# Statistical Models and Methods for Reliability and Survival Analysis

# Statistical Models and Methods for Reliability and Survival Analysis

Edited by
Vincent Couallier
Léo Gerville-Réache
Catherine Huber-Carol
Nikolaos Limnios
Mounir Mesbah

*Series Editor*
*Nikolaos Limnios*

iSTE

WILEY

First published 2014 in Great Britain and the United States by ISTE Ltd and John Wiley & Sons, Inc.



ISTE Ltd
27-37 St George's Road
London SW19 4EU
UK

www.iste.co.uk

John Wiley & Sons, Inc.
111 River Street
Hoboken, NJ 07030
USA

www.wiley.com



Library of Congress Control Number: 2013951426

British Library Cataloguing-in-Publication Data
A CIP record for this book is available from the British Library
ISBN: 978-1-84821-619-8

FSC
www.fsc.org
MIX
Paper from responsible sources
FSC® C013604

Printed and bound in Great Britain by CPI Group (UK) Ltd., Croydon, Surrey CR0 4YY

# Table of Contents

# Preface

*Statistical Models and Methods for Reliability and Survival Analysis* is a volume of contributions by specialists in statistical theory and their applications that provide up-to-date developments in methods used in survival analysis, statistical goodness of fit, stochastic processes for system reliability among others. Many of them are related to the work of Professor M. Nikulin in statistics which has spanned 30 years. The contributors accepted this challenging project to gather various contributions with a wide range of techniques and results, all of them on the topics of the past S2MRSA conference, 4–6th July 2012, dedicated to M. Nikulin for his 20th anniversary as a professor in the Bordeaux Segalen University (http://www.sm.u-bordeaux2.fr/Stat.Bordeaux.2012). The book is intended for researchers interested in statistical methodology and models useful in survival analysis, system reliability and statistical testing for censored and non-censored data.

Vincent COUALLIER
Léo GERVILLE-RÉACHE
Catherine HUBER-CAROL
Nikolaos LIMNIOS
Mounir MESBAH
October 2013

# Biography of Mikhail Stepanovitch Nikouline

From his native city of St. Petersburg, whose name was Leningrad when he was born there on April 29, 1944, Mikhail Stepanovitch Nikulin (Nikouline by French passport) visited many foreign countries due to his interest in several different domains in his favorite disciplines: mathematics and statistics. It was Professor Login N. Bolshev who allowed him to be aware of his talent in mathematics and directed his thesis in the field of "Probability Theory and Mathematical Statistics".

He obtained his Master's degree in mathematics from the State University in St. Petersburg in 1966. In the meanwhile, he had met Valia, a young lady who had many admirers, and he was proud and happy to be the one she chose to be her husband. They married in early 1963 when they were both under 20. Valia was a scientist like her husband and they had a daughter named Hélène at the end of the same year.

Then they decided to learn French intensively in order to go to Brazzaville, capital of the "République Populaire du Congo", where they spent three years. There, both were professors of mathematics in the *Collège d'Enseignement Général* from 1966 to 1969. Just after leaving Congo, their second child Alexis was born; both children were very promising. Then they went back to their native country. There, M.S. Nikouline became a PhD student at the Steklov Mathematical Institute in Moscow, Academy of Science of Russia, while Valia was doing research in mathematics at the Department of Applied Mathematics at the State University in St. Petersburg. Under the supervision of Professor L.N. Bolshev, for whom Nikouline always had a great admiration for both his mathematical skill and his personal human qualities, and whose early death at the age of 56 saddened Nikouline greatly, he obtained his PhD thesis entitled "A generalization of chi square tests".

---

Chapter written by Vincent COUALLIER, Léo GERVILLE-RÉACHE, Catherine HUBER-CAROL, Nikolaos LIMNIOS and Mounir MESBAH.

From 1974 to 2006, he was a member of the famous Ildar Ibragimov's Laboratory of Statistical Methods at the Mathematical Institute of V. Steklov (St. Petersburg).

From 1988 to 1992, M.S. Nikouline was a docent in the Department of Statistics and Probability at the State University in St. Petersburg. In 1992, he was hired as an Associate Professor at the University Victor Segalen-Bordeaux 2, France, where he spent his career serving in various positions: Professor in 1996, Full Professor in 2006 and Distinguished Professor in 2008 until 2012 when he became Emeritus. During the same time, from 1996 to 2001, he was the head of the Department "SCIMS- Science et Modélisation", and, from 1999 to 2006, he was the head of the Laboratory "Equipe d'Accueil 2961: Mathematical Statistics and its Applications".

He published a large number of papers (more than 250), alone for most of them, or jointly, with several colleagues and friends from the USSR, Europe and Canada, such as V. Solev, V. Bagdonavicius, V.G. Voinov, N.N. Lyashenko, K. Dhzapridze, P. Greenwood, L. Gerville-Réache, J. Kruopis and many others.

His first field of interest, following the theme of his PhD thesis, was goodness-of-fit tests and best unbiased estimators for parametric models. Between 1973 and 1992, he published more than 30 papers on this subject.

From 1993 to 2013, his subjects of interest became more diversified. We can see from the references of Nikulin's papers that although he was interested in parametric models, goodness of fit for them and unbiased estimators, he also tackled semi-parametric and non-parametric models that are more general. In particular, he was interested in human survival data analysis and he was one of the few people who made the link with reliability whose traditional field is industry that seemed for a long time to be very much apart from the medical field. From a mathematical point of view, many advances in any of these two domains could be provided by the other domain. With this perspective in mind, M.S. Nikouline organized with three French colleagues M. Mesbah of the University of Vannes (Britanny) and then University of Paris VI, N. Limnios of the University of Technology of Compiègne and C. Huber-Carol of the University of Paris Descartes, a European Seminar devoted to the "Survival Analysis and Reliability Theory".

During his career in Bordeaux, Nikouline has directed 11 PhD theses and most of his students have obtained good positions either in universities or in industry.

Moreover, Nikouline was the author and editor of a number of volumes on mathematical statistics. He has also written several articles for the *Encyclopaedia* and was part of several editorial boards and scientific societies such as the International Statistical Institute (ISI) and the Mathematical Society of St. Petersburg. Throughout his long career, he has met a large number of colleagues and as he is a friendly and generous person, many of them became his friends and were always very faithful to him.

# PART 1

# Statistical Models and Methods

Chapter 1

# Unidimensionality, Agreement and Concordance Probability

The evaluation and comparison of various methods often arise in medical research. For example, the evaluation of reproducibility of a new measurement technique often needs a comparison with the established technique, and image interpretation is often read by two or more observers. In this chapter, we provide a review of the measures of agreement and association, describe the statistical models underlying the Cronbach's alpha coefficient (CAC) and the backward reliability curve (BRC), the kappa coefficient, and present a general approach based on the concept of concordance probability. In particular, we illustrate the relationship between the concordance probability and various existing measures of agreement and association, namely Kendall's $\tau$, Somer's $D$, area under receiver operating characteristic (ROC) curve and Harrell's $c$-index. In addition, we review the estimation of concordance probability and present its large sample properties. Recent developments in the analysis of right censored data are also presented.

## 1.1. Introduction

The evaluation and comparison of various methods often arise in medical research. For example, the evaluation of reproducibility of a measurement technique often needs a comparison with the established technique, and the interpretation of a computerized tomography (CT) or magnetic resonance imaging (MRI) scan is often read by two or more observers. There is considerable literature on the measure of

---

Chapter written by Zhezhen JIN and Mounir MESBAH.

agreement (see [CHO 04], [BAR 07], [WAT 10], [SHO 04] and [LIN 10]). The methods vary with different types of measurement, i.e. continuous or categorical measurements. When the response variable is continuous, there are several intuitive approaches, namely comparison of means, Cronbach's coefficient alpha (CAC), various correlation coefficients and the test of slope being 1 in a simple linear regression, as well as alternative methods, the limits of agreement [BLA 86, BLA 99], the concordance correlation coefficient [LIN 89], mean squared deviation and total deviation index [LIN 00], and coverage probability approach [LIN 02]. When the response variable is categorical, kappa statistic, Somer's D-statistic and logistic regression are commonly used. When one measure is binary and the other measure is continuous, the methods of the ROC curve and logistic regression approach are often applied. These methods are related to typical concordance correlation between repeated measurements through an underlying linear or nonlinear parametric model. Recently developed concordance probability is a non-parametric approach. The concordance probability is commonly used as a measure of discriminatory power and predictive accuracy of statistical models. We show that the concordance probability also provides a unified measure of agreement for different types of measurement.

In this chapter, we present a review of the statistical models underlying the CAC and the BRC in section 1.2, and the kappa coefficient in section 1.3. In section 1.4, we introduce the concordance probability and describe its relationship with Kendall's $\tau$, Somer's $D$ and area of ROC curve of sensitivity and 1–specificity for different cutoffs. In section 1.5, we review the estimation of concordance probability and present its large sample properties. In section 1.6, we present recent developments on how to use the concordance probability to assess the agreement among different measures. We present the extension of the approach to the right censored data in section 1.7 and conclude with some discussion in section 1.8.

## 1.2. From reliability to unidimensionality: CAC and curve

### 1.2.1. *Classical unidimensional models for measurement*

Latent variable models involve a set of observable variables $A = \{\mathbf{X}_1, \mathbf{X}_2, \ldots, \mathbf{X}_k\}$ and a latent (unobservable) variable $\theta$ of dimension $d \leq k$. In such models, the dimensionality of $A$ is captured by the dimension of $\theta$, the value of $d$. When $d = 1$, the dimensionality of set $A$ is called unidimensional.

In a health-related quality of life (HrQoL) study, measurements are taken with an instrument: the questionnaire, which consists of questions (or items). In such cases, the $\mathbf{X}_{ij}$ represents the random response of the $j$th question by the $i$th subject and the $\mathbf{X}_j$ denotes the random variable generating responses to the $j$th question.

The parallel model is a classical latent variable model describing the unidimensionality of a set $A = \{\mathbf{X}_1, \mathbf{X}_2, \ldots, \mathbf{X}_k\}$ of quantitative observable variables. Let $\mathbf{X}_{ij}$ be the measurement of subject $i$, given by a variable $\mathbf{X}_j$, $i = 1, \ldots, n, j = 1, \ldots, k$, then:

$$\mathbf{X}_{ij} = \tau_{ij} + \varepsilon_{ij}, \qquad [1.1]$$

where $\tau_{ij}$ is the unknown true measurement corresponding to the observed measurement $\mathbf{X}_{ij}$ and $\varepsilon_{ij}$ a measurement error. The model is called a parallel model if the $\tau_{ij}$ can be divided as:

$$\tau_{ij} = \beta_j + \theta_i,$$

where $\beta_j$ is an unknown fixed parameter (non-random) representing the effect of the $j$th variable, and $\theta_i$ is an unknown random parameter effect of the $i$th subject.

It is generally assumed that $\theta_i$ has zero mean and unknown standard deviation $\sigma_\theta$. It should be noted that the zero-mean assumption is an arbitrary identifiability constraint with consequence on the interpretation of the parameter: its value must be interpreted comparatively to the mean population value. *In HrQoL setting, $\theta_i$ is the true latent HrQoL that the clinician or health scientist wants to measure and analyze.* It is a zero mean individual random part of all observed subject responses $\mathbf{X}_{ij}$, the same whatever the variable $\mathbf{X}_j$ (in practice, a question $j$ of an HrQoL questionnaire). It is also generally assumed that $\varepsilon_{ij}$ are independent random errors with zero mean and standard deviation $\sigma$ corresponding to the additional measurement error. Moreover, the true measure and the error are assumed to be uncorrelated, i.e. $cov(\theta_i, \varepsilon_{ij}) = 0$. This model is known as the parallel model, because the regression lines relating any observed item $\mathbf{X}_j$, $j = 1, \ldots, k$, and the true unique latent measure $\theta_i$ are parallel.

Model [1.1] can be obtained in an alternative way through modeling the conditional moments of the observed responses. Specifically, the conditional mean of $\mathbf{X}_{ij}$ can be specified as:

$$E[\mathbf{X}_{ij}|\theta_i; \beta_j] = \beta_j + \theta_i, \qquad [1.2]$$

where $\beta_j$, $j = 1, \ldots, k$, are fixed effects and $\theta_i$, $i = 1, \ldots, n$, are independent random effects with zero mean and standard deviation $\sigma_\theta$. The conditional variance of $\mathbf{X}_{ij}$ is specified as:

$$Var[\mathbf{X}_{ij}|\theta_i; \beta_j] = Var(\varepsilon_{ij}) = \sigma^2. \qquad [1.3]$$

Assumptions [1.2] and [1.3] are classical in experimental design. The model defines relationships between different kinds of variable: the observed score $\mathbf{X}_{ij}$, the true score $\tau_{ij}$ and the measurement error $\varepsilon_{ij}$. It is significant to make some remarks about the assumptions underlying this model. The random part of the true measure given by response by the $i$th individual does not vary with the question number $j$ as the $\theta_i$ does not depend on $j$, $j = 1, \ldots, k$. The model is unidimensional in the sense that the random part of all observed variables (questions $\mathbf{X}_j$) is generated by the common unobserved variable ($\theta_i$). More precisely, let $\mathbf{X}^*_{ij} = \mathbf{X}_{ij} - \beta_j$ be the calibrated version of the response to the $j$th item by the $i$th subject, then models [1.2] and [1.3] can be rewritten as:

$$E[X^*_{ij}|\theta_i; \beta_j] = \theta_i, \text{ for } \forall j, \qquad [1.4]$$

along with the same assumptions on $\beta$ and $\theta$ and the conditional variance model [1.3].

When both $\theta_i$ and $\varepsilon_{ij}$ are normally distributed, then we have the so-called conditional independence property: whatever $j$ and $j'$, two observed items $\mathbf{X}_j$ and $\mathbf{X}_{j'}$ are independent conditional to the latent $\theta_i$.

### 1.2.2. *Reliability of an instrument: CAC*

A measurement instrument yields values that we call the observed measure. The reliability $\rho$ of an instrument is defined as the ratio of two variances of the true over the observed measure. Under the parallel model, we can show that the reliability of any variable $\mathbf{X}_j$ (as an instrument to measure the true value) is given by:

$$\rho = \frac{\sigma^2_\theta}{\sigma^2_\theta + \sigma^2}. \qquad [1.5]$$

This coefficient is also known as the intra-class coefficient. The reliability coefficient, $\rho$, can easily be interpreted as a correlation coefficient between the true measure and the observed measure. When the parallel model is assumed, the reliability of the sum of $k$ variables is:

$$\tilde{\rho}_k = \frac{k\rho}{k\rho + (1-\rho)}. \qquad [1.6]$$

This formula is known as the Spearman–Brown formula [BRO 10, SPE 10].

The Spearman–Brown formula shows a simple relationship between $\tilde{\rho}_k$ and $k$, the number of variables. It is easy to see that $\tilde{\rho}_k$ is an increasing function of $k$.

The maximum likelihood estimator of $\tilde{\rho}_k$, under the parallel model with normal distribution assumptions, is known as CAC [CRO 51, BLA 97], which is denoted as $\alpha$:

$$\alpha = \frac{k}{k-1}\left(1 - \frac{\sum\limits_{j=1}^{k} S^2_j}{S^2_{tot}}\right), \qquad [1.7]$$

where

$$S_j^2 = \frac{1}{n-1}\sum_{i=1}^{n}(X_{ij} - \overline{X_j})^2$$

and

$$S_{tot}^2 = \frac{1}{nk-1}\sum_{i=1}^{n}\sum_{j=1}^{k}(X_{ij} - \overline{X})^2.$$

Under the parallel model, the variance–covariance matrix of the observed items $X_j$ and the latent trait $\theta$ is:

$$V_{X,\theta} = \begin{pmatrix} \sigma_\theta^2 + \sigma^2 & \sigma_\theta^2 & \cdots & \cdots \sigma_\theta^2 & \sigma_\theta^2 \\ \sigma_\theta^2 & \sigma_\theta^2 + \sigma^2 & \sigma_\theta^2 & \cdots \sigma_\theta^2 & \sigma_\theta^2 \\ \vdots & \vdots & \vdots & \vdots & \vdots \\ \sigma_\theta^2 & \cdots & \sigma_\theta^2 & \sigma_\theta^2 + \sigma^2 & \sigma_\theta^2 \\ \sigma_\theta^2 & \cdots & \cdots & \sigma_\theta^2 & \sigma_\theta^2 \end{pmatrix},$$

and the corresponding correlation matrix of the observed items $X_j$ and the latent trait $\theta$ is:

$$R_{X,\theta} = \begin{pmatrix} 1 & \rho & \cdots & \cdots \rho & \sqrt{\rho} \\ \rho & 1 & \rho & \cdots \rho & \sqrt{\rho} \\ \vdots & \vdots & \vdots & \vdots & \vdots \\ \rho & \cdots & \rho & 1 & \sqrt{\rho} \\ \sqrt{\rho} & \cdots & \cdots & \sqrt{\rho} & 1 \end{pmatrix}.$$

The *marginal* covariance $V_X$ and correlation matrix $R_X$ of the $k$ observed variables $X_j$, under the parallel model, are:

$$V_X = \begin{pmatrix} \sigma_\theta^2 + \sigma^2 & \sigma_\theta^2 & \cdots & \cdots \sigma_\theta^2 \\ \sigma_\theta^2 & \sigma_\theta^2 + \sigma^2 & \sigma_\theta^2 & \cdots \sigma_\theta^2 \\ \vdots & \vdots & \vdots & \vdots \\ \sigma_\theta^2 & \cdots & \sigma_\theta^2 & \sigma_\theta^2 + \sigma^2 \end{pmatrix}$$

and

$$R_X = \begin{pmatrix} 1 & \rho & \cdots & \cdots \rho \\ \rho & 1 & \rho & \cdots \rho \\ \vdots & \vdots & \vdots & \vdots \\ \rho & \cdots & \rho & 1 \end{pmatrix}.$$

This structure is known as a *compound symmetry*-type structure. It is easy to show that the reliability of the sum of $k$ items given in [1.7] can be expressed as:

$$\tilde{\rho}_k = \frac{k}{k-1}\left[1 - \frac{trace(V_X)}{J'V_X J}\right], \qquad [1.8]$$

with $J$ a vector with all components being 1, and

$$\alpha = \frac{k}{k-1}\left[1 - \frac{trace(S_X)}{J'S_X J}\right], \qquad [1.9]$$

where $S_X$ is the observed variance, empirical estimation of $S_X$. There is, even in the recent literature, an understanable confusion between Cronbach's alpha as a population parameter (theoretical reliability of the sum of items) or its sample estimate.

In addition, it is easy to show a direct connection between the CAC and the percentage of variance of the first component in principal component analysis (PCA), which is often used to assess unidimensionality. The PCA is mainly based on the analysis of the latent roots of $V_X$ or $R_X$ (or, in practice their sample estimate). The matrix $R_X$ has only two different latent roots, the greater root is $\lambda_1 = (k-1)\rho + 1$, and the other multiple roots are $\lambda_2 = \lambda_3 = \lambda_4 = \cdots = 1 - \rho = \frac{k-\lambda_1}{k-1}$. So, using the Spearman–Brown formula, we can express the reliability of the sum of the $k$ variables as $\tilde{\rho}_k = \frac{k}{k-1}(1 - \frac{1}{\lambda_1})$.

This clearly indicates a monotonic relationship between $\tilde{\rho}_k$, which can be consistently estimated by the CAC, and the first latent root $\lambda_x$, which in practice is naturally estimated by the corresponding observed sample correlation matrix and thus the percentage of variance of the first principal component in a PCA. So, CAC can also be considered as a measure of unidimensionality.

Nevertheless such a measure is not very useful, because it is easy to show, using the Spearman–Brown formula [BRO 10, SPE 10], that under the parallel model assumption, the reliability of the total score is an increasing function of the number of variables.

Therefore, *if the parallel model is true*, increasing the number of items will increase the reliability of a questionnaire. Moreover, the coefficient lies between 0 and 1. Zero value indicates a totally unreliable scale, while unit value means that the scale is perfectly reliable. Of course, in practice, these two scenarios never occur.

The CAC is an estimate of the reliability of the raw-score (sum of item responses) of a person *if the model generating those responses is a parallel model.*

The result can be used as a criterion for checking the unidimensionality of such responses when those item responses are *generated by a parallel model.*

In the next section, we show how to build and to use a more operational and more valid criterion to measure the unidimensionality of a set of items: the BRC (the $\alpha$-curve).

### 1.2.3. *Unidimensionality of an instrument: BRC*

Statistical validation of unidimensionality can be performed through a goodness-of-fit test of the parallel model or Rasch model. There is a vast literature on the subject, see [MES 12]. The goodness-of-fit tests generally do not have power because their null hypotheses do not focus on unidimensionality: this includes indirectly other additional assumptions (the normality for parallel models, local independence for Rasch models etc.) As a result, the departure from the null hypotheses is not necessarily an indication of the departure from a unidimensionality.

In the following, we describe a graphical tool, which is helpful for checking the unidimensionality of a set of variables. It draws a curve in a stepwise manner, using estimates of reliability of subscores (total of a subset included in the starting set).

In the first step, the CAC will be calculated with all the variables. Then, at every successive step, the CAC will be calculated by deleting one variable each time, and the variable whose deletion yields the maximum CAC value among those CAC values will be removed. This procedure is repeated until only two variables remain. If the parallel model is true, increasing the number of variables increases the reliability of the total score, which can be consistently estimated by Cronbach's alpha. The number of variables and the CAC values can be plotted, which would yield a curve. This procedure is called the backward reliability curve (BRC). If there is a decrease of such a curve after adding a variable, it would strongly indicate that the added variable does not constitute a unidimensional set with variables already in the curve.

Drawing the BRC of a set of *unidimensional* items is an essential tool in the validation process of an HrQoL questionnaire. When we develop an HrQoL questionnaire, the main goal is generally to measure some unidimensional latent subjective traits (such as sociability and mobility). The use of the BRC in empirical data is very helpful for detection of non-unidimensional subsets of items. When the BRC is not an increasing curve, we can remove one or more items to obtain an increasing curve [MES 13]. If the reduced set gives an increasing curve, it is in some sense *more valid in terms of unidimensionality* than the previous set.

## 1.3. Agreement between binary outcomes: the kappa coefficient

### 1.3.1. *The kappa model*

The CAC in the previous section is a natural estimate of a monotonic function of the intra-class coefficient. In the case of multiple items, the parallel model leads to equal variances among marginal distributions of the items and equal covariances between any pairs of items. It can be shown that the formula of an intra-class coefficient is a correlation coefficient under such a constraint [MAK 88]. We illustrate this with the simple case of two binary items $X$ and $Y$. Let us assume that $E(X) = E(Y) = \pi$. Under such an assumption, it is easy to derive the correlation coefficient between $X$ and $Y$: $\rho_{X,Y} = \frac{E(XY)-\pi^2}{\pi(1-\pi)}$. The $\rho_{X,Y}$ is often denoted as the coefficient $\kappa$. It follows that:

$$
\begin{aligned}
p_{11} &= Prob(X=1, Y=1) = E(XY) = \pi^2 + \pi(1-\pi)\kappa, \\
p_{10} &= Prob(X=1, Y=0) = \pi(1-\pi) - \pi(1-\pi)\kappa, \\
p_{01} &= Prob(X=0, Y=1) = \pi(1-\pi)(1-\kappa),
\end{aligned}
$$

and

$$p_{00} = Prob(X=0, Y=0) = (1-\pi)^2 + \pi(1-\pi)\kappa.$$

### 1.3.2. *The kappa coefficient*

The probability of concordance between the two items is:

$$p_c = p_{11} + p_{00} = 2\pi^2 + 1 - 2\pi + 2\pi(1-\pi)\kappa.$$

The probability of concordance due to chance, i.e. when the two items are independent ($\kappa = 0$), is:

$$p_h = 1 - 2\pi(1-\pi).$$

Consequently:

$$\kappa = \frac{p_c - p_h}{1 - p_h}.$$

### 1.3.3. *Estimation of the kappa coefficient*

When we observe a $2 \times 2$ contingency table $\{n_{ij} : i, j = 0, 1\}$, a natural estimation of $p_c$ is:

$$\hat{p_c} = \frac{n_{11} + n_{00}}{n}.$$

A natural estimation of $p_h$ is:

$$\hat{p_h} = \frac{n_{1.}n_{.1} + n_{0.}n_{.0}}{n}.$$

Cohen's kappa coefficient [COH 60] is:

$$\hat{\kappa} = \frac{\hat{p_c} - \hat{p_h}}{1 - \hat{p_h}} = \frac{n_{11} + n_{00} - n_{1.}n_{.1} - n_{0.}n_{.0}}{n - n_{1.}n_{.1} - n_{0.}n_{.0}},$$

which is a natural estimator of the $\kappa$ coefficient.

The $\kappa$ can also be estimated by the maximum likelihood estimation method. Specifically, the likelihood of the observations is:

$$L = (p_{11})^n_{11} \times (p_{10})^n_{10} \times (p_{01})^n_{01} \times (p_{00})^n_{00}.$$

So, we easily find:

$$\begin{aligned} L(\kappa) &= (\pi^2 + \pi(1-\pi)\kappa)^n_{11} \times (\pi(1-\pi) - \pi(1-\pi)\kappa)^n_{10} \\ &\quad \times (\pi(1-\pi)(1-\kappa))^n_{01} \times ((1-\pi)^2 + \pi(1-\pi)\kappa)^n_{00}. \end{aligned}$$

The maximum likelihood estimator of $\kappa$ can be obtained by maximizing $L(\kappa)$ over $\kappa$.

## 1.4. Concordance probability

For a pair of bivariate observations $(X_1, Y_1)$ and $(X_2, Y_2)$, the concordance probability is defined as:

$$C_{X,Y} = P\{Y_2 > Y_1 | X_2 > X_1\}.$$

As it is defined as a conditional probability, the concordance probability can be used to assess the relationship between two variables that have a natural ordering. In particular, it is useful for the assessment of monotonic correlation between two variables. The concordance probability is invariant to rank-preserving transformation on either $X$ or $Y$. It takes values between 0 and 1. If $C_{X,Y} = 0$, then it means that $X$ and $Y$ are inversely related. If $C_{X,Y} = 1$, then it means that $X$ and $Y$ are related. If $X$ and $Y$ are independent, then $C_{X,Y} = 0.5$. In addition, if $Y$ is continuous, then it is easy to see that:

$$1 - C_{X,Y} = P\{Y_1 > Y_2 | X_2 > X_1\} = C_{-X,Y},$$

as $P(Y_1 = Y_2) = 0$.

### 1.4.1. *Relationship with Kendall's $\tau$ measure*

Kendall's $\tau$ measure has been used for the assessment of the relationship between two random variables. The concordance probability is closely related to Kendall's $\tau$ measure.

For a pair of bivariate observations $(X_1, Y_1)$ and $(X_2, Y_2)$, Kendall's $\tau$ is defined as:

$$\tau_{X,Y} = E[\text{sign}(X_2 - X_1)\text{sign}(Y_2 - Y_1)].$$

It is easy to see that $\tau_{X,Y} = \tau_{Y,X}$, which shows that Kendall's $\tau$ is a symmetric measure. In addition:

$$\begin{aligned}\tau_{X,Y} &= P(X_2 > X_1, Y_2 > Y_1) + P(X_2 < X_1, Y_2 < Y_1) \\ &\quad -P(X_2 > X_1, Y_2 < Y_1) - P(X_2 < X_1, Y_2 > Y_1),\end{aligned}$$

which means that Kendall's $\tau$ is the difference between two probabilities: the probability of concordance and the probability of discordance. Equivalently, it follows that:

$$\tau_{X,Y} = 4P(X_2 > X_1, Y_2 > Y_1) - 2P(X_2 > X_1) - 2P(Y_2 > Y_1) + 1.$$

If $(X_1, Y_1)$ and $(X_2, Y_2)$ are independent and identically distributed, then:

$$\tau_{X,Y} = 2P(Y_2 > Y_1 | X_2 > X_1) - 1 = 2C_{X,Y} - 1,$$

which shows that there is 1–1 correspondence between Kendall's $\tau$ and the concordance probability, namely a linear relationship.

### 1.4.2. *Relationship with Somer's $D$ measure*

Somer's $D$ is defined as:

$$D_{X,Y} = E[\text{sign}(X_2 - X_1)\text{sign}(Y_2 - Y_1) | X_2 \neq X_1].$$

It is the difference between two conditional probabilities: the conditional probability of concordance and the conditional probability of discordance given that two $X$ values are not equal. In light of the relationship between the concordance probability and Kendall's $\tau$, it follows that $D_{X,Y} = \frac{2C_{X,Y}-1}{P(X_2 \neq X_1)}$.

Somer's $D$ is often used as a measure of association for ordinal variables. For ordinal $X$, it is clear that $D_{X,Y} \neq D_{Y,X}$, which shows that Somer's $D$ is an asymmetric measure.

If $X$ is a continuous variable, then $P(X_2 \neq X_1) = 1$ and Somer's $D$ is the same as Kendall's $\tau$, which implies that $D_{X,Y} = 2C_{X,Y} - 1$.

If $X$ is not a continuous variable, then $P(X_2 \neq X_1) < 1$ and Somer's $D$ is larger than Kendall's $\tau$, and $D_{X,Y} > 2C_{X,Y} - 1$.

### 1.4.3. *Relationship with ROC curve*

The area under the curve (AUC) is defined as the area under the ROC curve of 1-specificity against sensitivity. If $X$ is binary with possible values 0 and 1, and $Y$ is continuous or ordinal, let $TPP(c) = P(Y \geq c|X = 1)$ be the true positive proportion or sensitivity, and $FPP(c) = P(Y \geq c|X = 0)$ the false positive proportion or 1-specificity, then the AUC will be the area under the set of points $\{(FPP(c), TPP(c)) : c \in (-\infty, \infty)\}$.

If $Y$ is continuous, then the AUC is the $C_{X,Y}$.

PROOF.– By definition $C_{X,Y} = P\{Y_2 > Y_1|X_2 > X_1\}$, therefore:

$$\begin{aligned} C_{X,Y} &= \int_{-\infty}^{\infty} P\{Y_2 > c|X_2 > X_1\} dP\{Y_1 \leq c|X_2 > X_1\} \\ &= \int_{-\infty}^{\infty} P\{Y_2 \geq c|X_2 = 1\} dP\{Y_1 < c|X_1 = 0\} \\ &= \int_{-\infty}^{\infty} TPP(c) d(1 - FPP(c)) \\ &= \int_{0}^{1} TPP(FPP^{-1}(u)) du. \end{aligned}$$

Here, $X_2 = 1$ and $X_1 = 0$ because both $X_1$ and $X_2$ are binary with values 0 and 1 and $X_2 > X_1$.

If $Y$ is not continuous, then the AUC is $C_{X,Y} + P(Y_2 = Y_1|X_2 = 1, X_1 = 0)/2$.

## 1.5. Estimation and inference

Suppose $(X_1, Y_1), \cdots, (X_n, Y_n)$ are a sample of $n$-independent and identically distributed bivariate random vectors. The concordance probability can be estimated by:

$$\hat{C}_{X,Y} = \frac{\sum_{i=1}^{n}\sum_{j=1}^{n} I\{X_i > X_j\}I\{Y_i > Y_j\}}{\sum_{i=1}^{n}\sum_{j=1}^{n} I\{X_i > X_j\}},$$

where $I(\cdot)$ is an indicator function taking values of 0 or 1.

The estimator is of the form of a $U$-statistic. Using the large sample theory of a $U$-statistic, it can be shown that $\hat{C}_{X,Y} \to C_{X,Y}$ as $n \to \infty$ and $\sqrt{n}(\hat{C}_{X,Y} - C_{X,Y}) \to N(0, \sigma_1^2)$ as $n \to \infty$. The variance $\sigma_1^2$ can be estimated with a plug-in method with corresponding empirical estimators.

## 1.6. Measure of agreement

The concordance probability can be used as a measure of agreement. Suppose $X$ is a latent variable, and $Y$ and $Z$ are two measures of the latent variable, we may want to assess if $Y$ or $Z$ is in agreement of measuring $X$. Liu *et al.* [LIU 12] suggested that this can be done by a comparison of concordance probabilities $C_{X,Y}$ and $C_{X,Z}$ with their difference, namely:

$$d_{YZ|X} = P(Y_2 > Y_1 | X_2 > X_1) - P(Z_2 > Z_1 | X_2 > X_1).$$

If the difference is close to 0, it means that $Y$ and $Z$ are in agreement. If the difference is different from 0, it means that $Y$ and $Z$ are not in agreement measuring the latent variable $X$: a positive value means that $Y$ is better than $Z$ in agreement with $X$, and a negative value means that $Z$ is better than $Y$ in agreement with $X$.

For independent and identically distributed $(X_i, Y_i, Z_i), i = 1, \cdots, n$, the difference can be estimated by:

$$\hat{d}_{YZ|X} = \frac{\sum_{i=1}^{n}\sum_{j=1}^{n}[I\{Y_i > Y_j, X_i > X_j\} - I\{Z_i > Z_j, X_i > X_j\}]}{\sum_{i=1}^{n}\sum_{j=1}^{n} I\{X_i > X_j\}},$$

which is a ratio of two $U$-statistics. By using the large sample theory of a $U$-statistic, it follows that $\hat{d}_{YZ|X} \to d_{YZ|X}$ and $\sqrt{n}(\hat{d}_{YZ|X} - d_{YZ|X}) \to N(0, \sigma_2^2)$.

The asymptotic results can be used to construct a statistical test for the hypotheses $H_0 : C_{X,Y} = C_{X,Z}$ versus $H_1 : C_{X,Y} \neq C_{X,Z}$, which is equivalent to the hypotheses $H_0 : d_{YZ|X} = 0$ versus $H_1 : d_{YZ|X} \neq 0$.

Liu *et al.* [LIU 12] noted that the denominator $\sum_{i=1}^{n}\sum_{j=1}^{n} I\{X_i > X_j\}$ in $\hat{d}_{YZ|X}$ does not play a significant role in the test and evaluation of the agreement, and the difference:

$$\hat{\Delta}_{YZ|X} = \sum_{i=1}^{n}\sum_{j=1}^{n}[I\{Y_i > Y_j, X_i > X_j\} - I\{Z_i > Z_j, X_i > X_j\}]$$

can be used. Obviously, $\hat{\Delta}_{YZ|X}$ is a $U$-statistic with kernel $U_{ij}$ where:

$$\begin{aligned} U_{ij} &= 0.5[I\{Y_i > Y_j, X_i > X_j\} + I\{Y_i < Y_j, X_i < X_j\} \\ &\quad -I\{Z_i > Z_j, X_i > X_j\} - I\{Z_i < Z_j, X_i < X_j\}] \\ &= 0.5\{\text{sign}^+[(X_i - X_j)(Y_i - Y_j)] - \text{sign}^+[(X_i - X_j)(Z_i - Z_j)]\}. \end{aligned}$$

Consequently,

$$Var(\hat{\Delta}_{YZ|X}) = \frac{n(n-1)}{2}Var(U_{12}) + \frac{n}{4}Cov(U_{12}, U_{13}),$$

which can be estimated by replacing $Var(U_{12})$ and $Cov(U_{12}, U_{13})$ by their corresponding moment estimators.

The inference can be carried out by examining the $1-\alpha$ confidence interval of $\Delta_{YZ|X}$ or constructing a Wald-type test statistic $\frac{\hat{\Delta}_{YZ|X}}{\sqrt{\widehat{Var}(\hat{\Delta}_{YZ|X})}}$ for testing null hypothesis $H_0 : d_{YZ|X} = 0$.

## 1.7. Extension to survival data

In survival analysis, the outcome variable is the time to event occurrence, where the event could be the initial diagnosis of a disease or death. It is likely that the time variable is right-censored due to dropout or study termination. Let $T$ be the length of time to event occurrence, $Q$ be the censoring time and $X$ be a predictor variable. Then, the observed data consist of $(Y, \delta, X)$, where $Y = \min(T, Q)$ and $\delta = I\{T \leq Q\}$. It is of significance to estimate the concordance probability $C_{T,X} = P(X_2 > X_1|T_2 > T_1)$ with the $n$-independent observations $(X_i,\ Y_i,\ \delta_i)$, $i = 1, \cdots, n$.

### 1.7.1. *Harrell's $c$-index*

Harrell *et al.* [HAR 82, HAR 96] developed Harrell's $c$-index to estimate the concordance probability for survival data. Harrell's $c$-index is defined as:

$$c_{Y,X} = \frac{\sum_{i=1}^{n}\sum_{j=1}^{n}[\delta_i I\{Y_i < Y_j\}I\{X_i < X_j\} + \delta_j I\{Y_j < Y_i\}I\{X_j < X_i\}]}{\sum_{i=1}^{n}\sum_{j=1}^{n}[\delta_i I\{Y_i < Y_j\} + \delta_j I\{Y_j < Y_i\}]}.$$

Nam and D'Agostino [NAM 02] developed a method to estimate the standard error of the estimator. Pencina and D'Agostino [PEN 04] derived alternative formulas for standard error estimation using the relationship between the $c$-index and the modified Kendall's $\tau$ for bivariate correlation and investigated how to construct its confidence intervals.

If there is no censoring, i.e. $\delta_i = 1$ for $i = 1, \cdots, n$, then by the large sample theory of a $U$-statistic, $c_{Y,X} = c_{T,X} \to C_{T,X} = C_{Y,X}$, where $C_{T,X}$ is the concordance probability $P(X_2 > X_1 | T_2 > T_1)$.

If there is censoring, and censoring random variable $Q$ is independent of $T$, $c_{Y,X}$ will converge to a quantity that will depend on censoring distribution [LIU 09] and the consistency to a concordance probability will become questionable. Specifically, for continuous $X$, the estimator $c_{Y,X}$ will converge to $P(X_2 > X_1 | T_2 > T_1, T_1 < Q_1, T_1 < Q_2)$ [LIU 09].

To estimate the concordance probability $C_{T,X} = P(X_2 > X_1 | T_2 > T_1)$ properly with the $n$-independent observations $(X_i,\ Y_i,\ \delta_i)$, $i = 1, \cdots,\ n$, Liu and Jin [LIN 09] proposed a modified estimator using the idea of inverse probability weighting:

$$\hat{c}^*_{Y,X} = \frac{\sum_{i=1}^n \sum_{j=1}^n I(X_i < X_j) I(Y_i < Y_j) \delta_i / G^2(Y_i)}{\sum_{i=1}^n \sum_{j=1}^n I(Y_i < Y_j) \delta_i / G^2(Y_i)},$$

where $G(t) = P(t < Q)$ for $t > 0$. If $G(t)$ is unknown, a consistent estimator $\hat{G}(t)$, constructed by the Kaplan–Meier product limit method, may be used. For $Y_{(n)} = \max_{1 \leq i \leq n} Y_i$, it is required to set $b_{(n)} = 0$ if $\delta_{(n)} = 0$ and $\hat{G}(Y_{(n)}) = 0$. The estimator is consistent to the concordance probability $C_{T,X}$ if censoring variable $Q$ is completely random and independent of $T$.

### 1.7.2. *Measure of discriminatory power*

Suppose that variables $X$ and $Z$ are quantitative predictors of the time to event variable $T$. To examine whether $X$ and $Z$ have the same discrimination accuracy for $T$, we may examine the difference between their concordance probabilities as $d_{XZ|T} = C_{T,X} - C_{T,Z}$, or, equivalently, the difference between the bivariate probabilities as:

$$\Delta_{XZ} = P(X_2 > X_1, T_2 > T_1) - P(Z_2 > Z_1, T_2 > T_1).$$

If there is no censoring, note that the difference between the bivariate probabilities $\Delta_{XZ}$ can be estimated by a $U$-statistic:

$$\begin{aligned}\hat{\Delta}_{XZ} &= \frac{1}{n(n-1)}\sum_{i=1}^{n}\sum_{j=1}^{n}\{\text{sign}^{+}[(X_i - X_j)(T_i - T_j)] \\ &\quad -\text{sign}^{+}[(Z_i - Z_j)(T_i - T_j)]\} \\ &= \frac{1}{n(n-1)}\sum_{i=1}^{n}\sum_{j=1}^{n} U_{ij}.\end{aligned}$$

By the central limit theorem of $U$-statistics, under some regularity conditions, $\hat{\Delta}_{XZ}$ converges to $\Delta_{XZ}$ in distribution that $\sqrt{n}(\hat{\Delta}_{XZ} - \Delta_{XZ}) \rightarrow N(0, V)$ as $n \rightarrow \infty$. The variance $V$ can be estimated empirically.

With random censoring, Liu *et al.* [LIU 12] have developed a modified estimator using the idea of inverse probability weighting:

$$\hat{\Delta}^{*}_{XZ} = \frac{1}{n(n-1)}\sum_{i=1}^{n}\sum_{j=1}^{n}\frac{\delta_i I(Y_i < Y_j)}{\hat{G}^2(Y_i)}[I(X_i < X_j) - I(Z_i < Z_j)],$$

where $\hat{G}(\cdot)$ is the Kaplan–Meier product limit estimator of censoring distribution $G(\cdot)$. Under regularity conditions, Liu *et al.* [LIU 12] showed that $\hat{\Delta}^{*}_{XZ}$ is consistent and $\sqrt{n}(\hat{\Delta}^{*}_{XZ} - \Delta_{XZ}) \rightarrow N(0, V^{*})$ as $n \rightarrow \infty$. The variance $V^{*}$ can also be estimated empirically. As a result, the inference can be carried out by examining the $1 - \alpha$ confidence interval of $\hat{\Delta}^{*}_{XZ}$ or constructing a Wald-type test statistic $\frac{\hat{\Delta}^{*}_{XZ}}{\sqrt{\hat{V}^{*}}}$ for testing null hypothesis $H_0 : \Delta_{XZ} = 0$.

## 1.8. Discussion

In this chapter, we have reviewed various concordance coefficient measures. In particular, we presented non-parametric concordance coefficient and concordance probability, and described its relationship with commonly used measures of association, Kendall's $\tau$, Somer's $D$, AUC and Harrell's $c$-index. In addition, we have illustrated how to use the concordance probability to assess agreement and to evaluate the discriminatory power of two different predictors.

For more than two covariates, Gönen and Hellner [GON 05] proposed model-based estimation of concordance probability using the Cox proportional hazards model for right-censored data. However, it is challenging to estimate the asymptotic variance of the resulting estimator of concordance probability as it involves a non-smooth and non-continuous function of unknown parameters. It is of significance to develop an alternative method to estimate the asymptotic variance efficiently.

It is also noted that the concordance probability is invariant to the monotone transformation as its estimator only depends on ordering. However, the difference between two concordance probabilities might not be sensitive for detecting small differences in two competitive measures; caution should be taken in applications.

## 1.9. Bibliography

[BAR 07] BARNHART H.X., HABER M.J., LIN L.I., "An overview on assessing agreement with continuous measurement", *Journal of Biopharmaceutical Statistics*, vol. 17, pp. 529–569, 2007.

[BLA 86] BLAND J.M., ALTMAN D.G., "Statistical methods for assessing agreement between two methods of clinical measurement", *Lancet*, vol. i, pp. 307–310, 1986.

[BLA 89] BLAND J.M., ALTMAN D.G., "Measuring agreement in method comparison studies", *Statistical Methods in Medical Research*, vol. 8, pp. 135–160, 1989.

[BLA 97] BLAND J.M., ALTMAN D.G., "Statistics notes: Cronbach's alpha", *British Medical Journal*, vol. 314, p. 572, 1997.

[BRO 10] BROWN W., "Some experimental results in the correlation of mental abilities", *British Journal of Psychology*, vol. 3, pp. 296–322, 1910.

[CHO 04] CHOUDHARY P.K., NAGARAJA H.N., "Measuring agreement in method comparison studies-a review", in BALAKRISHNAN N., KANNAN N., NAGARAJA H.N. (eds), *Advances in Ranking and Selection, Multiple Comparisons, and Reliability*, Birkhauser, Boston, MA, pp. 215–244, 2004.

[COH 60] COHEN J., "A coefficient of agreement for nominal scales", *Educational and Psychological Measurement*, vol. 20, pp. 37–46, 1960.

[CRO 51] CRONBACH L.J., "Coefficient alpha and the internal structure of tests", *Psychometrika*, vol. 16, pp. 297–334, 1951.

[GON 05] GONON M., HELLER G., "Concordance probability and discriminatory power in proportional hazards regression", *Biometrika*, vol. 92, pp. 965–970, 2005.

[HAR 84] HARRELL F.E., LEE K.L., CALIFE R.M., *et al.*, "Regression modeling strategies for improved prognostic prediction", *Statistics in Medicine*, vol. 3, no. 2, pp. 143–152, 1984.

[HAR 96] HARRELL F.E., LEE K.L., MARK D.B., "Tutorial in biostatistics multivariable prognostic models: issues in developing models, evaluating assumptions and adequacy, and measuring and reducing errors", *Statistics in Medicine*, vol. 15, no. 4, pp. 361–387, 1996.

[KEN 38] KENDALL M.G., "A new measure of rank correlation", *Biometrika*, vol. 30, pp. 81–93, 1938.

[KEN 49] KENDALL M.G., "Rank and product-moment correlation", *Biometrika*, vol. 36, pp. 177–193, 1949.

[LIN 89] LIN L.I., "A concordance correlation coefficient to evaluate reproducibility", *Biometrics*, vol. 45, pp. 255–268, 1989.

[LIN 00] LIN L.I., "Total deviation index for measuring individual agreement with applications in laboratory performance and bioequivalence", *Statistics in Medicine*, vol. 19, pp. 255–270, 2000.

[LIN 02] LIN L.I., HEDAYAT A.S., SINHA B., *et al* "Statistical methods in assessing agreement: models, issues and tools", *Journal of American Statistical Association*, vol. 97, pp. 257–270, 2002.

[LIN 10] LIN L.I., HEDAYAT A.S., WU W., *Statistical Tools for Measuring Agreement*, Springer, New York, 2010.

[LIU 09] LIU X., JIN Z., "A non-parametric approach to scale reduction for uni-dimensional screening scales", *International Journal of Biostatistics*, vol. 5, no. 1, pp. 1–22, 2009.

[LIU 12] LIU X., JIN Z., GRAZIANO J.H., "Comparing paired biomarkers in predicting quantitative health outcome subject to random censoring", *Statistical Methods in Medical Research*, 2012.

[MAK 88] MAK T.K., "Analysing intraclass correlation for dichotomous variables", *Applied Statistics*, vol. 37, pp. 344–352, 1988.

[MES 12] MESBAH M., "Measurement and analysis of quality of life in epidemiology", in CHAKRABORTY R., RAO C.R., SEN P.K. (eds), *Handbook of Statistics, BioInformatics in Human Health and Heredity*, vol. 2, Chapter 15, Amsterdam, North Holland, 2012.

[MES 13] MESBAH M., "From measurement to analysis", in CHRISTENSEN K.B., KREINER S., MESBAH M. (eds), *Rasch Models in Health*, Chapter 13, ISTE, London, John Wiley & Sons, New York, 2013.

[NAM 02] NAM B.H., D'AGOSTINO R.B., *Discrimination Index, the Area under the ROC Curve, in Goodness-of-Fit tests and Model Validity*, Chapter 20, Birkhauser, Boston, MA, 2002.

[PEN 04] PENCINA M.J., D'AGOSTINO R.B., "Overall C as a measure of discrimination in survival analysis: model specific population value and confidence interval estimation", *Statistics in Medicine*, vol. 23, pp. 2109–2123, 2004.

[SHO 04] SHOUKRI M.M., *Measures of Interobserver Agreement*, Chapman & Hall/CRC, New York, 2004.

[SOM 62] SOMERS R.H., "A new asymmetric measure of association for ordinal variables", *American Sociological Review*, vol. 27, pp. 799–811, 1962.

[SPE 10] SPEARMAN C., "Correlation calculated from faulty data", *British Journal of Psychology*, vol. 3, pp. 271–295, 1910.

[WAT 10] WATSON P.F., PETRIE A., "Method agreement analysis: a review of correct methodology", *Theriogenology*, vol. 73, pp. 1167–1179, 2010.

Chapter 2

# A Universal Goodness-of-Fit Test Based on Regression Techniques

## 2.1. Introduction

Model fitting is an important part of all statistical research. However, model fitting must be accompanied by an assessment of the fit of the model. The development of the first numerical techniques to test goodness of fit can probably be attributed to Karl Pearson who developed the well-known chi-squared test. The chi-squared test groups a given data set into intervals and then compares the observed cell counts to the expected cell counts. Chi-squared goodness-of-fit tests were followed by tests of the EDF type (tests based on the empirical distribution function); the well-known Kolmogrov–Smirnov test and the Cramér–von Mises statistic being part of this family. EDF tests are based on measures of distance between the empirical cumulative distribution function of the data and the hypothesized cumulative distribution function. Recently, there has been an explosion in the number of procedures available to test distributional assumptions, for example tests based on regression and correlation, tests based on the characterizing properties of specific distributions and tests based on empirical Laplace transforms (see [DAG 86] for more details).

This chapter presents an omnibus test to test any distributional assumption and is based on the regression test of Brain and Shapiro [BRA 83] to test for an underlying exponential distribution. The techniques of Brain and Shapiro [BRA 83] were used by Gulati and Shapiro [GUL 06] and Gulati [GUL 11] to develop goodness-of-fit

---

Chapter written by Florence GEORGE and Sneh GULATI.

procedures for the Pareto, the Rayleigh and the Laplace distributions. The data were transformed to an exponential random sample using simple transformations and then tested for exponentiality using the Brain and Shapiro test. This chapter attempts to extend the results of Gulati and Shapiro [GUL 06] and Gulati [GUL 11] to develop a universal test for all single-parameter (belonging to the scale family of distributions) or two-parameter distributions (characterized by a scale and a shape parameter) based on the Brain and Shapiro test. The data are transformed to an exponential distribution either directly or indirectly (through the probability integral transform) and then tested for exponentiality.

Certainly, both the chi-squared-type tests and EDF tests are universal tests and can be used to test any distribution for underlying family. The main drawback of the chi-squared test is its dependency on the grouping intervals and the fact that it tends to be conservative. Tests based on the EDF statistics are more powerful but when the null hypothesis is composite, the exact distribution of the test statistic becomes very difficult to find. For the tests of the Anderson darling type, asymptotic theory is available; however, the computation of the null percentiles still needs to be done separately for each family being tested. For the K–S type statistics, even asymptotic theory is not available and the percentiles need to be computed (see [DAG 86]). In comparison, the test statistic proposed in this chapter is extremely simple to compute and under the null has an asymptotic chi-squared distribution. Simulation studies indicate that the null distribution of the proposed test statistic for hypothesized distributions with a single parameter is a chi-squared distribution with two degrees of freedom (exponential) even for sample sizes as small as 10, while for distributions with both a scale and a shape parameter, the test has a chi-squared distribution with one degree of freedom. Thus, a $p$-value can be obtained directly from a chi-squared table and there is no need for extensive computations for the calculation of the null percentiles.

The remaining chapter is organized as follows. Section 2.2 discusses the procedure proposed by Brain and Shapiro [BRA 83], while section 2.3 describes the application of this test to any underlying distribution. Sections 2.4 and 2.5 deal with simulation studies to assess the null distribution of the statistic and corresponding power studies for various distributions characterized by a scale parameter or a scale and a shape parameter. We present some real examples in section 2.6, while section 2.7 contains some concluding remarks.

## 2.2. The Brain and Shapiro procedure for the exponential distribution

Brain and Shapiro [BRA 83] developed a test for exponentiality based on the Laplace test, suggested by Cox and Lewis [COX 66]. The procedure determines exponentiality of the given data by testing whether the underlying failure rate function is constant and is based on the fact that the ordered weighted spacings of i.i.d. exponential random variables are independent and identically distributed (i.i.d.) exponential. The test statistic is a combination of two test statistics: the first test statistic tests for a monotone hazard rate and is a function of the linear regression of the ordered weighted spacings versus the order number. The second test statistic tests for a non-monotone and a non-constant hazard rate and is a function of a quadratic regression of the order number versus the weighted ordered spacings. Under the null hypothesis, the two statistics are distributed as independent normal variates. Since the final test statistic is obtained by adding the squares of the two statistics, its null distribution is that of a chi-squared random variable with two degrees of freedom. The test is summarized as follows:

1) Let the ordered data set be given by $X_{(1)} \leq X_{(2)} \leq ... \leq X_{(n)}$.

2) The null hypothesis of interest is given by:

$H_0$ : The data come from an exponential distribution.

3) Compute the weighted spacings as $Y_i = (n - i + 1)(X_{(i)} - X_{(i-1)})(i = 1, 2, ..., n)$ and $X_0 = 0$. As it is well known, the spacings will be generated as i.i.d exponential variates under the null hypothesis.

4) Now compute $t_i = \sum_{j=1}^{i} Y_j$, i = 1, 2, . . ., n and $u_i = t_i/t_n$, i = 1, 2, . . ., n-1; then the test statistic for the linear component is given by:

$$Z_1 = \sqrt{12(n-1)}(\bar{u} - 1/2), \qquad [2.1]$$

where $\bar{u} = \sum_{i=1}^{n-1} \frac{u_i}{n-1}$.

5) Next compute the test statistic for the quadratic component given by:

$$Z_2 = \sqrt{\frac{5(n-1)}{(n+1)(n-2)}} \times \left(n - 2 + 6n\bar{u} - 12\sum_{i=1}^{n-1} \frac{iu_i}{n-1}\right). \qquad [2.2]$$

6) Finally, the test statistic is given by:

$$Z = Z_1^2 + Z_2^2. \qquad [2.3]$$

As mentioned earlier, the null distribution of $Z$ is in the limit a chi-squared distribution with two degrees of freedom or an exponential distribution with a mean equal to 2. The test is an upper tail test and thus the $p$-value is given by:

$$p - value = \exp^{-Z/2}. \qquad [2.4]$$

A Monte Carlo analysis conducted of the null distribution of $Z$ in [BRA 83] showed that for small to moderate samples, a sample size correction factor resulted in an exact percentile but the correction factor was small and approached zero fairly quickly.

## 2.3. Applications of the Brain and Shapiro test

The Brain and Shapiro test can be used to test for any underlying distribution since every distribution can be transformed to an exponential distribution either directly or indirectly. For example, the Weibull, the Pareto, the Gumbel and the extreme value distributions are easily transformed to the exponential through straightforward transformations. For distributions where the transformation is not straightforward, the probability integral transform leads to a uniform distribution that can then be transformed to an exponential distribution. As mentioned in section 2.1, Gulati and Shapiro [GUL 06] and Gulati [GUL 11] used the Brain and Shapiro test to develop testing procedures for the Pareto, the Rayleigh and the Laplace distributions by transforming them and then testing the transformed data for exponentiality. Here, we extend this concept to include all distributions and test it for some specific distributions characterized by either a scale parameter or both a scale and a shape parameter.

The details of the testing procedure are similar to that discussed in section 2.2 with the exception of the fact that we work with transformed data rather than the actual observations. Assume once again that the ordered observations are given by $X_{(1)} \leq X_{(2)} \leq ... \leq X_{(n)}$. The null hypothesis of interest is $H_0$: the underlying cumulative distribution function (c.d.f) is given by $F_0(x)$, where $F_0$ is a continuous distribution whose form is specified (although the parameters may not be completely specified).

1) Transform the data so that the transformed observations are exponential under the null hypothesis. Denote the ordered transformed observations by $W_{(1)} \leq W_{(2)} \leq ... \leq W_{(n)}$. As an example, note that if $X$ is a Weibull random variable with scale parameter $\theta$ and shape parameter $\alpha$, then $X^{\alpha}$ has an exponential distribution with scale parameter = $\alpha$; or if, for example, X has the type I Pareto distribution with density function $f(x) = \frac{\alpha\sigma^{\alpha}}{x^{\alpha+1}}$, then the variable $Y = \frac{ln(X}{\sigma}$ has an exponential distribution with a mean of $1/\alpha$ .

Here, we distinguish between the following two cases:

i) The underlying parameters are completely specified, so that the transformed observations have the exponential distribution under the null hypothesis.

ii) The underlying parameters are unknown. In this case, we estimate the parameters through the maximum likelihood estimators and then transform the data. In this situation, the transformed variables will have an asymptotic exponential distribution.

2) If the data cannot be transformed directly, then transform the data to the uniform distribution, using the probability integral transform and then transform the resulting uniform random variable to an exponential variable. Thus, the transformation will

be $W_i = -ln(F_0(X_i))$ (or $-ln\hat{F}_0(X_i)$, where $\hat{F}_0$ represents $F_0$ with estimated parameters.)

3) Compute the $i$th weighted spacing as before by $Y_i = (n - i + 1)(W_{(i)} - W_{(i-1)})(i = 1, 2, ..., n)$ and $W_0 = 0$.

4) Compute $t_i$, $Z_1$ and $Z_2$ as in section 2.2.

5) Then, as in section 2.2, the test statistic is given by:

$$Z = Z_1^2 + Z_2^2. \quad [2.5]$$

When the underlying distribution is completely specified, the statistic $Z$ has an asymptotic chi-squared distribution with two degrees of freedom. Note also that since maximum likelihood estimates converge to the true value of the underlying parameters, the test statistic is always parameter free, although our simulation results show that the test statistic is not distribution free. We found that when the underlying distribution is characterized only by a scale parameter, the test statistic $Z$ does have an asymptotic chi-squared distribution with two degrees of freedom. However, when the underlying distribution is characterized by two or more parameters, according to our simulation results, the test statistic has an asymptotic chi-squared distribution with one degree of freedom. Thus, for case (i) mentioned above and for distributions characterized by a scale parameter only, the asymptotic $p$-value of the test will be given by:

$$p - value = \exp^{-Z/2}. \quad [2.6]$$

In case of distributions characterized by both a scale and a shape parameter, the approximate $p$-value of the test is given by:

$$p - value = P(Z \geq T), \quad [2.7]$$

where $T$ is a chi-squared random variable with one degree of freedom.

## 2.4. Small sample null distribution of the test statistic for specific distributions

As a preliminary study, we conducted Monte Carlo simulations to determine the small sample null distribution of $Z$. We concentrated on distributions characterized by a single parameter only (usually a scale parameter) or two parameters (both a scale and a shape parameter.) The purpose of the simulations was twofold. We wanted to determine how much (if any) information was lost by using parameter estimates instead of the true parameter values and if the transformation still preserved exponentiality for small samples. The simulations were conducted for the uniform distribution on (0,1), the Weibull with known shape, the Weibull with unknown shape, the Gumbel distribution and the generalized Pareto distribution (GPD). The sample size varied from 10 to 100 (in increments of 10) and the number of simulation runs was fixed at 10,000. We found that for the uniform and the Weibull distributions with known shape (but unknown scale), the empirical percentiles were fairly close to the percentiles of the chi-squared distribution with two degrees of freedom. The approximation improved with increasing sample size and was generally independent of the true parameter value. However, as mentioned before, for the two-parameter distributions, the null percentiles of $Z$ were closely approximated by a chi-squared distribution with one degree of freedom. This was not surprising since we had observed the same phenomenon in the case of the two-parameter Pareto distribution [GUL 06] and it adheres to the principle of "lose one degree of freedom for every estimated parameter".

Tables 2.1–2.4 give the simulated percentiles for the uniform, the Weibull, the Gumbel and the extreme value distributions. For the Weibull distribution and the GPD, we set the scale parameter as 1.0 without loss of generality and varied the shape parameter only. For the Gumbel distribution, we considered various location and scale parameters.

| $N$ | 90th Percentile | 95th Percentile |
|---|---|---|
| 10 | 4.433 | 6.050 |
| 20 | 4.483 | 5.877 |
| 30 | 4.472 | 5.958 |
| 40 | 4.517 | 5.933 |
| 50 | 4.503 | 5.908 |
| 60 | 4.467 | 5.836 |
| 70 | 4.513 | 5.879 |
| 80 | 4.583 | 5.879 |
| 90 | 4.469 | 5.928 |
| 100 | 4.488 | 5.757 |

**Table 2.1.** *Percentiles for the U(0,1) distribution*

| Shape | 0.5 | | 1.0 | | 1.5 | |
|---|---|---|---|---|---|---|
| Sample Size | P90 | P95 | P90 | P95 | P90 | P95 |
| 10 | 2.924 | 3.758 | 2.981 | 3.766 | 2.902 | 3.740 |
| 20 | 2.839 | 3.797 | 2.936 | 3.837 | 2.811 | 3.759 |
| 30 | 2.819 | 3.826 | 2.849 | 3.878 | 2.820 | 3.824 |
| 40 | 2.834 | 3.854 | 2.817 | 3.876 | 2.839 | 3.832 |
| 50 | 2.809 | 3.932 | 2.769 | 3.779 | 2.807 | 3.910 |
| 60 | 2.775 | 3.833 | 2.776 | 3.790 | 2.847 | 3.801 |
| 70 | 2.838 | 3.934 | 2.696 | 3.755 | 2.750 | 3.847 |
| 80 | 2.751 | 3.854 | 2.812 | 3.886 | 2.782 | 3.906 |
| 90 | 2.772 | 3.959 | 2.858 | 3.958 | 2.843 | 3.941 |
| 100 | 2.816 | 3.914 | 2.797 | 3.908 | 2.813 | 3.922 |
| P90 = 90th percentile & P95 = 95th percentile | | | | | | |

**Table 2.2.** *Percentiles for the Weibull distribution, scale = 1*

| Shape | 0.5 | | 1.0 | | 1.5 | |
|---|---|---|---|---|---|---|
| Sample size | P90 | P95 | P90 | P95 | P90 | P95 |
| 10 | 4.433 | 6.050 | 4.433 | 6.050 | 4.433 | 6.050 |
| 20 | 4.483 | 5.877 | 4.483 | 5.877 | 4.483 | 5.877 |
| 30 | 4.472 | 5.958 | 4.472 | 5.958 | 4.472 | 5.958 |
| 40 | 4.517 | 5.933 | 4.517 | 5.933 | 4.517 | 5.933 |
| 50 | 4.503 | 5.908 | 4.503 | 5.908 | 4.503 | 5.908 |
| 60 | 4.467 | 5.836 | 4.467 | 5.836 | 4.467 | 5.836 |
| 70 | 4.513 | 5.879 | 4.513 | 5.879 | 4.513 | 5.879 |
| 80 | 4.583 | 5.879 | 4.583 | 5.879 | 4.583 | 5.879 |
| 90 | 4.469 | 5.928 | 4.469 | 5.928 | 4.469 | 5.928 |
| 100 | 4.488 | 5.757 | 4.488 | 5.757 | 4.488 | 5.757 |
| P90 = 90th percentile and P95 = 95th percentile. | | | | | | |

**Table 2.3.** *Percentiles for the Weibull distribution with known shape and scale = 1*

| Shape | 0.0 | | 0.2 | | −0.2 | |
|---|---|---|---|---|---|---|
| Sample Size | P90 | P95 | P90 | P95 | P90 | P95 |
| 10 | 26.70 | 28.09 | 26.16 | 27.82 | 26.62 | 28.00 |
| 20 | 4.98 | 22.36 | 2.92 | 4.89 | 22.02 | 25.77 |
| 30 | 2.80 | 3.81 | 2.65 | 3.67 | 3.24 | 5.12 |
| 40 | 2.75 | 3.67 | 2.57 | 3.53 | 2.96 | 4.03 |
| 50 | 2.70 | 3.71 | 2.57 | 3.47 | 2.84 | 3.85 |
| 60 | 2.71 | 3.75 | 2.62 | 3.67 | 2.83 | 3.80 |
| 70 | 2.75 | 3.71 | 2,72 | 3.76 | 2.80 | 3.84 |
| 80 | 2.78 | 3.76 | 2.69 | 3.83 | 2.85 | 3.92 |
| 90 | 2.72 | 3.86 | 2.73 | 3.76 | 2.90 | 3.96 |
| 100 | 2.75 | 3.84 | 2.65 | 3.76 | 2.90 | 3.92 |
| P90 = 90th percentile and P95 = 95th percentile. | | | | | | |

**Table 2.4.** *Percentiles for the generalized Pareto distribution, scale = 1*

## 2.5. Power studies

Monte Carlo studies were conducted to determine the power of the procedure. Previous studies have shown that this procedure generally has high power, see [BRA 83], [GUL 06] and [GUL 11]. Here, we studied the power of the procedure for the uniform distribution, for the Gumbel distribution, for the Weibull distribution with known and unknown shapes and for the GPD. All these studies were based on 10,000 simulation runs with sample sizes generally ranging between 10 and 100 (in increments of 10). In all cases, we chose alternate distributions that were skewed as well as symmetric. We found that the power of the procedure was extremely high for the uniform distribution for left skewed and symmetric alternatives and for the Weibull distribution with known shape. For the Gumbel distribution, the power depends on the alternate distribution but tends to be high in general. However, for the Weibull distribution and the GPD, the procedure tends to have low power. Some of these results are presented in Tables 2.5–2.9 for $n = 20$ and 50 and for $\alpha = 0.1$. The power results for the GPD are displayed for $n$= 30, 50 and 80 instead. This is because, as indicated in Table 2.3, the test statistic for this distribution is not approximated by the chi-squared distribution for $n \leq 30$.

## 2.6. Some real examples

Choulakian and Stephens [CHO 01] examined $n = 72$ exceedances of the Wheaton River near Carcross in Yukon Territory, Canada. They tested if the data came from the GPD using both the Cramér–von Mises test as well as the Anderson–Darling test and found that both tests reject the GPD as a model for the data. Using the $Z$ test for the GPD as detailed in this chapter, we found that the maximum likelihood estimates (MLES) for the shape and scale parameters were given by 0.00932 and 12.1928 with a $Z$ value of 5.954. Comparing that to a chi-squared distribution with one degree of freedom, we find that our test rejects the GPD as a possible model for the exceedances as well. Besides testing the data for the GPD, we also decided to test to see if the data came from the Weibull or the Gumbel distribution. We found that the test statistic for the Weibull distribution was given by 4.207, with a corresponding $p - value$ of 0.0402, while that for the Gumbel was given by 80.00196 with a corresponding $p - value$ of 0.00. Thus, none of the three possible tail distributions fit the data. As another example, we considered the daily maximum temperature in Miami during the month of January, 2011. The data are available at http://climatecenter.fsu.edu/climate-data-access-tools/ downloadable-data and are given by: 75 75 77 77 75 76 79 66 73 73 76 80 63 64 71 75 74 73 80 81 78 78 69 68 76 74 75 70 64 68 72. Since the data represent daily maxima, we decided to try to fit them with a Gumbel distribution first. The $Z$ statistic for the Gumbel distribution was given by 141.132 (corresponding $p$ -value 0.0). Thereafter, as in the previous example, we tried to fit the data with the GPD and the Weibull distribution. The $Z$ statistic for the GPD is given by 54.31153 (corresponding $p$-value 0.0) but the test statistic for the Weibull distribution is given by 0.3453 giving a $p$-value of 0.5567. Thus, it seems that the Weibull distribution might provide an adequate fit for the above data.

| Shape | 0.5 | | 0.0 | | 3.0 | |
|---|---|---|---|---|---|---|
| Sample size | P90 | P95 | P90 | P95 | P90 | P95 |
| 10 | 2.87 | 3.70 | 2.95 | 3.74 | 2.96 | 3.77 |
| 20 | 2.81 | 3.73 | 2.94 | 3.86 | 2.80 | 3.75 |
| 30 | 2.82 | 3.87 | 2.77 | 3.83 | 2.78 | 3.84 |
| 40 | 2.77 | 3.82 | 2.77 | 3.85 | 2.74 | 3.80 |
| 50 | 2.90 | 3.93 | 2.70 | 3.66 | 2.77 | 3.83 |
| 60 | 2.83 | 3.86 | 2.82 | 3.83 | 2.75 | 3.75 |
| 70 | 2.78 | 3.80 | 2.78 | 3.83 | 2.88 | 4.01 |
| 80 | 2.84 | 3.83 | 2.80 | 3.97 | 2.85 | 3.97 |
| 90 | 2.76 | 3.92 | 2.79 | 3.83 | 2.82 | 3.95 |
| 100 | 2.74 | 3.98 | 2.79 | 3.79 | 2.82 | 3.91 |
| P90 = 90th percentile and P95 = 95th percentile. | | | | | | |

**Table 2.5.** *Percentiles for the Gumbel distribution with scale = 1*

| Alternate distribution | $n = 20$ | $n = 50$ |
|---|---|---|
| Beta (2,1) – skewed right | 0.0908 | 0.0954 |
| Beta (2,2) – symmetric | 0.5563 | 0.9292 |
| Beta (1,2) – skewed left | 0.5030 | 0.9040 |

**Table 2.6.** *Power of the uniform distribution,* $\alpha = 0.1$

| Alternate distribution | $n = 20$ | $n = 50$ |
|---|---|---|
| Gamma (2, 0.5) | 0.2377 | 0.4597 |
| U(0,1) | 0.1526 | 0.1719 |
| LN(0,1) | 0.7307 | 0.9823 |
| Weibull(0.5,1) | 0.9956 | 1.000 |
| N(0,1) | 0.3177 | 0.6685 |
| t(1) | 0.8674 | 0.9981 |
| t(4) | 0.5097 | 0.8672 |
| HN | 0.2952 | 0.5910 |
| Cauchy | 0.8707 | 0.9960 |

Gamma(a,b) = gamma with shape a and scale b.
Weibull(a, b) = Weibull with shape and scale b.
LN(0, 1) = standard lognormal.
HN = standard half-normal.
t(k) = t with k degrees of freedom.

**Table 2.7.** *Power of the test for the Gumbel distribution,* $\alpha = 0.1$

| Alternate distribution | $n = 20$ | $n = 50$ |
|---|---|---|
| Gamma (1,1) | 0.1105 | 0.0988 |
| Gamma (0.5, 1) | 0.1545 | 0.2140 |
| Gamma (1.5, 1) | 0.1176 | 0.1343 |
| Chisq(1) | 0.1546 | 0.2252 |
| Chisq(4) | 0.1291 | 0.1774 |
| LN | 0.3163 | 0.6726 |
| HN | 0.1516 | 0.2115 |

Gamma(a,b) = gamma with shape a and scale b.
Chisq(k) = chi Square with k degrees of freedom.
LN = standard lognormal.
HN = standard half-normal.

**Table 2.8.** *Power of the test for the Weibull distribution (shape and scale unknown),* $\alpha = 0.1$

| Shape | Alternate distribution | $n = 20$ | $n = 50$ |
|---|---|---|---|
| 0.5 | Gamma (0.5, 1) | 0.2392 | 0.5606 |
| 0.5 | Gamma (1.5, 1) | 0.9990 | 1.000 |
| 0.5 | Chi(1) | 0.2392 | 0.5606 |
| 0.5 | Chi(4) | 1.000 | 1.000 |
| 0.5 | HN | 0.9974 | 1.000 |
| 0.5 | LN | 0.9988 | 1.000 |
| 1.5 | Gamma (0.5, 1) | 0.9738 | 0.9999 |
| 1.5 | Gamma (1.5, 1) | 0.2413 | 0.4241 |
| 1.5 | Chi(1) | 0.9738 | 0.9999 |
| 1.5 | Chi(4) | 0.1291 | 0.1653 |
| 1.5 | HN | 0.1781 | 0.3235 |
| 1.5 | LN | 0.6851 | 0.9530 |
| 2.0 | Gamma (0.5, 1) | 0.9983 | 1.000 |
| 2.0 | Gamma (1.5, 1) | 0.7065 | 0.9729 |
| 2.0 | Chi(1) | 0.9983 | 1.000 |
| 2.0 | Chi(4) | 0.4627 | 0.8017 |
| 2.0 | HN | 0.6211 | 0.9407 |
| 2.0 | LN | 0.9412 | 0.9996 |

**Table 2.9.** *Power of the test for the Weibull distribution (unknown scale),* $\alpha = 0.1$

| Alternate Distribution | $n = 30$ | $n = 50$ | $n = 80$ |
|---|---|---|---|
| Weibull (1.25, 1) | 0.2286 | 0.2872 | 0.3874 |
| Weibull (0.75, 1) | 0.2640 | 0.3666 | 0.5302 |
| Gamma (2.0, 1) | 0.5954 | 0.8100 | 0.9476 |
| LN | 0.3512 | 0.5874 | 0.8248 |

**Table 2.10.** *Power of the test for the generalized Pareto distribution (known scale),* $\alpha = 0.1$

## 2.7. Conclusions

This chapter extends the results of Gulati and Shapiro [GUL 06] and Gulati [GUL 11] to develop a universal test. The procedure transforms the underlying data and tests the transformed data for exponentiality using the regression test of Brain and Shapiro [BRA 83]. The procedure is best suited to distributions that can easily be transformed to the exponential distribution (either directly or indirectly). The test statistic is simple to compute and the $p$-values of the test can be obtained easily from the chi-squared distribution. The test generally has high power for completely specified distributions and for all distributions characterized by a single unknown parameter (usually a scale parameter). The test also exhibits good power results for most two-parameter distributions, with the exception of the Weibull distribution.

## 2.8. Acknowledgment

The authors would like to extend their sincere gratitude to Alan Kimber at the University of Southampton who suggested the idea of a universal test to the second author at a conference.

## 2.9. Bibliography

[BRA 83] BRAIN C.W., SHAPIRO S.S., "A regression test for exponentiality: censored and complete samples", *Technometrics*, vol. 25, pp. 69–76, 1983.

[COX 66] COX D.R., LEWIS, P.A., *The Statistical Analysis of a Series of Events*, Methune, London, 1966.

[CHO 01] CHOULAKIAN V., STEPHENS M., "Goodness-of-fit tests for the generalized Pareto distribution", *Technometrics*, vol. 43, no. 4, pp. 478–484, 2001.

[DAG 86] D'AGASTINO R.B., STEPHENS M., *Goodness of Fit Techniques*, Marcel Dekker, New York, 1986.

[GUL 11] GULATI S., "Goodness of fit for the Rayleigh and the Laplace distributions", *International Journal of Applied Mathematics and Statistics*, vol. 24, pp. 74–85, 2011.

[GUL 06] GULATI S., SHAPIRO S., "Goodness of fit tests for the Pareto distribution", in VONTA F., NIKULIN M., LIMNIOS N., HUBER C. (eds), *Statistical Models and Methods for Biomedical and Technical Systems*, Birkhauser, Boston, pp. 263–277, 2006.

## Chapter 3

# Entropy-type Goodness-of-Fit Tests for Heavy-Tailed Distributions

### 3.1. Introduction

The maximum entropy principle [JAY 57] is a criterion for selecting *a priori* probabilities. For a given amount of information, the probability distribution that best describes our knowledge is the distribution that maximizes the Shannon entropy subject to a given evidence as constraints. Maximum entropy modeling has been successfully applied to computer vision, spatial physics, natural language processing and many other fields.

Forte [FOR 84] provides the Boltzmann–Shannon entropy as:

$$H(f) = -\int_{-\infty}^{\infty} f(x)\log(f(x))dx \qquad [3.1]$$

which measures the amount of uncertainty of the value $x$ of a real-valued random variable $X$ given its probability density function $f(x)$.

For continuous distributions, Jaynes [JAY 63, JAY 68, JAY 03] gave the following formula, which is better known as Kullback–Leibler divergence between the densities

---

Chapter written by Andreas MAKRIDES, Alex KARAGRIGORIOU and Filia VONTA.

$q$ and $p$:

$$H_c = -\int p(x)\log\left(\frac{p(x)}{q(x)}\right)dx. \qquad [3.2]$$

The Shannon entropy for a discrete random variable is not considered to be the discrete analogue of [3.1]. Instead, Forte and Hughes [FOR 88] proposed, as a good candidate for a discrete analogue of [3.1], the quantity:

$$-\sum p_i \log(p_i/(x_i - x_{i-1}))$$

since for a variable defined in $[a, b]$:

$$\lim_{\max_i |x_i - x_{i-1}| \to 0} -\sum_{i=1}^{n} p_i \log(p_i/(x_i - x_{i-1})) = H(f), \qquad [3.3]$$

where $p_i = P[x_{i-1} < X \leq x_i] = \int_{x_{i-1}}^{x_i} f(x)dx$, $i = 1, \ldots, n-1$ and $a = x_0 < \ldots < x_n = b$.

Goodness-of-fit (gof) tests measure the degree of agreement between the distribution of an observed random sample and a theoretical statistical distribution. The problem of gof to any distribution on the real line is frequently treated by partitioning the range of data in $m$ disjoint intervals. In all cases, a test statistic is compared against a known critical value to accept or reject the hypothesis that the sample is from the postulated distribution. Over the years, numerous nonparametric gof methods including the chi-squared test and various empirical distribution function (EDF) tests [D'AG 86] have been developed. At the same time, measures of entropy, divergence and information such as the ones given in [3.1–3.3] are quite popular in gof tests. Over the years, several measures such as [3.2] and [3.3] have been suggested to reflect the fact that some probability distributions are *closer together* than others. Many of the currently used tests, such as the likelihood ratio, the chi-squared, the score and the Wald tests are defined in terms of appropriate measures. Other such tests include those based on entropy [VAS 76, DUD 81, GOK 83] and the $\phi$−family of Csiszar measures [CSI 63, REA 88, ZOG 90, PAR 06] and its generalizations [MAT 10].

In this chapter, we propose an alternative information measure that generalizes [3.3] and is designed for gof tests for heavy-tailed distributions. In such cases, the focus is on the tail part of the distribution and, therefore, a special weight should be placed on observations lying on the tail part. In section 3.2, we provide the test statistic and its asymptotic distribution under the null hypothesis with a discussion in section 3.2.2. In section 3.3, we provide some simulation studies in order to explore the capabilities of the proposed test statistic.

### 3.2. The entropy test for heavy-tailed distributions

#### 3.2.1. *Development and asymptotic theory*

Let $Y_i$, $i = 1, \ldots, n$ be a random sample from a distribution with unknown cumulative distribution function $F$ and consider the following test of fit: For continuous distributions, we propose the following generalization of Forte and Hughes [FOR 88] entropy:

$$S^w(F) = -\sum_{i=1}^{m} w_i \left(F(s_i) - F(s_{i-1})\right) \log \left( \frac{F(s_i) - F(s_{i-1})}{s_i - s_{i-1}} \right), \qquad [3.4]$$

where the $w$s are appropriate weight functions with $0 \leq w_i \leq 1$ and $\sum_{i=1}^{m} w_i = 1$, $m$ is the number of disjoint intervals for partitioning the data range, and $-\infty < a \leq s_1 \leq \ldots \leq s_m \leq b < \infty$ are preassigned partition points. For a properly selected constant $c$, the null hypothesis will be rejected if $|S^w(F_n) - S^w(F_0)| \geq c$ or, even more stringently, if $\sup_w |S^w(F_n) - S^w(F_0)| \geq c$, where $F_n$ is the empirical distribution based on the sample, namely $F_n(x) = n^{-1} \sum_{i=1}^{n} I(Y_i \leq x)$. The test has been proposed in [LEE 11] as a general gof test with general (random) weights. As expected, the above test statistic with general, random weights is appropriate for typical distributions but fails when we are dealing with distributions with heavy tails. Our intention here is to adjust the weights appropriately so that the test will perform satisfactorily for heavy-tailed distributions.

The proposed test statistic is justified by the fact that it is closely related to the entropy measure $H(f)$ given in [3.1]. Indeed, observe that the unweighted form of [3.4] takes the form:

$$\begin{aligned} S_{\max}(F) \quad &\approx \quad -\sum_{i=1}^{m} f(s_i)\,(s_i - s_{i-1}) \log f(s_i) \\ &\overset{m\to\infty}{\longrightarrow} \quad -\int_{-\infty}^{\infty} f(x) \log f(x) dx \equiv H(f), \end{aligned}$$

where $f$ is the probability density function.

As mentioned earlier, there are often cases when it is necessary to test whether a distribution fits the data well mainly in the upper tail and the fit elsewhere is of limited importance. Examples include operational risk and insurance claims modeling, in which gof in the tails determines the value-at-risk (or conditional value-at-risk) and the ruin probabilities. Given the Basel II Capital Accords recommendations, under the loss distribution approach the operational risk capital charge is derived from the value-at-risk measure, which requires an accurate estimate of the upper tail of the loss distribution. Similarly, in insurance science, the upper tail of the claim size distribution is vital to obtain the right estimate of ruin probability In such case, an appropriate weigh function $w_i$ should be defined so that observations in the upper tail are assigned a much higher weight than those in the lower tail. For this purpose, we use the empirical distribution for each of the $m$ subintervals. In particular, let:

$$w_i = \frac{1}{1 - F(\xi_i)}, \qquad [3.5]$$

where $\xi_i$ is any random point within the $i$th subinterval. For the selection of weights, see section 3.2.2.

Observe that if $F_0$ is the uniform distribution in $[0, 1]$, then $S^w(F_0) = 0$. Note that without loss of generality, we can concentrate on the uniform distribution on $[0, 1]$ so that the hypothesis problem becomes:

$$H_0 : F = F_0 \equiv U[0, 1] \text{ vs. } H_1 : F \neq F_0 \equiv U[0, 1]. \qquad [3.6]$$

Indeed, with the use of the probability integral transform we can provide the basis for testing whether the $Y_i$s can reasonably be modeled as arising from *any* specified continuous distribution $F_0$. Specifically, the probability integral transform can be applied to construct the equivalent set of values $F_0(Y_i) = U_i$, and a test then examines whether a uniform distribution is appropriate for the $U_i$s. Note that this transformation is applied for the well-known P-P plots and the Kolmogorov–Smirnov tests.

Although the simple null hypothesis appears frequently in practice, it is common to test the composite null hypothesis that the unknown distribution belongs to a parametric family $\{F_{\boldsymbol{\theta}}\}_{\boldsymbol{\theta} \in \Theta}$, where $\Theta$ is an open subset in $R^k$. In this case, we can again consider a partition of the original sample space with $m$ disjoint intervals. Observe though that in this case, the probability integral transformation depends on the unknown $k-$dimensional parameter $\boldsymbol{\theta}$. Indeed, in this case, a consistent estimator $\hat{\boldsymbol{\theta}}$ is required so that if the null hypothesis is $H_0 : F = F_{\boldsymbol{\theta}}$, then the probability integral transformation is applied for the construction of the values $F_{\hat{\boldsymbol{\theta}}}(Y_i) = U_i$. In this case, the limiting distribution, given in theorem 3.1, can be affected by the estimation of $\boldsymbol{\theta}$. The effect, though, may diminish when $m$ is large and $\max_i(s_i - s_{i-1})$ is small, as in the case of the chi-squared test.

In regard to the estimating method applied for obtaining the estimator of $\boldsymbol{\theta}$, the traditional maximum likelihood estimator (MLE), under the null distribution, can be evaluated and implemented. Note though that one may alternatively consider a wider class of estimators, known as $\Phi-$divergence estimators. More specifically, let the partition $\{E_i\}_{i=1,\ldots,m}$ of the original sample space. Then, the minimum $\Phi-$divergence estimator of $\boldsymbol{\theta}$ is any $\hat{\boldsymbol{\theta}}_\Phi \in \Theta$ satisfying:

$$d_a(\hat{\boldsymbol{\theta}}_\Phi) = \min_{\boldsymbol{\theta}\in\Theta} d_a(\boldsymbol{\theta}) = \min_{\boldsymbol{\theta}\in\Theta} \sum_{i=1}^{m} p_{i0}(\boldsymbol{\theta})\Phi_\alpha\left(\frac{\widehat{p}_i}{p_{i0}(\boldsymbol{\theta})}\right), \; \Phi \in \Phi^*, \; a > 0 \quad [3.7]$$

with

$$p_{i0}(\boldsymbol{\theta}) = \int_{E_i} dF_{\boldsymbol{\theta}}, \; i = 1, \ldots, m,$$

where $\hat{p_i}$ is the MLE of the probability of the partition and $\Phi^*$ is the class of all convex functions $\Phi$ on $[0, \infty)$ such that $\Phi(1) = \Phi'(1) = 0$ and $\Phi''(1) \neq 0$. We also assume the conventions $0\Phi(0/0) = 0$ and $0\Phi(u/0) = \lim_{u\to\infty} \Phi(u)/u, \; u > 0$. Obviously, the resulting estimator depends on the $\Phi$-function chosen. Observe that for $\Phi$ having the special form:

$$\Phi_a(u) = u^{1+a} - \left(1 + \frac{1}{a}\right) u^a + \frac{1}{a}, \; a > 0 \quad [3.8]$$

or $\Phi^1_\alpha(u) = \Phi_a(u)/(1+a)$ and for $a \to 0$ the resulting estimator is the usual MLE for the grouped data. Note that for $\Phi$ as in [3.8], the measure [3.7] reduces to the Basu-Harris-Hjort-Jones (BHHJ) measure of divergence of Basu *et al.* [BAS 98], which was proposed for the development of a minimum divergence estimating method for robust parameter estimation. Basu *et al.* [BAS 98] established that an ideal range for the index $a$ in [3.7] is the interval $[0, 1]$ where the robust features of the estimator are better preserved.

It should be pointed out that as long as the original data are available, it is preferable to rely on the maximum likelihood estimating procedure that provides efficient estimators that are usually simpler and easier to obtain. Furthermore, the associated tests are more powerful than those based on the grouped data. It is for this reason we have chosen to use the MLE procedure for the Monte Carlo simulations in section 3.3.

Theorem 3.1 provides the asymptotic distribution of the proposed test statistic, which for arbitrary weights was established by Lee *et al.* [LEE 11] but continues to hold in our setting without any alterations.

THEOREM 3.1.– Let $U_1, \ldots, U_n$ be a random sample from a continuous distribution with cumulative distribution function $F$. Under $H_0$ given in [3.6], as $n \to \infty$, we have:

$$\sqrt{n} \sup_{\{w\in W\}} |S^w(F_n)| \xrightarrow{d} \sup_{\{w\in W\}} \left|\sum_{i=1}^{m} w_i \left(B(s_i) - B(s_{i-1})\right)\right|, \quad [3.9]$$

where $B(s)$ is the Brownian bridge on [0,1] and $W$ denotes the space of bounded weight functions $w_i : [0,1] \to [0,1]$ with $\sum_{i=1}^{m} w_i = 1$, and $0 = s_0 \leq s_1 \leq \ldots \leq s_m = 1$.

To implement our test in practice, we consider $w_i^{(l)}$, $l = 1, \ldots, L$ different random variables associated with the weights defined in [3.5]. For this purpose, we simulate independent and identically distributed random variables from an $m$-dimensional multinomial distribution with:

$$p_i = \frac{1}{2}[F(x_{max_i}) + F(x_{min_i})]$$

$i = 1, \ldots, m$ where $x_{max_i}$ and $x_{min_i}$ represent the largest and smallest observation in the $i$th interval. Then, if we put:

$$w_i = \frac{1}{1 - p_i}, \quad i = 1, \ldots, m$$

and we have, as $L \to \infty$,

$$\max_{1 \leq l \leq L} \Big| \sum_{i=1}^{m} w_{li} \left( B(s_i) - B(s_{i-1}) \right) \Big| \xrightarrow{d} \sup_{\{w \in W\}} \Big| \sum_{i=1}^{m} w_i \left( B(s_i) - B(s_{i-1}) \right) \Big|. \quad [3.10]$$

Subsequently, by taking $s_i = i/m$, $i = 1, \ldots, m$ for convenience, we can use as the entropy test statistic the quantity:

$$S_{\max}^{w} = \max_{1 \leq l \leq L} \left| \sum_{i=1}^{m} w_{li} \left( E_n\left(\frac{i}{m}\right) - E_n\left(\frac{i-1}{m}\right) \right) \right| \log \left( \frac{E_n\left(\frac{i}{m}\right) - E_n\left(\frac{i-1}{m}\right)}{1/m} \right)$$

$$\overset{d}{\approx} \sup_{\{w \in W\}} \left| \sum_{i=1}^{m} w_i \left( B\left(\frac{i}{m}\right) - B\left(\frac{i-1}{m}\right) \right) \right|. \quad [3.11]$$

Note that the argument in [3.10] is true since

$$P(\overline{\{(w_{l1}, \ldots, w_{lm}) : l \geq 1\}} = W) = 1;$$

otherwise, there exists an open ball $V$ in the hyperplane $W$ such that $p := P((w_{l1}, \ldots, w_{lm}) \in V^c \text{ for all } l \geq 1) > 0$, which is in fact impossible since $p \leq \rho^L$ for any $L$ with $\rho = P((w_{11}, \ldots, w_{1m}) \in V^c) < 1$ (or equivalently $P((w_{11}, \ldots, w_{1m}) \in V) > 0$). This indicates:

$$\sup_{l \geq 1} \Big| \sum_{i=1}^{m} w_{li} \left( B(s_i) - B(s_{i-1}) \right) \Big| = \sup_{\{w \in W\}} \Big| \sum_{i=1}^{m} w_i \left( B(s_i) - B(s_{i-1}) \right) \Big| \text{ a.s.},$$

which immediately implies [3.10].

### 3.2.2. *Discussion*

In regard to the choice of the value of $L$ for practical purposes, we recommend the use of $L = 1,000$. We have run a number of simulations with the uniform distribution and some non-uniform distributions (Weibull, beta, gamma, inverse Gaussian [IG]) with values of $L$ ranging from as low as 100 to as high as 10,000. Both the size and the power of the test have been evaluated with various alternatives. The results clearly show that a value of $L$ between 500 and 2,000 is sufficiently good for both the size and the power of the test in all cases to be stabilized. As a result, the choice of $L = 1,000$ appears to be satisfactory for all practical purposes. The results presented in the simulation section are based on this choice of $L$.

In regard to the number of classes $m$ required in tests of fit, we should point out that even for the case of the $\chi^2$, in spite of its wide use, it is still not clear how many partition points should be used and how the class intervals should be formed. However, it is generally recommended that researchers use equiprobable partitions [STU 99].

The problem of determining the optimum number of classes dates back to the 1940s and 1950s with the work of Mann and Wald [MAN 42] and Cochran [COC 52] who proposed similar techniques for the case of the chi-squared test. More recently, Harrison [HAR 85] proposed a generalization of the Mann and Wald method. These methods, however, arrive at diverse conclusions. Although Mann and Wald's recommendation for at most 24 classes is based on asymptotic theory, they suggest that the results hold approximately for sample sizes as low as 200 and may be true for considerably smaller samples. Harrison [HAR 85] found that at least 23 classes can ensure a power equal to 0.5. However, various numerical studies have presented empirical evidence to show that such values of $m$ are too large, resulting in loss of power in many situations. Williams [WIL 50] indicates that the value of $m$ may be halved for practical purposes, without relevant loss of power. See also [DAH 73], who suggest values of $m$ between 3 and 12 for several different alternatives in testing normality for sample sizes of $n = 50$ and 100. Finally, Cochran suggested a fixed number of classes equal to 5. In almost all these cases, the optimum number of classes depends on various equally important factors such as the type of alternative hypothesis under consideration, the "distance" between the alternative hypothesis and the data, the minimum power to be achieved, the significance level of the test and the size of the sample.

We should be always aware of the danger associated with the use of too many classes in cases where the observations are spread too thinly over the data range. Another important factor is the gain anticipated when the number of classes increases. The power of the test together with the size is the key factor in deciding if the increase in the number of classes is useful. It should be noted that even in cases where the number of classes is allowed to increase as the sample size increases, specific assumptions should be imposed in order to secure satisfactory asymptotic results. One such assumption is the well-known sparseness assumption according to which the limit as $n \to \infty$, of $n/m$, where $m$ depends on $n$, is finite. Results of this type have been reported by Holst [HOL 72], Dale [DAL 86], and Read and Cressie [REA 88]. In our analysis where we dealt with samples with less than 100 observations, we have considered 3–20 classes and concluded that for classes at most equal to 10, the results are quite satisfactory as compared with other tests commonly used in the literature.

### 3.3. Simulation study

In this section, we report simulation results for the proposed entropy test for small, medium and large sample sizes with $L = 1{,}000$. For this task, we use samples of size 10, 20, 30, 50, 100, 300, 500 and 1,000 and values of $m$ equal to 3, 4, 5 and 10. The Monte Carlo simulations are based on 10,000 repetitions. The results presented are for $n = 10, 20, 30, 50$ and 100.

For our investigation, we consider various continuous distributions and apply the probability integral transformation. In this case, we have:

$$H_0 : Y \sim F = F_0 \Leftrightarrow H_0 : U = F_0(Y) \sim U[0,1].$$

We choose to focus on distributions such as the exponential, lognormal, gamma, Pareto, IG and Weibull, which frequently appear in biomedicine, engineering and reliability. For instance, the family of the two-parameter IG distribution is one of the basic models for describing positively skewed data that arise in a variety of fields of applied research such as cardiology, hydrology, demography, linguistics, employment service, etc. Recently, Huberman *et al.* [HUB 98] have argued and demonstrated the appropriateness of the IG family for studying the Internet traffic and, in particular, the number of pages visited per user within an Internet site. Furthermore, distributions such as the Weibull are frequently encountered in survival modeling where the existence of censoring schemes makes the determination of the proper distribution an extremely challenging problem. Finally, distributions such as the exponential, the gamma, the lognormal and others are very common in lifetime problems. For the null distributions, we focus on the gamma, Weibull and IG distributions. Since there are often limitations regarding the number of alternatives a test can detect [JEN 00], we have chosen to investigate the performance of the tests of fit in relation to the degree of skewness of the distribution. For instance, the shape parameter of the gamma, the Weibull, the lognormal and the IG distributions corresponding to a skewness equal to 1.414 and 2.000 are 2 and 1, 1.26 and 1, 0.42 and 0.551, and 4.5 and 2.25, respectively. Observe that the gamma and the Weibull distributions with shape parameter equal to 1 coincide with the exponential distribution. Distributions with the same skewness values have been used as possible alternatives in each case. The scale parameter is considered to be equal to 1 in all cases, since all three distributions are scale invariant. Other skewness values considered are 0.707, 1.000 and 2.828 but due to space limitations the results are not presented here. The number of intervals used are $m = 3, 4, 5$ and 10 and we focus on small and medium sample sizes, namely $n = 10, 20, 30, 50$ and 100. For completeness, we also present, for $n = 100$, the best alternative test, among known tests in the literature, for the IG. The best alternative tests appear to be the Anderson-Darling (AD) test and the Φ-test of Vonta *et al.* [VON 12].

Note that although results are not presented, the size of the test is quite satisfactory and very close to the nominal level. Of course, as expected, when the sample size is very small and the number of classes is very large, the results should be treated with caution, since the intervals constructed may contain a very small number of observations.

| | | Maximum entropy test | | | | | BA test |
|---|---|---|---|---|---|---|---|
| $m$ | Shape | 10 | 20 | 30 | 50 | 100 | 100 |
| 3 | | 0.043 | 0.036 | 0.058 | 0.083 | 0.129 | |
| 4 | Gamma | 0.094 | 0.105 | 0.177 | 0.249 | 0.469 | |
| 5 | 2 | 0.107 | 0.163 | 0.206 | 0.367 | 0.650 | 0.888(AD) |
| 10 | | 0.130 | 0.304 | 0.449 | 0.692 | 0.894 | |
| 3 | | 0.352 | 0.590 | 0.801 | 0.947 | 0.999 | |
| 4 | Wei | 0.448 | 0.703 | 0.886 | 0.988 | 0.999 | |
| 5 | 1.26 | 0.448 | 0.797 | 0.930 | 0.994 | 1.000 | 0.992(AD) |
| 10 | | 0.547 | 0.873 | 0.973 | 1.000 | 1.000 | |
| 3 | | 0.302 | 0.412 | 0.683 | 0.872 | 0.998 | |
| 4 | Ln | 0.427 | 0.690 | 0.851 | 0.973 | 1.000 | |
| 5 | 0.42 | 0.488 | 0.771 | 0.916 | 0.994 | 1.000 | 0.4767 (Φ**) |
| 10 | | 0.588 | 0.902 | 0.980 | 1.000 | 1.000 | |

*Φ-test [VON 12].

**Table 3.1.** *Inverse Gaussian null model: power and size of the maximum entropy test for* $n = 10, 20, 30, 50$ and 100 *and best alternative (BA) test for* $n = 100$

According to the selective results presented in Tables 3.1 and 3.2, we conclude that, as expected, the power of the test increases as the sample size increases for all settings considered. When the sample size is small, it is natural to choose a relatively small number of classes/intervals otherwise some intervals may be empty. However, for large sample sizes, the number of intervals is of limiting importance, although larger rather than smaller values appear to give better results. In addition, we conclude that the power study reveals excellent discriminatory ability of the proposed entropy test when the null distribution is the IG and the alternative distribution is Weibull or lognormal. This last result appears to be quite important since it is known that the IG and lognormal distributions are indistinguishable under standard testing techniques (Kolmogorov, AD, etc.). In other situations, we observe relatively poor discriminatory ability, especially when the two competing distributions are extremely close to each other.

| | | Maximum entropy test | | | | |
|---|---|---|---|---|---|---|
| $m$ | Shape | 10 | 20 | 30 | 50 | 100 |
| 3 | | 0.000 | 0.005 | 0.006 | 0.138 | 0.796 |
| 4 | Gamma | 0.000 | 0.001 | 0.007 | 0.063 | 0.690 |
| 5 | 2 | 0.000 | 0.002 | 0.014 | 0.080 | 0.682 |
| 10 | | 0.000 | 0.000 | 0.022 | 0.427 | 0.986 |
| 3 | | 0.092 | 0.092 | 0.216 | 0.886 | 1.000 |
| 4 | Wei | 0.073 | 0.230 | 0.410 | 0.788 | 1.000 |
| 5 | 1.26 | 0.061 | 0.123 | 0.308 | 0.804 | 1.000 |
| 10 | | 0.027 | 0.131 | 0.372 | 0.899 | 1.000 |
| 3 | | 0.037 | 0.037 | 0.049 | 0.081 | 0.092 |
| 4 | Pareto | 0.065 | 0.106 | 0.172 | 0.284 | 0.395 |
| 5 | 8 | 0.081 | 0.166 | 0.238 | 0.382 | 0.615 |
| 10 | | 0.114 | 0.226 | 0.388 | 0.592 | 0.838 |

**Table 3.2.** *Lognormal null model: power and size of the maximum entropy test for* $n = 10, 20, 30, 50$ *and* 100

### 3.4. Conclusions

In conclusion, we have proposed a maximum entropy test for heavy-tailed distributions and established its asymptotic distribution. Taking into consideration the size and the power results, the Monte Carlo simulation study for various distributions and a variety of values of $m$ has clearly shown a satisfactory performance of the proposed entropy test for small, medium and large sample sizes. Finally, note the applicability of the proposed method in various classes of lifetime distributions for describing several aging criteria such as the classes of the increasing failure rate (IFR), the decreasing failure rate (DFR) and the new better than used (NBU) where the exponential null model is examined against all other members of the class.

### 3.5. Bibliography

[AND 54] ANDERSON T.W., DARLING D.A., "A test of goodness of fit", *Journal of American Statistical Association*, vol. 49, pp. 765–769, 1954.

[BAS 98] BASU A., HARRIS I.R., HJORT N.L., *et al.*, "Robust and efficient estimation by minimising a density power divergence", *Biometrika*, vol. 85, pp. 549–559, 1998.

[CHH 77] CHHIKARA R.S., FOLKS J.L., "The inverse Gaussian distribution as a lifetime model", *Technometrics*, vol. 19, no. 4, pp. 461–468, 1977.

[COC 52] COCHRAN W.C., "The $\chi^2$ test of goodness of fit", *Annals of Mathematical Statistics*, vol. 23, pp. 315–345, 1952.

[CSI 63] CSISZAR I., *Eine Informationstheoretische Ungleichung und ihre Anwendung auf den Beweis der Ergodizität von Markoffschen Ketten*, Mathematical Institute of the Hungarian Academy of Sciences, vol. 8, pp. 84–108, 1963.

[D'AG 86] D'AGOSTINO R.B., STEPHENS M.A., *Goodness-of-Fit Techniques*, Marcel Dekker, New York, 1986.

[DAH 73] DAHIYA R.C., GURLAND J., "How many classes in the Pearson chi-square test?", *Journal of the American Statistical Association*, vol. 68, pp. 678–689, 1973.

[DAL 86] DALE J.R., "Asymptotic normality of goodness-of-fit statistics for sparse product multinomials", *Journal of the Royal Statistical Society, Series B*, vol. 48, pp. 48–59, 1986.

[DUD 81] DUDEWICZ E.J., VAN DER MEULEN E.C., "Etropy-based tests of uniformity", *Journal of the American Statistical Association,* vol. 76, pp. 967–974, 1981. Review, *Journal of the Royal Statistical Society*, vol. 40, pp. 263–289.

[FOR 84] FORTE B., "Entropies with and without probabilities. Applications to questionnaires", *Information Processing and Management*, vol. 20, pp. 397–405, 1984.

[FOR 88] FORTE B., HUGHES W., "The maximum entropy principle: a tool to define new entropies", *Reports on Mathematical Physics*, vol. 26, pp. 227–235, 1988.

[GOK 83] GOKHALE D.V., "On entropy-based goodness of fit test", *Computational Statistics and Data Analysis* vol. 1, pp. 157–165, 1983.

[HAR 85] HARRISON R.H., "Choosing the optimum number of classes in the chi-square test for arbitrary power", *Sankhya B*, vol. 47, pp. 319–324, 1985.

[HOL 72] HOLST L., "Asymptotic normality and efficiency for certain goodness-of-fit tests", *Biometrika,* vol. 59, pp. 137–145, 1972. [*Correction*, vol. 59, pp. 699, 1972].

[HUB 98] HUBERMAN B.A., PIROLLI P.L.T., PITKOW J.E., *et al.*, "Strong regularities in world wide web surfing", *Science*, vol. 280, pp. 95–97, 1998.

[JAY 57] JAYNES E.T., "Information theory and statistical mechanics", *Physical Review*, vol. 106, pp. 620–630, 1957.

[JAY 63] JAYNES E.T., "Information theory and statistical mechanics", in FORD K. (ed.), *Statistical Physics*, Benjamin, New York, pp. 181–218, 1963.

[JAY 68] JAYNES E.T., "Prior probabilities", *IEEE Transactions on Systems Science and Cybernetics*, vol. SSC-4, pp. 227–241, 1968.

[JAY 03] JAYNES E.T., "Note on thermal heating efficiency", *American Journal of Physics*, vol. 71, pp. 180–182, 2003.

[JEN 00] JENSSEN A., "Global power functions of goodness of fit tests", *Annals of Statistics*, vol. 28, pp. 239–253, 2000.

[KOU 12] KOUTROUVELIS I., KARAGRIGORIOU A., "Cumulant plots and goodness of fit tests for the inverse Gaussian distribution", *Journal of Statistical Computation and Simulation*, vol. 82, no. 3, pp. 343–358, 2012.

[LEE 11] LEE S., VONTA I., KARAGRIGORIOU A., "A maximum entropy type test of fit", *Computational Statistics & Data Analysis*, vol. 55, pp. 2635–2643, 2011.

[MAN 42] MANN H.B., WALD A., "On the choice of the number of class intervals in the application of the chi-square test", *Annals of Mathematical Statistics*, vol. 13, pp. 306–317, 1942.

[MAT 10] MATTHEOU K., KARAGRIGORIOU A., "A new family of divergence measures for tests of fit", *Australian & New Zealand Journal of Statistics*, vol. 52, pp. 187–200, 2010.

[PAR 06] PARDO L., *Statistical Inference Based on Divergence Measures*, Chapman & Hall, 2006.

[REA 88] READ R.C., CRESSIE N., *Goodness-of-Fit Statistics for Discrete Multivariate Data*, Springer, New York, 1988.

[STU 99] STUART A., ORD J.K., ARNOLD S., *Kendall's Advanced Theory of Statistics*, vol. 2A, 6th ed., Edward Arnold, London, 1999.

[VAS 76] VASICEK O., "A test of normality based on sample entropy", *Journal of the Royal Statistical Society: Series B*, vol. 38, pp. 54–59, 1976.

[VON 31] VON MISES R., *Wahrscheinlichkeitsrechnung*, Leipzig-Vienna, 1931.

[VON 12] VONTA F., MATTHEOU K., KARAGRIGORIOU A., "On properties of the $(\Phi, a)$-power divergence family with applications in goodness of fit tests", *Methodology and Computing in Applied Probability*, vol. 14, pp. 335–356, 2012.

[WAT 61] WATSON G.S., "Goodness-of-fit tests on the circle", *Biometrika*, vol. 48, pp. 109–114, 1961.

[WIL 50] WILLIAMS JR. C.A., "On the choice of the number and width of classes for the chi-square test of goodness-of-fit", *Journal of the American Statistical Association*, vol. 45, pp. 77–86, 1950.

[ZHA 02] ZHANG J., "Powerful goodness-of-fit tests based on likelihood ratio", *Journal of the Royal Statistical Society: Series B*, vol. 64, no. 2, pp. 281–294, 2002.

[ZOG 90] ZOGRAFOS K., FERENTINOS K., PAPAIOANNOU T., "$\phi$-divergence statistics: sampling properties, multinomial goodness of fit and divergence tests", *Communications in Statistics Theory and Methods*, vol. 19, pp. 1785–1802, 1990.

# Chapter 4

# Penalized Likelihood Methodology and Frailty Models

## 4.1. Introduction

The availability of massive data along with new scientific problems have reshaped statistical thinking and data analysis in recent years. To this end, variable selection is fundamental, especially in high-dimensional statistical modeling, and has received much attention in the literature. In recent years we have seen an increased interest in penalized least squares and penalized likelihood methods, to which most of the well-known variable selection criteria are closely related. Obviously, parsimonious models are always desirable. They provide simple and easily interpretable relations among variables/features. Furthermore, forecasting errors are reduced. As a result, innovative variable selection procedures are needed which can be applied to diverse fields of sciences, such as medicine, biology, public health, epidemiology, engineering, geology, finance and economics among others.

Traditional variable selection methods require subset selection, such as the best subset selection and stepwise deletion. Although they seem useful, these procedures ignore stochastic errors. Moreover, the best subset selection in combination with the commonly used Akaike Information Criterion (AIC) (Akaike [AKA 73] and [AKA 74]) and Bayesian Information Criterion (BIC) (Schwartz [SCH 78]), suffers from lack of stability, as is noted in Breiman [BRE 96]. However, these two criteria were the first which essentially suggested a unified approach to variable and model

---

Chapter written by Emmanouil ANDROULAKIS, Christos KOUKOUVINOS and Filia VONTA.

selection, namely, the use of a form of a penalized log-likelihood, which must be maximized with respect to the parameter of interest $\boldsymbol{\beta}$:

$$l(\boldsymbol{\beta}) - \lambda||\boldsymbol{\beta}||_0 \quad [4.1]$$

where the $L_0$–norm counts the number of non-vanishing components in $\boldsymbol{\beta}$ and $\lambda \geq 0$ is a regularization parameter. Specifically, suppose that the data are of the form $(\mathbf{x}_i^T, y_i)_{i=1}^n$ where $\mathbf{x}_i$ is the $d$-dimensional vector of covariates and $y_i$ is the $i$th observation of the response vector $\mathbf{y}$. Akaike [AKA 73], [AKA 74] suggested that a penalty of $d$ should be deducted from the log-likelihood. As a result, Akaike proposed the use of:

$$-l(\hat{\boldsymbol{\beta}}) + \lambda dim(\hat{\boldsymbol{\beta}}). \quad [4.2]$$

The use of $\lambda = 1$ leads to the AIC. Moreover, $dim(\hat{\boldsymbol{\beta}})$ is the number of predictors or the dimension of the covariate vector in the model and $\hat{\boldsymbol{\beta}}$ is the maximum likelihood estimator (MLE) of $\boldsymbol{\beta}$. The above expression is minimized rather than maximized, due to the change of sign. On the other hand, Schwartz [SCH 78] proposed the BIC which has the form:

$$-l(\hat{\boldsymbol{\beta}}) + (\log n/2) dim(\hat{\boldsymbol{\beta}}). \quad [4.3]$$

It has a similar form as that of [4.2], but with the use of $\lambda = \log n/2$. As a result, many traditional methods can be regarded as penalized likelihood methods, with different choices of $\lambda$. For example, the $C_p$ criterion of Mallows [MAL 73], and the criteria of Risannen [RIS 78] and Hannan and Quinn [HAN 79] have similar expressions. However, it must be noted that the computation of the penalized $L_0$ problem [4.1] is a combinatorial problem with NP-complexity. Therefore, the computation is infeasible in high dimensional setups.

In order to retain the virtues of subset selection and to avoid its instability, many authors use a more general form of penalized log-likelihood. This is achieved through the use of different penalty functions. Observe that [4.1] can be generalized as:

$$l(\boldsymbol{\beta}) - n \sum_{j=1}^{d} p_\lambda(|\beta_j|) \quad [4.4]$$

where $p_\lambda(|\beta_j|)$ is an appropriate penalty function indexed by the regularization parameter $\lambda$. The aim is to maximize [4.4] so as to select the significant variables and estimate the regression coefficients simultaneously. As a result, insignificant variables are automatically deleted by estimating their coefficients as zero.

An example of the aforementioned approach is the bridge regression or equivalently the penalized $L_p$−regression of Frank and Friedman [FRA 93], in which $p_\lambda(|\beta_j|) = \lambda|\beta_j|^p$ for $0 < p \leq 2$. Bridge regression intermediates the best subset ($L_0$ penalty) and the ridge regression ($L_2$ penalty). Concerning the latter, it must be noted that the ridge estimator regularizes and stabilizes the estimator, but introduces biases and also does not shrink any coefficient directly to zero [FAN 06]. Furthermore, the non-negative garrote of Breiman [BRE 95] shares the same spirit as that of bridge regression. Another example is the penalized $L_1$−regression, known as the Least Absolute Shrinkage and Selection Operator (LASSO) method and proposed by Tibshirani [TIB 96] in the ordinary regression setting, for which $p_\lambda(|\beta|) = \lambda|\beta|$, for $\beta \in R$. Moreover, we must also mention the hard thresholding penalty $p_\lambda(|\beta|) = \lambda^2 - (|\beta| - \lambda)^2 I(|\beta| < \lambda)$ (Antoniadis [ANT 97]), where $I(\cdot)$ is an indicator function. In the same spirit as LASSO, the penalized likelihood with nonconcave penalty functions has been proposed to select significant variables for various parametric models, including generalized linear and robust linear models [FAN 01] and [FAP 04], along with some semiparametric models, such as the Cox model and partially linear models [FAN 02], [FAN 04] and [CAI 05].

The previous discussion gives rise to an obvious question: How can we choose an appropriate penalty function for variable selection? To this end, Antoniadis and Fan [ANT 01] discussed how to choose penalty functions for wavelet regression. In addition, Fan and Li [FAN 01] proposed a family of new variable selection methods based on a nonconcave penalized likelihood approach and advocated penalties with the following properties:

– Unbiasedness: the resulting estimator is nearly unbiased, when the true unknown parameter is large, in order to avoid modeling bias.

– Sparsity: the resulting estimator is a thresholding rule, namely, it automatically sets small estimated coefficients to zero. As a result, model complexity is reduced.

– Continuity: the resulting estimator is continuous in the data in order to reduce instability in model prediction.

We must stress that the $L_p$ penalty with $0 \leq p < 1$ does not satisfy the continuity condition, whereas for $p > 1$ it does not satisfy the sparsity condition. Moreover, the $L_1$ penalty does not satisfy the unbiasedness condition. To this end, Fan [FAN 97] and Fan and Li [FAN 01] introduced the Smoothly Clipped Absolute Deviation (SCAD) penalty, the first derivative of which is defined by $p'_\lambda(\beta) = \lambda\Big\{I(\beta \leq \lambda) + \frac{(\alpha\lambda-\beta)_+}{(\alpha-1)\lambda} I(\beta > \lambda)\Big\}$, for some $\beta > 0$ and $\alpha > 2$, with $p_\lambda(0) = 0$. For the choice of $\alpha$, according to the relevant literature [FAN 01], the value $\alpha \approx 3.7$ appears to perform quite satisfactorily in numerous variable selection problems. This penalty satisfies all the previously mentioned properties. Another penalty similar to SCAD, is the minimax concave penalty (MCP) of Zhang [ZHA 07], whose derivative is given by $p'_\lambda(\beta) = (\alpha\lambda - \beta)_+/\alpha$. We must also mention the family of concave penalties that bridge the $L_0$ and $L_1$ of Lv and Fan [LV 09], used for model selection and sparse recovery. In addition, the elastic net proposed by Zou and Hastie [ZOU 05] is a linear combination of the $L_0$ and $L_1$ penalties.

The maximization of [4.4] is an optimization problem, for which several algorithms have been proposed. Among others, the following are the most well known. Firstly, Fan and Li [FAN 01] proposed the local quadratic approximation (LQA) algorithm for optimizing nonconcave penalized likelihood. However, maximizing the penalized likelihood function is computationally challenging, because the objective function is nondifferentiable and nonconcave. In order to overcome this problem, Zou and Li [ZOU 08] developed a unified algorithm based on the local linear approximation (LLA) for maximizing the penalized likelihood, which reduces the computational cost. Another known algorithm is the least angle regression (LARS) proposed by Efron *et al.* [EFR 04]. Furthermore, Zhang [ZHA 07] proposed the penalized linear unbiased selection (PLUs) algorithm which can be used for the computation of the solution path in the penalized least squares problem. For the optimization of the same problem, a coordinate descent algorithm was proposed by Fu [FU 98], Daubechies *et al.* [DAU 04] and Wu and Lange [WU 08]. However, it can also be used in the more general case of penalized likelihood. See for example Fan and Lv [FAN 11], who developed the iterative coordinate ascent (ICA) algorithm. A similar approach was studied in Zhang and Li [ZHL 11], who introduced the ICM algorithm.

## 4.2. Penalized likelihood in frailty models for clustered data

The penalized likelihood approach of Fan and Li [FAN 01] was extended to the penalized Cox proportional hazards model and the Gamma frailty model [FAN 02], for the case where we have survival data with censored observations and censored clustered data, respectively. It must be noted that similarly to the Cox regression model there are some challenges in parameter estimation for the frailty model. Specifically, the profile and penalized likelihood do not have a closed form solution and this poses a problem in finding the estimates of the parameters of interest.

We now put the emphasis on the penalized frailty models framework. It is commonly known that frailty models are very effective in describing dependency within a cluster of observations or heterogeneity between clusters. They are mainly used when the proportional hazard assumption might be violated in some situations. As a special case we could be dealing with univariate frailty models in which each individual has his/her own frailty. To this end, a frailty term which follows a specific distribution, is added, which corresponds to a random block effect that acts multiplicatively on the hazard rates of all subjects in a group [FAN 02]. As a result, it is frequently assumed that given the frailty $u_i$, the data in the $i$th group are independent.

Various frailty distributions are discussed in several papers in the literature. Among others, the interested reader is referred to Hougaard [HOU 84], [HOU 86], McGilchrist and Aisbett [MCG 91], Aalen [AAL 98], Lee and Klein [LEE 88], Lindley and Singpurwalla [LIN 86] and the references therein. For the estimation of parameters in a class of frailty models in the univariate data case, see Vonta [VON 96], Slud and Vonta [SLU 04] and Kosorok *et al.* [KOS 04].

The methodology of Fan and Li [FAN 02], which was limited to the Gamma distribution case, was extended in Androulakis *et al.* [AND 12a], [AND 12b] to a whole class of frailty models, in which a generalized form of the penalized likelihood function designed for clusters is given (see also Hougaard [HOU 00] for the non-penalized case). This approach allows the direct use of many continuous distributions for the frailty parameter. As a result, this procedure can be considered very effective for parameter estimation and variable selection purposes. Three theoretical examples were analyzed, concerning the cases of the Gamma frailty (with a likelihood that slightly corrects the corresponding likelihood of Fan and Li [FAN 02]) and the Inverse Gaussian frailty [AND 12a], as well as the Uniform frailty case [AND 12b]. Note that, in practice, it is difficult to choose the right frailty distribution. It is better to examine each problem separately since the choice of frailty distribution depends on the problem considered. For example, the Gamma frailty model is mostly used to describe high late dependence while the Inverse Gaussian frailty model describes moderate dependence along the whole range of times. The Uniform frailty case could be considered for example when one analyzes data coming from the reliability of two-component systems exposed to the same environmental random stress factor. When the operating environment of the system differs from the laboratory environment, it makes sense to restrict the support of the random stress factor to a fixed interval $[a, b]$ [LEE 88].

We now provide a description of the work of Androulakis *et al.* [AND 12a] and [AND 12b]. Consider the frailty model in the case of clusters where the subjects in the same cluster share the same frailty. The frailty term is a positive random variable. Let $n$ be the number of clusters, $J_i$ the size of the cluster $i$, $i = 1, \ldots, n$ and $u_i$ the shared frailty in the cluster $i$ with c.d.f. $F_{u_i}$. The data are described by the triplet $(z_{ij}, \delta_{ij}, \mathbf{x}_{ij})$ where $z_{ij} = \min\{T_{ij}, C_{ij}\}$, $T_{ij}$ is the survival time, $C_{ij}$ the censoring time, $\delta_{ij}$ the censoring indicator and $\mathbf{x}_{ij}$ the $d-$dimensional vector of covariates of the $j$th individual belonging to the cluster $i$. Also assume that the hazard rate of the $j$th subject in the $i$th subgroup-cluster is given conditionally on the covariates $\mathbf{x}_{ij}$ and the shared frailty $u_i$ as:

$$h_{ij}(t|\mathbf{x}_{ij}, u_i) = u_i h_0(t) e^{\mathbf{x}_{ij}^T \boldsymbol{\beta}} \qquad [4.5]$$

where $\beta \in R^d$ is the parameter of interest and $h_0(t)$ the baseline hazard function. The corresponding survival function is:

$$S_{ij}(t|\mathbf{x}_{ij}, u_i) = e^{-u_i H_0(t) e^{\mathbf{x}_{ij}^T \beta}}, \quad i = 1, ..., n, \quad j = 1, ..., J_i \qquad [4.6]$$

where $H_0(t)$ is the cumulative baseline hazard function.

Given the frailty $u_i$, all individuals in the group $i$ are considered to be independent. Therefore, the joint survival of the group $i$ given $u_i$ is:

$$\prod_{j=1}^{J_i} S(z_{ij}|\mathbf{x}_{ij}, u_i) = e^{-u_i \sum_{j=1}^{J_i} e^{\mathbf{x}_{ij}^T \beta} H_0(z_{ij})}.$$

In the censored case, the contribution of the $i$th cluster in the likelihood, by integrating out the frailty $u_i$, is:

$$\int_0^\infty \prod_{j=1}^{J_i} h_{ij}(z_{ij}|\mathbf{x}_{ij}, u_i)^{\delta_{ij}} S_{ij}(z_{ij}|\mathbf{x}_{ij}, u_i)\, dF_{u_i}(u_i)$$

which by [4.5] and [4.6] is equal to:

$$\left(\prod_{j=1}^{J_i} (h_0(z_{ij}) e^{\mathbf{x}_{ij}^T \boldsymbol{\beta}})^{\delta_{ij}}\right) \int_0^\infty \exp(-u_i \sum_{j=1}^{J_i} e^{\mathbf{x}_{ij}^T \boldsymbol{\beta}} H_0(z_{ij}))\, u_i^{A_i}\, dF_{u_i}(u_i)$$

where $A_i = \sum_{j=1}^{J_i} \delta_{ij}$.

The full likelihood for all clusters is equal to:

$$\prod_{i=1}^{n} \left\{ \left(\prod_{j=1}^{J_i} (h_0(z_{ij}) e^{\mathbf{x}_{ij}^T \boldsymbol{\beta}})^{\delta_{ij}}\right) \int_0^\infty \exp(-u_i \sum_{j=1}^{J_i} e^{\mathbf{x}_{ij}^T \boldsymbol{\beta}} H_0(z_{ij}))\, u_i^{A_i}\, dF_{u_i}(u_i) \right\}$$

while the full loglikelihood is equivalently written as:

$$\sum_{i=1}^{n} \sum_{j=1}^{J_i} \delta_{ij} \ln(h_0(z_{ij})) + \sum_{i=1}^{n} \sum_{j=1}^{J_i} \delta_{ij} \mathbf{x}_{ij}^T \boldsymbol{\beta} + \sum_{i=1}^{n} \ln\left( \left| L^{(A_i)}\Big(\sum_{j=1}^{J_i} e^{\mathbf{x}_{ij}^T \boldsymbol{\beta}} H_0(z_{ij})\Big) \right| \right) \qquad [4.7]$$

where:

$$L^{(A_i)}(x) = \int_0^{\infty} e^{-ux} \, (-1)^{A_i} u^{A_i} \, dF_u(u) \qquad [4.8]$$

denotes the $A_i$th derivative of the Laplace transform $L(x) = \int_0^{\infty} e^{-ux} \, dF_u(u)$ of the distribution function $F_u$ of the frailty. It must be stressed here that different distributions of the frailty term give rise to different Laplace transforms and therefore a class of frailty models arises with corresponding loglikelihood given in [4.7].

Consider now the "least informative" nonparametric modeling for $H_0(\cdot)$ in which $H_0(t)$ has a possible jump of size $\mu_l$ at the observed failure time $z_l$. Then,

$$H_0(t) = \sum_{l=1}^{N} \mu_l I(z_l \leq t), \qquad [4.9]$$

where $z_1, ..., z_N$ are pooled observed failure times. Substituting [4.9] into [4.7] we obtain the following profile likelihood:

$$\sum_{i=1}^{n}\sum_{j=1}^{J_i} \delta_{ij}\mathbf{x}_{ij}^T\boldsymbol{\beta} + \sum_{l=1}^{N} \ln \mu_l +$$

$$\sum_{i=1}^{n} \ln \left( \int_0^{\infty} \exp\Big( - u_i \sum_{j=1}^{J_i} e^{\mathbf{x}_{ij}^T\boldsymbol{\beta}} \sum_{l=1}^{N} \mu_l I(z_l \leq z_{ij})\Big) \, u_i^{A_i} \, dF_{u_i}(u_i) \right). \qquad [4.10]$$

The derivative of the above loglikelihood with respect to $\mu_l$, $l = 1, ..., N$ gives:

$$\frac{1}{\mu_l} = \sum_{i=1}^{n} \frac{\left| L^{(A_i+1)}\left(\sum_{j=1}^{J_i} e^{\mathbf{x}_{ij}^T\boldsymbol{\beta}} \sum_{l=1}^{N} \mu_l I(z_l \leq z_{ij})\right)\right|}{\left| L^{(A_i)}\left(\sum_{j=1}^{J_i} e^{\mathbf{x}_{ij}^T\boldsymbol{\beta}} \sum_{l=1}^{N} \mu_l I(z_l \leq z_{ij})\right)\right|} \sum_{j=1}^{J_i} e^{\mathbf{x}_{ij}^T\boldsymbol{\beta}} I(z_l \leq z_{ij}). \qquad [4.11]$$

The first and second gradient derivatives of [4.10] with respect to $\boldsymbol{\beta} = (\beta_1, \ldots, \beta_d)^T$ are given in Androulakis *et al.* [AND 12a]. Note that the loglikelihood given in [4.10] is of a general character and can accommodate any continuous distribution of the frailty term. In the case of independent observations it reduces to the corresponding loglikelihood of the Cox model. Moreover, the aforementioned procedure can also be used for the case of independent univariate data. The corresponding loglikelihood function can be obtained by [4.10] by setting all the sizes $J_i = 1$ while $n$ represents the number of individuals in the sample.

In what follows we present three examples of commonly used distributions of the frailty term, namely, the Gamma, which has also been treated in Fan and Li [FAN 02], the Inverse Gaussian and the Uniform distribution.

### 4.2.1. *Gamma distributed frailty*

Consider a frailty term which is distributed as Gamma$(\alpha, 1/\alpha)$, which for identifiability purposes is assumed to have mean 1 and variance $1/\alpha$. The corresponding Laplace transform is given as:

$$L(x) = \left(\frac{\alpha}{\alpha + x}\right)^{\alpha} \qquad [4.12]$$

while the $n$th derivative of the above Laplace transform is given as:

$$L^{(n)}(x) = (-1)^n \frac{\Gamma(a+n)}{a^n \Gamma(a)} \left(\frac{\alpha}{\alpha + x}\right)^{a+n} \qquad [4.13]$$

and the loglikelihood [4.10], along with a penalty function, becomes:

$$\sum_{i=1}^{n}\sum_{j=1}^{J_i} \delta_{ij}\mathbf{x}_{ij}^T\boldsymbol{\beta} + \sum_{l=1}^{N} \ln \mu_l - \sum_{i=1}^{n} (\alpha + A_i) \ln \left(\alpha + \sum_{j=1}^{J_i} e^{\mathbf{x}_{ij}^T\boldsymbol{\beta}} \sum_{l=1}^{N} \mu_l I(z_l \leq z_{ij})\right)$$

$$+ \sum_{i=1}^{n} \alpha \ln \alpha + \sum_{i=1}^{n} \ln \left(\frac{\Gamma(A_i + \alpha)}{\Gamma(\alpha)}\right) - n \sum_{j=1}^{d} p_\lambda(|\beta_j|). \qquad [4.14]$$

The derivative with respect to $\mu_l$, $l = 1, ..., N$ from [4.11] and [4.13] gives:

$$\frac{1}{\mu_l} = \sum_{i=1}^{n} \frac{A_i + \alpha}{a + \sum_{j=1}^{J_i} e^{\mathbf{x}_{ij}^T\boldsymbol{\beta}} \sum_{l=1}^{N} \mu_l I(z_l \leq z_{ij})} \sum_{j=1}^{J_i} e^{\mathbf{x}_{ij}^T\boldsymbol{\beta}} I(z_l \leq z_{ij}), \qquad [4.15]$$

which coincides with formula [3.12] in Fan and Li [FAN 02]. Analogous expressions hold for the first and second derivatives with respect to $\boldsymbol{\beta}$.

### 4.2.2. *Inverse Gaussian distributed frailty*

Consider an Inverse Gaussian distributed frailty $N^-(b_1, b_2)$ which for identifiability purposes is assumed to have $b_1 = b_2$ so that the mean is 1 and the variance is equal to $1/2b$. The corresponding Laplace transform function $L(x)$ takes the form:

$$L(x) = e^{2b-\sqrt{4b(b+x)}}. \quad [4.16]$$

The absolute value of the $A_i$th derivative of the Laplace transform $L(x)$ given in [4.8], takes the form:

$$|L^{(A_i)}(x)| = e^{2b-\sqrt{4(b+x)b}} E_{g_1}(U_i^{A_i}) = L(x)E_{g_1}(U_i^{A_i}) \quad [4.17]$$

where the latter expectation is with respect to a density $g_1$ of an $N^-(b+x, b)$ random variable where $x$ stands for the term $\sum_{j=1}^{J_i} e^{\mathbf{x}_{ij}^T\boldsymbol{\beta}} \sum_{l=1}^{N} \mu_l I(z_l \leq z_{ij})$.

Therefore the loglikelihood [4.10] takes the form

$$\sum_{l=1}^{N} \ln \mu_l + \sum_{i=1}^{n}\sum_{j=1}^{J_i} \delta_{ij}\mathbf{x}_{ij}^T\boldsymbol{\beta} + \sum_{i=1}^{n} \ln\Big(L(x)\Big) + \sum_{i=1}^{n} \ln(E_{g_1}(U_i^{A_i})). \quad [4.18]$$

Since the $n$th moment of a $N^-(b_1, b_2)$ random variable is given as:

$$e^{\sqrt{4b_1b_2}}\sqrt{4b_2/\pi}\left(\sqrt{\frac{b_2}{b_1}}\right)^{n-(1/2)} K_{1/2-n}(\sqrt{4b_1b_2}\,)$$

where $K_{1/2-n}(z)$ is a modified Bessel function of the second type, the loglikelihood [4.18] for $b_1 = b + x$ and $b_2 = b$ along with a penalty term, becomes:

$$\sum_{l=1}^{N} \ln \mu_l + \sum_{i=1}^{n}\sum_{j=1}^{J_i} \delta_{ij}\mathbf{x}_{ij}^T\boldsymbol{\beta} + \sum_{i=1}^{n} 2b + \sum_{i=1}^{n} \frac{1}{2}(A_i - \frac{1}{2}) \ln\left(\frac{b}{b+x}\right)$$

$$+\sum_{i=1}^{n} \frac{1}{2}\ln(\frac{4b}{\pi}) + \sum_{i=1}^{n} \ln\left(K_{1/2-A_i}\left(\sqrt{4(b+x)b}\right)\right) - n\sum_{j=1}^{d} p_\lambda(|\beta_j|). \quad [4.19]$$

The derivative with respect to $\mu_l$, $l = 1, ..., N$ by [4.11] and [4.17] becomes:

$$\frac{1}{\mu_l} = \sum_{i=1}^{n} \sqrt{\frac{b}{b+x}} \frac{K_{1/2-A_i-1}(\sqrt{4(b+x)b}\,)}{K_{1/2-A_i}(\sqrt{4(b+x)b})} \sum_{j=1}^{J_i} e^{\mathbf{x}_{ij}^T\boldsymbol{\beta}} I(z_l \leq z_{ij}) \quad [4.20]$$

while analogous expressions hold for the first and second gradients with respect to $\boldsymbol{\beta}$.

### 4.2.3. *Uniform distributed frailty*

Consider now a Uniform distributed frailty on $[b, a]$ where $a > b > 0$ and the Laplace transform is given as

$$L(x) = \frac{1}{(a-b)x}(e^{-bx} - e^{-ax}). \qquad [4.21]$$

From [4.8] by a series of integration by parts we get that $L^{(A_i)}(x)$ is equal to:

$$(-1)^{A_i}\left\{\frac{e^{-bx}}{(a-b)x}\sum_{j=0}^{A_i} j!\binom{A_i}{j}\left(\frac{1}{x}\right)^j b^{A_i-j} - \frac{e^{-ax}}{(a-b)x}\sum_{j=0}^{A_i} j!\binom{A_i}{j}\left(\frac{1}{x}\right)^j a^{A_i-j}\right\}$$

$$= \frac{(-1)^{A_i}}{(a-b)x}\frac{2x^{-A_i} - x^{-A_i+1}}{-2+x}\{\Gamma(A_i+1, bx) - \Gamma(A_i+1, ax)\}$$

$$= \frac{(-1)^{A_i+1}}{(a-b)x^{A_i+1}}\{\Gamma(A_i+1, bx) - \Gamma(A_i+1, ax)\} \qquad [4.22]$$

where $\Gamma(a, x)$ is the incomplete Gamma function defined as $\int_x^\infty t^{a-1}e^{-t}dt$.

The loglikelihood [4.10] along with a penalty function becomes:

$$\sum_{l=1}^{N} \ln \mu_l + \sum_{i=1}^{n}\sum_{j=1}^{J_i} \delta_{ij}\mathbf{x}_{ij}^T\beta + \sum_{l=1}^{n} \ln\left(\Gamma(A_i+1, bx) - \Gamma(A_i+1, ax)\right)$$

$$-\sum_{l=1}^{n} \ln(a-b) - \sum_{l=1}^{n}(A_i+1)\ln x - n\sum_{j=1}^{d} p_\lambda(|\beta_j|). \qquad [4.23]$$

The derivative with respect to $\mu_l$, $l = 1, ..., N$ by [4.11] and [4.22] becomes:

$$\frac{1}{\mu_l} = \sum_{l=1}^{n}\left\{\frac{\Gamma(A_i+2, bx) - \Gamma(A_i+2, ax)}{(\Gamma(A_i+1, bx) - \Gamma(A_i+1, ax))x}\right\}\sum_{j=1}^{J_i} e^{\mathbf{x}_{ij}^T\beta} I(z_l \le z_{ij}). [4.24]$$

Analogous expressions hold for the first and second derivatives with respect to $\beta$ [AND 12b].

### 4.3. Simulation results

A thorough simulation study was conducted to support theoretical findings in Androulakis *et al.* [AND 12a] and [AND 12b]. Here we present only a small part of them. The simulation scheme considered was the same as in Fan and Li [FAN 02], therefore we simulated 100 data sets consisting of $n$ groups and $J$ subjects in each group from the exponential hazard frailty model:

$$h(t|\mathbf{x}, u) = u \exp(\mathbf{x}^T \boldsymbol{\beta}),$$

where the true parameter $\boldsymbol{\beta}_0 = (0.8, 0, 0, 1, 0, 0, 0.6, 0)$ and the $x_i$ are marginally standard normal with the correlation between $x_i$ and $x_j$ equal to $\rho^{|i-j|}$ with $\rho = 0.5$. The distribution of the censoring time was assumed to be exponential with mean $U \exp(\mathbf{x}^T \boldsymbol{\beta}_0)$, where $U$ is randomly generated from the Uniform distribution over [1,3] for each simulated data set so that about 30% data are censored. Here $\boldsymbol{\beta}_0$ is regarded as a known constant so that the censoring scheme is noninformative. Concerning the frailty $u$, we present three cases. First, we provide the results for a Gamma frailty with $a = 4$ and secondly an Inverse Gaussian frailty with $b = 2$ so that in both cases the variance is equal to $1/4$. We also examined the effects of the misspecification of the frailty distribution on the estimation of the regression coefficients and on variable selection. In addition, we present the results for the Uniform frailty with $b = 0.25$ and $a = 0.75$. The second case concerns $b = 1.25$ and $a = 1.75$ while in the last case, the values of $b$ and $a$ were $0.75$ and $1.25$, respectively. The performance of variable selection via the proposed generalized penalized likelihood is compared in terms of their model errors, model complexity and accuracy. Model errors of the proposed procedures are compared to those of the maximum profile likelihood estimators. Specifically, we present the median of relative model errors (MRME) over the 100 simulated data sets, for $n = 100$ and $J = 2$. Moreover, the average number of zero coefficients are also reported in the following tables, in which the column labeled "correct" presents the average restricted only to the true zero coefficients, while the column labeled "incorrect" depicts the average of coefficients erroneously set to 0. The penalized likelihood approach with the SCAD penalty (SCAD), the $L_1$ penalty (LASSO) and the hard thresholding penalty (Hard) were used, along with the generalized cross validation as a tuning parameter selection method. We also present the results for the Oracle estimator, obtained by fitting the ideal model consisting of only the variables $X_1$, $X_4$ and $X_7$.

Tables 4.1 and 4.2 contain the results for the Gamma and the Inverse Gaussian frailty case, respectively. In Tables 4.3–4.5, the results concerning the Uniform frailty are depicted.

A general remark is that all methods select about the same number of correct significant variables with the SCAD being the best one. The SCAD penalty generally outperforms the LASSO and Hard penalties, and moreover it provides estimators that perform as well as the oracle estimator in terms of MRME. Note also that the misspecification of the frailty distribution does not seem to significantly affect the selection of variables. However, for more details concerning the simulations, along with some real data examples, the interested reader is referred to Androulakis *et al.* [AND 12a] and [AND 12b].

| Method | MRME (%) | Aver. no. of 0 coeff. | |
|---|---|---|---|
| | | correct | incorrect |
| $n = 100, J = 2$ | | | |
| SCAD | 0.6432(0.5277) | 4.6900(4.4000) | 0(0) |
| LASSO | 1.0945(1.3630) | 4.1000(4.0500) | 0(0) |
| Hard | 0.9470(0.9389) | 4.1700(4.1000) | 0(0) |
| Oracle | 0.6272(0.4844) | 5(5) | 0(0) |

**Table 4.1.** *Simulation results when the data are generated from the gamma frailty model. Estimation of the model parameters is conducted through the gamma frailty model and the Inverse Gaussian frailty model-misspecification case (in parentheses)*

| Method | MRME (%) | Aver. no. of 0 coeff. | |
|---|---|---|---|
| | | correct | incorrect |
| $n = 100, J = 2$ | | | |
| SCAD | 0.5592(0.5711) | 4.4800(4.7600) | 0(0) |
| LASSO | 1.0795(0.6610) | 4.0800(4.3200) | 0(0) |
| Hard | 0.8918(0.7378) | 4.2200(4.3300) | 0(0) |
| Oracle | 0.5275(0.5183) | 5(5) | 0(0) |

**Table 4.2.** *Simulation results when the data are generated from the Inverse Gaussian frailty model. Estimation of the model parameters is conducted through the Inverse Gaussian frailty model and the gamma frailty model-misspecification case (in parentheses)*

| Method | MRME (%) | Aver. no. of 0 coeff. | |
|---|---|---|---|
| | | correct | incorrect |
| $n = 100, J = 2$ | | | |
| SCAD | 0.6193 | 4.7200 | 0 |
| LASSO | 1.1136 | 3.8900 | 0 |
| Hard | 0.6737 | 4.8700 | 0.0200 |
| Oracle | 0.5720 | 5 | 0 |

**Table 4.3.** *Case 1: $b = 0.25$ and $a = 0.75$. Simulation results when the data are generated from the Uniform frailty model*

| Method | MRME (%) | Aver. no. of 0 coeff. | |
|---|---|---|---|
| | | correct | incorrect |
| $n = 100, J = 2$ | | | |
| SCAD | 0.5435 | 4.5700 | 0.0100 |
| LASSO | 1.0674 | 3.8200 | 0.0100 |
| Hard | 0.4759 | 4.7400 | 0.0600 |
| Oracle | 0.4614 | 5 | 0 |

**Table 4.4.** *Case 2: $b = 1.25$ and $a = 1.75$. Simulation results when the data are generated from the Uniform frailty model*

| Method | MRME (%) | Aver. no. of 0 coeff. | |
|---|---|---|---|
| | | correct | incorrect |
| $n = 100, J = 2$ | | | |
| SCAD | 0.5767 | 4.9800 | 0.0100 |
| LASSO | 0.8588 | 4.1400 | 0 |
| Hard | 0.6065 | 4.9200 | 0.0100 |
| Oracle | 0.5662 | 5 | 0 |

**Table 4.5.** *Case 3: $b = 0.75$ and $a = 1.25$. Simulation results when the data are generated from the Uniform frailty model*

## 4.4. Concluding remarks

We provided a selective overview of the most known methods used for variable selection purposes, which are generally based on the penalized likelihood approach. Emphasis was given to the frailty models framework, but only for continuous distributions for the frailty term. Discrete frailty models could also be considered with the appropriate modifications in the likelihood. It could also be very interesting to obtain asymptotic results for the penalized estimators for the case of censored clustered data and general frailty distribution.

## 4.5. Bibliography

[AAL 98] AALEN O.O., "Heterogeneity in survival analysis", *Statistics in Medicine*, vol. 7, pp. 1121–1137, 1998.

[AKA 73] AKAIKE H., "Information theory and an extension of the maximum likelihood principle", in PETROV B.N., CSAKI F. (eds), *Second International Symposium on Information Theory*, Akademiai Kiado, Budapest, pp. 267–281, 1973.

[AKA 74] AKAIKE H., "A new look at the statistical model identification", *IEEE Transactions on Automatic Control*, vol. 19, pp. 16–23, 1974.

[AND 12a] ANDROULAKIS E., KOUKOUVINOS C., VONTA F., "Estimation and variable selection via frailty models with penalized likelihood", *Statistics in Medicine,* vol. 31, pp. 2223–2239, 2012.

[AND 12b] ANDROULAKIS E., KOUKOUVINOS C., VONTA F., "On the uniform frailty model with penalized likelihood and clustered data", *Journal of Reliability and Statistical Studies,* vol. 5, pp. 97–106, 2012.

[ANT 97] ANTONIADIS A., "Wavelets in statistics: a review" (with discussion), *Journal of the Italian Statistical Association*, vol. 6, pp. 97–144, 1997.

[ANT 01] ANTONIADIS A., FAN J., "Regularization of wavelets approximations" (with discussion), *Journal of the Italian Statistical Association*, vol. 96, pp. 939–967, 2001.

[BRE 95] BREIMAN L., "Better subset regression using the non-negative garrote", *Technometrics,* vol. 37, pp. 373–384, 1995.

[BRE 96] BREIMAN L., "Heuristics of instability and stabilization in model selection", *The Annals of Statistics*, vol. 24, pp. 2350–2383, 1996.

[CAI 05] CAI J., FAN J., LI R., *et al.*, "Variable selection for multivariate failure time data", *Biometrika,* vol. 92, pp. 303–316, 2005.

[DAU 04] DAUBECHIES I., DEFRISE M., DE MOL C., "An iterative thresholding algorithm for linear inverse problems with a sparsity constraint", *Communications on Pure and Applied Mathematics*, vol. 57, pp. 1413–1457, 2004.

[EFR 04] EFRON B., HASTIE T., JOHNSTONE I., *et al.*, "Least angle regression" (with discussion), *The Annals of Statistics*, vol. 32, pp. 407–499, 2004.

[FAN 97] FAN J., "Comments on 'Wavelets in statistics: a review' by A. Antoniadis", *Journal of the Italian Statistical Association*, vol. 6, pp. 131–138, 1997.

[FAN 01] FAN J., LI R., "Variable selection via nonconcave penalized likelihood and its oracle properties", *Journal of the Italian Statistical Association*, vol. 96, pp. 1348–1360, 2001.

[FAN 02] FAN J., LI R., "Variable selection for Cox's proportional hazards model and frailty model", *The Annals of Statistics*, vol. 30, pp. 74–99, 2002.

[FAN 04] FAN J., LI R., "New estimation and model selection procedures for semiparametric modeling in longitudinal data analysis", *Journal of the Italian Statistical Association*, vol. 99, pp. 710–723, 2004.

[FAN 06] FAN J., LI R., "Statistical challenges with high dimensionality: feature selection in knowledge discovery", SANZ-SOLE M., SORIA J., VARONA J.L., VERDERA J. (eds), *Proceedings of the International Congress of Mathematicians*, vol. III, European Mathematical Society, Zurich, pp. 595–622, 2006.

[FAN 11] FAN J., LV J., "Non-concave penalized likelihood with NP-dimensionality", *IEEE Transactions on Information Theory*, vol. 57, pp. 5467–5484, 2011.

[FAN 04] FAN J., PENG H., "On non-concave penalized likelihood with diverging number of parameters", *The Annals of Statistics*, vol. 32, pp. 928–961, 2004.

[FRA 93] FRANK I.E., FRIEDMAN J.H., "A statistical view of some chemometrics regression tools" (with discussion), *Technometrics,* vol. 35, pp. 109–148, 1993.

[FU 98] FU W.J., "Penalized regression: the bridge versus the LASSO", *Journal of Computational and Graphical Statistics,* vol. 7, pp. 397–416, 1998.

[HAN 79] HANNAN E.J., QUINN B.G., "The determination of the order of an autoregression", *Journal of the Royal Statistical Society, Ser. B*, vol. 41, pp. 190–195, 1979.

[HOU 84] HOUGAARD P., "Life table methods for heterogeneous populations: distributions describing the heterogeneity", *Biometrika*, vol. 71, pp. 75–83, 1984.

[HOU 86] HOUGAARD P., "Survival models for heterogeneous populations derived from stable distributions", *Biometrika*, vol. 73, pp. 387–396, 1986.

[HOU 00] HOUGAARD P., *Analysis of Multivariate Survival Data*, Springer, New York, 2000.

[KOS 04] KOSOROK M., LEE B., FINE J., "Robust inference for univariate proportional hazards frailty regression models", *The Annals of Statistics*, vol. 32, pp. 1448–1491, 2004.

[LEE 88] LEE S., KLEIN J.P., "Bivariate models with a random environmental factor", *Industrial Journal of Productivity, Reliability, and Quality Control*, vol. 13, pp. 1–18, 1988.

[LIN 86] LINDLEY D.V., SINGPURWALLA N.A., "Multivariate distributions for the life lengths of components of a system sharing a common environment", *Journal of Applied Probability*, vol. 23, pp. 418–431, 1986.

[LV 09] LV J., FAN Y., "A unified approach to model selection and sparse recovery using regularized least squares", *The Annals of Statistics*, vol. 37, pp. 3498–3528, 2009.

[MAL 73] MALLOWS C.L., "Some comments on $C_p$", *Technometrics,* vol. 15, pp. 661–675, 1973.

[MCG 91] MCGILCHRIST C.A., AISBETT C.W., "Regression with frailty in survival analysis", *Biometrics*, vol. 47, pp. 461–466, 1991.

[RIS 78] RISANNEN J., "Modeling by shortest data description", *Automatica,* vol. 14, pp. 465–471, 1978.

[SCH 78] SCHWARTZ G., "Estimating the dimension of a model", *The Annals of Statistics*, vol. 6, pp. 461–464, 1978.

[SLU 04] SLUD E.V., VONTA F., "Consistency of the NPML estimator in the right-censored transformation model", *Scandinavian Journal of Statistics*, vol. 31, pp. 21–41, 2004.

[TIB 96] TIBSHIRANI R., "Regression shrinkage and selection via the LASSO", *Journal of the Royal Statistical Society, Ser. B*, vol. 58, pp. 267–288, 1996.

[VON 96] VONTA F., "Efficient estimation in a non-proportional hazards model in survival analysis", *Scandinavian Journal of Statistics*, vol. 23, pp. 49–61, 1996.

[WU 08] WU T.T., LANGE K., "Coordinate descent algorithms for LASSO penalized regression", *The Annals of Applied Statistics*, vol. 2, pp. 224–244, 2008.

[ZHA 07] ZHANG C.H., Penalized linear unbiased selection, Technical Report 2007–003, Dept. Statistics, Rutgers University, 2007.

[ZHL 11] ZHANG Y., LI R., "Iterative conditional maximization algorithm for nonconcave penalized likelihood", *Nonparametric statistics and mixture models*, HUNTER D.R., RICHARDS, ST. D., ROSENBERGER J. (eds), a festschrift in honor of THOMAS P. HETTMANSPERGER (pp. 336–352), Singapore, World Scientific Publishing, 2011.

[ZOU 05] ZOU H., HASTIE T., "Regularization and variable selection via the elastic net", *Journal of the Royal Statistical Society, Ser. B*, vol. 67, pp. 301–320, 2005.

[ZOU 08] ZOU H., LI R., "One-step sparse estimates in nonconcave penalized likelihood models" (with discussion), *The Annals of Statistics*, vol. 36, pp. 1509–1566, 2008.

Chapter 5

# Interactive Investigation of Statistical Regularities in Testing Composite Hypotheses of Goodness of Fit

In this chapter[1], a "real-time" ability to simulate and research distributions of test statistics in the course of testing a complex hypothesis of goodness of fit for distributions with estimated parameters is implemented by means of parallel computing. It makes it possible to make correct statistical inferences even in those situations when the distribution of test statistic is unknown before the testing procedure starts.

## 5.1. Introduction

In composite hypotheses testing in the form $H_0 : F(x) \in \{F(x,\theta), \theta \in \Theta\}$, when an estimate $\hat{\theta}$ of scalar or vector distribution parameter $\theta$ is calculated from sample under testing, the non-parametric goodness-of-fit Kolmogorov (K), $\omega^2$ Cramér–von Mises–Smirnov (CMS) and $\Omega^2$ Anderson–Darling (AD) tests lose their property of being distribution free.

The value:

$$D_n = \sup_{|x|<\infty} |F_n(x) - F(x,\theta)|$$

Chapter written by Boris LEMESHKO, Stanislav LEMESHKO and Andrey ROGOZHNIKOV.

1 This chapter was prepared at Novosibirsk State Technical University.

is used in the Kolmogorov test as a distance between empirical and theoretical laws (where $F_n(x)$ is an empirical distribution function, and $n$ is the size of the sample). When testing hypotheses, this statistic should be used with Bolshev's correction [BOL 87] in the form [BOL 83]:

$$S_K = \frac{6nD_n + 1}{6\sqrt{n}}, \qquad [5.1]$$

where $D_n = \max(D_n^+, D_n^-)$,

$$D_n^+ = \max_{1 \leq i \leq n} \left\{ \frac{i}{n} - F(x_i, \theta) \right\}, D_n^- = \max_{1 \leq i \leq n} \left\{ F(x_i, \theta) - \frac{i-1}{n} \right\},$$

$n$ is the size of the sample, and $x_1, x_2, \ldots, x_n$ are sample values in an increasing order. The statistic [5.1] in testing a simple hypothesis follows the Kolmogorov distribution $K(s) = \sum\limits_{k=-\infty}^{\infty} (-1)^k e^{-2k^2 s^2}$.

In the $\omega^2$ CMS test, we use a statistic in the form:

$$S_\omega = \frac{1}{12n} + \sum_{i=1}^{n} \left\{ F(x_i, \theta) - \frac{2i-1}{2n} \right\}^2, \qquad [5.2]$$

and in a test of $\Omega^2$ AD type [AND 52, AND 54], a statistic in the form:

$$S_\Omega = -n - 2\sum_{i=1}^{n} \left\{ \frac{2i-1}{2n} \ln F(x_i, \theta) + \left(1 - \frac{2i-1}{2n}\right) \ln(1 - F(x_i, \theta)) \right\}. \qquad [5.3]$$

In testing a simple hypothesis, [5.2] has the distribution with CDF $a1(\cdot)$ (see [BOL 83]):

$$\begin{aligned} a1(s) &= \frac{1}{\sqrt{2s}} \sum_{j=0}^{\infty} \frac{\Gamma(j+1/2)\sqrt{4j+1}}{\Gamma(1/2)\Gamma(j+1)} \exp\left\{ -\frac{(4j+1)^2}{16s} \right\} \\ &\times \left\{ I_{-\frac{1}{4}} \left[ \frac{(4j+1)^2}{16s} \right] - I_{\frac{1}{4}} \left[ \frac{(4j+1)^2}{16s} \right] \right\}, \end{aligned}$$

where $I_{-\frac{1}{4}}(\cdot)$ and $I_{\frac{1}{4}}(\cdot)$ are the modified Bessel functions:

$$I_\nu(z) = \sum_{k=0}^{\infty} \frac{(z/2)^{\nu+2k}}{\Gamma(k+1)\Gamma(k+\nu+1)}, \quad |z| < \infty, \quad |\arg z| < \pi,$$

and [5.3] has the distribution with CDF $a2(\cdot)$ [BOL 83]:

$$\begin{aligned} a2(s) &= \frac{\sqrt{2\pi}}{s} \sum_{j=0}^{\infty} (-1)^j \frac{\Gamma(j+1/2)(4j+1)}{\Gamma(1/2)\Gamma(j+1)} \exp\left\{-\frac{(4j+1)^2\pi^2}{8s}\right\} \times \\ &\times \int_0^{\infty} \exp\left\{\frac{s}{8(y^2+1)} - \frac{(4j+1)^2\pi^2 y^2}{8s}\right\} dy. \end{aligned}$$

## 5.2. Distributions of the test statistics in the case of testing composite hypotheses

Let us denote the distribution of a statistic $S$ when hypothesis $H_0$ is true as $G(S|H_0)$. In our case, we will consider the statistics of Kolmogorov $S_K$, CMS $S_\omega$ and AD $S_\Omega$ in place of $S$. In composite hypothesis testing, $G(S|H_0)$ is affected by a number of factors: the form of the distribution $F(x, \theta)$ that corresponds to the true hypothesis $H_0$; types and number of parameters to be estimated; the method of parameter estimation; value of a specific parameter (e.g. for gamma distribution, inverse Gaussian distribution (IGD), generalized Weibull distribution (GWD), generalized Gaussian distribution (GGD) and beta-distribution families).

The investigation of statistic distributions of the non-parametric goodness-of-fit tests for composite hypotheses began from the paper [KAC 55]. Then, various approaches to solve this problem were used [DAR 55, DAR 57, DUR 73, DUR 75, DUR 76, GIH 53, MAR 78, MAR 11, PEA 72, STE 70, STE 74, CHA 81, TYU 84a, TYU 84b, DZH 82, NIK 92a, NIK 92b].

In our research [LEM 98, LEM 01, LEM 02, LEM 04, LEM 07a, LEM 07b, LEM 09a, LEM 09b, LEM 10a, LEM 10b, LEM 10c, LEM 11a, LEM 11b], the distributions of statistics of the non-parametric goodness-of-fit tests have been investigated by means of methods of statistical simulation, and approximate models of the distributions have been found on the basis of the obtained empirical distributions. The most complete list of the constructed models of distributions of statistics and tables of percentage points for the non-parametric goodness-of-fit tests is provided in [LEM 09a], [LEM 09b], [LEM 10c], [LEM 11a], [LEM 11b] and [LEM 11f]. These models and tables are applicable when maximum likelihood (ML) estimators (MLEs) are used.

The distributions of statistics of non-parametric goodness-of-fit tests are affected by a number of factors: the form of the observed law $F(x, \theta)$ that corresponds to the true hypothesis $H_0$; types and number of parameters to be estimated; the method of parameter estimation. In these cases, there are no impediments for studying test statistic distributions by means of statistical simulation and further construction of approximate models for them when testing composite hypothesis [LEM 10a, LEM 10b, LEM 10c].

Complications arise when distributions $G(S|H_0)$ of non-parametric goodness-of-fit tests depend on the value of specific parameter(s) of $F(x,\theta)$ (for gamma distribution, GGD, IGD, GWD and beta-distribution families).

The existing dependence on parameters values should not be neglected. For example, in composite hypotheses testing subject to GWD with the density function:

$$f(x;\theta_0,\theta_1,\theta_2) = \frac{\theta_0}{\theta_1}{\theta_2}^{\theta_0}x^{\theta_0-1}\left(1+\left(\frac{x}{\theta_2}\right)^{\theta_0}\right)^{\frac{1}{\theta_1}-1} \times \exp\left(1-\left(1+\left(\frac{x}{\theta_2}\right)^{\theta_0}\right)^{\frac{1}{\theta_1}}\right), \qquad [5.4]$$

where $x \geq 0$ and $\theta_0$, $\theta_1$, $\theta_2$ are positive, the limiting distributions of statistics of the non-parametric goodness-of-fit tests depend on value of the form parameter $\theta_1$. In Figures 5.1 and 5.2 we can see the behavior of the distribution of statistic [5.3] in testing composite hypotheses for family [5.4]. In the case when three parameters are estimated by the ML method (Figure 5.1), we can see the following: when the value of the shape parameter increases up to $\theta_1 \approx 2$, the distribution $G(S|H_0)$ shifts to the left. With $\theta_1$ continuing to increase further, the distribution $G(S|H_0)$ shifts into the opposite direction. In a case of two parameters estimated by MLE (Figure 5.2), we can see the following: as value of $\theta_1$ increases, the distribution $G(S|H_0)$ shifts to the right.

Percentage points have been obtained by simulations, and models of marginal statistic distributions of Kolmogorov, CMS and AD tests have been computed for the following values of shape parameter $\theta_1$: 0.5, 1.0, 3.0, 4.0, 5.0, 6.0, 7.0, 8.0. The points and models are presented in Tables 5.1–5.6. The distribution parameters were estimated with ML method.

Distributions $G(S|H_0)$ of the Kolmogorov, the CMS and the AD statistics are best of all approximated by the family of the type III beta-distributions with the density function:

$$B_3(\theta_0,\theta_1,\theta_2,\theta_3,\theta_4) = \frac{\theta_2^{\theta_0}}{\theta_3 B(\theta_0,\theta_1)} \frac{\left(\frac{x-\theta_4}{\theta_3}\right)^{\theta_0-1}\left(1-\frac{x-\theta_4}{\theta_3}\right)^{\theta_1-1}}{\left[1+(\theta_2-1)\frac{x-\theta_4}{\theta_3}\right]^{\theta_0+\theta_1}},$$

or by the family of the *Sb*-Johnson distributions:

$$Sb(\theta_0,\theta_1,\theta_2,\theta_3) = \frac{\theta_1\theta_2}{(x-\theta_3)(\theta_2+\theta_3-x)}\exp\left\{-\frac{1}{2}\left(\theta_0-\theta_1\ln\frac{x-\theta_3}{\theta_2+\theta_3-x}\right)^2\right\}.$$

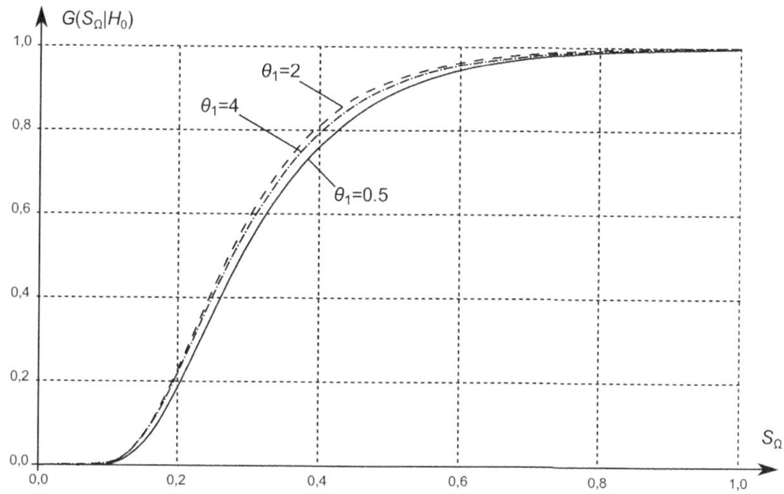


**Figure 5.1.** *Distributions of the Anderson–Darling statistic in testing goodness of fit of family [5.4]. ML method is used to estimate parameters $\theta_0$, $\theta_1$ and $\theta_2$*

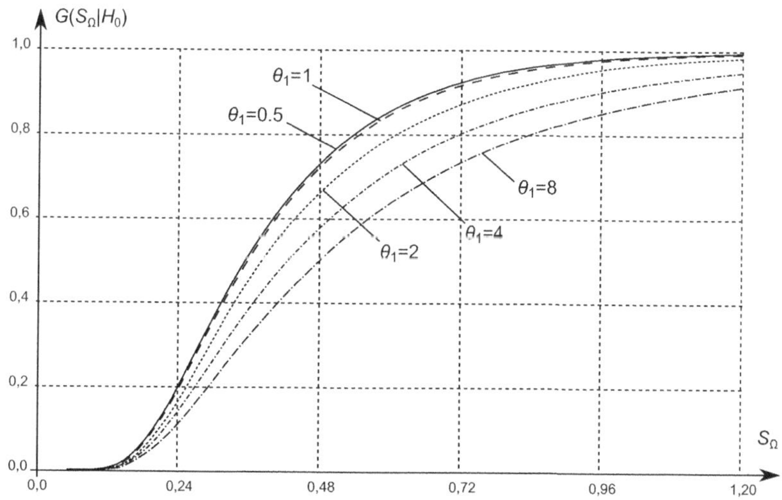


**Figure 5.2.** *Distributions of the Anderson–Darling statistic in testing goodness of fit of family [5.4]. ML method is used to estimate parameters $\theta_0$ and $\theta_1$*

The tables of percentage points and the models of distributions of statistics have been constructed from simulated samples of statistics of the size $N = 10^6$. Under this size, the deviation the empirical cumulative distribution function (CDF) $G_N(S|H_0)$

from the theoretical counterpart is less than $10^3$. The values of statistics of goodness-of-fit tests have been calculated using samples of size $n = 10^3$ of pseudorandom variables following the given $F(x,\theta)$. For this value of $n$, CDF. $G(S_n|H_0)$ of statistics is almost the same as the marginal CDF. $G(S|H_0)$.

| Parameters estimated | Percentage points | | | Model |
|---|---|---|---|---|
| | 0.9 | 0.95 | 0.99 | |
| | | The Kolmogorov test | | |
| $\theta_0$ | 1.001 | 1.102 | 1.309 | $B_3$(6.5294, 6.8315, 3.5901, 2.0446,0.2801) |
| $\theta_1$ | 1.084 | 1.199 | 1.427 | $B_3$(5.4860, 5.9744, 3.4348, 2.1402, 0.3000) |
| $\theta_2$ | 1.038 | 1.144 | 1.360 | $B_3$(4.7833, 6.1285, 3.0596, 2.0214, 0.3200) |
| $\theta_0,\theta_1$ | 0.849 | 0.922 | 1.071 | $B_3$(6.2332, 6.0259,2.8200, 1.3000, 0.2800) |
| $\theta_0,\theta_2$ | 0.837 | 0.909 | 1.054 | *Sb*(2.1787, 1.8756, 1.5259, 0.2567) |
| $\theta_1,\theta_2$ | 0.848 | 0.922 | 1.076 | *Sb*(2.4861, 1.8758, 1.7026, 0.2664) |
| $\theta_0,\theta_1,\theta_2$ | 0.780 | 0.845 | 0.979 | *Sb*(2.3507, 1.9291, 1.4629, 0.2495) |
| | | The Cramer–von Mises–Smirnov test | | |
| $\theta_0$ | 0.181 | 0.232 | 0.359 | $B_3$(5.1297, 2.5959, 22.9591,0.8000, 0.0081) |
| $\theta_1$ | 0.227 | 0.296 | 0.466 | $B_3$(7.4650, 2.6576, 44.4162, 1.3633, 0.0000) |
| $\theta_2$ | 0.198 | 0.255 | 0.395 | $B_3$(5.4489, 2.7019, 31.5609, 1.1500, 0.0062) |
| $\theta_0,\theta_1$ | 0.110 | 0.135 | 0.192 | $B_3$(6.3779, 4.6451, 27.3376, 1.0000, 0.0050) |
| $\theta_0,\theta_2$ | 0.106 | 0.129 | 0.183 | *Sb*(3.7541, 1.5434, 0.5800, 0.0058) |
| $\theta_1,\theta_2$ | 0.112 | 0.138 | 0.200 | $B_3$(10.3369, 4.0734, 25.8270, 0.5802, 0.0000) |
| $\theta_0,\theta_1,\theta_2$ | 0.086 | 0.103 | 0.145 | $B_3$(6.7252, 4.6508, 16.7920, 0.4800, 0.0050) |
| | | The Anderson–Darling test | | |
| $\theta_0$ | 1.125 | 1.415 | 2.140 | $B_3$(4.9800, 4.1685, 17.0454, 7.1000, 0.0500) |
| $\theta_1$ | 1.279 | 1.625 | 2.478 | $B_3$(4.7602, 5.1000, 9.8527, 6.8675, 0.0000) |
| $\theta_2$ | 1.157 | 1.454 | 2.186 | $B_3$(3.0331, 4.0598, 9.3429, 5.9880, 0.1000) |
| $\theta_0,\theta_1$ | 0.673 | 0.806 | 1.120 | $B_3$(5.7172, 5.0419, 10.1641, 3.0044, 0.0550) |
| $\theta_0,\theta_2$ | 0.655 | 0.781 | 1.079 | *Sb*(3.8953, 1.6481, 3.5052, 0.0513) |
| $\theta_1,\theta_2$ | 0.743 | 0.902 | 1.290 | *Sb*(4.1462, 1.6136, 4.6254, 0.0535) |
| $\theta_0,\theta_1,\theta_2$ | 0.523 | 0.617 | 0.839 | *Sb*(3.9313, 1.6905, 2.7078, 0.0530) |

**Table 5.1.** *Percentage points and models of the limiting distributions of statistics of the non-parametric goodness-of-fit tests when ML method is used for parameter estimation ($\theta_1$= 0.5)*

The most serious impediment to a complete solution of the problem of testing composite hypotheses by means of the non-parametric goodness-of-fit tests is that the distributions of the test statistics depend on specific values of shape parameters of the observed laws. In [LEM 07a, LEM 09a, LEM 09b, LEM 10a, LEM 10b, LEM 10c, LEM 11a and LEM 11b], models of distributions of statistics were obtained for a limited set of combinations of (integer) values of shape parameters (for gamma distribution, two-sided exponential law, IGD and beta-distribution families). It is unrealistic to build the models for an infinite set of combinations of the parameter values.

| Parameters estimated | Percentage points | | | Model |
|---|---|---|---|---|
| | 0.9 | 0.95 | 0.99 | |
| The Kolmogorov test | | | | |
| $\theta_0$ | 1.181 | 1.316 | 1.585 | $B_3$(6.9734, 4.8247, 5.3213, 2.3800, 0.2690) |
| $\theta_1$ | 1.083 | 1.196 | 1.425 | $B_3$(4.6425, 6.6688, 2.8491, 2.2246, 0.3200) |
| $\theta_2$ | 0.994 | 1.092 | 1.290 | $B_3$(6.2635, 7.1481, 3.2059, 2.0000, 0.2800) |
| $\theta_0, \theta_1$ | 0.874 | 0.954 | 1.117 | *Sb*(2.4299, 1.8866, 1.7504, 0.2598) |
| $\theta_0, \theta_2$ | 0.823 | 0.893 | 1.033 | $B_3$(5.8989, 7.5040, 2.4180, 1.3724, 0.2800) |
| $\theta_1, \theta_2$ | 0.815 | 0.883 | 1.023 | *Sb*(2.4499, 1.9720, 1.6016, 0.2486) |
| $\theta_0, \theta_1, \theta_2$ | 0.758 | 0.820 | 0.946 | *Sb*(2.3012, 1.9386, 1.3863, 0.2464) |
| The Cramer–von Mises–Smirnov test | | | | |
| $\theta_0$ | 0.320 | 0.431 | 0.706 | $B_3$(2.2422, 2.2970, 16.4663, 1.6500, 0.0130) |
| $\theta_1$ | 0.227 | 0.295 | 0.464 | $B_3$(5.3830, 2.6954, 40.5199, 1.6450, 0.0050) |
| $\theta_2$ | 0.174 | 0.221 | 0.336 | $B_3$(3.6505, 3.2499, 16.5445, 1.0000, 0.0100) |
| $\theta_0, \theta_1$ | 0.117 | 0.144 | 0.209 | *Sb*(3.8667, 1.4603, 0.7583, 0.0059) |
| $\theta_0, \theta_2$ | 0.102 | 0.123 | 0.174 | $B_3$(12.2776, 4.1107, 27.2069, 0.4875, 0.0000) |
| $\theta_1, \theta_2$ | 0.103 | 0.127 | 0.182 | $B_3$(4.7144, 4.6690, 10.8816, 0.5261, 0.0059) |
| $\theta_0, \theta_1, \theta_2$ | 0.080 | 0.097 | 0.135 | *Sb*(4.1842, 1.6587, 0.4794, 0.0061) |
| The Anderson–Darling test | | | | |
| $\theta_0$ | 1.724 | 2.280 | 3.639 | $B_3$(4.8106, 2.6855, 35.5593,11.8700, 0.0500) |
| $\theta_1$ | 1.275 | 1.617 | 2.468 | $B_3$(3.6999, 3.9108, 16.4841, 9.0300, 0.0740) |
| $\theta_2$ | 1.056 | 1.314 | 1.953 | $B_3$(4.9871, 4.1479, 16.5432, 6.4500, 0.0600) |
| $\theta_0, \theta_1$ | 0.687 | 0.827 | 1.161 | $B_3$(4.6368, 6.6727, 7.1680, 3.6356, 0.0521) |
| $\theta_0, \theta_2$ | 0.633 | 0.753 | 1.037 | $B_3$(3.0467, 5.9239, 5.0944, 2.7870, 0.1000) |
| $\theta_1, \theta_2$ | 0.696 | 0.842 | 1.194 | $B_3$(6.9638, 4.5238, 17.7792, 3.8000, 0.0522) |
| $\theta_0, \theta_1, \theta_2$ | 0.494 | 0.582 | 0.786 | *Sb*(3.9578, 1.6861, 2.5760, 0.0547) |

**Table 5.2.** *Percentage points and models of the limiting distributions of statistics of the non-parametric goodness-of-fit tests when ML method is used for parameter estimation ($\theta_1$= 1)*

In this chapter, a "real-time" ability is implemented by the use of parallel computing to simulate and research the distributions of test statistics in the case of testing a composite goodness-of-fit hypothesis (for distributions with estimated parameters). It makes it possible to make correct statistical inferences even in those situations when the distribution of the test statistic is unknown (before the testing procedure starts).

| Parameters estimated | Percentage points | | | Model |
|---|---|---|---|---|
| | 0.9 | 0.95 | 0.99 | |
| The Kolmogorov test | | | | |
| $\theta_0$ | 1.150 | 1.279 | 1.538 | $B_3$(5.0155, 5.4869, 3.3992, 2.2476, 0.3000) |
| $\theta_1$ | 1.084 | 1.199 | 1.427 | $B_3$(5.4860, 5.9744, 3.4348, 2.1402,0.3000) |
| $\theta_2$ | 0.892 | 0.968 | 1.121 | $B_3$(4.6527, 7.8624, 1.8636, 1.4770, 0.3110) |
| $\theta_0, \theta_1$ | 0.992 | 1.095 | 1.301 | $B_3$(37.6836, 9.6249, 24.7703, 4.2400, 0.1000) |
| $\theta_0, \theta_2$ | 0.823 | 0.895 | 1.041 | $B_3$(6.6694, 6.5961, 3.0264,1.3700, 0.2650) |
| $\theta_1, \theta_2$ | 0.807 | 0.877 | 1.020 | $B_3$(5.3859, 8.4947, 2.3199, 1.4900, 0.2850) |
| $\theta_0, \theta_1, \theta_2$ | 0.751 | 0.811 | 0.932 | $B_3$(5.7236, 7.0743, 2.3212, 1.1488, 0.2714) |
| The Cramer–von Mises–Smirnov test | | | | |
| $\theta_0$ | 0.286 | 0.383 | 0.620 | *Sb*(3.4745, 1.1215, 2.1611, 0.0065) |
| $\theta_1$ | 0.227 | 0.296 | 0.466 | $B_3$(8.0420, 2.6222, 50.1417, 1.3950, 0.0000) |
| $\theta_2$ | 0.135 | 0.167 | 0.240 | $B_3$(9.8988, 3.6331, 27.2342, 0.6611, 0.0000) |
| $\theta_0, \theta_1$ | 0.167 | 0.215 | 0.334 | *Sb*(3.6343, 1.2549, 1.1752, 0.0074) |
| $\theta_0, \theta_2$ | 0.100 | 0.121 | 0.172 | $B_3$(4.9109, 4.8805, 11.3991, 0.5400, 0.0058) |
| $\theta_1, \theta_2$ | 0.099 | 0.121 | 0.174 | $B_3$(9.7955, 5.0455, 35.0176, 0.9000, 0.0000) |
| $\theta_0, \theta_1, \theta_2$ | 0.078 | 0.094 | 0.133 | $B_3$(4.2414, 3.7719, 8.6839, 0.2744, 0.0087) |
| The Anderson–Darling test | | | | |
| $\theta_0$ | 1.556 | 2.040 | 3.234 | $B_3$(4.3943, 2.4670, 38.0035, 10.7000, 0.0900) |
| $\theta_1$ | 1.279 | 1.625 | 2.478 | $B_3$(5.3689, 3.2667, 21.3222, 6.8675, 0.0535) |
| $\theta_2$ | 0.900 | 1.096 | 1.572 | $B_3$(3.5132, 4.3501, 8.8168, 4.2500, 0.1000) |
| $\theta_0, \theta_1$ | 0.871 | 1.088 | 1.623 | $B_3$(5.6254, 3.7452, 20.0868, 4.9237, 0.0588) |
| $\theta_0, \theta_2$ | 0.619 | 0.737 | 1.015 | $B_3$(7.1939, 6.8828, 3.2613, 1.5626, 0.2598) |
| $\theta_1, \theta_2$ | 0.640 | 0.769 | 1.072 | $B_3$(30.1793 ,4.4373, 60.5986, 3.2000, 0.0000) |
| $\theta_0, \theta_1, \theta_2$ | 0.483 | 0.568 | 0.773 | $B_3$(5.2772, 4.4958, 7.9102, 1.5891, 0.0664)) |

**Table 5.3.** *Percentage points and models of the limiting distributions of statistics of the non-parametric goodness-of-fit tests when the ML method is used for parameter estimation ($\theta_1 = 3$)*

## 5.3. Testing composite hypotheses in "real-time"

In this chapter, we propose an approach that is based upon the software, that is being developed by authors, and the use of simulation [LEM 11e, LEM 11]. Computational processes in the simulation of distributions of statistics of various tests can be parallelized rather easily by the use of available resources of nearby computer network. This makes it possible to dramatically reduce the time required to simulate (study) an unknown distribution $G(S|H_0)$ of statistic. Statistical analysis is carried out by the scheme in Figure 5.3 in the case of the use of non-parametric goodness-of-fit tests for testing composite hypotheses in regard to the distributions, for which the statistic distributions depend on parameter values. Such an approach was used in [LEM 10b] and [LEM 11d]. Here, the study of $G(S|H_0)$ is carried out in "real time" of testing the hypothesis [LEM 11c].

| Parameters estimated | Percentage points | | | Model |
|---|---|---|---|---|
| | 0.9 | 0.95 | 0.99 | |
| The Kolmogorov test | | | | |
| $\theta_0$ | 1.131 | 1.256 | 1.506 | $B_3$(5.0752, 5.5757, 3.3089, 2.1797, 0.3000) |
| $\theta_1$ | 1.084 | 1.199 | 1.427 | $B_3$(3.8892, 6.2974, 2.5413, 2.1402, 0.3400) |
| $\theta_2$ | 0.890 | 0.966 | 1.119 | *Sb*(2.1569,1.8555, 1.6361, 0.2661) |
| $\theta_0, \theta_1$ | 1.024 | 1.133 | 1.352 | $B_3$(14.6423, 5.3789, 9.0355, 2.1287, 0.2000) |
| $\theta_0, \theta_2$ | 0.839 | 0.914 | 1.068 | $B_3$(5.1515, 6.1071, 2.8573, 1.3900, 0.3000) |
| $\theta_1, \theta_2$ | 0.833 | 0.909 | 1.065 | $B_3$(7.3590, 7.0743, 3.0755, 1.4500, 0.2450) |
| $\theta_0, \theta_1, \theta_2$ | 0.769 | 0.834 | 0.970 | $B_3$(4.0431, 7.9330, 1.6664, 1.2059, 0.3007) |
| The Cramer–von Mises–Smirnov test | | | | |
| $\theta_0$ | 0.269 | 0.359 | 0.578 | *Sb*(3.4774, 1.1443, 1.9761, 0.0066) |
| $\theta_1$ | 0.227 | 0.296 | 0.466 | $B_3$(7.2936, 2.6369, 40.7763, 1.2800, 0.0000) |
| $\theta_2$ | 0.135 | 0.166 | 0.240 | $B_3$(6.9544, 4.2952, 17.0098, 0.7100, 0.0000) |
| $\theta_0, \theta_1$ | 0.186 | 0.242 | 0.379 | $B_3$(10.0457, 2.7234, 74.1688, 1.4000, 0.0000) |
| $\theta_0, \theta_2$ | 0.104 | 0.127 | 0.182 | $B_3$(10.3993, 4.2771, 25.5455, 0.5600, 0.0000) |
| $\theta_1, \theta_2$ | 0.104 | 0.128 | 0.186 | $B_3$(5.2006, 4.4814, 13.7165, 0.5770, 0.0050) |
| $\theta_0, \theta_1, \theta_2$ | 0.082 | 0.099 | 0.141 | $B_3$(4.3747, 3.2066, 9.2236, 0.2479, 0.0088) |
| The Anderson–Darling test | | | | |
| $\theta_0$ | 1.484 | 1.934 | 3.047 | $B_3$(12.5725, 2.7914, 75.0000, 9.6500, 0.0000) |
| $\theta_1$ | 1.279 | 1.625 | 2.478 | $B_3$(6.9691, 2.9121, 32.3978, 6.8675, 0.0535) |
| $\theta_2$ | 0.895 | 1.090 | 1.556 | $B_3$(16.0792, 4.1280, 41.0115, 4.9000, 0.0000) |
| $\theta_0, \theta_1$ | 0.958 | 1.213 | 1.838 | $B_3$(5.9821, 3.4306, 23.7037, 5.4000, 0.0500) |
| $\theta_0, \theta_2$ | 0.632 | 0.754 | 1.043 | $B_3$(19.4692, 4.7303, 32.4566, 2.8950, 0.0000) |
| $\theta_1, \theta_2$ | 0.645 | 0.776 | 1.087 | $B_3$(19.2831, 4.8148, 37.5002, 3.4100, 0.0000) |
| $\theta_0, \theta_1, \theta_2$ | 0.496 | 0.587 | 0.805 | $B_3$(5.9771,4.3144,9.7987,1.7085,0.0619) |

**Table 5.4.** *Percentage points and models of the limiting distributions of statistics of the non-parametric goodness-of-fit tests when ML method is used for parameter estimation ($\theta_1 = 4$)*

When the composite hypothesis $H_0 : F(x) \in \{F(x,\theta), \theta \in \Theta\}$ is tested by an existing sample $x_1, x_2, \ldots, x_n$, the parameter vector estimate $\hat{\theta}$ for the distribution $F(x,\theta)$ is found in accordance with the selected method (MLE, in the case). Then, the value of statistic $S^*$ of the goodness-of-fit test in use is calculated in accordance with the estimate $\hat{\theta}$ found. To make an inference on whether to reject or to accept the hypothesis $H_0$ under test, it is necessary to know the distribution $G(S|H_0)$ of the test statistic that corresponds to the parameter value $\hat{\theta}$.

After that, a statistical simulation procedure is started that results in obtaining empirical distribution $G_N(S_n|H_0)$ of the test statistic for the corresponding sample size $n$, the given number of simulations $N$ and the parameters vector $\theta = \hat{\theta}$ of $F(x,\theta)$. We can find an estimate of an achieved significance level $P\{S_n > S^*\}$ or estimates of percentage points by the use of empirical distribution $G_N(S_n|H_0)$. The hypothesis is not rejected if $P\{S_n > S^*\} > \alpha$, where $\alpha$ is the given type I error probability.

| Parameters estimated | Percentage points | | | Model |
|---|---|---|---|---|
| | 0.9 | 0.95 | 0.99 | |
| The Kolmogorov test | | | | |
| $\theta_0$ | 1.104 | 1.223 | 1.461 | $B_3$(5.0453,5.6018,3.3300, 2.1145, 0.3100) |
| $\theta_1$ | 1.084 | 1.199 | 1.427 | $B_3$(5.3655, 6.0543, 3.3092, 2.1402, 0.3000) |
| $\theta_2$ | 0.955 | 1.047 | 1.231 | $B_3$(8.8643, 20.9468, 7.9001, 9.1000, 0.2300) |
| $\theta_0, \theta_1$ | 1.066 | 1.180 | 1.409 | *Sb*(2.4625, 1.7390, 2.3814, 0.2668) |
| $\theta_0, \theta_2$ | 0.895 | 0.980 | 1.153 | $B_3$(4.2520, 7.5684, 2.1829, 1.6786, 0.3100) |
| $\theta_1, \theta_2$ | 0.902 | 0.991 | 1.169 | $B_3$(4.5096, 5.6482, 3.0218, 1.6000, 0.3100) |
| $\theta_0, \theta_1, \theta_2$ | 0.839 | 0.918 | 1.079 | $B_3$(8.5291, 6.5470, 4.4062, 1.6000, 0.2400) |
| The Cramer–von Mises–Smirnov test | | | | |
| $\theta_0$ | 0.246 | 0.323 | 0.516 | $B_3$(7.5042, 2.4317, 48.3146, 1.4000, 0.0000) |
| $\theta_1$ | 0.227 | 0.296 | 0.466 | $B_3$(6.2641, 2.8729, 33.7742, 1.3750, 0.0000) |
| $\theta_2$ | 0.156 | 0.196 | 0.290 | $B_3$(4.1621, 3.9072, 14.0226, 0.8986, 0.0059) |
| $\theta_0, \theta_1$ | 0.213 | 0.278 | 0.441 | *Sb*(3.4488, 1.2020, 1.4196, 0.0061) |
| $\theta_0, \theta_2$ | 0.122 | 0.151 | 0.223 | $B_3$(7.9405, 3.8743, 23.4697, 0.6700, 0.0000) |
| $\theta_1, \theta_2$ | 0.124 | 0.155 | 0.234 | $B_3$(7.5192, 4.0675, 25.1497, 0.7945, 0.0000) |
| $\theta_0, \theta_1, \theta_2$ | 0.010 | 0.123 | 0.181 | $B_3$(5.5784, 3.2913, 17.0579, 0.4290, 0.0067) |
| The Anderson–Darling test | | | | |
| $\theta_0$ | 1.380 | 1.778 | 2.770 | *Sb*(3.7593, 1.3295, 9.6362, 0.0552) |
| $\theta_1$ | 1.279 | 1.625 | 2.478 | $B_3$(4.8031, 3.4732, 17.6302, 6.8675, 0.0535) |
| $\theta_2$ | 0.972 | 1.193 | 1.724 | $B_3$(3.4890, 4.7102, 10.2828, 5.9597, 0.0800) |
| $\theta_0, \theta_1$ | 1.111 | 1.420 | 2.184 | $B_3$(5.4232, 3.1894, 26.3229, 6.7000, 0.0539) |
| $\theta_0, \theta_2$ | 0.692 | 0.835 | 1.178 | $B_3$(11.9769, 4.7144, 19.2233, 3.0000, 0.0000) |
| $\theta_1, \theta_2$ | 0.701 | 0.852 | 1.225 | $B_3$(22.3537, 4.1744, 51.2639, 3.6000, 0.0000) |
| $\theta_0, \theta_1, \theta_2$ | 0.563 | 0.677 | 0.955 | $B_3$(8.0353, 3.9949, 18.1724, 2.4000, 0.0500) |

**Table 5.5.** *Percentage points and models of the limiting distributions of statistics of the non-parametric goodness-of-fit tests when ML method is used for parameter estimation ($\theta_1$= 7)*

The value of $N$ defines the required accuracy of simulation of $G(S_n|H_0)$. However, the time spent for simulation increases along with the growth of $N$; therefore, we can determine $N$ during parallelization of simulation process on the basis of available computer resources (number of processors and cores) that could be used for the problem under solution.

The probability that elements of $\hat{\theta}$ are integer is zero. Thus, we should cautiously use models and percentage points of test statistic distributions for values of parameters close to integer values provided in [LEM 07a, LEM 09a, LEM 09b,

LEM 10a, LEM 10b and LEM 10c] as, with interpolation applied, results obtained can be far from the true distribution $G(S|H_0)$ with the given $\hat{\theta}$.

| Parameters estimated | Percentage points | | | Model |
|---|---|---|---|---|
| | 0.9 | 0.95 | 0.99 | |
| The Kolmogorov test | | | | |
| $\theta_0$ | 1.100 | 1.218 | 1.454 | $B_3$(8.0781, 4.8128, 5.8094, 2.0960, 0.2735) |
| $\theta_1$ | 1.084 | 1.199 | 1.428 | *Sb*(2.4326, 1.7778, 2.3797, 0.2673) |
| $\theta_2$ | 0.978 | 1.074 | 1.266 | $B_3$(8.4485, 5.1812, 5.5890, 1.8364, 0.2700) |
| $\theta_0, \theta_1$ | 1.072 | 1.186 | 1.417 | $B_3$(5.7833, 6.1641, 3.2903, 2.1269, 0.2699) |
| $\theta_0, \theta_2$ | 0.911 | 0.999 | 1.179 | *Sb*(2.6863, 1.8734, 2.0545, 0.2559) |
| $\theta_1, \theta_2$ | 0.929 | 1.012 | 1.198 | *Sb*(2.6357, 1.8244, 2.0497, 0.2612) |
| $\theta_0, \theta_1, \theta_2$ | 0.863 | 0.948 | 1.117 | $B_3$(11.1281, 6.1031, 6.0962, 1.7021, 0.2200) |
| The Cramer–von Mises–Smirnov test | | | | |
| $\theta_0$ | 0.242 | 0.319 | 0.507 | $B_3$(4.5895, 2.5584, 15.2153, 0.8500, 0.0000) |
| $\theta_1$ | 0.228 | 0.296 | 0.467 | $B_3$(6.0112, 2.5379, 23.0339, 0.8900, 0.0000) |
| $\theta_2$ | 0.166 | 0.209 | 0.314 | $B_3$(5.8877, 4.0329, 23.3907, 1.2150, 0.0000) |
| $\theta_0, \theta_1$ | 0.217 | 0.284 | 0.450 | *Sb*(3.4552, 1.1997, 1.4606, 0.0061) |
| $\theta_0, \theta_2$ | 0.128 | 0.160 | 0.239 | *Sb*(4.6035, 1.4434, 1.3182, 0.0060) |
| $\theta_1, \theta_2$ | 0.131 | 0.165 | 0.252 | *Sb*(4.4612, 1.4003, 1.3183, 0.0059) |
| $\theta_0, \theta_1, \theta_2$ | 0.107 | 0.134 | 0.201 | $B_3$(6.9845, 2.7596, 2.6920, 0.4000, 0.0060) |
| The Anderson–Darling test | | | | |
| $\theta_0$ | 1.363 | 1.752 | 2.722 | $B_3$(5.6824, 4.0065, 18.9636, 8.4000, 0.0000) |
| $\theta_1$ | 1.279 | 1.624 | 2.477 | *Sb*(3.4000, 1.3163, 7.4752, 0.0535) |
| $\theta_2$ | 1.005 | 1.240 | 1.799 | $B_3$(3.4843, 5.3032, 9.1592, 6.2767, 0.0800) |
| $\theta_0, \theta_1$ | 1.139 | 1.457 | 2.243 | $B_3$(6.3736, 2.8599, 35.0312, 6.7458, 0.0538) |
| $\theta_0, \theta_2$ | 0.713 | 0.864 | 1.227 | $B_3$(4.6820, 5.7296, 7.8880, 3.4597, 0.0523) |
| $\theta_1, \theta_2$ | 0.722 | 0.881 | 1.275 | $B_3$(4.3613, 6.0352, 7.6499, 3.8116, 0.0520) |
| $\theta_0, \theta_1, \theta_2$ | 0.589 | 0.714 | 1.009 | $B_3$(7.9147, 4.0088, 21.3294, 2.9120, 0.0500) |

**Table 5.6.** *Percentage points and models of the limiting distributions of statistics of the non-parametric goodness-of-fit tests when ML method is used for parameter estimation ($\theta_1$= 8)*

Let us consider an example where a composite hypothesis is tested in regard to the IGD with the density function:

$$f(x) = \left(\frac{\theta_1}{2\pi x^3}\right)^{1/2} \exp\left(-\frac{\theta_1(x-\theta_0)^2}{2\theta_0^2 x}\right).$$

In this case, distributions $G(S|H_0)$ of the non-parametric tests depend on specific values of $\theta_0$ and $\theta_1$.

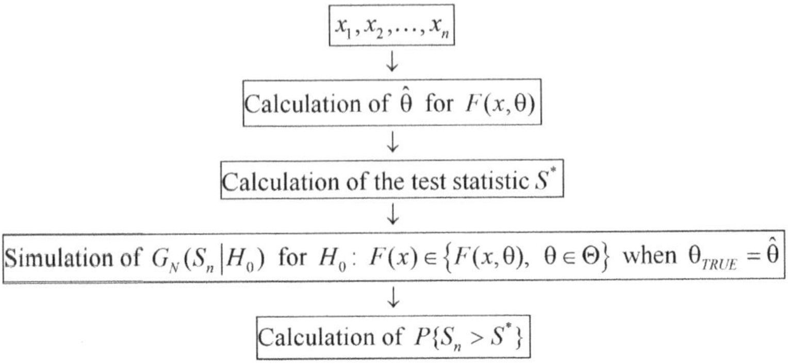


**Figure 5.3.** *Testing the composite hypothesis* $H_0 : F(x) \in \{F(x,\theta), \theta \in \Theta\}$

The following is the sample under analysis:

| | | | | | | | | | |
|---|---|---|---|---|---|---|---|---|---|
| 0.278 | 0.633 | 0.928 | 1.078 | 1.334 | 1.937 | 2.297 | 2.630 | 3.554 | 5.674 |
| 0.312 | 0.686 | 0.933 | 1.080 | 1.497 | 1.965 | 2.362 | 2.919 | 3.593 | 5.989 |
| 0.358 | 0.716 | 0.936 | 1.089 | 1.612 | 1.991 | 2.364 | 2.995 | 3.948 | 6.284 |
| 0.361 | 0.776 | 0.938 | 1.113 | 1.671 | 2.012 | 2.417 | 3.002 | 3.996 | 6.863 |
| 0.362 | 0.777 | 0.956 | 1.119 | 1.680 | 2.026 | 2.467 | 3.120 | 4.053 | 7.580 |
| 0.374 | 0.789 | 0.996 | 1.159 | 1.687 | 2.027 | 2.566 | 3.149 | 4.141 | 7.644 |
| 0.403 | 0.796 | 1.038 | 1.165 | 1.731 | 2.069 | 2.577 | 3.166 | 4.363 | 7.874 |
| 0.590 | 0.805 | 1.053 | 1.166 | 1.735 | 2.146 | 2.599 | 3.224 | 4.597 | 9.236 |
| 0.597 | 0.822 | 1.060 | 1.192 | 1.763 | 2.210 | 2.621 | 3.278 | 5.022 | 11.704 |
| 0.599 | 0.849 | 1.066 | 1.245 | 1.898 | 2.213 | 2.628 | 3.528 | 5.201 | 20.069 |

ML estimates of parameters calculated are $\hat{\theta}_0 = 2.4706$ and $\hat{\theta}_1 = 2.5769$. Values of test statistics and achieved significance levels ($p$-values) obtained by simulated (in "real time") test statistic distributions under different values of $N$ are given in Table 5.7.

| Parameters estimated | $S^*$ | $P\{S_n > S^*\}$ | | | | |
|---|---|---|---|---|---|---|
| | | $N = 1{,}000$ | $N = 5{,}000$ | $N = 10{,}000$ | $N = 100{,}000$ | $N = 1{,}000{,}000$ |
| K | 0.59361 | 0.6380 | 0.6532 | 0.6535 | 0.6552 | 0.6556 |
| CMS | 0.05380 | 0.5390 | 0.5550 | 0.5555 | 0.5582 | 0.5582 |
| AD | 0.35021 | 0.5280 | 0.5498 | 0.5496 | 0.5485 | 0.5480 |

**Table 5.7.** *P-values of different tests under different volumes of simulation*

It should be noted that distributions of non-parametric goodness-of-fit test statistics [5.1–5.3] for $\hat{\theta}_0 = 2.4706$ and $\hat{\theta}_1 = 2.5769$ differ substantially from corresponding distributions under different combinations of integer values of $\theta_0$ and $\theta_1$.

## 5.4. Conclusions

In this chapter, implementation of a software has been discussed, which makes it possible to test composite hypotheses with the use of non-parametric goodness-of-fit tests in the cases when statistic distributions depend on specific values of the observed distributions.

The interactive mode is implemented for the Kolmogorov, the CMS and the AD tests, as well as the Kuiper test [KUI 60], the Watson test [WAT 61] and the test by Zhang *et al.* [ZHA 02, ZHA 05, ZHA 06].

## 5.5. Acknowledgment

This work was supported by the Ministry of Education and Science of the Russian Federation within the state assignment (project 8.1274.2011) and the Federal target programme "Research and Pedagogical Personnel for Innovative Russia" (agreement # 14.B37.21.0860).

## 5.6. Bibliography

[AND 52] ANDERSON T., DARLING D., "Asymptotic theory of certain goodness of fit criteria based on stochastic processes", *The Annals of Mathematical Statistics*, vol. 23, no. 2, pp. 193–212, 1952.

[AND 54] ANDERSON T., DARLING D., "A test of goodness of fit", *Journal of the American Statistical Association*, vol. 49, no. 268, pp. 765–769, 1954.

[BOL 83] BOLSHEV L., SMIRNOV N., *Tables of Mathematical Statistics*, Nauka, Moscow, 1983. [In Russian]

[BOL 87] BOLSHEV L., "On the question on testing some composite statistical hypotheses", *Theory of Probability and Mathematical Statistics*, Selected Works, Nauka, Moscow, pp. 5–63, 1987. [In Russian]

[CHA 81] CHANDRA M., SINGPURWALLA N., STEPHENS M., "Kolmogorov statistics for tests of fit for the extreme value and Weibull distribution", *Journal of the American Statistical Association*, vol. 76, no. 375, pp. 729–731, 1981.

[DAR 55] DARLING D., "The Cramer-Smirnov test in the parametric case", *Annals of Mathematical Statistics*, vol. 26, no. 1, pp. 1–20, 1955.

[DAR 57] DARLING D., "The Cramer-Smirnov test in the parametric case", *Annals of Mathematical Statistics*, vol. 28, no. 1, pp. 823–938, 1957.

[DUR 73] DURBIN J., "Weak convergence of the sample distribution function when parameters are estimated", *Annals of Statistics*, vol. 1, no. 2, pp. 279–290, 1973.

[DUR 75] DURBIN J., "Kolmogorov-Smirnov tests when parameters are estimated with applications to tests of exponentiality and tests of spacings", *Biometrika*, vol. 62, no. 1, pp. 5–22, 1975.

[DUR 76] DURBIN J., "Kolmogorov-Smirnov test when parameters are estimated", *Empirical Distributions and Processes*, vol. 566 of Lecture Notes in Mathematics, Springer, Berlin-Heidelberg, pp. 33-44, 1976.

[DZH 82] DZHAPARIDZE K., NIKULIN M., "Probability distribution of the Kolmogorov and omega-square statistics for continuous distributions with shift and scale parameters", *Journal of Soviet Mathematics*, vol. 20, no. 3, pp. 2147–2164, 1982.

[GIH 53] GIHMAN I., "Some remarks on the consistency criterion of A.N. Kolmogorov", *Doklady Akademii nauk SSSR*, vol. 91, no. 4, pp. 715–718, 1953.

[KAC 55] KAC M., KIEFER J., WOLFOWITZ J., "On tests of normality and other tests of goodness of fit based on distance methods", *Annals of Mathematical Statistics*, vol. 26, no. 2, pp. 189–211, 1955.

[KUI 60] KUIPER N., "Tests concerning random points on a circle", *Nederlandse Akademie van Wetenschappen, Proc. Ser. A*, vol. 63, pp. 38–47, 1960.

[LEM 98] LEMESHKO B., POSTOVALOV S., "Statistical distributions of nonparametric goodness-of-fit tests as estimated by the sample parameters of experimentally observed laws", *Industrial Laboratory*, vol. 64, no. 3, pp. 197–208, 1998.

[LEM 01] LEMESHKO B., POSTOVALOV S., "Application of the nonparametric goodness-of-fit tests in testing composite hypotheses", *Optoelectronics, Instrumentation and Data Processing*, vol. 37, no. 2, pp. 76–88, 2001.

[LEM 02] LEMESHKO B., POSTOVALOV S., "The nonparametric goodness-of-fit tests about fit with Johnson distributions in testing composite hypotheses", *News of the SB AS HS*, vol. 1, no. 5, pp. 65–74, 2002. [In Russian]

[LEM 04] LEMESHKO B., MAKLAKOV A., "Nonparametric test in testing composite hypotheses on goodness of fit exponential family distributions", *Optoelectronics, Instrumentation and Data Processing*, vol. 40, no. 3, pp. 3–18, 2004.

[LEM 07a] LEMESHKO B., LEMESHKO S., "Statistic distributions of the nonparametric goodness-of-fit tests in testing hypotheses relative to beta-distributions", *News of the SB AS HS*, vol. 2, no. 19, pp. 6–16, 2007. [In Russian]

[LEM 07b] LEMESHKO S., Expansion of applied opportunities of some classical methods of mathematical statistics, PhD Thesis, Novosibirsk State Technical University, 2007.

[LEM 09a] LEMESHKO B., LEMESHKO S., "Distribution models for nonparametric tests for fit in verifying complicated hypotheses and maximum-likelihood estimators. Part I", *Measurement Techniques*, vol. 52, no. 6, pp. 555–565, 2009.

[LEM 09b] LEMESHKO B., LEMESHKO S., "Models for statistical distributions in nonparametric fitting tests on composite hypotheses based on maximum-likelihood estimators. Part II", *Measurement Techniques*, vol. 52, no. 8, pp. 799–812, 2009.

[LEM 10a] LEMESHKO B., LEMESHKO S., AKUSHKINA K., "Models of statistical distributions of nonparametric goodness-of-fit tests in testing composite hypotheses of the generalized Weibull distribution", *Proceedings of the 3rd International Conference on Accelerated Life Testing, Reliability-based Analysis and Design*, Clermont-Ferrand, France, pp. 125–132, May 2010.

[LEM 10b] LEMESHKO B., LEMESHKO S., NIKULIN M., *et al.*, "Modeling statistic distributions for nonparametric goodness-of-fit criteria for testing composite hypotheses with respect to the inverse Gaussian law", *Automation and Remote Control*, vol. 71, no. 7, pp. 1358–1373, 2010.

[LEM 10c] LEMESHKO B., LEMESHKO S., POSTOVALOV S., "Statistic distribution models for some nonparametric goodness-of-fit tests in testing composite hypotheses", *Communications in Statistics – Theory and Methods*, vol. 39, no. 3, pp. 460–471, 2010.

[LEM 11a] LEMESHKO B., LEMESHKO S., "Construction of statistic distribution models for nonparametric goodness-of-fit tests in testing composite hypotheses: the computer approach", *Quality Technology & Quantitative Management*, vol. 8, no. 4, pp. 359–373, 2011.

[LEM 11b] LEMESHKO B., LEMESHKO S., "Models of statistic distributions of nonparametric goodness-of-fit tests in composite hypotheses testing for double exponential law cases", *Communications in Statistics – Theory and Methods*, vol. 40, no. 16, pp. 2879–2892, 2011.

[LEM 11c] LEMESHKO B., LEMESHKO S., AKUSHKINA K., "Real-time studying of statistic distributions of non-parametric goodness-of-fit tests when testing composite hypotheses", *Proceedings of the International Workshop Äpplied Methods of Statistical Analysis. Simulations and Statistical Inference, AMSA'2011*, Novosibirsk, Russia, pp. 19–27, September 2011.

[LEM 11d] LEMESHKO B., LEMESHKO S., AKUSHKINA K., *et al.*, "Inverse Gaussian model and its applications in reliability and survival Analysis", in RYKOV V., BALAKRISHNAN N., NIKULIN M. (eds), *Mathematical and Statistical Models and Methods in Reliability. Applications to Medicine, Finance, and Quality Control*, Birkhaüser, Boston, MA, pp. 433–453, 2011.

[LEM 11e] LEMESHKO B., LEMESHKO S., CHIMITOVA E., *et al.*, "Software system for simulation and research of probabilistic regularities and statistical data analysis in reliability and quality control", in RYKOV V., BALAKRISHNAN N., NIKULIN M. (eds), *Mathematical and Statistical Models and Methods in Reliability. Applications to Medicine, Finance, and Quality Control*, Birkhaüser, Boston, MA, pp. 417–432, 2011.

[LEM 11f] LEMESHKO B., LEMESHKO S., POSTOVALOV S., *et al.*, *Statistical Data Analysis, Simulation and Study of Probability Regularities. Computer Approach*, NSTU Monographs, NSTU Publishing House, Novosibirsk, 2011. [In Russian]

[MAR 78] MARTYNOV G., *Omega-Square Tests*, Nauka, Moscow, 1978. [In Russian]

[MAR 11] MARTYNOV G., "Weighted Cramer-von Mises test with estimated parameters", *Communications in Statistics – Theory and Methods*, vol. 40, nos. 19–20, pp. 3569–3586, 2011.

[NIK 92a] NIKULIN M., "Gihman statistic and goodness-of-fit tests for grouped data", *Mathematical Reports of the Academy of Science of the Royal Society of Canada*, vol. 14, no. 4, pp. 151–156, 1992.

[NIK 92b] NIKULIN M., "A variant of the generalized omega-square statistic", *Journal of Soviet Mathematics*, vol. 61, no. 4, pp. 1896–1900, 1992.

[PEA 72] PEARSON E., HARTLEY H., *Biometrica Tables for Statistics*, vol. 2, University Press, Cambridge, 1972.

[STE 70] STEPHENS M., "Use of the Kolmogorov-Smirnov, Cramer-Von Mises and related statistics without extensive tables", *Journal of the Royal Statistical Society*, vol. 32, no. 1, pp. 115–122, 1970.

[STE 74] STEPHENS M., "EDF statistics for goodness of fit and some comparisons", *Journal of the American Statistical Association*, vol. 69, no. 347, pp. 730–737, 1974.

[TYU 84a] TYURIN Y., "On the limiting Kolmogorov-Smirnov statistic distribution for composite hypothesis", *News of the AS USSR. Ser. Math.*, vol. 48, no. 6, pp. 1314–1343, 1984. [In Russian]

[TYU 84b] TYURIN Y., SAVVUSHKINA N., "Goodness-of-fit test for Weibull-Gnedenko distribution", *News of the AS USSR. Ser. Techn. Cybernetics*, vol. 3, pp. 109–112, 1984. [In Russian]

[WAT 61] WATSON G., "Goodness-of-fit tests on a circle", *Biometrika*, vol. 48, nos. 1–2, pp. 57–63, 1961.

[ZHA 02] ZHANG J., "Powerful goodness-of-fit tests based on the likelihood ratio", *Journal of the Royal Statistical Society: Series B*, vol. 64, no. 2, pp. 281–294, 2002.

[ZHA 05] ZHANG J., WU Y., "Likelihood-ratio tests for normality", *Computational Statistics & Data Analysis*, vol. 49, no. 3, pp. 709–721, 2005.

[ZHA 06] ZHANG J., "Powerful two-sample tests based on the likelihood ratio", *Technometrics*, vol. 48, no. 1, pp. 90–103, 2006.

Chapter 6

# Modeling of Categorical Data

## 6.1. Introduction

In multivariate statistics, several random variables need to be investigated simultaneously. We will find connections between variables or the response of some variables on others. These variables can be metric variables or categorical variables. We will explain the influence of the endogenous variables on the response variables. If all variables are metric variables, then often a regression model will be used. If the response is metric, but the endogenous factors are categorical variables, then an analysis of variance can help. For categorical response variables, a discriminant analysis or logistic regression is used. In [AGR 02] and [EVE 01], many methods for statistical analysis are described. We will concentrate on statistical modeling and scaling of categorical variables. Mostly, as scaling is connected with proximities or distances, we consider a statistical approach as in [LÄU 13] or [RAM 10]. In this chapter, a scaling will be based on the estimation of parameters in conditional distributions. We will see which parameters in such families are estimable and find estimates for those. The conditional parameters will be used for methods of scaling of categorical data and can be used for the interpretation of the results of the analysis of the data. Some other methods for statistical scaling are based on tests and lead to most separating scales, as was considered by Ahrens and Läuter [AHR 81] and Ramadan [RAM 10].

In general, we consider a paired random variable $(Z_1, Z_2)$ where the variable $Z_1$ is the endogenous variable and $Z_2$ is the response variable. We are looking at the conditional variable $\tilde{U} := Z_2 \,|\, Z_1 = t$ and we will find the influence of $t$ in this

---

Chapter written by Henning LÄUTER.

variable. We admit both metric and categorical variables for $Z_1$. If $Z_1$ is a metric variable and $t$ is known, then often the regression:

$$r(t) := \mathsf{E}(Z_2 \,|\, Z_1 = t) \qquad [6.1]$$

is studied, non-parametric or parametric regression. If $Z_1$ is a categorical variable, then we will find tools for representing the analyzed data. Here, we consider the statistical scaling and find points in metric spaces equivalent to the categorical values. These results for scaling are the basis for some other statistical analyses of categorical data as for discrimination and classification.

## 6.2. Continuous conditional distributions

### 6.2.1. *Conditional normal distribution*

We consider an $(m_1 + m_2)$ dimensional, normally distributed random variable $Z$ with

$$Z = \begin{pmatrix} Z_1 \\ Z_2 \end{pmatrix} \sim \mathsf{N}_{m_1+m_2}\left(\begin{pmatrix} \mu_1 \\ \mu_2 \end{pmatrix}, \begin{pmatrix} \Sigma_{11} & \Sigma_{12} \\ \Sigma_{21} & \Sigma_{22} \end{pmatrix}\right). \qquad [6.2]$$

Here, we have $Z_1, \mu_1 \in R^{m_1}, Z_2, \mu_2 \in R^{m_2}$ and $\Sigma_{11}$ is an $m_1 \times m_1$ matrix and the others correspondingly. The variance matrix $\Sigma = varZ$ should be non-singular. Then, the conditional distribution of $Z_2$ given $Z_1 = t$ is well known as:

$$Z_2 \,|\, Z_1 = t \sim \mathsf{N}_{m_2}\left(\mu_2 + \Sigma_{21}\Sigma_{11}^{-1}(t - \mu_1), \Sigma_{2\cdot 1}\right) \qquad [6.3]$$

with $\Sigma_{2\cdot 1} = \Sigma_{22} - \Sigma_{21}\Sigma_{11}^{-1}\Sigma_{12}$.

#### 6.2.1.1. *Estimation of parameters*

This conditional distribution depends on the parameters $\mu_1, \mu_2, \Sigma, t$. Obviously, these parameters are not identifiable after observing these conditional variables. Assuming for positive integers $n_1, ..., n_L$ and values $t_1, ..., t_L$, we have for $l = 1, ..., L$ variables $U_{l1}, ..., U_{ln_l}$ with:

$$U_{lj} \sim \mathsf{N}_{m_2}\left(\mu_2 + \Sigma_{21}\Sigma_{11}^{-1}(t_l - \mu_1), \Sigma_{2\cdot 1}\right), \quad j = 1, ..., n_l. \qquad [6.4]$$

All these variables should be independent, the sample sizes $n_l$ should be known. Then, we see that the parameters $\mu_1, \mu_2, \Sigma$ determine the distribution of $Z$, but the parameters $t_1, ..., t_L$ determine the localizations where the first $m_1$-dimensional subvectors of the $U_{lj}$ are fixed. Therefore, the whole parameter $(\mu_1, \mu_2, \Sigma, t_1, ..., t_L)$

consists of global distributional parameters, global for all variables, and localized parameters. Although, e.g., $t_l$ and $\mu_1$ influence the distribution of $U_{lj}$, in the same way their meaning is completely different. $\mu_1$ describes the expectation of the first subvector and $t_l$ describes the point where $Z_1$ is fixed. We see immediately that the parameters:

$$\tau_l = \mu_2 + \Sigma_{21}\Sigma_{11}^{-1}(t_l - \mu_1) \qquad [6.5]$$

are estimable but not any of the parameters $\mu_1, \mu_2, t_1, ..., t_L$ separately. Moreover all the differences:

$$\delta_{ls} := \Sigma_{21}\Sigma_{11}^{-1}(t_l - t_s) \qquad [6.6]$$

are estimable. This means that we can determine relations between the points where the first subvectors are to be observed. In the special case $m_1 = 1$, we can estimate the values $t_2 - t_1, ..., t_L - t_1$ up to a factor and these estimations can be ordered on a line. For $m_2 = 1$, we represent the estimations of the real values $\Sigma_{21}\Sigma_{11}^{-1}(t_2 - t_1), ..., \Sigma_{21}\Sigma_{11}^{-1}(t_L - t_1)$.

THEOREM 6.1.– Assume [6.2] and let the variables $\{U_{lj}, j = 1, ..., n_l;\ l = 1, ..., L.\}$ be independent; the variables $\{U_{lj}, j = 1, ..., n_l\}$ are identically distributed and it is known that $U_{lj}$ has a conditional distribution as in [6.3] and [6.4] for unknown $t_l$. There is no connection between the $t_l$s. Then, the best unbiased estimate for $\delta_{ls} = \Sigma_{21}\Sigma_{11}^{-1}(t_l - t_s)$ is given by:

$$\hat{\delta}_{ls} = \bar{U}_{l\cdot} - \bar{U}_{s\cdot}$$

with

$$\bar{U}_{l\cdot} = \frac{1}{n_l}\sum_{j=1}^{n_l} U_{lj} = \frac{1}{n_l} U_{l+} \quad .$$

The values $\nu = \mu_2 - \Sigma_{21}\Sigma_{11}^{-1}\mu_1$ and $t_1, ..., t_L$ are not estimable but all differences $\delta_{ls}$. From this, it follows that we can choose in an arbitrary way the origin, i.e. an arbitrary $\hat{t}_1$, say $\hat{t}_1 = 0$ and then all other $\Sigma_{21}\Sigma_{11}^{-1}\hat{t}_2, ..., \Sigma_{21}\Sigma_{11}^{-1}\hat{t}_L$ are calculable.

We see that for normal distributions, the localized parameters $\delta_{ls}$ can be estimated from observations of the conditioned distributions. It is clear from [6.4] that the $t_1, ..., t_L$ cannot be estimated but relations between them can be. We remember that the values of the localized parameters can be identified with the levels under which the observations are taken.

From the distribution, it follows that $\Sigma_{11}$ and $\Sigma_{21}$ are not estimable. Therefore, these parameters have to be assumed known or it is sufficient to estimate the $\delta_{ls}$.

If the conditions $t_1, ..., t_L$ are functionally completely independent, then the estimates for the $\delta_{ls}$ are given in theorem 6.1. But there are important cases of different $t_1, ..., t_L$ where in parts connections exist between them. We assume that $\tau = (t'_1, ..., t'_L)'$ varies in a $p$-dimensional subspace of $R^{Lm_1}$, $p \leq Lm_1$ and $p$ is as small as possible. Then, we know $m_1 \times p$ matrices $C_1, ..., C_L$ with:

$$t_l = C_l\beta, \quad l = 1, ..., L, \beta \in R^p \qquad [6.7]$$

and we obtain

$$U_{lj} = \nu + \Sigma_{21}\Sigma_{11}^{-1}t_l + \varepsilon_{lj} = \nu + \Sigma_{21}\Sigma_{11}^{-1}C_l\beta + \varepsilon_{lj}. \qquad [6.8]$$

With the Kronecker product $A \otimes B$ of matrices $A, B$ and $1_k$ the $k$-dimensional vector of ones, $I_k$ the $k \times k$ identity matrix and:

$$E_n = \begin{pmatrix} 1_{n_1} & 0 & 0 & \dots & 0 \\ 0 & 1_{n_2} & 0 & \dots & \vdots \\ \vdots & & \ddots & & \vdots \\ 0 & \dots & & 0 & 1_{n_L} \end{pmatrix},$$

we get

$$U = (1_n \otimes I_{m_2})\nu + (E_n \otimes \Sigma_{21}\Sigma_{11}^{-1})\begin{pmatrix} C_1 \\ \vdots \\ C_L \end{pmatrix}\beta + \varepsilon = X\begin{pmatrix} \nu \\ \beta \end{pmatrix} + \varepsilon \qquad [6.9]$$

with $U = (U'_{11}, ..., U'_{Ln_L})'$, $\varepsilon = (\varepsilon'_{11}, ..., \varepsilon'_{Ln_L})'$ and

$$X = \left(1_n \otimes I_{m_2} \vdots (E_n \otimes \Sigma_{21}\Sigma_{11}^{-1})\begin{pmatrix} C_1 \\ \vdots \\ C_L \end{pmatrix}\right).$$

Then, for a matrix $F$ with $m_2 + p$ columns, the best unbiased estimation for an estimable $F(\nu', \beta')'$ is:

$$F\begin{pmatrix} \hat{\nu} \\ \hat{\beta} \end{pmatrix} = F\left(X'(I_n \otimes \Sigma_{2\cdot 1}^{-1})X\right)^+ X'(I_n \otimes \Sigma_{2\cdot 1}^{-1})u$$

for a realization $u$ from [6.9]. As a result, the optimal estimate for $\delta_{ls}$ is:

$$\hat{\delta}_{ls} = \Sigma_{21}\Sigma_{11}^{-1}(C_l - C_s)\,\hat{\beta}, \quad l = 1, ..., L. \qquad [6.10]$$

THEOREM 6.2.– Assume [6.2] and let the variables:

$$U_{lj}, \quad j = 1, ..., n_l; l = 1, ..., L$$

be independently distributed. The variables $U_{lj}, \quad j = 1, ..., n_l$ are identically distributed and it is known that $U_{lj}$ has a conditional distribution as in [6.3] and [6.4] for unknown $t_l$. There exist such $m_1 \times p$ matrices $C_1, ..., C_L$ that [6.7] is fulfilled. Then, $\hat{\delta}_{ls}$ from [6.10] is the maximum likelihood estimator for $\delta_{ls}$.

Summarizing in this standard model where the random variables are normally distributed and the conditional variable will be observed at unknown points, we see that the expectations of the variables $Z_1$ and $Z_2$ are not estimable, the conditional points $t_1, ..., t_L$ are not estimable, and the variance of $Z_1$ and the correlation between $Z_1$ and $Z_2$ are not estimable. The contrasts $\delta_{ls} = \Sigma_{21}\Sigma_{11}^{-1}(C_l - C_s)\beta$ are estimable.

We cannot expect that in general distributional families, the possibilities for estimated parameters are completely better. We consider general continuous random variables in section 6.2.2, where the principle will be repeated in an obvious way.

### 6.2.2. *More general continuous conditional distributions*

Let us assume that the $(m_1 + m_2)$ dimensional variable $Z = (Z_1', Z_2')'$ has any continuous distribution. The one-dimensional marginal densities of the components of $Z_1$ and $Z_2$ are denoted by $f_{11}, ..., f_{1m_1}$ and $f_{21}, ..., f_{2m_2}$ and the distribution functions by $F_{1i}$ and $F_{2j}$, respectively. The dependence between $Z_1$ and $Z_2$ is described by a copula function $C$. We denote $F_1(\xi_1) = \Big(F_{11}(\xi_{11}), ..., F_{1m_1}(\xi_{1m_1})\Big)$ and $F_2(\xi_2) = \Big(F_{21}(\xi_{21}), ..., F_{2m_2}(\xi_{2m_2})\Big)$. Then, the joint distribution function of $(Z_1, Z_2)$ is:

$$F(\xi_1, \xi_2) = C\Big(F_1(\xi_1), F_2(\xi_2)\Big).$$

For any distribution $F$ with the marginal distributions $F_{1i}$, $F_{2j}$, such a copula function exists and is uniquely determined. Assuming that the densities of $F, F_{1i}, F_{2j}$ and the copula $C$ exist, then for the joint density $f$ we have the representation:

$$f(\xi_1, \xi_2) = c\Big(F_1(\xi_1), F_2(\xi_2)\Big) f_{11}(\xi_{11}) \cdot ... \cdot f_{1m_1}(\xi_{1m_1})\, f_{21}(\xi_{21}) \cdot ... \cdot f_{2m_2}(\xi_{2m_2}).$$

6.2.2.1. *Conditional distribution*

As a result, the density of $Z_2 \mid Z_1 = t$ with $t = (t_{(1)}, ..., t_{(m_1)})'$ is:

$$\begin{aligned} f^{Z_1=t}(\xi_2) &= c\Big(F_1(t), F_2(\xi_2)\Big) f_{11}(t_{(1)}) \cdot ... \cdot f_{1m_1}(t_{(m_1)})\, f_{21}(\xi_{21}) \\ &\quad \cdot ... \cdot f_{2m_2}(\xi_{2m_2}) / f^{Z_1}(t). \end{aligned} \qquad [6.11]$$

Therefore, the copula function determines the dependence of $Z_1$ and $Z_2$ and hence the copula function is an essential part of the statistical model.

Now, we assume that for positive integers $n_1, ..., n_L$ and values $t_1, ..., t_L$ and for all $l = 1, ..., L$, the variables $U_{l1}, ..., U_{ln_l}$ are identically distributed with the density $f^{Z_1=t_l}$ from [6.11]. All these variables should be independent, the sample sizes are known. We assume that the distribution function $F_1$ of $Z_1$ depends on a parameter $\vartheta_1 \in R^{p_1}$ and the distribution function $F_2$ depends on $\vartheta_2 \in R^{p_2}$ for some $p_1, p_2$. Then, the global parameter for the distribution of the sample is $\vartheta = (\vartheta_1, \vartheta_2)$. We observe $u = (u_{11}, ..., u_{Ln_L})$ with $u_{lj} = (u_{lj,1}, ..., u_{lj,m_2})'$. The likelihood function for the whole sample is with $t_l = (t_{l1}, ..., t_{lm_1})'$

$$\begin{aligned} \mathcal{L}(\vartheta, t_1, ..., t_L; u) &= \prod_{l=1}^{L} \prod_{j=1}^{n_l} c\Big(F_1(t_l; \vartheta_1), F_2(u_{lj}; \vartheta_2)\Big) f_{11}(t_{l1}; \vartheta_1) \cdot ... \\ &\quad ... \cdot f_{1m_1}(t_{lm_1}; \vartheta_1)\, f_{21}(u_{lj,1}, \vartheta_2) \cdot ... \\ &\quad ... \cdot f_{2m_2}(u_{lj,m_2}, \vartheta_2) / f^{Z_1}(t_l; \vartheta_1). \end{aligned} \qquad [6.12]$$

Now, we can define the maximum likelihood estimation of the parameters. Of course, maximization of the likelihood is taken over the possible parameter set. If the $t_1, ..., t_L$ are unrestricted, then we maximize $\mathcal{L}$ without additional conditions. If there are restrictions as in [6.7], then we substitute $t_l = C_l\, \beta$ and consider $\mathcal{L}$ as a function of $\vartheta, \beta$. Then, we proceed as in the preceding section.

In the special case $m_1 = 1$, the likelihood function has a simple representation. We see that:

$$\begin{aligned} \mathcal{L}(\vartheta, t_1, ..., t_L; u) &= \prod_{l=1}^{L} \prod_{j=1}^{n_l} c\Big(F_1(t_l; \vartheta_1), F_2(u_{lj}; \vartheta_2)\Big) f_{21}(u_{lj,1}, \vartheta_2) \cdot ... \\ &\quad ... \cdot f_{2m_2}(u_{lj,m_2}, \vartheta_2). \end{aligned} \qquad [6.13]$$

6.2.2.2. *Normal copula*

Obviously, the maximum likelihood estimation depends on the copula, i.e. it clearly depends on the model that we have assumed. In the special case of a normal copula, we will look at the properties. Let $\Phi$ be the distribution of $\mathsf{N}(0,1)$ and $\Phi_R$ be the distribution function of $\mathsf{N}_m(0,R)$ where $m = m_1 + m_2$ and the variance matrix $R$ at the same time is a correlation matrix. Then, the normal copula has the form:

$$C_R^N(u_1, ..., u_m) = \Phi_R(\Phi^{-1}(u_1), ..., \Phi^{-1}(u_m))$$

$$= \frac{1}{(2\pi)^{m/2}|R|^{1/2}} \int_{-\infty}^{\Phi^{-1}(u_1)} \cdots \int_{-\infty}^{\Phi^{-1}(u_m)} \exp(-\frac{1}{2}\,\xi' R^{-1}\xi)\, \mathrm{d}\xi.$$

Denoting the copula density of $C_R^N$ by $c_R^N$, the conditional density in [6.11] then has the form:

$$f^{Z_1=t}(\xi_2) = c_R^N\Big(F_1(t;\vartheta_1), F_2(\xi_2;\vartheta_2)\Big) f_{11}(t_{(1)};\vartheta_1) \cdot ... \cdot f_{1m_1}(t_{(m_1)};\vartheta_1) \cdot ...$$

$$... \cdot f_{21}(\xi_{21},\vartheta_2) \cdot ... \cdot f_{2m_2}(\xi_{2m_2},\vartheta_2)/f^{Z_1}(t;\vartheta_1).$$

The copula density has the form:

$$c_R^N(u_1, ..., u_m) = \frac{1}{|R|^{1/2}} \exp\Big(-\frac{1}{2}[Q_R(u) - \Phi^{-1}(u_1)^2 - ... - \Phi^{-1}(u_m)^2]\Big)$$

with the quadratic form:

$$Q_R(u) = \Big(\Phi^{-1}(u_1), ..., \Phi^{-1}(u_m)\Big) R^{-1} \Big(\Phi^{-1}(u_1), ..., \Phi^{-1}(u_m)\Big)'.$$

Under $m_2 = 1$, the correlation matrix $R$ in a special standardized form is significant. This is:

$$R = R_0 := \begin{pmatrix} & & & & \rho_1 \\ & & I_{m_1} & & \vdots \\ & & & & \rho_{m_1} \\ \rho_1 & \rho_2 & \cdots & \rho_{m_1} & 1 \end{pmatrix}$$

and $R^{-1}$ can be calculated in a direct way.

## 6.3. Discrete conditional distributions

We consider $(m_1 + m_2)$ dimensional variables $Z$ where $Z_1$ is $m_1$-dimensional and $Z_2$ is a $m_2$-dimensional discrete random variables. For given $t$, the distribution of $Z_2|Z_1 = t$ depends on a parameter $\mu$ and is denoted by $\mathsf{P}_{\mu,t}$. The support of $Z_2|Z_1 = t$ should be independent on $t$.

### 6.3.1. *Parametric conditional distributions*

The parameter $\mu$ is a distributional parameter and $t$ can be considered as a localized parameter. We consider measures $\mathsf{P}_{\mu,t}$ in the following way. The support points of $Z_2$ are $c_1, c_2, ..., c_k$ with $\mathsf{P}(Z_2 = c_i) = \pi_i, \quad i = 1, 2, ...$ and let us take:

$$\mathsf{P}(Z_1 \in B, Z_2 = c_i) = \pi_i \int_B \varphi_i(\xi)\eta(\,\mathrm{d}\xi), \quad B \subset \mathbb{R}^{m_1}, i = 1, 2, ...$$

for a $\sigma$-finite measure $\eta$. As a result, $Z_1$ has the density $\varphi$ with:

$$\varphi(\xi) = \sum_i \pi_i \varphi_i(\xi)$$

and we have the conditional probability:

$$\mathsf{P}(Z_2 = c_i \,|\, Z_1 = t) = \frac{\pi_i \varphi_i(t)}{\sum_j \pi_j \varphi_j(t)}. \qquad [6.14]$$

We assume that the densities $\varphi_i$ are known up to a parameter $\vartheta_i$. Then, the distribution of $Z_2|Z_1 = t$ depends on $\mu = (\vartheta_1', ..., \vartheta_k')'$ and $t$. We use the notation:

$$p_i(t; \mu) := \frac{\pi_i \varphi_i(t)}{\sum_j \pi_j \varphi_j(t)} \qquad [6.15]$$

for the conditional probability [6.14] and the corresponding measure is $\mathsf{P}_{\mu,t}$.

Assuming for known positive integers $n_1, ..., n_L$ and values $t_1, ..., t_L$, we have for $l = 1, ..., L$ variables $U_{l1}, ..., U_{ln_l}$ with $U_{lj} \sim \mathsf{P}_{\mu,t_l}$. All these variables should be independently distributed. The observation $u_{lj}$ is one of the points $c_1, ..., c_k$ and all observations can be presented in Table 6.1 with the frequencies of the support points.

Here, we see in the column under $t_l$ the values $h_{1l}, ..., h_{kl}$ are realizations of a multinomial distributed variable $W_l$ with:

$$W_l \sim \mathsf{Mult}\Big(h_{+l}, p_1(t_l; \mu), ..., p_k(t_l; \mu)\Big) \qquad [6.16]$$

if

$$h_{+l} := \sum_{i=1}^{k} h_{il} = n_l$$

is the sample size of observations in the point $t_l$.

| | $t_1$ | $t_2$ | $t_3$ | $\cdots$ | $t_L$ |
|---|---|---|---|---|---|
| $c_1$ | $h_{11}$ | $h_{12}$ | $h_{13}$ | $\cdots$ | $h_{1L}$ |
| | | | | $\ddots$ | |
| $c_k$ | $h_{k1}$ | $h_{k2}$ | $h_{k3}$ | $\cdots$ | $h_{kL}$ |

**Table 6.1.** *Frequencies in a contingency table*

As a result, the frequencies in Table 6.1 can be considered as a realization of a multivariate multinomial distributed variable $W = (W_1, ..., W_L)$ for independent $W_l$ and [6.16]. Sometimes the table of the relative frequencies $f_{il} = h_{il}/h_{+l}$ will be used instead of Table 6.1. Examples for joint distributions can be constructed in various way. For instance, for a binomial distribution $B(m, \vartheta_i)$, we have:

$$\mathsf{P}(Z_1 = j, Z_2 = c_i) = \pi_i \binom{m}{j} \vartheta_i^j (1 - \vartheta_i)^{m-j}$$

and therefore

$$\mathsf{P}(Z_2 = c_i \mid Z_1 = j) = \frac{\pi_i \vartheta_i^j (1 - \vartheta_i)^{m-j}}{\sum_{s=1}^{k} \pi_s \vartheta_s^j (1 - \vartheta_s)^{m-j}} = p_i(j; \mu)$$

with $\mu = (\vartheta_1, ..., \vartheta_k)$. Similarly, we can use a Poisson distribution $\mathsf{Poi}(\lambda_i)$ and obtain:

$$\mathsf{P}(Z_2 = c_i \mid Z_1 = j) = \frac{\pi_i \lambda_i^j \mathrm{e}^{-\lambda_i}}{\sum_{s=1}^{k} \pi_s \lambda_s^j \mathrm{e}^{-\lambda_s}} = p_i(j; \mu)$$

with $\mu = (\lambda_1, ..., \lambda_k)$.

If we consider for the joint distribution of $(Z_1, Z_2)$ the normal distribution:

$$\mathsf{P}(Z_1 \le t, Z_2 = c_i) = \pi_i \frac{1}{(2\pi)^{m_1/2} |\Sigma|^{1/2}} \int_{-\infty}^{t} \exp(-\frac{1}{2}(\xi - \vartheta_i)' \Sigma^{-1} (\xi - \vartheta_i) \,\mathrm{d}\xi,$$

then we have $\mu = (\vartheta_1, ..., \vartheta_k, \Sigma)$. As a result, we obtain:

$$\mathsf{P}(Z_2 = c_i \mid Z_1 = t) = \frac{\pi_i \exp(-\frac{1}{2}(t - \vartheta_i)' \Sigma^{-1} (t - \vartheta_i)}{\sum_{j=1}^{k} \pi_j \exp(-\frac{1}{2}(t - \vartheta_j)' \Sigma^{-1} (t - \vartheta_j)} = p_i(t; \mu).$$

### 6.3.2. *Estimation of parameters*

In the model, described by $W = (W_1, ..., W_L)$ with [6.16], the unknown parameter:

$$(\mu', t_1', ..., t_L') = (\vartheta_1', ..., \vartheta_k', t_1', ..., t_L')$$

is to be estimated. The likelihood function corresponding to the model and the observation $w = (h_{11}, ..., h_{k1}, h_{12}, ..., h_{kL})$ is given by:

$$\mathcal{L}(\mu, t_1, ..., t_L \mid w) = \prod_{l=1}^{L} \frac{h_{+l}!}{h_{1l}! \cdot \cdots \cdot h_{kl}!} p_1(t_l; \mu)^{h_{1l}} \cdot \cdots \cdot p_k(t_l; \mu)^{h_{kl}},$$

the log-likelihood is determined up to a constant by:

$$\tilde{l}(\mu, t_1, ..., t_L \mid w) = \sum_{l=1}^{L} \sum_{i=1}^{k} h_{il} \log p_i(t_l; \mu). \qquad [6.17]$$

DEFINITION 6.1.– Any parameter $(\hat{\mu}^*, \hat{t}_1^*, ...., \hat{t}_L^*)$ maximizing $\mathcal{L}$ is called (multinomial) maximum likelihood estimate ($m.l.e.$) for $(\mu, t_1, ..., t_L)$. If some of the parameters are known or determined by relations, then $m.l.e.$ is used for the solution of optimization with respect to the unknown parameters.

## 6.4. Goodness of fit

In discrete distributional families as a measure for the goodness of fit, we could choose a statistic that we use for testing in contingency tables. The relative frequencies for the cells were denoted by $f_{il}$ and we use:

$$\hat{X}^2 = \sum_{l=1}^{L} h_{+l} \sum_{i=1}^{k} \frac{\left(f_{il} - p_i(\hat{t}_l; \hat{\mu})\right)^2}{p_i(\hat{t}_l; \hat{\mu})} \qquad [6.18]$$

as a measure for the distance between observed frequencies and estimated frequencies under the model. Here, $(\hat{\mu}, \hat{t}_1, ..., \hat{t}_L)$ is an estimation for the unknown parameters. Usually, the minimum-$\chi^2$-estimates or the maximum likelihood estimates are to be chosen.

DEFINITION 6.2.– The minimum-$\chi^2$-estimate (M$\chi^2$) minimizes the functional:

$$X^2(\mu, t_1, ...t_L) := \sum_{l=1}^{L} h_{+l} \sum_{i=1}^{k} \frac{\left(f_{il} - p_i(t_l; \mu)\right)^2}{p_i(t_l; \mu)}. \qquad [6.19]$$

If some of the parameters are known or determined by relations, then $M\chi^2$ is used for the solution of optimization with respect to the unknown parameters.

Here, we again distinguish two cases. If the parameters $t_1, ..., t_L$ are any points in a given set $\mathcal{M}$, then we maximize the likelihood over this set. If we have linear conditions as in section 6.2.1.1, e.g. in [6.7], then we apply the same technique used there. In general, we cannot get an explicit representation for the maximum likelihood estimation, we have to use some numerical procedure to determinate the m.l.e.

### 6.4.1. *Distribution of* $\hat{X}^2$

The distribution of $\hat{X}^2$ depends on the choice of the estimates. As is known from the properties of the $\chi^2$-statistic for some estimates, the asymptotic distribution of $\hat{X}^2$ is distribution free, but it is not distribution free for some other estimates. With the theorem of Cramér, it follows that the asymptotic distribution of $\hat{X}^2$ together with the minimum-$\chi^2$-statistic or with the (multinomial) maximum likelihood estimate is distribution free, the limit distribution is a $\chi^2$-distribution.

ASSUMPTION 6.1.– Let $a$ be a function, $a : \Gamma \to \mathbb{R}^k, \quad \Gamma \subset \mathbb{R}^{\tilde{k}}$.

1) $\nabla a$ exists and is continuous on $\Gamma$.

2) The matrix $\nabla a$ has the rank $k$.

ASSUMPTION 6.2.– Let us be given a $q \times q$ matrix $F$ and $m_1 \times m_1$ matrices $F_1, ..., F_L$. The parameter $(\mu, t_1, ..., t_L)$ is identifiable if there is a unique solution of:

$$F\mu = \nu; \quad F_l t_l = \nu_l, l = 1, ..., L$$

for any given $\nu \in \mathbb{R}^q, \quad \nu_l \in \mathbb{R}^{m_1}, l = 1, ..., L$.

ASSUMPTION 6.3.– Put $\pi(\mu, t) = \Big(p_1(t; \mu), ..., p_k(t; \mu)\Big)'$,

$$\tilde{\pi}(\mu, t_1, ..., t_L) = \begin{pmatrix} \pi(\mu, t_1) \\ \vdots \\ \pi(\mu, t_L) \end{pmatrix}, \quad \mu \in \mathbb{R}^q, t_l \in \mathbb{R}^{m_1}.$$

Under assumption 6.2, a function $b : \Gamma \to \mathbb{R}^{kL}$ exists that for any $\mu, t_1, ..., t_L$ a $\gamma \in \Gamma$ with.

$$\tilde{\pi}(F\mu, F_1 t_1, ..., F_L t_L) = b(\gamma), \quad \gamma \in \Gamma \subset \mathbb{R}^h$$

and $b$ satisfies assumption 6.1 for $b, h$ instead of $a, \tilde{k}$, respectively.

THEOREM 6.3.– Let $W = (W_1, ..., W_L)$ with independent variables:

$$W_l \sim \text{Mult}\Big(h_{+l}, p_1(t_l;\mu), ..., p_k(t_l;\mu)\Big).$$

We assume assumptions 6.2 and 6.3. If $M\chi^2$ denotes the minimum-$\chi^2$-estimate then.

$$\mathfrak{L}(X^2(M\chi^2)) \xrightarrow{\text{d}} \chi^2_{L(k-1)-h}.$$

PROOF.– The proof is based on the results given in [WIT 95] and also in [NIK 73] and [DZA 74].

THEOREM 6.4.– Let $W = (W_1, ..., W_L)$ with independent variables:

$$W_l \sim \text{Mult}\Big(h_{+l}, p_1^{(l)}, ..., p_k^{(l)}\Big).$$

Assuming that there does not exist a $(\mu, t_1, ..., t_L)$ in such a way that:

$$(p_1^{(l)}, ..., p_k^{(l)}) = \Big(p_1(t_l;\mu), ..., p_k(t_l;\mu)\Big), \quad l = 1, ..., L.$$

then

$$X^2(M\chi^2) \to \infty \quad a.s.$$

PROOF.– The proof is based on the results given in [WIT 95].

## 6.5. Modeling of categorical data

Some technical or medical products and some behaviors or properties are described by categorical data. The perception of the objects or categories should be represented in a metric space in such a way that distances are expressive. A geometric nearness should reflect a proximity or similarity of objects.

In a statistical background, the data are realizations of random variables. For the categories or conditions, probabilities exist and data are observed in each category. Conditions should be interpreted to be near if observed probabilities under these conditions are near. We will find a family of distributions over a space $\mathcal{Y}$ that the observed probabilities can be explained by these distributions.

DEFINITION 6.3.– Given a paired random variable $Z = (Z_1, Z_2)$. Let $\mathcal{P}$ be a statistical model for $Z$. If the observed data in each level can be considered as realizations of conditional variables $Z_2|Z_1 = t$ for appropriate $t$, then the family $\mathcal{P}$ is called a statistical model for the observed data.

Modeling depends strongly on the special data sets. We will concentrate on models of analysis of variance or contingency tables. Here, some levels of factors are observed and these levels are categorized. The response variables can be metric or categorical. The levels are mapped into a metric space and will be estimated there. Here, we will use methods that were discussed in preceding sections.

### 6.5.1. *Contingency tables*

Contingency tables are to be used for finding a relation or connection between variables where at least one of these is a categorical variable. In general, the data describe a frequency distribution, multivariate or univariate. Observations are frequencies of categories $c_1, ..., c_k$ observed at several levels $b_1, ..., b_L$.

#### 6.5.1.1. *General tables*

The observations are presented in a contingency table, Table 6.2.

| | $b_1$ | $b_2$ | $b_3$ | $\cdots$ | $b_L$ |
|---|---|---|---|---|---|
| $c_1$ | $h_{11}$ | $h_{12}$ | $h_{13}$ | $\cdots$ | $h_{1L}$ |
| | | | | $\ddots$ | |
| $c_k$ | $h_{k1}$ | $h_{k2}$ | $h_{k3}$ | $\cdots$ | $h_{kL}$ |

**Table 6.2.** *Contingency table*

Here, $h_{jl}$ is the frequency of the category $c_j$ under the level $b_l$. The responses are the frequencies for the categories, depending on the levels $t_1, ..., t_L$ that are related to the exploratory variables. Under the level $b_l$, we have $h_{+l}$ observations. Hence, the data at any level give observations for the given categories, the relative frequency of $c_i$ is $f_{il} = h_{il}/h_{+l}$. This $f_{il}$ is an estimation for the probability to observe the $c_i$ at level $t_l$.

We assume that the levels can be identified or described with $m_1$ parameters. Hence, we identify the level $b_l$ by a vector $t_l = (x_{l1}, ..., x_{lm_1})$. Furthermore, we assume that $k$ distributions with the densities $\varphi_1, ..., \varphi_k$ over $\mathbb{R}^{m_1}$ are given and at any point $t \in \mathbb{R}^{m_1}$ the conditional probability for observing $c_j$ is defined by [6.15]. In section 6.3, it was discussed that $h_{1l}, ..., h_{kl}$ are realizations of a multinomially distributed variable $W_l$ with:

$$W_l \sim \mathsf{Mult}\Big(h_{+l}, p_1(t_l; \mu), ..., p_k(t_l; \mu)\Big) \qquad [6.20]$$

from [6.16]. Then, we use the notion of a statistical model in the following sense.

DEFINITION 6.4.– For the data in the contingency table shown in Table 6.2, the multivariate multinomial distribution of $W = (W_1, ..., W_L)$ with [6.20] for independent $W_1, ..., W_L$ is called the statistical model. The unknown parameters are the distributional parameter $\mu = (\vartheta_1, ..., \vartheta_k)$ and the parameters $t_1, ..., t_L$.

From the data, the parameters are to be estimated. The estimates $\hat{\vartheta}_1, ..., \hat{\vartheta}_k$ can be used for estimating the densities for the classes $G_1, ..., G_k$. The estimates $\hat{t}_1, ..., \hat{t}_L$ are the estimated levels. These values give hints about the ordering of the levels or about the local relations between the levels.

The relative frequencies $f_{il}$ are estimations for $p_i(t_l; \mu)$. With these frequencies, the parameter $(\mu, t_1, ..., t_L)$ will be estimated. It turns out that, in general, $(\mu, t_1, ..., t_L)$ is not estimable, we need restrictions for $\mu$ and the levels. Some of these reparametrizations are given for the levels, e.g. if we have cross-classified designs. We use restrictions described by [6.7]. In section 6.3.2, the maximum likelihood estimation $(\hat{\mu}, \hat{t}_1, ..., \hat{t}_L)$ for $(\mu, t_1, ..., t_L)$ was introduced.

Modeling includes the choice of the probabilities $p_i(t; \mu)$. Mostly, we use the multivariate multinomial $W = (W_1, ..., W_L)$ with [6.20] for independent $W_1, ..., W_L$ and:

$$p_i(t; \mu) := \frac{\varphi_i(t; \vartheta_i)}{\sum_j \varphi_j(t; \vartheta_j)}$$

as the model. Here, $\varphi_i$ are densities over $\mathbb{R}^{m_1}$ for the distribution in the category $c_i$.

EXAMPLE 6.1.– We consider a one-way classification design. An example is given in Table 6.3 concerning the parliamentary elections of the Bundestag in Germany for the year 2009.

We have the factor $A$ (state) with $L = 16$ levels. For any state, we observe the percentages for the parties CDU, SPD, FDP, Linke and Grüne. For the level $l$ with $l = 1, ..., 16$, the percentage for the party $i$ is denoted by $q_{i,l}$. The number of voices for the party $i$ is:

$$h_{il} = v_l q_{il}/100, \quad l = 1, ..., L$$

if $v_l$ is the number of valid voices. Hence, we have data as in Table 6.2. There is $k = 5$ and $c_1, ..., c_5$ are the parties.

For modeling, the idea is the following. There is no natural space where the levels are mapped in. We can imagine that the states are characterized by $m_1$ variables or values and instead of the names of the states we use these $m_1$-dimensional observations. We assume that in $\mathbb{R}^{m_1}$ $k$ densities $\varphi_1, ..., \varphi_k$ are given. Here, $\varphi_i$

describes the distribution of values $t \in \mathbb{R}^{m_1}$ for taking the result $c_i$. In section 6.3, $Z_1$ takes values in $\mathbb{R}^{m_1}$ and $Z_2$ has a connection to $Z_1$ and takes the different values $c_1, .., c_k$. $Z_1$ takes values $t$ and these values are identified with the states. With [6.14] the distribution of $Z_2|Z_1 = t$ is described if prior probabilities $\pi_1, ..., \pi_k$ for the results $c_1, ..., c_k$ are given.

| | State | Percentage of voices | Valid voices |
|---|---|---|---|
| $l$ | | CDU \| SPD \| FDP \| Linke \| Grüne | $v_l$ |
| 1 | Baden-Württemberg | 42.5 \| 23.7 \| 11.9 \| 6.4 \| 12.8 | 5424349 |
| 2 | Bayern | 48.2 \| 20.1 \| 10.6 \| 5.8 \| 10.2 | 6625797 |
| 3 | Berlin | 26.3 \| 24.3 \| 7.2 \| 20.7 \| 17.4 | 1718759 |
| 4 | Brandenburg | 24.8 \| 28.7 \| 7.2 \| 29.7 \| 5.5 | 1381965 |
| 5 | Bremen | 27.4 \| 36.0 \| 7.2 \| 12.7 \| 14.4 | 337779 |
| 6 | Hamburg | 32.6 \| 33.6 \| 7.9 \| 9.7 \| 14.5 | 883658 |
| 7 | Hessen | 39.4 \| 32.3 \| 9.6 \| 7.1 \| 9.0 | 3172039 |
| 8 | Mecklenburg-Vorpommern | 34.9 \| 18.6 \| 7.1 \| 29.5 \| 5.4 | 865709 |
| 9 | Niedersachsen | 38.5 \| 35.8 \| 7.9 \| 7.5 \| 8.4 | 4422903 |
| 10 | Nordrhein-Westfalen | 39.6 \| 35.1 \| 8.9 \| 7.1 \| 7.8 | 9366765 |
| 11 | Rheinland-Pfalz | 41.4 \| 29.4 \| 10.6 \| 8.1 \| 8.0 | 2180560 |
| 12 | Saarland | 35.7 \| 31.3 \| 7.2 \| 18.4 \| 5.3 | 582528 |
| 13 | Sachsen | 39.5 \| 15.1 \| 10.2 \| 24.2 \| 6.0 | 2249678 |
| 14 | Sachsen-Anhalt | 32.3 \| 19.8 \| 7.8 \| 32.0 \| 4.6 | 1200777 |
| 15 | Schleswig-Holstein | 38.4 \| 32.4 \| 10.6 \| 7.1 \| 10.1 | 1604890 |
| 16 | Thüringen | 33.1 \| 20.7 \| 7.7 \| 29.1 \| 5.1 | 1229844 |

**Table 6.3.** *Percentage of voices for the parties and valid voices in the states*

We use the conditional probabilities $p_i(t; \mu)$ as in [6.15] with $\pi_i = 1/k$. Here, $\mu$ includes the unknown distributional parameters. In this context, we remember that $p_i(t; \mu)$ gives the probability for $Z_2 = c_i$ given $Z_1 = t$.

Modeling in such one-way tables takes the following idea. We map the levels of the factor $A$ into $\mathbb{R}^{m_1}$ and we will find such densities $\varphi_1, ..., \varphi_k$ and $t_1, ..., t_L$ that the conditional probabilities $p_i(t_l; \mu)$ are near to the relative frequencies $f_{il}$. Then, we can plot the points $t_1, ..., t_L$ and can interpret the relations between these points as relations between the states. We put $m_1 = 1$. $h_{il}$ is the number of voices for the party $i$ in the state $l$. According to [6.16], we consider $(h_{1l}, ..., h_{kl})$ as realization of $W_l$ with:

$$W_l \sim \mathsf{Mult}\Big( h_{+l}, p_1(t_l; \mu), ..., p_k(t_l; \mu)\Big).$$

A maximum likelihood estimation is calculated by maximizing the log-likelihood in [6.17], i.e. with the observation $w = (h_{11}, ..., h_{k1}, h_{12}, ..., h_{kL})$ we maximize:

$$\tilde{l}(\mu, t_1, ..., t_L \mid w) = \sum_{l=1}^{L} v_l \sum_{i=1}^{k} q_{il} \log p_i(t_l; \mu).$$

If we use for $\varphi_i$ the density of $\mathsf{N}_1(\vartheta_i, 1)$, then the maximum likelihood estimation for:

$$\vartheta = (\vartheta_1, ..., \vartheta_k)'$$

is given by:

$$\hat{\vartheta} = (\hat{\vartheta}_1, ..., \hat{\vartheta}_k)' = (0, 0.06608, 0.2803, 0.2869, 0.288)'.$$

For the estimated levels, we obtain:

$$\hat{t} = (\hat{t}_1, ..., \hat{t}_L)' = (-4.606, -5.54, -1.82, -2.174, -3.476, -4.026, -5.314, ..$$

$$.., -2.626, -5.603, -5.651, -5.246, -4.345, -3.03, -2.243, -4.951, -2.571)'.$$

In the plot it is visible which states have a similar result in elections and which others are very different.

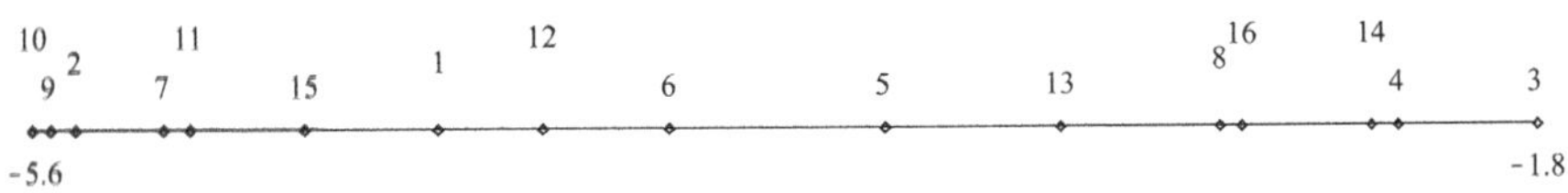


**Figure 6.1.** *Estimated levels for the states*

In Figure 6.1, we have the locations of the states and we can see that, e.g. the Eastern states have similarities, their distance to the Western states is relatively large.

For checking the model, the statistic $\hat{X}^2$ from [6.18] with the calculated estimations is used. With theorem 6.3, the limit distribution of $\hat{X}^2$ is a $\chi^2$-distribution with 44 degrees of freedom under the hypothesis that the model is adequate. The significance point for $\alpha = 0.05$ is $c = 60.481$ and the observed value for $\hat{X}^2$ is $\hat{x}^2 = 28.778$. Hence, the hypothesis is not to be rejected.

#### 6.5.1.2. *Further examples*

The preceding example can be extended for two-way (or $k-$way) contingency tables in a direct procedure. We consider the same idea for modeling as in the previous one-way tables. We map the levels of factor $A$ into an $m_{11}$-dimensional space and those of factor $B$ into an $m_{12}$-dimensional space. We assume, e.g., that there is a connection between states and years, on the one hand, and the votes for the different parties, on the other hand, in our representation a connection between $Z_2$ and $Z_1$. $Z_1$ takes values in $\mathbb{R}^{m_1}$ for $m_1 = m_{11} + m_{12}$ and $Z_2$ takes the values for $c_1, ..., c_k$. We assume that we are given densities $\varphi_1, ..., \varphi_k$ in $\mathbb{R}^{m_1}$ and again $\varphi_i$ describes the distribution of values $t \in \mathbb{R}^{m_1}$ for taking the result $c_i$. With [6.14], the distribution of $Z_2|Z_1 = t$ is described if prior probabilities $\pi_1, ..., \pi_k$ for the results $c_1, ..., c_k$ are given. We use the conditional probabilities as in [6.15]. Here, $\mu$ includes the unknown distributional parameters. In this context, we remember that $p_i(t; \mu)$ gives the probability for $Z_2 = c_i$ given $Z_1 = t$.

For univariate or multivariate analysis of variance tables, we observe data under some levels $b_1, ..., b_L$. Mostly, the levels are categorized. Then, for modeling, we use the conditional distributions discussed in section 6.2. In this way, we get geometric relations between the levels and there is a good possibility for interpreting the whole given data set.

## 6.6. Bibliography


[AGR 02] AGRESTI A., *Categorical Data Analysis*, Wiley, New York, 2002.

[AHR 81] AHRENS H., LÄUTER J., *Mehrdimensionale Varianzanalyse*, Akademie-Verlag, Berlin, 1981.

[DZA 74] DZAPARIDSE K.O., NIKULIN M.S., "On a modification of the standard statistics of Pearson", *Theory of Probability and Its Applications*, vol. 19, pp. 851–852, 1974.

[EVE 01] EVERITT B.S., DUNN G., *Applied Multivariate Data Analysis*, Hodder Education, London, 2001.

[LÄU 13] LÄUTER H., RAMADAN A.M., "New approach for multidimensional scaling with categorical data", *J. Applied Functional Anal.*, vol. 8, pp. 246–252, 2013.

[NIK 73] NIKULIN M.S., "Chi-square test for continuous distributions", *Theory of Probability and Its Applications*, vol. 18, pp. 559–568, 1973.

[RAM 10] RAMADAN A.M., Statistical model for categorical data, PhD Thesis, University of Potsdam, 2010.

[WIT 95] WITTING H., MÜLLER-FUNK U., *Mathematische Statistik II*, Teubner, Stuttgart, 1995.

Chapter 7

# Within the Sample Comparison of Prediction Performance of Models and Submodels: Application to Alzheimer's Disease

Our objective is to compare the predictive ability of several nested models. It derives from the following problem in epidemiology: the occurrence of a certain disease is to be predicted to happen within a fixed period of time based on the values of a number of items measured on the observed patients. It may happen that one or several items, proved to be relevant for the best fitting model, have a non-significant contribution to the prediction of who is at risk of developing the disease. The indices we use to compare the respective predictive ability of two models are the integrated discrimination improvement (IDI) and the Brier's score improvement (BRI). Estimation of the models and their relative IDI and BRI are conducted on the same sample, and their respective asymptotic properties are proved. We apply the results to Alzheimer's disease.

## 7.1. Introduction

When the objective of modeling a data set is explanatory, it is most appropriate to choose the best fitting model using the usual model selection procedures. But if the objective is to predict and not to explain the facts, and some of the factors selected by the "best fitting model" are not available for all subjects, as is sometimes the case for

---

Chapter written by Catherine HUBER-CAROL, Shulamith T. GROSS and Annick ALPÉROVITCH.

certain genetic markers, we can compare models by their prediction qualities rather than by their goodness of fit to the data. In this chapter, we consider the case of two competing, nested, probability predicting models, similar to Whittemore's [WHI 10] example, and consistent with the Pencina *et al.* setting [PEN 08]. The nested model contains traditional factors, and the larger model contains some expensive, or generally hard to obtain, often genetic, relevant markers.

## 7.2. Framework

### 7.2.1. *General description of the data set and the models to be compared*

Let $X = (Y, \mathbf{Z})$ be a random variable such that the response variable $Y$ is binary with values in $\{0, 1\}$, and $\mathbf{Z}$ is a k-dimensional real variable. The data consists of $\mathbf{X} = (X_1, \cdots, X_n)$, $n$ independent and identically distributed (i.i.d.) observations of $X$, and two models for predicting $Y$ on the basis of $\mathbf{Z}$ are to be compared:

$$\begin{array}{lrcl} \text{Model 1} & P(Y = 1|\mathbf{Z} = \mathbf{z}) & = & p_1(\mathbf{z}) \\ \text{Model 2} & P(Y = 1|\mathbf{Z} = \mathbf{z}) & = & p_2(\mathbf{z}) \end{array}$$

while the true distribution of $Y$ given $Z = z$, which will remain unknown all along, is given by:

$$P(Y = 1|\mathbf{Z} = \mathbf{z}) \quad = \quad p(\mathbf{z})$$

This setting originates from a problem in epidemiology: $Y_i$ is the indicator of the occurrence of a specfic disease for subject $i$ within a given period of time. The prediction of occurrence of this event is based on the value $\mathbf{z_i}$ of $\mathbf{Z}$ observed on subject $i$. In the special case of linear logistic models, $p_1$ and $p_2$ are denoted by $g_1$ and $g_2$ in the following. While $g_1$ includes all $k$ components of $\mathbf{Z}$, $g_2$ is obtained by dropping $k"$ components of $\mathbf{Z}$, thus keeping only $k' = k - k" < k$ of them. Without loss of generality, we treat in detail the case when one drops only one factor. The theoretical aim of this chapter is to derive the asymptotic properties of the estimators of the IDI and the BRI in order to obtain confidence intervals for them within the sample used to estimate the two models.

### 7.2.2. *Definition of the performance prediction criteria: IDI and BRI*

From Pencina *et al.* [PEN 08], the IDI of model 2 with respect to model 1 is:

$$IDI_{2/1} = E[p_2(\mathbf{Z}) - p_1(\mathbf{Z})|Y = 1] - E[p_2(\mathbf{Z}) - p_1(\mathbf{Z})|\mathbf{Y} = \mathbf{0}]$$

where $E$ denotes the expectation with respect to the distribution of $X$. We denote:

$$\pi \quad = \quad P(Y=1) \quad := \quad E[p(\mathbf{Z})]$$

the population prevalence of the event under study. We derive a simpler expression for $IDI_{2/1}$ to be used later:

$$IDI_{2/1} = E\left[(p_2(\mathbf{Z}) - p_1(\mathbf{Z}))\left(\frac{Y-\pi}{\pi\,(1-\pi)}\right)\right] \qquad [7.1]$$

Gu and Pepe [GU 09] define the proportion of explained variance (PEV) of a model:

$$PEV = \frac{var(P(Y=1|Z))}{\pi\,(1-\pi)} \qquad [7.2]$$

from which it is clear that:

$$IDI_{2/1} = PEV_2 - PEV_1 \qquad [7.3]$$

For a single model $p$, the Brier score is defined as:

$$BR(p) = E[(Y - p(\mathbf{Z}))^2] \qquad [7.4]$$

As the bigger the Brier's score, the worst the model; we define the Brier's score improvement for model 2 with respect to model 1 as:

$$\begin{aligned} BRI_{2/1} &= BR(p_1) - BR(p_2) \\ &= E[(p_1(\mathbf{Z}) - p_2(\mathbf{Z})) \times \\ &\quad (p_1(\mathbf{Z}) + p_2(\mathbf{Z}) \; - \; 2Y)] \end{aligned} \qquad [7.5]$$

A negative $IDI_{2/1}$ as well as a negative $BRI_{2/1}$ means that the predictive properties of model 2 are not as good as those of model 1. The ranges of IDI and BRI are, respectively, $-2\,,+2$ and $-1\,,+1$.

### 7.3. Estimation of IDI and BRI

Now we assume that we have a sample of size $n$ of $X$, and the two prediction models $p_j(\theta_j, \mathbf{z})$, defined for $j = 1, 2$ through the respective parameters $\theta_1, \theta_2$, as:

$$\begin{aligned} P(Y=1|\text{ model 1}) &= p_1(\theta_1, \mathbf{z}) \\ P(Y=1|\text{ model 2}) &= p_2(\theta_2, \mathbf{z}) \end{aligned}$$

### 7.3.1. *General estimating equations for IDI and BRI*

Using [7.1] and [7.5], and denoting for simplicity

$$\widehat{p_{ji}} := p_j(\widehat{\theta}_j, \mathbf{z_i}) \quad \text{for } \mathbf{j} = \mathbf{1}, \mathbf{2}, \mathbf{i} = \mathbf{1}, \cdots, \mathbf{n} \qquad [7.6]$$

where $\widehat{\theta}_j, j = 1, 2$ are maximum likelihood estimates for the parameters of models 1 and 2, natural estimates of $IDI_{2/1}$ and $BRI_{2/1}$ are, respectively,

$$\widehat{IDI_{2/1}} = \frac{1}{n}\sum_{i=1}^{n}(\widehat{p_{2i}} - \widehat{p_{1i}})\frac{y_i - \overline{y}}{\overline{y}(1-\overline{y})} \qquad [7.7]$$

$$\widehat{BRI_{2/1}} = \frac{2}{n}\sum_{i=1}^{n}\left[(\widehat{p_{1i}} - \widehat{p_{2i}})\left(\frac{(\widehat{p_{1i}} + \widehat{p_{2i}})}{2} - y_i\right)\right] \qquad [7.8]$$

Under usual regularity conditions on models 1 and 2, their parameter estimates are asymptotically consistent and normal.

### 7.3.2. *Estimation of IDI and BRI in the logistic case*

Let $u =< \theta, \mathbf{z} >$ be the scalar product of two $(k+1)$-dimensional real vectors $\theta = (\theta_0, \cdots, \theta_k)$ , $\mathbf{z} = (1, z_1, \cdots, z_k))$ and $g$ the function:

$$g(u) = \frac{e^u}{1 + e^u} \qquad [7.9]$$

Models 1 and 2 are logistic: $p_1 \equiv g_1$ and $p_2 \equiv g_2$:

$$\begin{aligned} g_1(\mathbf{z}) &= g(< \boldsymbol{\theta_1}, \mathbf{z} >) \\ g_2(\mathbf{z}) &= g(< \boldsymbol{\theta_2}, \mathbf{z} >) \end{aligned}$$

where some components of $\theta_1$ and some (possibly different) components of $\theta_2$ are predefined. If we refer to the motivating example of the introduction:

$$\begin{aligned} \boldsymbol{\theta_1} &= (\theta_0, \theta_1, \cdots, \theta_{k-1}, \theta_k) \\ \boldsymbol{\theta_2} &= (\theta'_0, \theta'_1, \cdots, \theta'_{k-1}, 0) \end{aligned}$$

Dropping index $j$ for simplicity of notation, we get the log-likelihood $L_n$:

$$\begin{aligned} L_n(\theta) &= \frac{1}{n}\sum_{i=1}^{n} y_i \log(g(<\theta, \mathbf{z_i}>)) \\ &\quad +(1-y_i)\log(1-g(<\theta,\mathbf{z_i}>)) \\ &= \frac{1}{n}\sum_{i=1}^{n} [y_i < \theta, \mathbf{z_i} > \; - \; \log(1+e^{<\theta,\mathbf{z_i}>})] \end{aligned} \qquad [7.10]$$

Then we estimate parameters $\theta_1$ and $\theta_2$ of the two logistic models. Those two estimators, obtained through classical maximum likelihood equations, are proved to be consistent and asymptotically normal.

7.3.2.1. *Asymptotics of $\widehat{IDI}_{2/1}$ for logistic predictors*

For simplicity, and since there is no ambiguity as we always consider how model 2 behaves with respect to model 1, we now drop the index $2/1$.

Define $\widetilde{IDI}$:

$$\widetilde{IDI} = \frac{1}{n}\sum_{i=1}^{n}(g_{2i}-g_{1i})\frac{y_i-\overline{y}}{\overline{y}(1-\overline{y})} \qquad [7.11]$$

which is the estimate of IDI when the two models are perfectly known, or more realistically, estimated "out of the sample", and $\widehat{g_{ji}}$:

$$\widehat{g_{ji}} = g(<\widehat{\theta}_j, z_i>) \ , \ j=1,2 \qquad [7.12]$$

the estimates of the two models. From equation [7.7], by considering $(\widehat{IDI}-\widetilde{IDI})+\widetilde{IDI}$, we get:

$$\begin{aligned} \widehat{IDI} &= \frac{1}{n}\sum_{i=1}^{n}[(\hat{g}_{2i}-\hat{g}_{1i})-(g_{2i}-g_{1i})]\frac{y_i-\overline{y}}{\overline{y}(1-\overline{y})} \\ &+ \frac{1}{n}\sum_{i=1}^{n}[g_{2i}-g_{1i}]\frac{y_i-\overline{y}}{\overline{y}(1-\overline{y})} \\ &:= T_{1n}+\widetilde{IDI} \end{aligned}$$

THEOREM 7.1.– [Consistency of $\widehat{IDI}$] $\widehat{IDI} \xrightarrow[n\to\infty]{a.s.} IDI$.

Theorem 7.1 results from the almost sure convergence of $T_{1n}$ to 0, due to the CLT for the model estimators $\widehat{g_{ji}}$ and the boundedness of the derivative of $g$, $(g(u)(1 - g(u))u')$, and the fact that

$$\widehat{IDI} \xrightarrow[n\to\infty]{a.s.} IDI$$

THEOREM 7.2.– [CLT of $\widehat{IDI}$]

$$\sqrt{n}(\widehat{IDI} - IDI) \xrightarrow[n\to\infty]{\mathscr{L}} \mathscr{N}(0, \sigma^2) \qquad [7.13]$$

where $\sigma^2 = (\frac{1}{(1-\pi)\,\pi})^2\; var(V)$ with $V$ defined as

$$\begin{aligned} V &= (g(\theta_2, \mathbf{Z_2}) - g(\theta_1, \mathbf{Z_1}) - E_\Delta)(Y - \pi) \\ &+ (Y - g(\theta_2, \mathbf{Z_2}))\; {}^t(\mathbf{Z_2})(I^{-1}(\theta_2))E_2 \\ &- (Y - g(\theta_1, \mathbf{Z_1}))^t(\mathbf{Z_1})(I^{-1}(\theta_1))E_1 \\ &+ IDI\,(2\pi - 1)\,(Y - \pi), \end{aligned}$$

where

$$\begin{aligned} E_\Delta &= E(g(\theta_2, \mathbf{Z_2}) - g(\theta_1, \mathbf{Z_1})) \\ E_j &= E[g_j(1 - g_j)(Y - \pi)\mathbf{Z_j}] \;\;,\;\; j = 1, 2 \\ \widehat{\sigma^2} &= \left(\frac{1}{\overline{Y}(1 - \overline{Y})}\right)^2 \widehat{var(V)} \end{aligned}$$

where $\widehat{\sigma^2}$ is a consistent estimator of $\sigma^2$.

#### 7.3.2.2. *Asymptotics of $\widehat{BRI_{2/1}}$ for logistic predictors*

Again, we drop the index 2/1, and consider the estimated BRI [7.8].

$$\begin{aligned} \widehat{BRI} &= \frac{2}{n}\sum_{i=1}^{n}\Bigg[(g(<\hat{\theta}_1, Z_i>) - g(<\hat{\theta}_2, Z_i>))\times \\ &\times \left(\frac{(g(<\hat{\theta}_1, Z_i>) + g(<\hat{\theta}_2, Z_i>))}{2} - Y_i\right)\Bigg] \end{aligned}$$

Using the same method as for IDI, we get

$$\sqrt{n}(\widehat{BRI} - BRI) = \frac{1}{\sqrt{n}}\sum_{i=1}^{n}(W_i - BRI) + o_P(1)$$

where the random variables $W_i$ are defined as

$$\begin{aligned} W_i &= 2\Big[(g(<\widehat{\theta_1}, \mathbf{Z}_i>) - g(<\widehat{\theta_2}, \mathbf{Z}_i>))\times \\ &\times \left(\frac{(g(<\widehat{\theta_2}, \mathbf{Z}_i>) + g(<\widehat{\theta_1}, \mathbf{Z}_i>))}{2} - Y_i\right)\Big] \\ &- (Y_i - g(<\widehat{\theta_2}, \mathbf{Z}_{2i}>))^t Z_{2i} I^{-1}(\widehat{\theta}_2)(2E_2 - E_4) \\ &+ (Y_i - g(<\widehat{\theta_1}, \mathbf{Z}_{1i}>))^t Z_{1i} I^{-1}(\widehat{\theta}_1)(2E_1 - E_3) \end{aligned}$$

where the expectations $E_j, j = 1, \cdots, 4$ are defned as

$$\begin{aligned} E_1 &= E\Big[g(<\theta_1, \mathbf{Z}_1>)(1 - g(<\theta_1, \mathbf{Z}_1>))\mathbf{Z}_1\times \\ &\times \left(\frac{g(<\theta_2, \mathbf{Z}_2>) + g(<\theta_1, \mathbf{Z}_1>)}{2} - Y\right)\Big] \\ E_2 &= E\Big[g(<\theta_2, \mathbf{Z}_2>)(1 - g(<\theta_2, \mathbf{Z}_2>))\mathbf{Z}_2\times \\ &\times \left(\frac{(g(<\theta_1, \mathbf{Z}_1>) + g(<\theta_2, \mathbf{Z}_2>))}{2} - Y)\right)\Big] \\ E_3 &= E[g(<\theta_1, \mathbf{Z}_1>)(1 - g(<\theta_1, \mathbf{Z}>))\mathbf{Z}_1(g(<\theta_2, \mathbf{Z}_1>) \\ &-g(<\theta_1, \mathbf{Z}_1>)) \\ E_4 &= E[g(<\theta_2, \mathbf{Z}>)(1 - g(<\theta_2, \mathbf{Z}_2>))\mathbf{Z}_2(g(<\theta_2, \mathbf{Z}_2>) \\ &-g(<\theta_1, \mathbf{Z}_2>)) \end{aligned}$$

Actually, we have as follows.

THEOREM 7.3.– [Consistency of the estimated $BRI$]

$$\widehat{BRI} \xrightarrow[n\to\infty]{a.s.} BRI \qquad [7.14]$$

THEOREM 7.4.– [CLT for $\widehat{BRI}$]

$$\sqrt{n}(\widehat{BRI} - BRI) \xrightarrow[n\to\infty]{\mathscr{L}} \mathscr{N}(0, \sigma_B^2). \qquad [7.15]$$

A consistent estimate of $\sigma_B^2 = var(W)$ is given by:

$$\widehat{\sigma_B^2} = \frac{1}{n-1}\sum_{i=1}^{n}(\widehat{W_i} - \hat{\overline{W}})^2$$

The estimated Ws are obtained by replacing all parameters by their maximum likelihood estimates and expectations by sample averages.

## 7.4. Simulation studies

We do two different simulations: one to check the behavior of our estimates of IDI and BRI, and the other parallel to Gu and Pepe's examples.

### 7.4.1. *First simulation*

Our basic model to simulate the data is a logistic one with two independent covariates: $Z_1 \in \{-1, 0, 1\}$ with respective probabilities $(.2, .4, .4)$, and $Z_2 \in \{0, 1\}$ with respective probabilities $(.2, .8)$, with $\theta = (0, 2, 1)$ so that

$$\mathscr{L}(Y_i|Z_i) = \text{Bernoulli}\left(\frac{\exp(2Z_{1i} + Z_{2i})}{1 + \exp(2Z_{1i} + Z_{2i})}\right)$$

This is the true model used to simulate the data.

We generate 1,000 samples of 1,000 triplets $(Y_i, Z_{1i}, Z_{2i})$. On each sample obtained, three logistic models are considered: model 1 including both covariates, $\theta_1 = (\theta_{10}, \theta_{11}, \theta_{12})$, model 2 including only $Z_1$, $\theta_2 = (\theta_{20}, \theta_{21}, 0)$ and model 2' including only $Z_2$, $\theta_{2'} = (\theta_{2'0}, 0, \theta_{2'2})$. We know the true values of $\theta_1$, $\theta_2$ and $\theta_2'$: the first one is equal to $\theta$ of the true model:

$$\theta_1 = (0, 2, 1)$$

and the two others are obtained by minimizing the Kullback distance between the true law and the logistic based on covariate $Z_1$ alone and $Z_2$ alone, respectively:

$$\begin{aligned} \theta_2 &= (0.096, 1.969, 0) \\ \theta_{2'} &= (0.307, 0, 0.674) \end{aligned}$$

The true values of $IDI_{2/1}$ and $IDI_{2'/1}$ can thus be computed:

$$\begin{aligned} IDI_{2/1} &= -0.01037. \\ IDI_{2'/1} &= -0.3282. \end{aligned}$$

For each sample, we estimated the three models and computed $\widehat{IDI_{2/1}}$ and $\widehat{IDI_{2'/1}}$, as well as our estimates of their standard errors. In Table 7.1, we compare the mean estimated asymptotic standard errors of the IDIs, and the empirical standard errors of the estimated IDIs. The true values of the Brier's score of the three models lead to the true values of the BRI of models 2 and 2' with respect to model 1, which can be seen in Table 7.3.

| | Asymptotic | Empirical |
|---|---|---|
| standard error of $IDI_{2/1}$ | 0.0057029 | 0.0052650 |
| standard error of $IDI_{2'/1}$ | 0.025556 | 0.026306 |

**Table 7.1.** *Comparison of the estimated and simulated standard errors of $\widehat{IDI_{2/1}}$ and $\widehat{IDI_{2'/1}}$*

| | 95% confidence | interval | True value |
|---|---|---|---|
| $IDI_{2/1}$ | −0.02140714 | −0.00076843 | −0.01037 |
| $IDI_{2'/1}$ | −0.38141 | −0.27829 | −0.3282 |

**Table 7.2.** *95% confidence intervals for $IDI_{2/1}$ and $IDI_{2'/1}$*

| | True Brier's score | True BRI with respect to model 1 |
|---|---|---|
| model 1 | 0.1603 | 0 |
| model 2 | 0.16281 | −0.002511 |
| model 2' | 0.23962 | −0.079316 |

**Table 7.3.** *True Brier's scores and true BRI for the three models*

It can be seen from the Q-Q plots below that the estimated IDI for model 2' with respect to model 1 is close to the normal while the estimated IDI for model 2 with respect to model 1, which is very small, is more erratic. The same occurs for the Q-Q plots of the BRIs.

| | 95% confidence | interval | True value |
|---|---|---|---|
| $BRI_{2/1}$ | −0.00535781 | −0.00067996 | −0.002511 |
| $BRI_{2'/1}$ | −0.090107 | −0.069466 | −0.079316 |

**Table 7.4.** *95% confidence intervals for $BRI_{2/1}$ and $BRI_{2'/1}$*

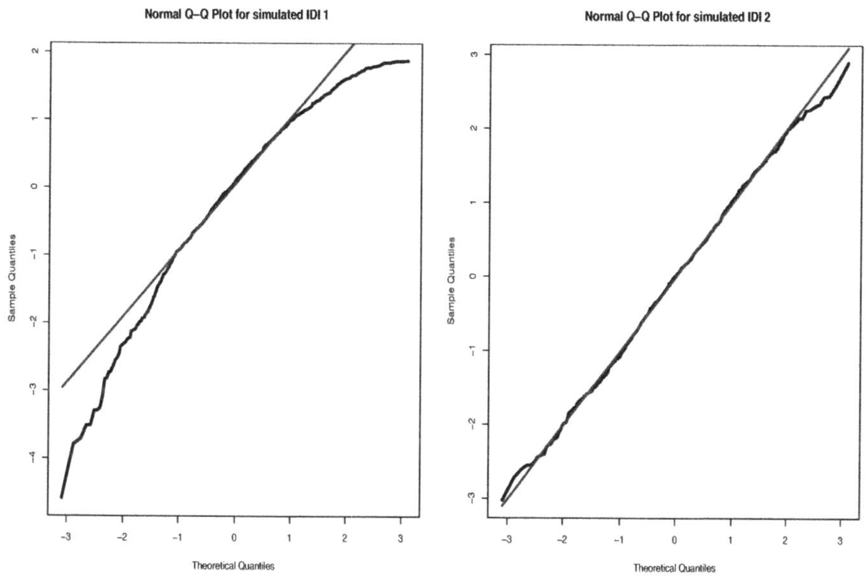


**Figure 7.1.** *Q-Q plot for $IDI_{2/1}$ and $IDI_{2'/1}$*

### 7.4.2. *Second simulation: Gu and Pepe's example*

Gu and Pepe's example is as follows: they assume that there is only one covariate so that $Z = (Z_0, Z_1)$, $Z_0$ being fixed and equal to 1, that the laws of $Z_1$, conditionally on $Y = 0$ and $Y = 1$, are normal and that $P(Y = 1) = 0.2$:

$$\begin{array}{rcl} P(Y=1) & = & 0.2 \\ \mathscr{L}(Z_1|Y=1) & = & \mathscr{N}(1,1) \\ \mathscr{L}(Z_1|Y=0) & = & \mathscr{N}(0,1) \end{array}$$

Computing the law of $Y$ conditional on $Z_1$, we get the following logistic model:

$$\begin{array}{rcccl} P(Y=1) & = & p_{i1} & = & 0.2 \\ \theta & = & (\theta_0, \theta_1) & = & (log(p_{i1}/(1-p_{i1})) - .5, 1) \end{array}$$

$$
\begin{aligned}
P & \quad (Y = 1|Z_1 = z_1) \\
& = \frac{\exp(z_1 + (log(p_{i1}/(1 - p_{i1})) - .5))}{1 + \exp(z_1 + (log(p_{i1}/(1 - p_{i1})) - .5)}
\end{aligned}
$$

$$
\begin{aligned}
P & \quad (Y = 0|Z_1 = z_1) \\
& = \frac{1}{1 + \exp(z_1 + (log(p_{i1}/(1 - p_{i1})) - .5)}
\end{aligned}
$$

Three different expressions are given by Gu and Pepe for the PEV:

$$
\begin{aligned}
PEV & = \frac{var(P(Y = 1|Z_1))}{p_{i1}(1 - p_{i1})} \\
& = \frac{var(Y) - E(var(Y|Z_1))}{var(Y)} \\
& = corr(Y, P(Y = 1|Z_1))
\end{aligned}
$$

It seems that the third definition is not correct as the result for the three definitions given by 1,000 simulations of a sample of size 1,000 is, respectively, 0.15433, 0.15503 and 0.39384.

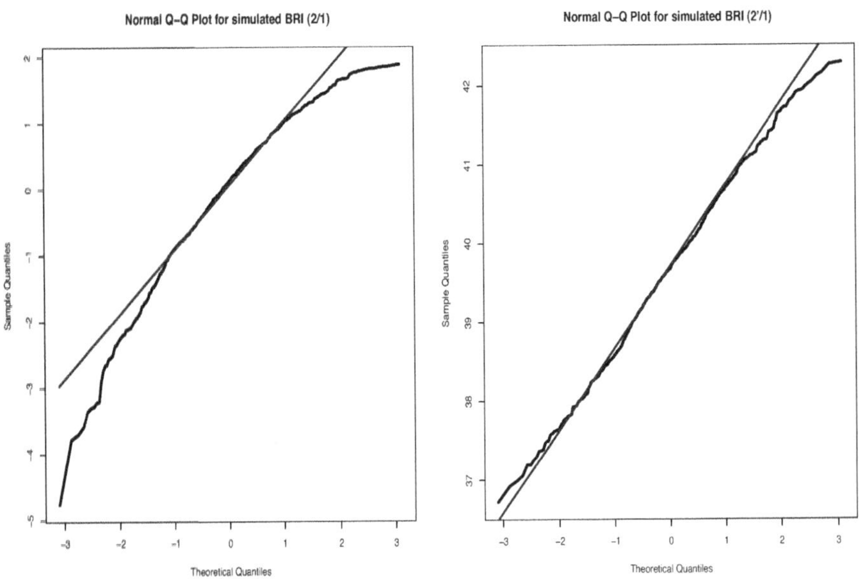


**Figure 7.2.** *Q-Q plot for* $BRI_{2/1}$ *and* $BRI_{2'/1}$

## 7.5. The three city study of Alzheimer's disease

The three city (3C) study is a cohort study conducted in three cities in France (Bordeaux, Dijon and Montpellier), aiming to estimate the risk of dementia and cognitive impairment attributable to vascular factors. A sample of non-institutionalized subjects aged 65 years and older was selected randomly from the electoral rolls of each city [3C 03]). Each follow-up examination included cognitive testing and dementia diagnosis.

We have obtained official permission to use the data for the purpose of studying the contribution of the genetic marker APOE4 for the prediction of developing dementia in the next four years beyond the traditional variables: age at recruitment into the study, gender, education, indicator for depression, indicator of previous cardiologic complications, indicator of consumption of psychotropic medication and indicator of incapacitation of any sort. We were given a subsample of n = 4,486 men and women among which 162 developed Alzheimer's within four years.

After removing records with missing data we ended up with n = 4,214 records. Based on these data, we fit the best model. It included the genetic marker APOE4:

1) age at recruitment in 3 classes: $65 \leq Age < 71, 71 \leq Age < 78, \geq 78$;

2) education level;

3) past history of vascular disease;

4) use of psychotropic drugs;

5) incapacities in activities of daily living;

6) the genetic marker APOE4.

| | Estimate | Standard error | $Pr(>\|z\|)$ |
|---|---|---|---|
| (Intercept) | −2.944 | 0.176 | <2e−16 |
| age.fac.31 | −2.089 | 0.330 | 2.3e−10 |
| age.fac.32 | −0.984 | 0.191 | 2.5e−07 |
| educ.level | −0.430 | 0.180 | 0.0167 |
| card | 0.616 | 0.233 | 0.0081 |
| depress | 0.786 | 0.201 | 9.5e−05 |
| incap | 1.180 | 0.206 | 1.1e−08 |
| APOE4 | 0.634 | 0.195 | 0.0012 |
| AIC | = | 1,094 | |

**Table 7.5.** *Logistic model 1 including the genetic marker APOE4*

All the covariates are highly significant.

For the model without the genetic marker:

| | Estimate | Standard error | $p$-value |
|---|---|---|---|
| (Intercept) | −2.797 | 0.168 | <2e−16 |
| age.fac.31 | −2.060 | 0.330 | 4.0e−10 |
| age.fac.32 | −0.963 | 0.190 | 4.2e−07 |
| educ.level | −0.434 | 0.179 | 0.0155 |
| card | 0.677 | 0.231 | 0.0034 |
| depress | 0.805 | 0.201 | 6.2e−05 |
| incap | 1.124 | 0.206 | 4.6e−08 |
| AIC | = | 1,102 | |

**Table 7.6.** *Model 2: logistic model without the genetic marker APOE4*

Clearly, model 1 that includes APOE4 is significantly a better fit to the data than model 2: it has a smaller AIC and the coefficient of APOE4 is highly significant. We turn now to check whether the inclusion of APOE4 improves significantly the prediction ability of model 2, by estimating the indices $IDI_{2/1}$ and $BRI_{2/1}$. We estimate the IDI and the BRI of model 2 with respect to model 1 directly from the data, and then run a bootstrap with 1,000 repetitions to obtain a competing estimate of the standard errors of the indices, as well as 95% confidence intervals for the true IDI and BRI for the dementia data.

Here AIC refers to the Akaike Information Criterion [AKA 73], a likelihood-based index widely used to evaluate the fit of a statistical model that penalizes the model for using a large number of parameters.

The area under the ROC curve has been widely used in biomedical and imaging research to help assess the discrimination ability of a binary predictor of disease or a tumor. Two patient groups whose disease status is known, 1 for diseased, 0 for disease-free, are evaluated on an ordinal or quantitative measure, then the ROC curve, which is numerically equivalent to the (one sample) Mann Whitney rank statistics [MAN45].

We placed the sample estimates and the bootstrap estimates in the same display.

| Sample Mean | Bootstrap Mean | Sample Standard error | Bootstrap Standard error |
|---|---|---|---|
| −.00298 | −.00374 | 0.00303 | 0.00328 |

**Table 7.7.** *Sample and bootstrap estimates for IDI*

| Sample Mean | Bootstrap Mean | Sample Standard error | Bootstrap Std error |
|---|---|---|---|
| −6.09e−05 | −9.02e−05 | 0.000115 | 0.000129 |

**Table 7.8.** *Sample and bootstrap estimates for BRI*

| | 95% Confidence interval for IDI | |
|---|---|---|
| Asymptotic | −0.008911 | +0.002958 |
| Bootstrap | −0.012012 | +0.000389 |

**Table 7.9.** *Asymptotic and percentile bootstrap* 95% *confidence intervals (CI) of IDI for the three cities study*

| | 95% Confidence interval for BRI | |
|---|---|---|
| Asymptotic | −2.87e−04 | 1.65e−04 |
| Bootstrap | −4.08e−04 | 9.03e−05 |

**Table 7.10.** *Asymptotic and percentile bootstrap* 95% *confidence intervals (CI)* of BRI for the three cities study*

These results certainly suggest that the dementia model that includes the genetic marker APOE4 is not significantly greater than the model without APOE4 in predicting the development of dementia. This result is obtained with regard to both the model discrimination as measured by the IDI and model calibration as measured by Brier's score difference for the two models.

The bootstrap distributions for both the IDI and the Brier's score difference remain somewhat of a mystery. Note that in both cases the bootstrap standard error is smaller than the asymptotic standard error as computed from the sample. More significantly, the Q-Q normal plots below show that in both cases the bootstrap distribution, although based on 1,000 repetitions, is markedly non-normal. This explains the fact that the bootstrap confidence intervals in both cases are wider than their corresponding asymptotic confidence intervals based on the indices and their standard errors as estimated from the sample.

## 7.6. Conclusion

We present results that enable researchers to make an inference on two important indices for measuring the relative effectiveness of two models in predicting the probabilities of future events. Most importantly, we allowed for model estimation prior to index computation on the same data by providing new standard errors for both the IDI and the BRI when the indices are computed on the same data that provided model parameter estimates. There are additional indices for comparison of models' predictiveness accuracy and discrimination. We mention, in particular, the difference in the area under the ROC curve of two models and Pencina *et al.*'s [PEN 08] net reclassification improvement (NRI). We are currently working on the asymptotic theory for these indices when parameters are estimated from the same data as the indices.

We first tested our asymptotic results and standard error formulas in simulation studies. We then applied them to Alzheimer's data from the French Three Cities study and did not find any evidence to support the effectiveness of the genetic marker APOE4 in predicting the occurrence of Alzheimer's beyond that achieved by standard non-genetic predictor variables such as age, education and additional health variables.

One unusual finding is the markedly non-normal bootstrap distribution for these data, based on 1,000 bootstrap samples. Although in simulations we found no unusual behavior of the bootstrap distribution, for the dementia data, the bootstrap distribution for the IDI, with in-sample estimation of the models, is visibly right-skewed, whereas the bootstrap distribution for the Brier difference, again with in-sample model estimation, is visibly left-skewed. We offer no explanation for these phenomena at this time.

## 7.7. Bibliography

[AKA 73] AKAIKE, H., "Information theory and an extension of the maximum likelihood principle", *Proc. 2nd International Symposium on Information Theory*, (eds). B.N. PETROV and F. CSAKI, Akademiai Kiado, Budapest, pp. 267–281, 1973. Reproduced in: *Breakthroughs in Statistics, I, Foundations and Basic Theory*, (eds) S. KOTZ and N.L. JOHNSON, Springer-Verlag, New York, pp. 610–624, 1992.

[GU 09] GU W., PEPE M., "Measures to summarize and compare the predictive capacity of markers", *The International Journal of Biostatistics*, vol. 5, no. 1, 2009.

[MAN 45] WILCOXON F., "Individual comparisons by ranking methods", *Biometrics Bulletin*, vol. 1, no. 6, pp. 80–83, December 1945.

[PEN 08] PENCINA M.J., D'AGUSTINO SR. R.B., D'AGUSTINO JR. R.B., VASRAN R.S., "Evaluating the added predictive ability of a new marker. From area under the ROC curve to reclassification and beyond", *Statistics in Medicine*, vol. 27, pp. 157–172, 2008.

[3C 03] THE 3C STUDY GROUP, "Vascular factors and risk of dementia. Design of the Three-City study and baseline characteristics of the study population", *Neuroepidemiology*, vol. 22, pp. 316–325, 2003.

[WHI 12] WHITTEMORE, A.S., "Evaluating health risk models", *Statistics in Medicine*, vol. 29, pp. 2438–2452, 2010.

# Chapter 8

# Durbin–Knott Components and Transformations of the Cramér-von Mises Test

## 8.1. Introduction

Durbin *et al.* ([DUR 72] and [DUR 75]) have defined the independent components of the Cramér-von Mises statistic by using the Karhunen–Loève expansion of the empirical process. Studying the properties of the component was continued by Schoenefeld ([SCH 77] and [SCH 80]). Here we consider the transformation of the weighted statistics using its components. These statistics allow us to change the test sensitivity of amplitudes and frequencies of deviations from the hypothetical distribution. The collection of Karhunen–Loève expansions is discussed in sections 8.2–8.4. In section 8.5, the statistics based on the components is discussed. Two new forms of the statistics are considered.

## 8.2. Weighted Cramér-von Mises statistic

We will study the components of the Cramér-von Mises statistic

$$\omega_n^2 = n \int_0^1 \psi^2(t)(F_n(t) - t)^2 dt, \qquad [8.1]$$

---

Chapter written by Gennady MARTYNOV.

where $F_n(t)$ is the empirical distribution function based on the sample $X_1, X_2, ..., X_n$ from the uniform distribution on $[0,1]$, and $\psi(t)$ is a weight function. If the condition

$$\int_0^1 \psi^2(t)\, t\,(1-t)dt < \infty$$

is fulfilled then the statistic $\omega_n^2$ converges in distribution to

$$\omega^2 = \int_0^1 \xi^2(t)dt, \quad [8.2]$$

where $\xi(t),\ t \in [0,1]$, is the Gaussian process with zero mean and the covariance function

$$K_\psi(t,\tau) = \psi(t)\psi(\tau)(\min(t,\tau) - t\tau)$$

(see, for example [VAN 96, p. 50]). The statistic [8.1] designed to test the hypothesis $H_0:\ \ F(x) = t$ against the alternative $H_1:\ \ F(x) \neq t$. The Gaussian process $\xi(t)$ can be developed in the Karhunen-Loève series

$$\xi(t) = \sum_{i=1}^{\infty} \frac{x_k \varphi_k(t)}{\sqrt{\lambda_k}},$$

where $x_k,\ k = 1, 2, ...$, are the independent standard normal random variables, and $\lambda_k$ and $\varphi_k(t)$, $i = 1, 2, ...$, are the eigenvalues and eigenfunctions of the linear operator with the kernel $K(t,\tau)$, i.e. solutions of the Fredholm integral equation:

$$\varphi(t) = \lambda \int_0^1 \psi(t)\psi(\tau)(\min(t,\tau) - t\tau)\varphi(\tau)d\tau. \quad [8.3]$$

By twice differentiation [8.3] with respect to $t$, we obtain the differential equation:

$$h''(t) + \lambda\psi^2(t)h(t) = 0$$

with the conditions $h(0) = h(1) = 0$. Here $h(t) = \varphi(t)/\psi(t)$.

Some partial solutions of equation [8.3] are presented in the following sections.

## 8.3. Examples of the Cramér-von Mises statistics

### 8.3.1. *Classical Cramér-von Mises statistic*

The classical Cramér-von Mises statistics is

$$\omega_n^2 = n \int_0^1 (F_n(t) - t)^2 dt$$

with $\psi(t) = 1$. The eigenvalues and eigenfunctions of the covariance function $K(t, \tau)$ are $\lambda_i = (\pi i)^2$ and $\varphi_i(t) = \sqrt{2}\sin(\pi i t)$, $i = 1, 2, \ldots$ . The limit in distribution $\omega^2$ of $\omega_n^2$ can be written as

$$\omega^2 = \int_0^1 B^2(t)dt = \sum_{k=1}^{\infty} \frac{x_k^2}{(\pi k)^2},$$

where $B(t)$ is the Brown bridge process, i.e. Gaussian process with zero mean and covariance function $K(t, \tau) = \min(t, \tau) - t\tau$.

### 8.3.2. *Anderson–Darling statistic*

The Anderson–Darling statistic is well known and widely used in practice. In [AND 52], Anderson and Darling proposed the variant of the Cramér-von Mises statistic with the weight function $\psi^2(t) = 1/t(1-t)$:

$$A_n^2 = n \int_0^1 \psi^2(t)(F_n(t) - t)^2 dt.$$

$A_n^2$ converges in distribution to

$$A^2 = \int_0^1 \xi^2(t)dt,$$

where $\xi(t)$ is the Gaussian process with the covariance function

$$K(t, \tau) = (\min(t, \tau) - t\tau)/\sqrt{t(1-t)\tau(1-\tau)}.$$

The variance of the process $\xi(t)$ does not depend on $t$ and equal to 1. The eigenvalues corresponding to $K(t, \tau)$ are $\lambda_k = k(k+1)$, $k = 1, 2, \ldots$, and the eigenfunctions are

$$\varphi_k(t) = t(1-t)P_k^{'}(2t-1), \quad k = 1, 2, \ldots ,$$

where $P_k(t)$ is the Legendre polynomials.

### 8.3.3. *Cramér-von Mises statistic with the power weight function*

Deheuvels and Martynov [DEH 03] described the following result. Let $\{B(t) : 0 \leq t \leq 1\}$ be the Brownian bridge. Then, for each $\beta = \frac{1}{2\nu} - 1 > -1$, the Karhunen-Loeve expansions of $\{t^\beta B(t) : 0 < t \leq 1\}$ are given by

$$t^\beta B(t) = \sum_{k=1}^{\infty} \sqrt{\lambda_{kB}}\, \omega_k e_{kB}(t).$$

Here, $\{\omega_k : k \geq 1\}$ are independent and identically distributed (i.i.d.) $N(0,1)$ random variables, and, for $k = 1, 2, \ldots$, the eigenvalues are $\lambda_k = (2\nu/z_{\nu,k})^2$, the corresponding eigenfunctions are

$$e_k(t) = \frac{t^{\frac{1}{2\nu}-\frac{1}{2}} J_\nu\left(z_{\nu,k}\, t^{\frac{1}{2\nu}}\right)}{\sqrt{\nu} J_{\nu-1}\left(z_{\nu,k}\right)}, \quad 0 < t \leq 1,$$

and $z_{\nu,k}$, $k = 1, 2, ...$, are zeros of the Bessel functions $J_\nu(z)$.

## 8.4. Weighted parametric Cramér-von Mises statistic

### 8.4.1. *Covariance functions of weighted parametric empirical process*

In [KAC 55], the following result is presented. We will test the complex hypothesis

$$H_0: \quad F(x) \in \mathcal{G} = \{G(x,\theta), \quad \theta = (\theta_1, \theta_2, ...\theta_k)^\top \in \Theta \subset R_k\},$$

where $\theta$ is a vector of unknown parameters. In this case, the Cramér-von Mises statistic takes the form

$$\omega_n^2 = n \int_{-\infty}^{\infty} \psi^2(G(x,\theta_n))(F_n(x) - G(x,\theta_n))^2\, dG(x,\theta_n),$$

where $\theta_n$ is the likelihood estimate of $\theta$ and $\psi(t)$ is a weight function. This statistic converges in distribution under $H_0$ and under some regularity conditions to the following integral (see Kac, Kiefer and Wolfowitz [KAC 55]):

$$\omega^2 = \int_0^1 \psi^2(t)\xi^2(t,\theta_0)dt.$$

Here $\psi(t)\xi(t,\theta_0)$ is the Gaussian process with the expectation $E\psi^2(t)\xi(t,\theta_0) = 0$ and with covariance function

$$K(t,\tau) = E(\psi(t)\xi(t,\theta_0)\psi(\tau)\xi(\tau,\theta_0))$$
$$= \psi(t)\psi(\tau)(\min(t,\tau) - t\tau - q^\top(t,\theta_0)I^{-1}(\theta_0)q(\tau,\theta_0)),\ t,\tau \in (0,\ 1),$$

$\theta_0$ is an unknown value of $\theta$,

$$q^\top(t,\theta) = (\partial G(x,\theta)/\partial\theta_1, ..., \partial G(x,\theta)/\partial\theta_k)|_{t=G(x,\theta)},$$

$I(\theta)$ is the Fisher information matrix,

$$I(\theta) = (E((\partial/\partial\theta_i)\log g(X,\theta)(\partial/\partial\theta_j)\log g(X,\theta)))_{1\le i,j\le k},$$

$$g(x,\theta) = (\partial G(x,\theta)/\partial x).$$

It must fulfill the condition

$$\int_0^1 \psi^2(t)K(t,t)dt < \infty.$$

The distribution of $\omega^2$ depends, in general, on unknown parameter $\theta_0$ and on the family $\mathcal{G}$.

It was stated in [KAC 55] that the limit distribution of the statistic $\omega_n^2$ is independent of the parameters for a family $\left\{G(x,\theta_1,\theta_2) = G\left(\frac{x-\theta_1}{\theta_2}\right)\right\}, \theta_2 > 0$. The same result was obtained in [MAR 10] for the related family $\left\{G(x,\theta_1,\theta_2) = G\left(\left(\frac{x}{\theta_2}\right)^{\theta_1}\right)\right\}, \theta_2 > 0$. Both classes of distribution family contain distribution functions most frequently used in practice. Further results on testing of the parametric hypothesis with the parameters of the location and the scale are discussed in [DZH 79]. It was considered in [KHM 79] a transformation of the Cramér-von Mises and Kolmogorov-Smirnov statistics to the form with the limit distributions independent of parameters.

### 8.4.2. *Eigenvalues and eigenfunctions for weighted parametric Cramér-von Mises statistic*

In [MAR 11], the following result was described. We have the following algorithm for searching the eigenvalues and eigenfunctions of equation

$$\phi(t) = \lambda \int_0^1 K(t,\tau)\phi(\tau)d\tau. \qquad [8.4]$$

(see Martynov [MAR 11]). ($K(t,\tau)$ has the form described in section 8.4.1. The dependence of $K(t,\tau)$ on the unknown parameter $\theta_0$ may occur here but is omitted for brevity. We denote $y(t) = \phi(t)/\psi(t)$. After double differentiation of equation [8.4], we get the equation

$$y''(t) + \lambda\psi^2(t)y(t) = -\lambda(q''(t))^\top C_0$$

with the conditions

$$y(0) = -\lambda q^\top(0)C_0 \text{ and } y(1) = -\lambda q^\top(1)C_0,$$

where

$$C_0 = (C_{01}, ..., C_{0k})^\top = I^{-1}\int_0^1 \psi^2(t)q(t)y(t)dt.$$

Here, we present the results only for the weight function $\psi(t) = t^\alpha$. Here, we used the results from section 8.3.3. The equation for eigenvalues is

$$\det\begin{pmatrix} z_0(\lambda) & z_1(\lambda) & Z(\lambda) \\ 0 & \frac{1}{\pi}\Gamma(\nu)(\nu\sqrt{\lambda})^{-\nu} & \lambda q^\top(0) \\ J_\nu\left(2\nu\sqrt{\lambda}\right) & Y_\nu\left(2\nu\sqrt{\lambda}\right) & \rho^\top(\lambda,1) + \lambda q^\top(1) \end{pmatrix} = 0, \qquad [8.5]$$

where $\nu = 2\beta + 2$, $J_\nu(x)$ and $Y_\nu(x)$ are the Bessel functions of the first and second order,

$$z_0(\lambda) = I^{-1}\int_0^1 t^{\alpha+1/2}q(t)J_\nu\left(2\nu\sqrt{\lambda}t^{1/2\nu}\right)dt,$$

$$z_1(\lambda) = I^{-1}\int_0^1 t^{\alpha+1/2}q(t)Y_\nu\left(2\nu\sqrt{\lambda}t^{1/2\nu}\right)dt,$$

$$Z(\lambda) = \int_0^1 [t^\alpha I^{-1} q(t) \rho^\top(\lambda, t) - E] dt,$$

$$\begin{aligned} r(\lambda,t) &= \pi\lambda\nu\left(\sqrt{t}J_\nu\left(2\nu\sqrt{\lambda}t^{1/2\nu}\right)\int_0^t \sqrt{\tau}q''(\tau)Y_\nu\left(2\nu\sqrt{\lambda}\tau^{1/2\nu}\right)d\tau\right. \\ &\quad \left. -\sqrt{t}Y_\nu\left(2\nu\sqrt{\lambda}t^{1/2\nu}\right)\int_0^t \sqrt{\tau}q''(\tau)J_\nu\left(2\nu\sqrt{\lambda}\tau^{1/2\nu}\right)d\tau\right). \end{aligned}$$

We finally obtain the solution in the form

$$y_i(t) = C_1^{(i)} y_1(t) + C_2^{(i)} y_2(t) + \lambda_i r^\top(\lambda_i, t) C_0^{(i)},$$

where

$$y_{1,\nu}(t) = \sqrt{t}J_\nu\left(2\nu\sqrt{\lambda}t^{1/2\nu}\right), \ t \in [0,1],$$

$$y_{2,\nu}(t) = \sqrt{t}Y_\nu\left(2\nu\sqrt{\lambda}t^{1/2\nu}\right), \ t \in [0,1].$$

The constants $C_1^i$, $C_2^i$ and $C_0^i$ are for each $\lambda_i$ the solutions of the homogeneous equation with the matrix used in the determinant [8.5].

## 8.5. Transformations of the Cramér-von Mises statistic

### 8.5.1. *Preliminary notes*

The Cramér-von Mises statistic

$$\omega_n^2 = n\int_0^1 \psi^2(t)(F_n(t) - t)^2 dt = \int_0^1 \xi_n^2(t) dt$$

was represented by Durbin *et al.* [DUR 72] (without the weight function) in the form

$$\sum_{i=1}^{\infty} V_{i,n}^2 = \sum_{i=1}^{\infty} \frac{D_{i,n}^2}{\lambda_i},$$

where

$$V_{i,n} = \int_0^1 \xi_n(t)\varphi_i(t) dt$$

are the Durbin–Knott components of the Cramér-von Mises statistic, and

$$D_{i,n} = \sqrt{\lambda_i} \int_0^1 \xi_n(t)\varphi_i(t)dt = \frac{1}{\sqrt{n}}\sqrt{\lambda_i}\sum_{j=1}^{n}\varphi_i(X_j)$$

are the normalized Durbin–Knott components, $\{X_1, X_2, ..., X_n\}$ is the sample. We have

$$EV_{i,n} = 0, \quad EV_{i,n}^2 = 1/\lambda_i, \quad i = 1, 2, ....$$

If $n \to \infty$, then the normalized components $D_i = \sqrt{\lambda_i}V_i, \quad i = 1, 2, ...,$ are independent, and they have the distribution $N(0,1)$. Here, the convergence of the components distributions to the limit distributions can be considered a similar manner as was done by Durbin *et al.* [DUR 72].

### 8.5.2. *Replacement of eigenvalues*

It was also considered a variant of the Cramér-von Mises test based on the Durbin–Knott components. We now consider the testing of standard hypothesis $H_0 : F(t) = t, \ t \in [0,1]$. Let the normalized components $D_{i,n}$ correspond to an arbitrary $\psi(t)$. Let $\mu_1, \mu_2, ...$ be the sequence of the positive numbers such that $1/\mu_1 + 1/\mu_2 + ... < \infty$. Then the statistic

$$\omega_n^{*2} = \sum_{i=1}^{\infty} \frac{D_{n,i}^2}{\mu_i}$$

is the double weighted statistic. One weighting realizes with the function $\psi(t)$ and the other weighting realizes with the sequence $\mu_1, \mu_2, ....$ Here, we have the "amplitude" and "frequency" weightings correspondingly.

EXAMPLE 8.1.– One form of the statistic $\omega_n^{*2}$ can be statistics, based on the finite number of the component: $\sum_{i=1}^{r} D_{n,i}^2$. This statistic has the $\chi^2$-distribution with the degrees of freedom number, equal to $k$.

EXAMPLE 8.2.– The classical Cramér-von Mises statistic

$$\omega_n^2 = n\int_0^1 (F_n(t) - t)^2 dt$$

has the following eigenvalues and eigenfunctions of the covariance function $K(t, \tau)$: $\lambda_i = (\pi i)^2$ and $\varphi_i(t) = \sqrt{2}\sin(\pi i t)$, $i = 1, 2, ...,$ correspondingly. Then

$$\omega_n^2 = \sum_{i=1}^{\infty} \frac{D_{n,i}^2}{(\pi i)^2}, \quad \text{where } D_{i,n} = \sqrt{2}\pi i \int_0^1 \xi_n(t)\sin(\pi i t)(t)dt.$$

If we change the denominators in the Durbin–Knott expansion of the classical Cramer-von Mises statistic on $k(k+1)$, than we have the new statistics

$$\omega_n^2 = \sum_{i=1}^{\infty} \frac{D_{n,i}^2}{k(k+1)}$$

with the limit distribution of the Anderson–Darling statistic. But this statistic does not coincide with the Adderson–Darling statistics.

### 8.5.3. *Transformed statistics*

It can be noted that the distribution function of the functional

$$\omega^2 = \int_0^1 w^2(t)dt$$

from a Gaussian process $w(t)$ depends only on eigenvalues, corresponding to its covariance function $K(t,\tau)$, but it does not depend on the eigenfunctions. Let the eigenvalues and eigenfunctions of an arbitrary covariance function $K(t,\tau)$ be $d_1,\ d_2, ...$ and $q_1(t),\ q_2(t), ...$, correspondingly. By Mercer's theorem, we deduce:

$$K(t,\tau) = \sum_{k=1}^{\infty} \frac{q_k(t)q_k(\tau)}{\lambda_i}.$$

First, we consider the series

$$\omega^{*2} = \sum_{i=1}^{\infty} \frac{D_i^2}{\mu_i}.$$

This series is the limit in distribution of the statistic

$$\omega_n^{*2} = \sum_{i=1}^{\infty} \frac{D_{n,i}^2}{\mu_i}.$$

This statistic is the functional of the empirical process $\xi_n(t)$.

We can transform $\omega^{*2}$ as follows:

$$\omega^{*2} = \sum_{i=1}^{\infty} \frac{\lambda_i}{d_i} \int_0^1 \xi(s)\varphi_i(s)ds \int_0^1 \xi(z)\varphi_i(z)dz = \int_0^1 \int_0^1 \xi(s)\widetilde{K}(s,z)\xi(z)dsdz,$$

where

$$\widetilde{K}(s,z) = \sum_{i=1}^{\infty} \frac{\lambda_i}{\mu_i} \varphi_i(s)\varphi_i(z).$$

It is clear that the integral and the sum can be changed in places if $\sum_{i=1}^{\infty} \mu_i/\lambda_i < \infty$. Corresponding statistics in general form can be written as

$$\Omega_n^2 = \int_0^1 \int_0^1 \xi_n(s)\Psi(s,z)\xi_n(z)dsdz,$$

where $\Psi(s,z)$ is a positively defined kernel with eigenvalues $\nu_1, \nu_2, ...$ and normalized eigenfunctions $\psi_1, \psi_2, ...$. In this section, we consider the processes $\xi(t)$ and $\xi_n(t)$ without the weight function because these can be included in weight kernels. We suppose that it can be found in the kernel $\Psi^{<1/2>}(s,z)$ so that

$$\Psi(s,z) = \int_0^1 \Psi^{<1/2>}(s,r)\Psi^{<1/2>}(r,z)dr.$$

We assume that the following inequalities are satisfied:

$$\sum_{i=1}^{\infty} \frac{1}{\nu_i} < \infty \text{ and } \sum_{i=1}^{\infty} \frac{1}{\sqrt{\nu_i}} < \infty$$

or, equivalently,

$$\int_0^1 \Psi(t,t)dt < \infty \text{ and } \int_0^1 \Psi^{<1/2>}(t,t)dt < \infty.$$

With simple transformations, it can be stated that

$$\Psi^{<1/2>}(s,z) = \left( \sum_{i=1}^{\infty} \frac{\psi_i(s)\psi_i(z)}{\nu_i} \right)^{<1/2>} = \sum_{i=1}^{\infty} \frac{\psi_i(s)\psi_i(z)}{\sqrt{\nu_i}}.$$

If $\Psi(s,z)$ exists, then $\Psi^{<1/2>}(s,z)$ may not exist. The mathematical expectation of

$$\Omega^2 = \int_0^1 \int_0^1 \xi(s)\Psi(s,z)\xi(z)dsdz$$

is

$$E\Omega^2 = \int_0^1 \int_0^1 \xi(s)\Psi(s,z)\xi(z)dsdz = \sum_{i=1}^{\infty}\sum_{j=1}^{\infty} \frac{1}{\lambda_i \mu_j} \left( \int_0^1 \varphi_i(t)\psi_j(t)dt \right)^2 .$$

Consider the following chain of transformations:

$$\Omega^2 = \int_0^1 \int_0^1 \xi(s)\Psi(s,z)\xi(z)dsdz$$

$$= \int_0^1 \int_0^1 \xi(s) \int_0^1 \Psi^{<1/2>}(s,r)\Psi^{<1/2>}(r,z)dr\xi(z)dsdz$$

$$= \int_0^1 \left[ \int_0^1 \xi(s)\Psi^{<1/2>}(s,r)ds \int_0^1 \Psi^{<1/2>}(r,z)\xi(z)dz \right] dr$$

$$= \int_0^1 \left[ \int_0^1 \xi(s)\Psi^{<1/2>}(s,t)ds \right]^2 dt.$$

Further, we denote

$$\Omega^{*2} = \int_0^1 \left[ \int_0^1 \xi(s)\Psi^{<1/2>}(s,t)ds \right]^2 dt = \int_0^1 \eta^2(t)dt,$$

where

$$\eta(t) = \int_0^1 \xi(s)\Psi^{<1/2>}(s,t)ds$$

is the Gaussian process with the covariance function

$$K_\eta(t,\tau) = \int_0^1 \int_0^1 \Psi^{<1/2>}(z,t)K(z,u)\Psi^{<1/2>}(u,\tau)dzdu.$$

Let $\Upsilon(s,t)$ be a positively defined kernel. In general, we can consider the corresponding statistic

$$\Omega_n^{**2} = \int_0^1 \left[ \int_0^1 \xi_n(s)\Upsilon(s,t)ds \right]^2 dt = \int_0^1 \eta_n^{*2}(t)dt,$$

converging in distribution to

$$\Omega^{**2} = \int_0^1 \eta^{*2}(t)dt.$$

This assertion follows from the formula

$$\int_0^1 \xi_n(s)\Upsilon(s,t)ds = \frac{1}{\sqrt{n}}\sum_{i=1}^{n}\int_0^1 \left(I_{X_{i\leq s}} - s\right)\Upsilon(s,t)ds.$$

Here, the sum on the right side is the sum of independent and identically distributed random processes with covariance function

$$K_{\eta^*}(t,\tau) = \int_0^1\int_0^1 \Upsilon(z,t)K(z,u)\Upsilon(u,\tau)dzdu.$$

The convergence condition is given by the inequality

$$\int_0^1 K_{\eta^*}(t,t)dt = \int_0^1\int_0^1\int_0^1 \Upsilon(z,t)K(z,u)\Upsilon(u,t)dzdudt < \infty.$$

EXAMPLE 8.3.– Let $\delta(x)$ be the generalized $\delta$-function with the unit mass concentrated in 0. If $\Psi(s,z) = \psi(s)\delta(s-z)\psi(z)$, then $\Omega^2$ is the classical weighted Cramér-von Mises statistic

$$\omega_n^2 = n\int_0^1 \psi^2(t)(F_n(t)-t)^2dt.$$

EXAMPLE 8.4.– If $\Upsilon = \sqrt{\psi(s)}\delta(s-z)\sqrt{\psi(z)}$, then $\Omega^{**2}$ is also the classical weighted Cramér-von Mises statistic.

## 8.6. Bibliography

[AND 52] ANDERSON T.W., DARLING D.A., "Asymptotic theory of certain "goodness of fit" criteria based on stochastic processes", *The Annals of Mathematical Statistics*, vol. 23, no. 2, pp. 193–212, 1952.

[DEH 03] DEHEUVELS P., MARTYNOV G., "Karhunen-Loève Expansions for weighted Wiener processes and Brownian bridges via Bessel functions", in HOFFMANN-JØRGENSEN J., WELLNER J.A., MARCUS M.B. (eds), *High Dimensional Probability III*, vol. 55, Progress in Probability, Birkhäuser, Basel, pp. 57–93, 2003.

[DUR 72] DURBIN J., KNOTT M., "Components of the Cramér-von-Mises statistics. I", *Journal of the Royal Statistical Society: Series B*, vol. 34, pp. 290–307, 1972.

[DUR 75] DURBIN J., KNOTT M., TAYLOR L., "Components of the Cramér-von-Mises statistics. II", *Journal of the Royal Statistical Society: Series B*, vol. 37, pp. 216–237, 1975.

[DZH 79] DZHAPARIDZE K.O., NIKULIN M.S., "The probability distributions of the Kolmogorov and omega square statistics for continuous distributions with shift and location parameters", *Investigations in the Theory of Probability Distributions. Part IV*, Zapiski Nauchnych Seminarov LOMI, Leningrad, vol. 85, pp. 46–74, 1979.

[KAC 55] KAC M., KIEFER J., WOLFOWITZ J., "On tests of normality and other tests of goodness-of-fit based on distance methods", *The Annals of Mathematical Statistics*, vol. 30, pp. 420–447, 1955.

[KHM 79] KHMALADZE E., "The use of omega square tests for testing parametric hypotheses", *Theory of Probability and Its Applications*, vol. 24, no. 2, pp. 283–301, 1979.

[MAR 10] MARTYNOV G., "Note on the Cramer-von Mises test with estimated parameters", *Publicationes Mathematicae Debrecen*, vol. 76, no. 3, pp. 341–346, 2010.

[MAR 11] MARTYNOV G., "Weighted Cramer-von Mises test with estimated parameters", *Communications in Statistics-Theory and Methods*, vol. 40, no. 19–20, pp. 3569–3586, 2011.

[SCH 77] SCHOENEFELD D.A., "Asymptotic properties of tests based on linear combination of the orthogonal components of the Cramér-von Mises statistic", *The Annals of Statistics*, vol. 5, no. 5, pp. 1017–1026, 1977.

[SCH 80] SCHOENEFELD D.A., "Tests based on linear combinations of the orthogonal components of the Cramér-von Mises statistic when parameters are estimated", *The Annals of Statistics*, vol. 8, no. 5, pp. 1017–1022, 1980.

[VAN 96] VAN DER VAART A.W., WELLNER J.A., *Weak Convergence and Empirical Processes*, Springer-Verlag, 1996.

# Chapter 9

# Conditional Inference in Parametric Models

## 9.1. Introduction and context

This chapter explores conditional inference in parametric models. A comprehensive overview on this area is the illuminating review paper by Reid [REI 95]. Our starting point is as follows: given a model $\mathcal{P}$ defined as a collection of continuous distributions $P_\theta$ on $\mathbb{R}^d$ with density $p_\theta$ where the parameter $\theta$ belongs to some subset $\Theta$ in $\mathbb{R}^s$ and given a sample of independent copies of a random variable with distribution $P_{\theta_T}$ for some unknown value $\theta_T$ of the parameter, we intend to provide some inference about $\theta_T$ conditioning on some observed statistics of the data. The situations that we have in mind are of two different kinds.

The first one is the Rao–Blackwellization of estimators, which amounts to reducing the variance of an unbiased estimator (for a large review on unbiased estimators, see [VOI 93] in the univariate case and [VOI 96] in the multivariate case) by conditioning on any statistics; when the conditioning statistics is complete and sufficient for the parameter, this procedure provides optimal reduction, as stated by Lehmann–Scheffé theorem. These facts yield the following questions:

1) Is it possible to provide good approximations for the density of a sample conditioned on a given statistic, and, when applied for a model where some sufficient statistic for the parameter is known, does sufficiency with respect to the parameter still holds for the approximating density?

---

Chapter written by Michel BRONIATOWSKI and Virgile CARON.

2) In the case when the first question has a positive answer, is it possible to simulate samples according to the approximating density, and to propose some Rao–Blackwellized version for a given preliminary estimator? Also, we would hope that the proposed method would be feasible, the programming burden would be light, the run time for this method be short and the involved techniques would keep in the range of globally known ones by the community of statisticians.

The second application of conditional inference pertains to the role of conditioning in models with nuisance parameters. There is a huge bibliography on this topic, some of which will be considered in detail in the following. The usual frame for this field of problems is the exponential families one, for reasons related both to the importance of these models in applications and the role of the concept of sufficiency when dealing with the notion of nuisance parameter. Conditioning on a sufficient statistic for the nuisance parameter produces a new exponential family, which gets free of this parameter, and allows for simple inference on the parameter of interest, at least in simple cases. This will also be discussed, since the reality, as known, is not that simple, and since so many complementary approaches have been developed over decades in this area. Using the approximation of the conditional density in this context and performing simulations yields Monte Carlo tests for the parameter of interest, free from the nuisance parameter; comparison with the parametric bootstrap will also be discussed. Also, conditional maximum likelihood estimators will be produced.

This chapter is organized as follows. Section 9.2 describes a general approximation scheme for the conditional density of long runs of subsamples conditioned on a statistics, with explicit formulas. The proof of the main result of this section is presented in [BRO 13a]. Discussion about implementation is provided. Section 9.3 presents two aspects of the approximating conditional scheme: first, we show an example that sufficiency is kept under the approximating scheme and second, this yields to an easy Rao–Blackwellization procedure. An illustration of the Lehmann–Scheffé theorem is presented. Section 9.4 deals with models with nuisance parameters in the context of exponential families. We have found it useful to spend a few paragraphs on bibliographical issues. We address Monte Carlo tests based on the simulation scheme; in simple cases, its performance is similar to that of parametric bootstrap; however, conditional simulation-based tests improve clearly over parametric bootstrap procedure when the test pertains to models for which the likelihood is multimodal with respect to the nuisance parameter; an example is provided. Finally, we consider conditioned maximum likelihood based on the approximation of the conditional density; in simple cases its performance is similar to that of estimators defined through global likelihood optimization; however, when the preliminary estimator of the nuisance is difficult to obtain, for example when it depends strongly on some initial point for a Newton–Raphson routine (this is indeed a very common situation), by the very nature of sufficiency, conditional inference based on the proxy of the conditional likelihood performs better; this is illustrated with examples.

## 9.2. The approximate conditional density of the sample

Most attempts that have been proposed for the approximation of conditional densities stem from arguments developed in [LEH 86] for inference on the parameter of interest in models with a nuisance parameter; however, the proposals in this direction hinge at the approximation of the distribution of sufficient statistics for the parameter of interest, given the observed value of the sufficient statistics of the nuisance parameter. We will present some of these proposals in the section devoted to exponential families. To our knowledge, no attempt has been made to approximate the conditional distribution of a sample (or a long subsample), given some observed statistics.

However, generating samples from the conditional distribution itself (such samples are often called co-sufficient samples, following [LOC 07]) has been considered by many authors; see, for example, [LIN 03], [LIN 05] and the references therein.

In [ENG 97], simulating exponential or normal samples under the given value of the empirical mean is proposed. For example, under the exponential distribution $Exp(\theta)$, the minimal sufficient statistics for $\theta$ is the sum of the observations, say $t_n$; a co-sufficient sample $x^*$ can be created by generating an $x^{'}$ sample from $Exp(1)$ and taking $x_i^* = x_i^{'} t_n / \overline{x'}$. However, this approach may be at odds in simple cases, as for the Gamma density in the nonexponential case.

Lockhart *et al.* [LOC 07] proposed a different framework based on the Gibbs sampler, simulating the conditioned sample one at a time through a sequential procedure. The example which is presented is for the Gamma distribution under the empirical mean; in these examples it seems to perform well for location parameter, when the true parameter is in some range, therefore not uniformly on the model. Their chapter contains a comparative study with the global maximum likelihood method. In a simple case, they argue favorably for both methods. We will turn back to global likelihood maximization in relation with conditional likelihood estimators, in the last section of this chapter.

### 9.2.1. *Approximation of conditional densities*

For sake of clarity we consider the case when the model $\mathcal{P}$ is a family of distributions on $\mathbb{R}$.

Denote $\mathbf{X}_1^n := (\mathbf{X}_1, .., \mathbf{X}_n)$ a set of $n$ independent copies of a real random variable $\mathbf{X}$ with density $p_{\mathbf{X},\theta_T}$ on $\mathbb{R}$. Let $\mathrm{x}_1^n := (\mathrm{x}_1, ..., \mathrm{x}_n)$ denote the observed values of the data, each $\mathbf{x}_i$ resulting from the sampling of $\mathbf{X}_i$. Define the r.v. $\mathbf{U} := u(\mathbf{X})$ and $\mathbf{U}_{1,n} := u(\mathbf{X}_1) + ... + u(\mathbf{X}_n)$ where $u$ is a real-valued measurable function on $\mathbb{R}$, and, accordingly, $u_{1,n} := u(\mathbf{x}_1) + ... + u(\mathbf{x}_n)$. Denote $p_{\mathbf{U},\theta_T}$ the density of the r.v. $\mathbf{U}$. We consider approximations of the density of the vector $\mathbf{X}_1^k = (\mathbf{X}_1, .., \mathbf{X}_k)$ on $\mathbb{R}^k$ when $\mathbf{U}_{1,n} = u_{1,n}$. It will be assumed that the observed value $u_{1,n}$ is "typical", in the sense that it satisfies the law of large numbers. Since the approximation scheme for the conditional density is validated through limit arguments, it will be assumed that the sequence $u_{1,n}$ satisfies

$$\lim_{n\to\infty} \frac{u_{1,n}}{n} = Eu(\mathbf{X}). \qquad [9.1]$$

We propose an approximation for

$$p_{u_{1,n},\theta_T}\left(x_1^k\right) := p_{\theta_T}\left(x_1^k | \mathbf{U}_{1,n} = u_{1,n}\right)$$

where $x_1^k := (x_1, .., x_k)$ and $k := k_n$ is an integer sequence such that

$$0 \leq \limsup_{n\to\infty} k/n \leq 1 \qquad [K1]$$

together with

$$\lim_{n\to\infty} n - k = \infty \qquad [K2]$$

in other words, we may approximate $p_{u_{1,n},\theta_T}\left(x_1^k\right)$ on long runs. The rule which defines the value of $k$ for a given accuracy of the approximation is stated in section 3.2 of [BRO 13a].

The hypotheses pertaining to the function $u$ and the r.v. $\mathbf{U} = u(\mathbf{X})$ are as follows:

1) $u$ is real valued and the characteristic function of the random variable $\mathbf{U}$ is assumed to belong to $L^r$ for some $r \geq 1$.

2) The r.v. $\mathbf{U}$ is supposed to fulfill the Cramer condition: its moment generating function satisfies

$$\phi_{\mathbf{U}}(t) := E \exp t\mathbf{U} < \infty$$

for $t$ in a non-void neighborhood of 0.

Define the functions $m(t)$, $s^2(t)$ and $\mu_3(t)$ as the first, second and third derivatives of $\log \phi_{\mathbf{U}}(t)$. Denote

$$\pi^{\alpha}_{u,\theta_T}(x) := \frac{\exp tu(x)}{\phi_{\mathbf{U}}(t)} p_{\mathbf{X},\theta_T}(x)$$

with $m(t) = \alpha$ and $\alpha$ belongs to the support of $P_{\mathbf{U},\theta_T}$, the distribution of $\mathbf{U}$. Also it is assumed that this latest definition of $t$ makes sense for all $\alpha$ in the support of $\mathbf{U}$. These kinds of conditions are called "steepness properties" and are exposed in [BAR 78], p153 and the following.

We introduce a positive sequence $\epsilon_n$ which satisfies

$$\lim_{n\to\infty} \epsilon_n \sqrt{n-k} = \infty \quad \text{[E1]}$$

$$\lim_{n\to\infty} \epsilon_n (\log n)^2 = 0. \quad \text{[E2]}$$

### 9.2.2. *The proxy of the conditional density of the sample*

The density $g_{u_{1,n},\theta_T}(x_1^k)$ on $\mathbb{R}^k$, which approximates $p_{u_{1,n},\theta_T}\left(x_1^k\right)$ sharply with a relative error smaller than $\epsilon_n (\log n)^2$, is defined recursively as follows.

Set

$$m_0 := u_{1,n}/n$$

and

$$g_0(x_1|\, x_0) := \pi_{u,\theta_T}^{m_0}(x_1)$$

with $x_0$ arbitrary, and for $1 \leq i \leq k-1$ define the density $g(x_{i+1}|\, x_1^i)$ recursively.

Set $t_i$ the unique solution of the equation

$$m_i := m(t_i) = \frac{u_{1,n} - u_{1,i}}{n-i} \quad [9.2]$$

where $u_{1,i} := u(x_1) + ... + u(x_i)$. The tilted adaptive family of densities $\pi_{u,\theta_T}^{m_i}$ is the basic ingredient of the derivation of approximating scheme. Let

$$s_i^2 := \frac{d^2}{dt^2}\left(\log E_{\pi_{u,\theta_T}^{m_i}} \exp tu\left(\mathbf{X}\right)\right)(0)$$

and

$$\mu_j^i := \frac{d^j}{dt^j}\left(\log E_{\pi_{u,\theta_T}^{m_i}} \exp tu\left(\mathbf{X}\right)\right)(0)\,,\ j = 3, 4$$

which are the second, third and fourth cumulants of $\pi^{m_i}_{u,\theta_T}$. Let

$$g(x_{i+1}|x_1^i) = C_i p_{\mathbf{X},\theta_T}(x_{i+1}) \mathfrak{n}\left(\alpha\beta + m_0, \beta, u\left(x_{i+1}\right)\right) \qquad [9.3]$$

where $\mathfrak{n}\left(\mu,\tau,x\right)$ is the normal density with mean $\mu$ and variance $\tau$ at $x$. Here

$$\beta = s_i^2\left(n-i-1\right) \qquad [9.4]$$

$$\alpha = t_i + \frac{\mu_3^i}{2s_i^4\left(n-i-1\right)} \qquad [9.5]$$

and $C_i$ is a normalizing constant.

Define

$$g_{u_{1,n},\theta_T}(x_1^k) := g_0(x_1|x_0)\prod_{i=1}^{k-1} g(x_{i+1}|x_1^i) \qquad [9.6]$$

The following theorem holds.

THEOREM 9.1.– Assume [K1,K2] together with [E1,E2]. Then

1)

$$p_{u_{1,n},\theta_T}(x_1^k) = g_{u_{1,n},\theta_T}(x_1^k)(1 + o_{P_{u_{1,n},\theta_T}}(\epsilon_n\left(\log n\right)^2))$$

2)

$$p_{u_{1,n},\theta_T}(x_1^k) = g_{u_{1,n},\theta_T}(x_1^k)(1 + o_{G_{u_{1,n},\theta_T}}(\epsilon_n\left(\log n\right)^2)).$$

3) The total variation distance between $P_{u_{1,n},\theta_T}$ and $G_{u_{1,n},\theta_T}$ goes to 0 as $n$ tends to infinity.

For the proof, see [BRO 13a].

Statement 1) means that the conditional likelihood of any long sample path $\mathbf{X}_1^k$ given $(\mathbf{U}_{1,n} = u_{1,n})$ can be approximated by $G_{u_{1,n},\theta_T}(\mathbf{X}_1^k)$ with a small relative error on typical realizations of $\mathbf{X}_1^n$.

The second statement states that simulating $\mathbf{X}_1^k$ under $g_{u_{1,n},\theta_T}$ produces runs which could have been sampled under the conditional density $p_{u_{1,n},\theta_T}$ since $g_{u_{1,n},\theta_T}$ and $p_{u_{1,n},\theta_T}$ coincide sharply on larger and larger subsets of $\mathbb{R}^k$ as $n$ increases.

COMMENT 9.1.– Theorem 9.1 states that the density $g_{u_{1,n},\theta_T}$ on $\mathbb{R}^k$ approximates $p_{u_{1,n},\theta_T}$ on the sample $\mathbf{X}_1^n$ generated under $\theta_T$. However, in some cases, the r.v.'s $\mathbf{X}_i$s in theorem 9.1 may at time be generated under some other parameters, say $\theta_0$. Indeed, for direct applications developed in this chapter, theorem 9.1 has to hold when the sample is generated under another sampling scheme. Theorem 11 in [BRO 13a] states that the approximation scheme holds true in this case. Indeed, let $\mathbf{Y}_1^n$ be independent and identically distributed (i.i.d.) copies with distribution $P_{\theta_0}$. Define

$$p_{u_{1,n},\theta_0}\left(y_1^k\right) := p_{\theta_0}\left(\mathbf{Y}_1^k = y_1^k \middle| \mathbf{U}_{1,n} = u_{1,n}\right)$$

with distribution $P_{u_{1,n},\theta_0}$. It then holds

THEOREM 9.2.– With the same hypotheses and notation as in theorem 9.1,

$$p_{\theta_T}\left(\mathbf{X}_1^k = Y_1^k \left| \mathbf{U}_{1,n} = u_{1,n}\right.\right) = g_{u_{1,n},\theta_T}(Y_1^k)(1 + o_{P_{u_{1,n},\theta_0}}(\epsilon_n (\log n)^2)).$$

### 9.2.3. *Comments on implementation*

The simulation of a sample $X_1^k$ with density $g_{u_{1,n},\theta_T}$ is fast and easy. Indeed, the r.v. $X_{i+1}$ with density $g\left(x_{i+1}|x_1^i\right)$ is obtained through a standard acceptance–rejection algorithm. When $\theta_T$ is unknown, a preliminary estimator may be used. When $\mathbf{U}_{1,n}$ is sufficient for $p_{u_{1,n},\theta}$, it is nearly sufficient for its proxy $g_{u_{1,n},\theta}$ (see the next section); indeed, changing the value of this preliminary estimator does not alter the value of the likelihood of the sample; as shown in the simulations developed hereafter, any value of $\theta$ can be used; call $\theta^*$ the value of $\theta$ chosen as initial value, using henceforth $p_{\mathbf{X},\theta^*}$ instead of $p_{\mathbf{X},\theta_T}$ in [9.3]. In exponential families, the values of the parameters which appear in the Gaussian component of $g\left(x_{i+1}|x_1^i\right)$ in [9.3] are easily calculated; note also that due to [9.1] the parameters in $\mathfrak{n}\left(\alpha\beta, \beta, u\left(x_{i+1}\right)\right)$ are such that the dominating density can be chosen for all $i$ as $p_{\mathbf{X},\theta^*}$. The constant in the acceptance–rejection algorithm is then $C_i/\sqrt{2\pi\beta}$. The constant $C_i$ need not be evaluated since it cancels in the ratio defining the acceptance–rejection rule. This is in contrast with the case when the conditioning value is in the range of a large deviation with respect to $p_{\mathbf{X},\theta_T}$; in this case, which appears in a natural way in importance sampling estimation for rare event probabilities, the simulation algorithm is more complex (see [BRO 13b]).

## 9.3. Sufficient statistics and approximated conditional density

### 9.3.1. *Keeping sufficiency under the proxy density*

The density $g_{u_{1,n},\theta_T}(y_1^k)$ is used in order to handle Rao–Blackwellization of estimators or statistical inference for models with nuisance parameters. The basic property is with respect to the nuisance parameter. We show in one example that the family of densities $g_{u_{1,n},\theta}(y_1^k)$ defined in [9.6], when indexed by $\theta$, inherits the invariance with respect to the parameter $\theta$ when conditioning on a sufficient statistics. In [CAR 12], the inverse Gaussian density is studied.

Consider the Gamma density

$$f_{r,\theta}(x) := \frac{\theta^{-(r+1)}}{\Gamma(r+1)} x^r \exp -x/\theta \quad \text{for } x > 0. \qquad [9.7]$$

As $r$ varies in $(-1, \infty)$ and $\theta$ is positive, the density runs in an exponential family with parameters $r$ and $\theta$, and sufficient statistics $t(x) := \log x$ and $u(x) := x$, respectively, for $r$ and $\theta$. Given a data set $\mathbf{x}_1, ..., \mathbf{x}_n$ obtained through sampling from i.i.d. r.v.'s $\mathbf{X}_1, ...\mathbf{X}_n$ with density $f_{r_T,\theta_T}$, the resulting sufficient statistics are, respectively, $t_{1,n} := \log \mathbf{x}_1 + ... + \log \mathbf{x}_n$ and $u_{1,n} := \mathbf{x}_1 + ... + \mathbf{x}_n$. We consider two parametric models $(f_{r_T,\theta}, \theta \geq 0)$ and $(f_{r,\theta_T}, r > -1)$, respectively, assuming $r_T$ or $\theta_T$ known.

We first consider sufficiency of $\mathbf{U}_{1,n} := \mathbf{X}_1 + ... + \mathbf{X}_n$ in the first model. The density $g_{u_{1,n},(r_T,\theta_T)}(y_1^k)$ should be free of the current value of the true parameter $\theta_T$ of the parameter under which the data are drawn. However, as appears in [9.6], the unknown value $\theta_T$ should be used in its very definition. We show by simulation that whatever the value of $\theta$ inserted in place of $\theta_T$ in [9.6], the value of the likelihood of $\mathbf{x}_1^k$ under $g_{u_{1,n},(r_T,\theta)}$ does not depend on $\theta$. We thus observe that $\mathbf{U}_{1,n}$ is "sufficient" for $\theta$ in the conditional density approximating $p_{u_{1,n},(r_T,\theta)}$, as should hold as a consequence of theorem 9.1. Say that $\mathbf{U}_{1,n}$ is quasi-sufficient for $\theta$ in $g_{u_{1,n},(r_T,\theta)}$ if this loose invariance holds.

Similarly, the same fact occurs in the model $(f_{r,\theta_T}, r > 0)$.

In both cases, whatever the value of the parameter $\theta$ (Figure 9.1) or $r$ (Figure 9.2), the likelihood of $\mathbf{x}_1^k$ remains constant.

### 9.3.2. *Rao–Blackwellization*

The Rao–Blackwell theorem holds regardless of whether biased or unbiased estimators are used, since conditioning reduces the MSE. Although its statement is rather weak, in practice the improvement is often enormous. New interest in Rao–Blackwellization procedures has arisen in the recent years, conditioning on ancillary variables (see [FRA 04] for a survey on ancillaries in conditional inference); specific Rao–Blackwellization schemes have been proposed in [CAS 96], [CAS 98], [PER 99], [DOU 11] and [IAC 10], whose purpose is to improve the variance of a given statistics (for example, a tail probability) under a *known* distribution, through a simulation scheme under this distribution; the ancillary variables used in the simulation process itself are used as conditioning ones for the Rao–Blackwellization of the statistics. The present approach is more classical in this respect, since we do not assume that the parent distribution is known; conditioning on a sufficient statistics $\mathbf{U}_{1,n}$ with respect to the parameter $\theta$ and simulating samples according to the approximating density $g_{u_{1,n},\theta}$ will produce the improved estimator.

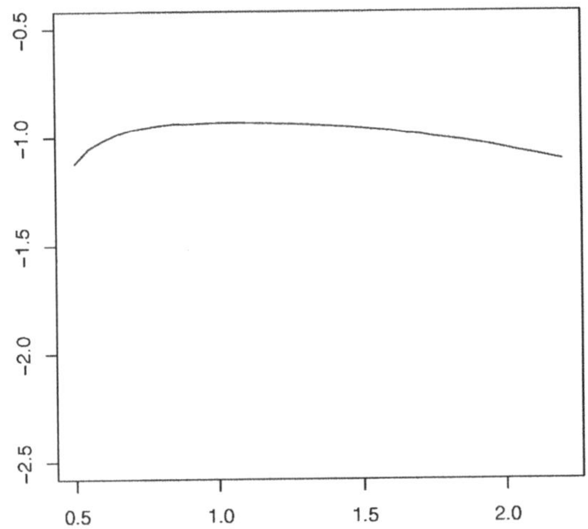


**Figure 9.1.** *Proxy of the conditional likelihood of $X_1^k$ under $g_{T_{1,n}}$ as a function of $\theta$ for $n = 100$ and $k = 80$ in the gamma case*

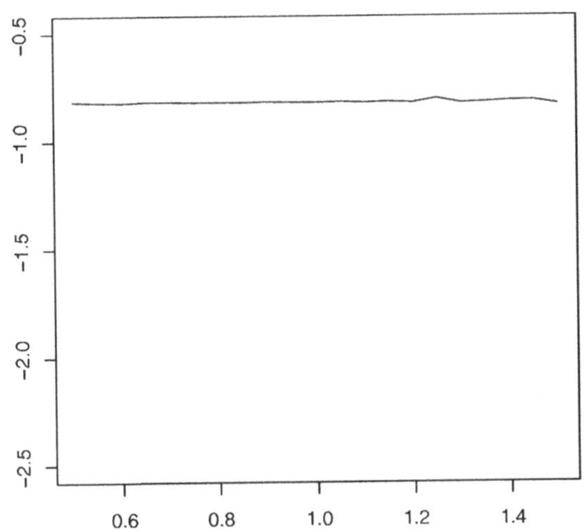


**Figure 9.2.** *Proxy of the conditional likelihood of $X_1^k$ under $g_{U_{1,n}}$ as a function of $r$ for $n = 100$ and $k = 80$ in the gamma case*

Since $\mathbf{U}_{1,n}$ is quasi-sufficient for the parameter $\theta$ in $g_{u_{1,n},\theta}$, it can be used in order to obtain improved estimators of $\theta_T$ through Rao–Blackwellization. We briefly illustrate the procedure and its results on some toy cases. Consider again the Gamma family defined above with canonical parameters $r$ and $\theta$.

First, the parameter to be estimated is $\theta_T$. A first unbiased estimator is chosen as

$$\widehat{\theta_2} := \frac{X_1 + X_2}{2r_T}$$

Given an i.i.d. sample $\mathbf{X}_1^n$ with density $f_{r_T,\theta_T}$, the Rao–Blackwellized estimator of $\widehat{\theta_2}$ is defined through

$$\theta_{RB,2} := E\left(\widehat{\theta_2}\middle| \mathbf{U}_{1,n}\right)$$

whose variance is less than $Var\widehat{\theta_2}$. Given the data set $\mathbf{x}_1, ...\mathbf{x}_n$, the estimate of $\theta_{RB,2}$ is produced through simulation of as many $\widehat{\theta_2}$s as wished, under $g_{u_{1,n},(r_T,\theta_T)}$. Denote by $\widehat{\theta}_{RB,2}$ the resulting Rao–Blackwellized estimator of $\widehat{\theta_2}$.

Consider $k = 2$ in $g_{u_{1,n},(r_T,\theta_T)}(y_1^k)$ and let $(Y_1, Y_2)$ be distributed according to $g_{u_{1,n},(r_T,\theta_T)}(y_1^2)$; note that any value $\theta$ can be used in practice instead of the unknown value $\theta_T$, by quasi-sufficiency of $\mathbf{U}_{1,n}$. Replications of $(Y_1, Y_2)$ produce an estimator $\widehat{\theta}_{RB,2}$ for fixed $u_{1,n}$; we have used 1,000 replications $(Y_1, Y_2)$. Iterating on 1,000 simulations of the runs $\mathbf{X}_1^n$ produces, for $n = 100$, an i.i.d. sample with size 1,000 of $\widehat{\theta}_{RB,2}$s and $Var\theta_{RB,2}$ is estimated. The resulting variance shows a net improvement with respect to the estimated variance of $\widehat{\theta_2}$. It is of some interest to confront this gain in variance as the number of terms involved in $\widehat{\theta_k}$ increases together with $k$. As $k$ approaches $n$, the variance of $\widehat{\theta_k}$ approaches the Cramer–Rao bound. The graph below shows the decay in variance of $\widehat{\theta_k}$. We note that whatever the value of $k$, the estimated value of the variance of $\theta_{RB,k}$ is constant. This is indeed an illustration of Lehmann–Scheffés theorem.

COMMENT 9.2.– Lockhart and O'Reilly [LOC 05] establish, under certain conditions and for *fixed* $k$, the asymptotic equivalence of the plug-in estimate for the distribution $P_{\theta_{ML}}\left(\mathbf{X}_1^k \in B\right)$ and the Rao–Blackwell estimate $P\left(\mathbf{X}_1^k \in B\middle| \mathbf{U}_{1,n}\right)$, where $\theta_{ML}$ is the maximum likelihood estimator of $\theta_T$ based on the whole sample $\mathbf{X}_1^n$ (this result is known as Moore's conjecture (see [MOO 73])). They also provide rates for this convergence.

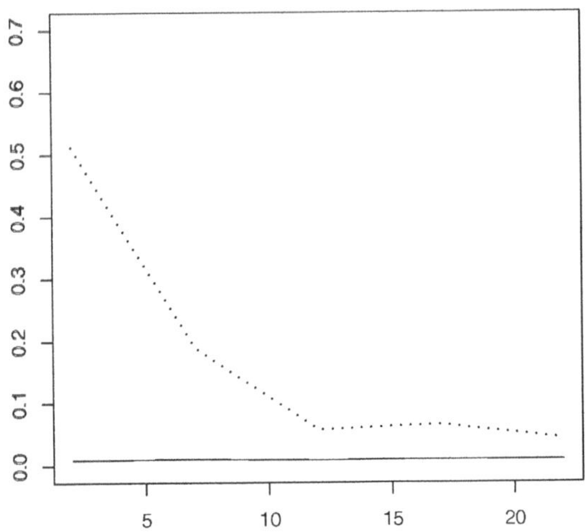


**Figure 9.3.** *Variance of $\widehat{\theta}_k$, the initial estimator (dotted line), along with the variance of $\theta_{RB,k}$, the Rao–Blackwellized estimator (solid line) with $n = 100$ as a function of k*

## 9.4. Exponential models with nuisance parameters

### 9.4.1. *Conditional inference in exponential families*

We consider the case when the parameter consists of two distinct subparameters: one of interest denoted by $\theta$ and a nuisance component denoted by $\eta$. As is well known, conditioning on a sufficient statistics for the nuisance parameter produces a new exponential family which is free of it. Assuming the observed data set $\mathbf{x}_1^n := (\mathbf{x}_1, ..., \mathbf{x}_n)$ resulting from sampling of a vector $\mathbf{X}_1^n := (\mathbf{X}_1, ..., \mathbf{X}_n)$ of i.i.d. random variables with distribution in the initial exponential model, and denoting $\mathbf{U}_{1,n}$ a sufficient statistics for $\eta$, simulation of samples under the conditional distribution of $\mathbf{X}_1^n$ given $\mathbf{U}_{1,n} = u_{1,n}$ and $\theta = \theta_0$ for some $\theta_0$ produces the basic ingredient for Monte Carlo tests with $H0 : \theta_T = \theta_0$, where $\theta_T$ stands for the true value of the parameter of interest. Changing $\theta_0$ for other values of the parameter of interest produces power curves as functions of the level of the test. This is the well-known principle of Monte Carlo tests, which are considered here. We consider a steep but not necessarily regular $\mathcal{P} := \{P_{\mathbf{X},(\theta,\eta)}, (\theta,\eta) \in \mathcal{N}\}$ defined on $\mathbb{R}$ with canonical parameterization $(\theta, \eta)$ and minimal sufficient statistics $(t, u)$ defined through the density

$$p_{\mathbf{X},(\theta,\eta)}(x) := \frac{dP_{\mathbf{X},(\theta,\eta)}(x)}{dx} = \exp\left[\theta t(x) + \eta u(x) - K(\theta,\eta)\right] h(x). \qquad [9.8]$$

For notational convenience and without loss of generality, both $\theta$ and $\eta$ belong to $\mathbb{R}$. Also the model can be defined on $\mathbb{R}^d$, $d > 1$, at the cost of similar but more evolved tools. The natural parameter space is $\mathcal{N}$ (which is a convex set in $\mathbb{R}^2$) defined as the effective domain of

$$k(\theta, \eta) := \exp\left[K(\theta, \eta)\right] = \int \exp\left[\theta t(x) + \eta u(x)\right] h(x) dx. \qquad [9.9]$$

As above, denote $\mathbf{x}_1^n := (\mathbf{x}_1, ..., \mathbf{x}_n)$ as the observed values of $n$ i.i.d. replications of a general random variable $\mathbf{X}$ with density [9.8]. Denote

$$t_{1,n} := \sum_{i=1}^{n} t(\mathbf{x}_i) \text{ and } u_{1,n} := \sum_{i=1}^{n} u(\mathbf{x}_i). \qquad [9.10]$$

Basu [BAS 77] discusses 10 different ways for eliminating the nuisance parameters, among which conditioning on sufficient statistics and consider UMPU tests pertaining to the parameter of interest. In most cases, the density of $\mathbf{T}_{1,n}$ given $\mathbf{U}_{1,n} = u_{1,n}$ is unknown. Two main processes have been developed to deal with this issue: approximating this conditional density of a statistics or simulating samples from the conditional density. These two approaches are combined here.

The classical technique is to approximate this conditional density using some expansion. Then integration produces critical values. For example, Pedersen [PED 79] defines the mixed Edgeworth-saddlepoint approximation, or the single saddlepoint approximation. However, the main problem of this technique is that the approximated density depends on the nuisance parameter. In order to obtain the expansion, some suitable values for the parameter of interest and for the nuisance parameter need to be chosen. In the method developed here, as seen before, the conditional approximated density inherits the invariance with respect to the nuisance parameter when conditioning on a sufficient statistics pertaining to this parameter.

Rephrasing the notation of section 9.2 in the present setting, the maximum likelihood estimator (MLE) $(\theta_{ML}, \eta_{ML})$ satisfies

$$\left.\frac{\partial K(\theta, \eta)}{\partial \eta}\right|_{\theta_{ML}, \eta_{ML}} = u_{1,n}/n$$

and therefore $u_{1,n}/n$ converges to $\left(\frac{\partial K(\theta_T, \eta)}{\partial \eta}\right)^{-1}(\eta_T)$.

For notational clarity, denote by $\mu$ the expectation of $u(\mathbf{X}_1)$ and by $\sigma^2$ its variance under $(\theta_T, \eta_T)$; hence

$$\mu := \mu_{(\theta_T, \eta_T)} := \partial K(\theta_T, \eta_T)/\partial \eta \quad \sigma^2 := \sigma^2_{\theta_T, \eta_T} := \frac{\delta K(\theta_T, \eta_T)}{\delta \eta^2}$$

Assume at present that $\theta_T$ and $\eta_T$ are known. It holds that

$$\phi(r) := E_{(\theta_T,\eta_T)} \exp[ru(\mathbf{X})] = \exp\left[K(\theta_T, \eta_T + r) - K(\theta_T, \eta_T)\right]$$

and

$$\begin{aligned} m(r) &= \mu_{(\theta_T,\eta_T+r)} \\ s^2(r) &= \sigma^2_{(\theta_T,\eta_T+r)} \\ \mu_3(r) &= \partial^3 K(\theta_T, \eta_T + r)/\partial r^3 . \end{aligned}$$

Furthermore,

$$\pi^{\alpha}_{u,\theta_T,\eta_T}(x) := \frac{\exp ru(x)}{\phi(r)} p_{\mathbf{X},(\theta_T,\eta_T)}(x) = p_{\mathbf{X},(\theta_T,\eta_T+r)}(x) \qquad [9.11]$$

for any given $\alpha$ in the range of $P_{\mathbf{X},(\theta_T,\eta_T)}$. In the above formula [9.11], parameter $r$ denotes the only solution of the equation

$$m(r) = \alpha.$$

For large $k$ depending on $n$, using Monte Carlo (MC) tests based on runs of length $k$ instead of $n$ does not affect the accuracy of the results.

### 9.4.2. *Application of conditional sampling to MC tests*

#### 9.4.2.1. *Context*

Consider a test defined through $H0 : \theta_T = \theta_0$ versus $H1 : \theta_T \neq \theta_0$. Aim at obtaining $p-$values through simulation when the distribution of the desired test statistics under $H0$ is either unknown or very cumbersome to obtain; a comprehensive reference is [JÖC 86].

Recall the principle of those tests: denote by $t$ the observed value of the studied statistic based on the data set and let $t_2, .., t_L$ the values of the resulting test statistics obtained through the simulation of $L - 1$ samples $\mathbf{X}_1^n$ under $H0$. If $t$ is the $M$th largest value of the sample $(t, t_2, ..., t_L)$, $H0$ will be rejected at the $\alpha = M/L$ significance level, since the rank of $t$ is uniformly distributed on the integer $2, ..., L$ when $H0$ holds. The present MC procedure uses simulated samples under the proxy of $p_{u_{1,n},(\theta_0,\eta_T)}$. Using quasi-sufficiency of $\mathbf{U}_{1,n}$, we may use any value in place of $\eta_T$; we have compared this simple choice with the common use, inserting the MLE $\hat{\eta}_{\theta_0}$ in place of $\eta_T$ in $g_{u_{1,n},(\theta_0,\eta_T)}$. This estimate $\hat{\eta}_{\theta_0}$ is the MLE of $\eta_T$ in the one-parameter family $p_{\mathbf{X},(\theta_0,\eta)}$ defined through [9.8]; this choice follows the commonly used one, as advocated, for instance, in [PED 79] and [PAC 92]. Innumerous simulation studies support this choice in various contexts; we found no difference in the resulting procedures.

Consider the problem of testing the null hypothesis $H0 : \theta_T = \theta_0$ against the alternative $H1 : \theta_T > \theta_0$ in model [9.8] where $\eta$ is the nuisance parameter.

When $p_{u_{1,n},(\theta_0,\eta_T)}$ is known, the classical conditional test $H0 : \theta_T = \theta_0$ versus $H1 : \theta_T > \theta_0$ with level $\alpha$ is UMPU.

Substituting $p_{u_{1,n},(\theta_0,\eta_T)}\left(\mathbf{X}_1^n = x_1^n \,|\mathbf{U}_{1,n} = u_{1,n}\right)$ by $g_{u_{1,n},(\theta_0,\eta_T)}\left(x_1^k\right)$ defined in [9.6], i.e. substituting the test statistics $\mathbf{T}_1^n$ by $\mathbf{T}_1^k$ and $p_{\theta_0}\left(\mathbf{X}_1^k = x_1^k \,|\mathbf{U}_{1,n} = u_{1,n}\right)$ by $g_{u_{1,n},(\theta_0,\eta_T)}\left(x_1^k\right)$, i.e. changing the model for a proxy while keeping the same parameter of interest $\theta$ yields the conditional test with level $\alpha$.

$$\psi_\alpha(x_1^k) := \begin{cases} 1 \text{ if } t_{1,k} > t_\alpha \\ \gamma \text{ if } t_{1,k} = t_\alpha \\ 0 \text{ if } t_{1,k} < t_\alpha \end{cases}$$

and

$$E_{G_{u_{1,n},(\theta_0,\eta_T)}}[\psi_\alpha(X_1^k)] = \alpha$$

i.e. $\alpha := \int \mathbb{1}_{t_{1,k}>t_\alpha} g_{u_{1,n},(\theta_0,\eta_T)}\left(x_1^k\right) dx_1 ... dx_k$. Its power under a simple hypothesis $\theta_T = \theta$ is defined through

$$\beta_{\psi_\alpha}(\theta|u_n) = E_{G_{u_{1,n},(\theta_0,\eta_T)}}[\psi_\alpha(\mathbf{X}_1^k)].$$

By quasi-sufficiency of $\mathbf{U}_{1,n}$ with respect to $\eta$, any value can be inserted in $g_{u_{1,n},(\theta_0,\eta_T)}$ in place of $\eta_T$.

Recall that the parametric bootstrap produces samples from a parametric model which is fitted to the data, often through maximum likelihood. In the present setting, the parameter $\theta$ is set to $\theta_0$ and the nuisance parameter $\eta$ is replaced by its estimator $\widehat{\eta}_{\theta_0}$ which is the MLE of $\eta_T$ when the parameter $\theta$ is fixed at the value $\theta_0$ defining $H0$. Comparing their exact conditional MC tests with parametric bootstrap ones for Gamma distributions, Lockhart *et al.* [LOC 05] conclude that no significant difference can be noticed in terms of level or in terms of power. We proceed in the same vein, comparing conditional sampling MC tests with the parametric bootstrap ones, again obtaining similar results when the nuisance parameter is estimated accurately (see [CAR 12]). However, the results are somehow different when the nuisance parameter cannot be estimated accurately, which may occur in various cases.

9.4.2.2. *Bimodal likelihood: testing the mean of a normal distribution in dimension 2*

In contrast to the above-mentioned examples, the following case study shows that estimation through the unconditional likelihood may fail to provide consistent estimators when the likelihood surface has multiple critical points.

Sundberg [SUN 10] proposes four examples that allow likelihood multimodality. Two of them can also be found in [EFR 75] and in [BAR 94, Chapter 2]. We consider the "Normal parabola" model which is a curved (2, 1) family (see example 2.35 in [BAR 94, Chapter 2]). Two independent Gaussian variates have unknown means and known variances; their means are related by a parabolic relationship.

Let $\mathbf{X}$ and $\mathbf{Y}$ be two independent Gaussian r.v.s with same variance $\sigma_T^2$ with expectation $\psi_T$ and $\psi_T^2$. In the present example, $\sigma_T^2 = 1$ and $\psi_T = 2$.

Let $(\mathbf{x}_i, \mathbf{y}_i)$ , $1 \leq i \leq n$, be i.i.d. realizations of $(\mathbf{X}_i, \mathbf{Y}_i)$ .

The parameter of interest is $\sigma^2$ while the nuisance parameters is $\psi$. Derivation of the likelihood function of the observed sample with respect to $\psi$ yields the following equation:

$$(u_{1,n} - \psi) + 2\psi \left(v_{1,n} - \psi^2\right) = 0$$

with $u_{1,n} := \mathbf{x}_1 + ... + \mathbf{x}_n$ and $v_{1,n} := \mathbf{y}_1 + ... + \mathbf{y}_n$. Define accordingly $\mathbf{U}_{1,n}$ and $\mathbf{V}_{1,n}$. The following table shows that the likelihood function is bimodal in $\psi$.

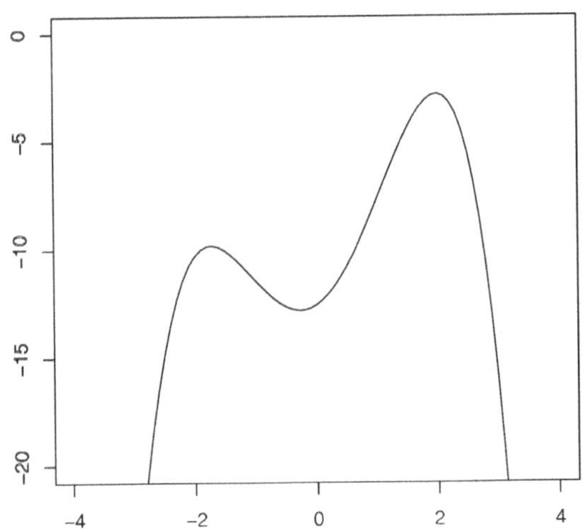


**Figure 9.4.** *Bimodal likelihood in* $\psi$

Estimation of the nuisance parameter $\psi$ is performed through the standard Newton–Raphson method. The Newton–Raphson optimizer of the likelihood function converges to the true value when the initial value is larger than 1 and fails to converge to $\psi_T = 2$ otherwise. Hence, the ML estimation based on this preliminary estimate of the nuisance parameter may lead to erroneous estimates of the parameter of interest. Indeed, according to the initial value, we obtained estimators of $\psi_T$ close to 2 or to $-2$. When the estimator of the nuisance parameter is close to its true value 2, parametric bootstrap yields Monte Carlo tests with power close to 1 for any $\alpha$ and any alternative close to $H0$. On the contrary, when this estimate is close to the second maximizer of the likelihood (i.e. close to $-2$), the resulting Monte Carlo test based on parametric bootstrap has power close to 0 irrespective of the value of $\alpha$ and of the alternative, when close to $H0$. In contrast to these results, Monte Carlo tests based on conditional sampling provides powers close to 1 for any $\alpha$; we have considered alternatives close to $H0$. This result is of course a consequence of the quasi-sufficiency of the statistics $(\mathbf{U}_{1,n}, \mathbf{V}_{1,n})$ for the parameter $(\psi, \psi^2)$ of the distribution of the sample $(\mathbf{x}_i, \mathbf{y}_i)_{i=1,\ldots,n}$; see the next section for a discussion of this point.

### 9.4.3. *Estimation through conditional likelihood*

Considering model [9.8], we intend to perform an estimation of $\theta_T$ irrespectively upon the value of $\eta_T$. Denote by $\widehat{\eta}_\theta$ the MLE of $\eta_T$ when $\theta$ holds; the model $p_{\mathbf{X},(\theta,\widehat{\eta}_\theta)}(x)$ is a one-parameter model which is fitted to the data for any peculiar choice of $\theta$. The optimizer in $\theta$ of the resulting likelihood function is the global MLE. Properties of the resulting estimators strongly rely on the constistency properties of $\widehat{\eta}_\theta$ at any given $\theta$.

Consider the consequence of theorem 9.2. Condition on the value of the sufficient statistics $\mathbf{U}_{1,n}$, and consider the conditional likelihood of the observed subsample $\mathbf{x}_1^k$ under parameter $(\theta, \widehat{\eta}_\theta)$; recall that $\mathbf{x}_1^k$ is generated under $(\theta_T, \eta_T)$. By theorem 9.2 this likelihood is approximated by $g_{u_{1,n},(\theta,\widehat{\eta}_\theta)}\left(\mathbf{x}_1^k\right)$ with a small relative error. Conditioned likelihood estimation is performed optimizing $g_{u_{1,n},(\theta,\widehat{\eta}_\theta)}\left(\mathbf{x}_1^k\right)$ upon $\theta$. Any value of the nuisance parameter $\eta$ can be used in place of $\widehat{\eta}_\theta$ as seen in section 3.1.

In most cases, as the normal, Gamma or inverse-Gaussian, estimations through the unconditional likelihood or conditional likelihood give a consistent estimator.

We consider the example of the bimodal likelihood from the above section, inheriting the notation, and explore the behavior of the proxy of the conditional likelihood of the sample $(\mathbf{x}_i, \mathbf{y}_i)$, $1 \le i \le n$, when conditioning on $u_{1,n}$ and $v_{1,n}$ as a function of $\sigma^2$. This likelihood becomes

$$L\left(\sigma^2 \middle| u_{1,n}, v_{1,n}\right) = p_{u_{1,n}\sigma^2}\left(\mathbf{x}_1^n\right) p_{v_{1,n}\sigma^2}\left(\mathbf{y}_1^n\right)$$

where we have used the independence of the r.v.'s $\mathbf{X}_i$s and $\mathbf{Y}_i$s.

Applying theorem 9.1 to the above expression, it appears that $\psi$ cancels in the resulting densities $g_{u_{1,n,\sigma^2}}$ and $g_{v_{1,n,\sigma^2}}$. This proves that the proxy of the conditional likelihood provides consistent estimation of $\sigma_T^2$ as shown in Figures 9.5 and 9.6 (see the solid lines).

In Figure 9.5, the dotted line is the likelihood function

$$L\left(\sigma^2\right) := \sum_{i=1}^{n} \log p_{\mathbf{X},\left(\sigma^2, \widehat{\psi}_{\sigma^2}\right)}(\mathbf{x}_i)$$

where $\widehat{\psi}_{\sigma^2}$ is a consistent estimator of the nuisance parameter; the resulting maximizer in variable $\sigma^2$ is close to $\sigma_T^2 = 1$. At the opposite in Figure 9.6, an inconsistent preliminary estimator of $\psi_T$ obtained through a bad tuning of the initial point in the Newton–Raphson procedure leads to inconsistency in the estimation of $\sigma_T^2$, the resulting likelihood function being unbounded.

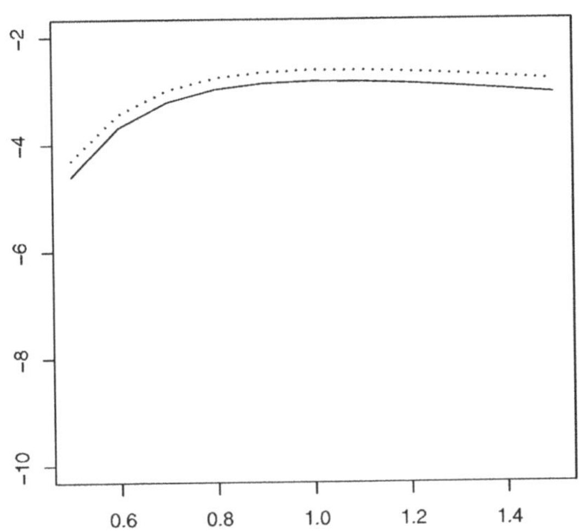


**Figure 9.5.** *Proxy of the conditional likelihood (solid line) along with the classical likelihood (dotted line) as a function of $\sigma^2$ for $n = 100$ and $k = 99$ in the case where a good initial point in Newton–Raphson procedure is chosen*

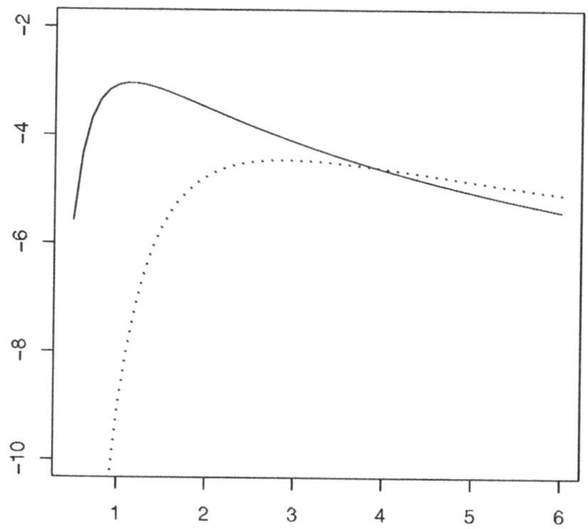


**Figure 9.6.** *Proxy of the conditional likelihood (solid line) along with the classical likelihood (dotted line) as a function of $\sigma^2$ for $n = 100$ and $k = 99$ in the case where a bad initial point in Newton–Raphson procedure is chosen*

## 9.5. Bibliography

[BAR 78] BARNDORFF-NIELSEN O., *Information and Exponential Families in Statistical Theory*, Wiley Series in Probability and Mathematical Statistics, John Wiley & Sons, Chichester, 1978.

[BAR 94] BARDNORFF-NIELSEN O., COX R., *Inference and Asymptotics*, Chapman & Hall, London, 1994.

[BAS 77] BASU D., "On the elimination of nuisance parameters", *Journal of the American Statistical Association*, vol. 72, no. 358, pp. 355–366, 1977.

[BRO 13a] BRONIATOWSKI M., CARON V., "Long runs under a conditional limit distribution", *Annals of Applied Probability*, 2013, forthcoming.

[BRO 13b] BRONIATOWSKI M., CARON V., "Towards zero variance estimators for rare event probabilities", *ACM TOMACS Special Issue on Monte Carlo Methods in Statistics*, vol. 23, no. 1, p. 7, 2013.

[CAR 12] CARON V., A limit conditional theorem. Applications to conditional inference and importance sampling schemes (in French), available at tel.archives-ouvertes.fr/docs/00/76/33/69/PDF/these.pdf, 2012.

[CAS 96] CASELLA G., ROBERT C.P., "Rao–Blackwellization of sampling schemes", *Biometrika*, vol. 83, no. 1, pp. 81–94, 1996.

[CAS 98] CASELLA G., ROBERT C.P., "Post-processing accept-reject samples: recycling and rescaling", *Journal of Computational and Graphical Statistics*, vol. 7, no. 2, pp. 139–157, 1998.

[DOU 11] DOUC R., ROBERT C., "A vanilla Rao–Blackwellization of Metropolis-Hastings algorithms", *Annals of Statistics*, vol. 39, no. 1, pp. 261–277, 2011.

[EFR 75] EFRON B., "Defining the curvature of a statistical problem (with applications to second order efficiency) (with discussion)", *Annals of Statistics*, vol. 3, no. 6, pp. 1189–1242, 1975.

[ENG 97] ENGEN S., LILLEGARD M., "Stochastic simulations conditioned on sufficient statistics", *Biometrika*, vol. 84, no. 1, pp. 235–240, 1997.

[FRA 04] FRASER D. A.S., "Ancillaries and conditional inference. With comments by Ronald W. Butler, Ib M. Skovgaard, Rudolf Beran and a rejoinder by the author.", *Statistical Science*, vol. 19, no. 2, pp. 333–369, 2004.

[IAC 10] IACOBUCCI A., MARIN J.-M., ROBERT C., "On variance stabilisation in population Monte Carlo by double Rao–Blackwellization", *Computational Statistics & Data Analysis*, vol. 54, no. 3, pp. 698–710, 2010.

[JÖC 86] JÖCKEL K.-H., "Finite sample properties and asymptotic efficiency of Monte Carlo tests", *The Annals of Statistics*, vol. 14, no. 1, pp. 336–347, 1986.

[LEH 86] LEHMANN E., *Testing Statistical Hypotheses*, Springer, New York, 1986.

[LIN 03] LINDQVIST B., TARALDSEN G., LILLEGÄRD M., *et al.*, "A counterexample to a claim about stochastic simulations", *Biometrika*, vol. 90, no. 2, pp. 489–490, 2003.

[LIN 05] LINDQVIST B., TARALDSEN G., "Monte Carlo conditioning on a sufficient statistic", *Biometrika*, vol. 92, no. 2, pp. 451–464, 2005.

[LOC 05] LOCKHART R., O'REILLY F., "A note on Moore's conjecture", *Statistics & Probability Letters*, vol. 74, no. 2, pp. 212–220, 2005.

[LOC 07] LOCKHART R.A., O'REILLY F.J., STEPHENS A., "Use of the Gibbs sampler to obtain conditional tests, with applications", *Biometrika*, vol. 94, no. 4, pp. 992–998, 2007.

[MOO 73] MOORE D.S., "A note on Srinivasan's goodness-of-fit test", *Biometrika*, vol. 60, no. 1, pp. 209–211, 1973.

[PAC 92] PACE L., SALVAN A., "A note on conditional cumulants in canonical exponential families", *Scandinavian Journal of Statistics*, vol. 19, no. 2, pp. 185–191, 1992.

[PED 79] PEDERSEN B., "Approximating conditional distributions by the mixed Edgeworth-saddlepoint expansion", *Biometrika*, vol. 66, no. 3, pp. 597–604, 1979.

[PER 99] PERRON F., "Beyond accept-reject sampling", *Biometrika*, vol. 86, no. 4, pp. 803–813, 1999.

[REI 95] REID N., "The roles of conditioning in inference.", *Statistical Science*, vol. 10, no. 2, pp. 138–157, 1995.

[SUN 10] SUNDBERG R., "Flat and multimodal likelihoods and model lack of fit in curved exponential families", *Scandinavian Journal of Statistics*, vol. 27, no. 4, pp. 632–643, 2010.

[VOI 93] VOINOV V., NIKULIN M., *Unbiased Estimators and Their Applications, V.1 Univariate Case*, Kluwer Academic Publishers, Dordrecht, 1993.

[VOI 96] VOINOV V., NIKULIN M., *Unbiased Estimators and Their Applications, V.2 Multivariate Case*, Kluwer Academic Publishers, Dordrecht, 1996.

# Chapter 10

# On Testing Stochastic Dominance by Exceedance, Precedence and Other Distribution-Free Tests, with Applications

We investigate the use of precedence and exceedance tests based on independent random samples for testing stochastic dominance. For homogeneous samples of equal size, we derive the exact finite sample distributions of the statistics. In general, the corresponding limit laws for large sample sizes are shown to be negative binomial. To illustrate the usefulness of these methods, we apply these tests to an experimental data set composed of two paired samples of size $n = 10$, observed by Seralini *et al.* [SER 12]. We show that exceedance tests compare well with other classical tests (Kolmogorov–Smirnov, Mann–Whitney and Wilcoxon). For either of these methods, in our example, the assumption that the samples are equally distributed is rejected with a level of significance varying between $0.5\%$ and $2.6\%$.

## 10.1. Introduction

Let $X := X_1, \ldots, X_{n_A}$ (sample $A$) and $Y := Y_1, \ldots, Y_{n_B}$ (sample $B$) be independent random samples, each composed of a sequence of independent and identically distributed (i.i.d.) random failure-times, with respective continuous distribution functions (DFs) $F(x) = \mathbb{P}(X \leq x)$ and $G(y) = \mathbb{P}(Y \leq y)$, for $x, y \in \mathbb{R}$. Our aim is to investigate *distribution-free* tests of the hypotheses $H_0 : F = G$ versus $H_1 : F > G$ (the *stochastic dominance* of $F$ over $G$). *Distribution-free* means the

---

Chapter written by Paul DEHEUVELS.

fact that the null distribution of the test statistics must be independent of the common DF, $F = G$. Also, when possible, we would want these tests to rely on partially observed data sets, where one only needs the knowledge of a limited number of *tail* (*upper* or *lower*) failure-times from each of the two samples. To be more explicit, for $1 \leq k \leq n$, denote by $N_k$ the number of observations from sample $A$ falling below the $k$th smallest observation from sample $B$, and by $N_k^*$ the number of observations from sample $B$ falling above the $k$-largest observation from sample $A$. For example, for $k = 1$, a test based on $N_1$ requires only the observation of the failure times of sample $A$ below $\min\{Y_1, \ldots, Y_{n_B}\}$, whereas a test based on $N_1^*$ requires only the observation of failures times of sample $B$ above $\max\{X_1, \ldots, X_{n_A}\}$. Both of these choices are of practical interest, and this will be illustrated below in a real-life example.

We shall concentrate on the comparison of samples of identical sizes, which is applicable to experiments where the effects of a treatment are compared to a control on *paired samples*. Therefore, we will assume that $n_A = n_B = n$, unless otherwise specified. Some of our forthcoming results will turn out to be specific to this case (see, e.g., section 10.2.3). We will consider unequal sample sizes in section 10.3, and also treat the case where $n_A$ and $n_B$ are random. Therefore, the results in this section will be applicable to the comparison of rates of partially observed *Poisson processes*. The extension of our results to the comparison of more than two samples of arbitrary sizes will be considered elsewhere, and we will limit ourselves to the present setup, which is of interest in and of itself.

Historically, one of the first practical solutions to this problem is the *slippage test* by Mosteller [MOS 48], which is based on rejection of $H_0$ for large values of the statistic $N_1$ (or $N_1^*$). *Slippage* (or *rigid translation*) generally means a parametric alternative to $H_0$, of the form $F(\cdot) = G(\cdot + \theta)$ for some $\theta \neq 0$. We will not restrict ourselves to such alternatives, and work, rather, in a purely non-parametric setup. The rationale of Mosteller's test is more or the less implicit in [WIL 42] through the notion of *tolerance limits* (see, e.g., the comments in [ROS 53]). Variants of Mosteller's slippage test have been discussed by Rosenbaum [ROS 53, ROS 54], Kamat [KAM 56], Haga [HAG 59], Conover [CON 69] and Neave [NEA 72], among others. A popular version of these tests is the thumb-rule recommendation (see, e.g., [SAK 82, p. 285], and [ROS 54]) to reject $H_0$ at a significance of 5%, if, for $5 < n \leq 25$, the $j = 5$ largest or smaller values (respectively, for $n \geq 25$, the $j = 6$ largest or smaller values), come from the same sample (this corresponds to reject $H_0$ when $\max\{N_1, N_1^*\} \geq j$). Another variant is the *Tukey–Duckworth statistic*, defined by Tukey [TUK 59] and based on the statistic $N_1 + N_1^*$. For $5 < n \leq 24$ (respectively, for $n \geq 25$), Tukey [TUK 59] recommends rejection of $H_0$ at a significance of 5%, whenever $N_1 + N_1^*$ is greater than or equal to 7 (respectively 8). Such two-sided *exceedance statistics* can also be used to test $H_0$ against *dispersion* alternatives. Let $M_1$ denote the number of observations from sample $B$ falling below the smallest observation from sample $A$, and let $M_1^*$ denote the number of observations from sample $A$ falling above the largest observation from sample $B$. Keeping in mind that $N_1$ and $M_1$ (and likewise $N_1^*$ and $M_1^*$) cannot be simultaneously positive, Haga's dispersion test (see, e.g., [HAG 59]), is based on the statistic $N_1 + M_1 - N_1^* - N_2^*$. This test is a minor variant of the Rosenbaum dispersion test (see, e.g., [ROS 53]), which is based on either $N_1 + M_1$ or $N_1^* + M_1^*$.

The basic version of Mosteller's slippage test is the *precedence test* (see, e.g., [NEL 63, NEL 93]), whose distribution was tabulated by Rosenbaum [ROS 54] and Balakrishnan and Tony Ng [BAL 06]. For a properly chosen $1 \leq j \leq n$, the *precedence test* rejects $H_0$, whenever $N_1 \geq j$. Another strategy (which is, under $H_0$, stochastically equivalent to the first one) is to reject $H_0$ when $N_1^* \geq j$. We will call this last decision rule an *exceedance test*. The exceedance tests with critical zone of the form $N_1^* \geq j$ were considered by Westlake [WES 71]. In the following, we will consider some extended versions of these tests with critical zones of the form, either $N_k \geq j$, or $N_k^* \geq j$. These tests have been discussed to some extent by Balakrishnan and Frattina [BAL 00] and Balakrishnan and Tony Ng [BAL 06]. They rely on counts of small (respectively, large) observations of sample $A$ (respectively, sample $B$), considered below (respectively, above) a specified order statistic from the other sample.

We will compare these tests with classical non-parametric procedures which may be used to test $H_0$ against $H_1$ when the whole sample is available. In the first place, we will consider the one-sided Kolmogorov–Smirnov test (see, e.g., section 10.2.3), and, second, the Wilcoxon [WIL 45] $W$-test, which is equivalent, in the present setup to the Mann and Whitney [MAN 47] $U$-test (see, e.g., section 10.2.2). The Wilcoxon–Mann–Whitney test is most often used to test $H_0$ against $H_1^* := \{\mathbb{P}(X > Y) < \frac{1}{2}\}$ (see, e.g., [MAN 47]). However, this procedure is also appropriate to test $H_0$ against $H_1$. To illustrate this point, we observe that, for non-negative observations, with $F(0) = G(0) = 0$, $H_1 \Rightarrow H_1^*$, since, under $H_1$,

$$\mathbb{P}(X > Y) = \int_0^\infty (1 - F(t))\mathrm{d}G(t) < \int_0^\infty (1 - G(t))\mathrm{d}G(t) = \frac{1}{2}.$$

The exact null distributions of the tests we consider are not always easily computed. This is specifically the case for the Mann–Whitney–Wilcoxon statistic. Fortunately, for this test, there is a huge literature exhibiting consistent evidence that the level of significance evaluated from the normal asymptotic distribution coincides remarkably well with that obtained from numerical tabulations. Further refinements will be found in [BUC 06, BUC 69], together with extensive tabulations of the distribution of the Mann–Whitney–Wilcoxon statistic under $H_0$. This property will be verified in our example, and we will not further discuss the corresponding numerical details. On the other hand, for one-sided Kolmogorov–Smirnov statistics, both exact and limiting distributions are available, but, for small values of $n$ (such as $n = 10$, which is considered in our example) there are some important discrepancies between limiting and exact distributions. To render the present chapter self-contained, we will provide in this case some easily implementable formulas, allowing us to compute the exact significance levels of these tests (see, e.g., section 10.2.3). With respect to the *exceedance* tests we consider (recall that, under $H_0$, *exceedance* tests are formally equivalent to *precedence* tests after a change of scale), the situation is more delicate. Some exact finite-sample formulas are given in [BAL 06]. However, they require delicate summations of products of binomial coefficients, which are not of easy use. The same results are mentioned, together with an erroneous formula, in [BAL 00]. We will combine these results, which are already present in [ROS 54], with a different combinatorial evaluation, provided by Westlake [WES 71] for $k = 1$. Thus, we will provide below the exact formulas for the level of significance of these tests, which are new to our best knowledge for an arbitrary $k \geq 1$. This will allow us to evaluate the limiting distributions of the precedence (or exceedance) statistics, which are also new to our best knowledge, for $k \geq 2$. Interestingly, we will show in section 10.3 that, under $H_0$, the large sample limiting laws of the test statistics $N_k$ and $N_k^*$, as $n \to \infty$, are *negative binomial*. Our finite-sample formulas turn out to be easy to implement and agree with the tabulated values given by Balakrishnan and Tony Ng [BAL 06] and Balakrishnan and Frattina [BAL 00]. The discussion of the distributional properties of these tests (exact and asymptotic) constitutes the main component of this chapter.

## 10.2. Results

### 10.2.1. *The experimental data set*

To illustrate the practical interest of these methods, we will treat a real-life data set taken from an experiment of Seralini *et al.* [SER 12]. From the raw experimental data of this paper, we extract two sets of $n = 10$ female rats. For each of these animals, one observes the time (on a daily unit) of first occurrence of sizeable tumors. Since the observations are limited to the total duration of the experiment, some of the animals do not develop tumors during their experimental life span, and the corresponding times of tumor occurrences are therefore not available. In this case, the (unknown) experimental occurrence times will still be supposed to exist, with values larger than the total duration of the study. We will assume here that sample $B$ corresponds to the control, whereas data from sample $A$ are generated by rats with a diet composed of 11% of a genetically modified (GM) maize (see, e.g., [SER 12]). There is no observation beyond 721 days, and, for this reason, as mentioned earlier, part of the data is not explicitly available. Table 10.1 gives the order statistics and pooled sample ranks of each observation. Since the four largest observations of the pooled sample are not observed and come from sample $A$, there is no loss of generality to define the ranks of these extreme data points to be equal to $17, 18, 19$ and $20$, respectively. This allows us to use, on this data set, statistical methods for which the total rank statistic should normally be needed (the Kolmogorov–Smirnov and Wilcoxon–Mann–Whitney tests).

| sample $A$ | sample $B$ | Ranks $A$ | Ranks $B$ |
|---|---|---|---|
| $X_{1,10} = 388$ | $Y_{1,10} = 427$ | $r_1 = 1$ | $s_1 = 2$ |
| $X_{2,10} = 490$ | $Y_{2,10} = 518$ | $r_2 = 3$ | $s_2 = 6$ |
| $X_{3,10} = 500$ | $Y_{3,10} = 546$ | $r_3 = 4$ | $s_3 = 9$ |
| $X_{4,10} = 507$ | $Y_{4,10} = 637$ | $r_4 = 5$ | $s_4 = 12$ |
| $X_{5,10} = 532$ | $Y_{5,10} = 714$ | $r_5 = 7$ | $s_5 = 15$ |
| $X_{6,10} = 542$ | $Y_{6,10} = 717$ | $r_6 = 8$ | $s_6 = 16$ |
| $X_{7,10} = 588$ | $Y_{7,10} > 720$ | $r_7 = 10$ | $s_7 = 17$ |
| $X_{8,10} = 609$ | $Y_{8,10} > 720$ | $r_8 = 11$ | $s_8 = 18$ |
| $X_{9,10} = 689$ | $Y_{9,10} > 720$ | $r_9 = 13$ | $s_9 = 19$ |
| $X_{10,10} = 696$ | $Y_{10,10} > 720$ | $r_{10} = 14$ | $s_{10} = 20$ |

**Table 10.1.**

We denote by $X_{1,n} < \ldots < X_{n,n}$ (respectively, $Y_{1,n} < \ldots < Y_{n,n}$) the order statistics of $X_1, \ldots, X_n$ (respectively, $Y_1, \ldots, Y_n$). We set $r_1, \ldots, r_n$ (respectively, $s_1, \ldots, s_n$) for the ranks of $X_1, \ldots, X_n$ (respectively, $Y_1, \ldots, Y_n$) within the pooled sample $X_1, \ldots, X_n, Y_1, \ldots, Y_n$. It will be convenient to set $U_i = F(X_i)$ and $V_i = G(Y_i)$ for $i = 1, \ldots, n$, so that $U_1, \ldots, U_n$ and $V_1, \ldots, V_n$ constitute independent random samples of a uniform distribution on $(0, 1)$. The corresponding order statistics, denoted by $0 < U_{1,n} < \ldots < U_{n,n} < 1$ and $0 < V_{1,n} < \ldots < V_{n,n} < 1$, fulfill, with probability 1, $U_{i,n} = F(X_{i,n})$ and $V_{i,n} = G(Y_{i,n})$, for $i = 1, \ldots, n$.

### 10.2.2. *An application of the Wilcoxon–Mann–Whitney statistics*

The Wilcoxon $W$ and Mann–Whitney $U$ statistics are readily evaluated from the data in Table 10.1. We compute, successively, for $n = 10$,

$$
\begin{aligned}
W_1 &:= \sum_{i=1}^{n} s_i = 134,\ W_2 := \sum_{i=1}^{n} r_i = 76 = \frac{2n(2n+1)}{2} - W_1, \\
U_1 &:= n^2 + \frac{n(n+1)}{2} - W_1 = 79,\ U_2 := n^2 + \frac{n(n+1)}{2} - W_2 = 21, \\
\widehat{z} &:= \frac{W_1 - W_2}{n\sqrt{(2n+1)/3}} \simeq 2.192194.
\end{aligned}
$$

It is noteworthy that if there are no ties in the observations (which is the case for the data set of Table 10.1), the Wilcoxon [WIL 45] $W$-test, rejecting $H_0$ when

$W := \min\{W_1, W_2\} \leq w$ for a suitable $w$, and the Mann–Whitney [MAN 47] $U$-test, rejecting $H_0$ for $U := \max\{W_1, W_2\} \leq u$, are equivalent. Presently, by either of these tests, the assumption $H_0$ is rejected at a significance level of 5% if $U \leq 27$. For the levels 2.5% and 1%, the critical regions are $U \leq 23$ and $U \leq 19$, respectively. Therefore, the $U$-test rejects $H_0$ at a significance level $\alpha$ between 1% and 2.5%. Since $\widehat{z}$ is close to a standard normal $N(0,1)$ law with DF $\Phi(z) = (2\pi)^{-1/2} \int_z^{\infty} \exp(-t^2/2)dt$, we can estimate the smallest $\alpha > 0$ for which this test rejects $H_0$ by $\widehat{\alpha} = 1 - \Phi(\widehat{z}) \simeq 1.418275\%$. This result is in agreement with the exact probability levels obtained by tabulation. According to [HOL 73, Table A.5, p. 282], (see, e.g., [KRA 68, Table B], and [BUC 06, BUC 69]), we have $\mathbb{P}(U_2 \leq 21|H_0) = \mathbb{P}(W_1 \leq 134|H_0) \simeq 1.40\%$.

### 10.2.3. *One-sided Kolmogorov-Smirnov tests*

Denote by $F_{n_A}(x) := n_A^{-1}\#\{X_i \leq x : 1 \leq i \leq n_A\}$ and $G_{n_B}(x) := n_B^{-1}\#\{Y_i \leq x : 1 \leq i \leq n_B\}$ the empirical DFs of samples $A$ and $B$. In the setup of [SER 12], $n_A = n_B = n$, and $nF_n(x)$ (respectively, $nG_n(x)$) denote the number of tumors observed prior to time $x$, within the control group (sample $A$) (respectively within the group fed with the GM maize (sample $B$)). Some classical tests of $H_0$ against $H_1$ are based on the one-sided *Kolmogorov–Smirnov statistics* $D^{\pm}_{n_A,n_B} := \sup_{x\in\mathbb{R}} \pm\{F_{n_A}(x) - G_{n_B}(x)\}$. Under $H_0$, the exact distribution of $D^{\pm}_{n_A,n_B}$ is known for arbitrary $n_A$ and $n_B$, but it involves, in general, tedious recursions (see, e.g., [HOD 58] and [DUR 73, pp. 42–43]). For $n_A = n_B = n$, the results simplify to some extent, and we will limit ourselves below to this case. We may write here:

$$D^{+}_{n,n} = \sup_{x\in\mathbb{R}}\{F_n(x) - G_n(x)\} = \max\left\{\max_{1\leq i\leq n}(2i - r_i), \max_{1\leq i\leq n}(s_i - 2i)\right\},$$

$$D^{-}_{n,n} = \sup_{x\in\mathbb{R}}\{G_n(x) - F_n(x)\} = \max\left\{\max_{1\leq i\leq n}(r_i - 2i), \max_{1\leq i\leq n}(2i - s_i)\right\}.$$

For tests of $H_0$ versus $H_1$, one makes use of critical regions, either of the form $\{nD^{+}_{n,n} \geq c\}$ or of the form $\{nD^{-}_{n,n} \leq d\}$, for appropriate choices of $c$ and $d$. Under $H_0$, $D^{+}_{n,n}$ and $D^{-}_{n,n}$ are identically distributed. Their common DF is given, for integer $c$, by (see, e.g., [GNE 51])

$$\mathbb{P}\left(nD^{\pm}_{n,n} \geq c\right) = 1 - \mathbb{P}\left(nD^{\pm}_{n,n} \leq c+1\right)$$
$$= \begin{cases} 1 & \text{if } c \leq 0, \\ \dfrac{\binom{2n}{n-c}}{\binom{2n}{n}} = \dfrac{(n!)^2}{(n-c)!(n+c)!} & \text{if } 1 \leq c \leq n, \\ 0 & \text{if } c > n. \end{cases} \qquad [10.1]$$

The following propositions are useful for numerical evaluations of $\mathbb{P}\left(nD^{\pm}_{n,n} \geq c\right)$ for large $n$.

PROPOSITION 10.1.– We have, for each integer $c \in \{0, \ldots, n\}$, the inequalities

$$\left(1+\frac{c}{n}\right)^{-(n+c)}\left(1-\frac{c}{n}\right)^{-(n-c)} \leq \mathbb{P}\left(nD^{\pm}_{n,n} \geq c\right)$$
$$\leq \left(1+\frac{c}{n+1}\right)^{-(n+1+c)}\left(1-\frac{c}{n+1}\right)^{-(n+1-c)}. \qquad [10.2]$$

PROOF.– [10.2] is trivial for $c = 0$. For $1 \leq c \leq n$, we may rewrite [10.1] into

$$\mathbb{P}\left(nD^{\pm}_{n,n} \geq c\right) = \prod_{k=1}^{c}\left(1-\frac{c}{n+k}\right) = \exp\left[\sum_{k=1}^{c}\log\left(1-\frac{c}{n+k}\right)\right] := \exp(-R_n).$$

Since the function $x \in [0,\infty) \to -\log(1-c/(n+x)) > 0$ is decreasing, we have

$$Q_{n+1} := -\int_1^{c+1}\log\left(1-\frac{c}{n+x}\right)\mathrm{d}x \leq R_n = -\sum_{k=1}^{c}\log\left(1-\frac{c}{n+k}\right)$$
$$\leq Q_n := -\int_0^{c}\log\left(1-\frac{c}{n+x}\right)\mathrm{d}x. \qquad [10.3]$$

Now, it is readily checked that, for $1 \leq c \leq n$,

$$Q_{n+1} \leq Q_n = (n+c)\log\left(1+\frac{c}{n}\right) + (n-c)\log\left(1-\frac{c}{n}\right), \qquad [10.4]$$

from which [10.2] is straightforward.

PROPOSITION 10.2.– We have, for each integer $c \in \{0, \ldots, n\}$, the inequalities:

$$\exp\left(-\frac{c^2}{n}-\frac{2c^4}{5n^3}\right) \leq \mathbb{P}\left(nD^{\pm}_{n,n} \geq c\right) \leq \exp\left(-\frac{c^2}{n+1}\right). \qquad [10.5]$$

PROOF.– Some calculus shows that, for $0 \leq c \leq n$ and $n \geq 1$, with $Q_n$ as in [10.4],

$$\frac{c^2}{n} \leq Q_n = \frac{c^2}{n} + \sum_{k=2}^{\infty}\frac{c^{2k}}{k(2k-1)n^{2k-1}} \leq \frac{c^2}{n} + \frac{c^4}{n^3}\sum_{k=2}^{\infty}\frac{1}{k(2k-1)}$$
$$= \frac{c^2}{n} + \frac{c^4}{n^3}(2\log 2 - 1) < \frac{c^2}{n} + \frac{2c^4}{5n^3},$$

where we have used the inequality $2 \log 2 - 1 \simeq 0.386294 < 0.4 = 2/5$. Given these inequalities, we readily infer [10.5] from [10.3] and [10.4].

COMMENT 10.1.– Making use of [10.5] it is readily checked that, for each $\lambda > 0$, $\mathbb{P}(n^{1/2} D^{\pm}_{n,n} \geq \lambda) \to \exp(-\lambda^2)$ as $n \to \infty$. In the practical example of the data of Table 10.1, where $n = 10$, the bounds in [10.2] and [10.5] turn out to be rather coarse, and it is necessary to use the exact formula [10.1] to properly evaluate the level of significance of tests based upon $D^{\pm}_{n,n}$. In particular, for $n = 10$ and $c = 6$, we infer from [10.2] that $\mathbb{P}(nD^{\pm}_{n,n} \geq c) \simeq 2.62238\%$ is in the interval $[2.11758\%, 3.14889\%]$, whereas [10.5] yields the broader interval $[1.62705\%, 3.79029\%]$. In practice, the numerical of the exact value of $\mathbb{P}(nD^{\pm}_{n,n} \geq c)$ is easily achieved by using a mathematical program in combination with [10.1].

EXAMPLE 10.1.– Set $n = 10$. We see that, for this value of $n$,

$$\mathbb{P}\left(nD^{+}_{n,n} \geq 5\right) \simeq 8.39161\% \quad \text{and} \quad \mathbb{P}\left(nD^{+}_{n,n} \geq 6\right) \simeq 2.62238\%,$$

$$\mathbb{P}\left(nD^{-}_{n,n} \leq 0\right) \simeq 9.09091\% \quad \text{and} \quad \mathbb{P}\left(nD^{-}_{n,n} \leq 1\right) \simeq 31.81818\%.$$

Therefore, an application of the Kolmogorov–Smirnov test, based on the statistic $D^{+}_{n,n}$, allows us to reject $H_0$ at the level $\alpha \simeq 2.62238\%$, whenever, for some $x$, we observe the event $nF_n(x) - nG_n(x) \geq 6$. On the other hand, if we seek to test $H_0$ against $H_1$ by rejecting $H_0$ for small values of $D^{-}_{n,n}$, we need to have, for all $x$, $nF_n(x) - nG_n(x) \geq 0$. If so, we should reject $H_0$ with level $\alpha \simeq 9.09091\%$.

For the data in Table 10.1, we obtain $nD^{+}_{n,n} = 6$ and $nD^{-}_{n,n} = 0$. Therefore, the test based on $D^{+}_{n,n}$ rejects $H_0$ at the level of significance of 2.62238%. The test based on $D^{-}_{n,n}$ reaches the same conclusion, but here, the (best possible) level of significance is 9.09091%, and this last test appears not to be sensitive enough for the sample size $n = 10$, which we consider.

#### 10.2.4. *Precedence and Exceedance Tests.*

We now concentrate on the *exceedance* (or *precedence*) tests (see, e.g., [NEL 63, NEL 93], [BAL 06]). To describe these tests, it is convenient to denote by $X_{1,n} < \ldots < X_{n,n}$ (respectively, $Y_{1,n} < \ldots < Y_{n,n}$) the order statistics of $X_1, \ldots, X_n$ (respectively, $Y_1, \ldots, Y_n$). For a specified integer value of $k \geq 1$, the $k$th precedence statistic $N_k$ is defined as the number of observations from the $X$-sample that precede $Y_{k,n}$. Namely,

$$N_k := nF_n(Y_{k,n}) = k + nF_n(Y_{k,n}) - nG_n(Y_{k,n}) \leq k + nD^{+}_{n,n}. \qquad [10.6]$$

PROPOSITION 10.3.– Under $H_0$, the DF of $N_k$ is given by, for each integer $j = 1, \ldots, n$,

$$\mathbb{P}(N_k \geq j) = \frac{1}{\binom{2n}{n}} \sum_{\ell=0}^{k-1} \binom{\ell + j - 1}{\ell} \binom{2n - j - \ell}{n - j}$$

$$= \frac{1}{\binom{2n}{n}} \sum_{i=j}^{n} \binom{k + i - 1}{k - 1} \binom{2n - i - k}{n - k}. \qquad [10.7]$$

PROOF.– The second half of [10.7] follows from the following argument. Under $H_0$, conditional on $V_{k,n} = t$, $N_k$ follows a binomial $\mathrm{B}(n,t)$ distribution. Since $V_{k,n}$ follows a $\beta(k, n-k+1)$ distribution, the unconditional law of $N_k$ is given by

$$P(N_k = i) = \int_0^1 \left\{ \binom{n}{j} t^i (1-t)^{n-i} \right\} \frac{t^{k-1}(1-t)^{n-k}}{\beta(k, n-k+1)} \, \mathrm{d}t$$

$$= \binom{n}{j} \frac{\beta(k+i, 2n-i-k+1)}{\beta(k, n-k+1)} = \frac{n!(k+i-1)!(2n-i-k)!n!}{j!(n-i)!(2n)!(k-1)!(n-k)!}$$

$$= \frac{1}{\binom{2n}{n}} \binom{k+i-1}{k-1} \binom{2n-i-k}{n-i}. \qquad [10.8]$$

We obtain the second half of [10.7] by summation of the probabilities in [10.8] for $i = j, \ldots, n$. An alternative derivation of $P(N_k = i)$ is as follows. The event $N_k = i$ is obtained when $i$ items from sample $A$ and $k-1$ items from sample $B$ are set on the $k-1$ lower ranks, 1 item from sample $B$ is set on rank $k$, and the $n - i$ remaining items from sample $A$ are set on the $2n - i - k$ remaining ranks. The number of distinct ways of obtaining this combination, out of the $\binom{2n}{n}$ possible cases, is therefore equal to

$$\binom{k+i-1}{k-1} \binom{2n-i-k}{n-i},$$

which readily implies [10.8]. A similar argument applies for the event $N_k \geq j$, for $j \geq 1$, which is obtained when $j-1$ items from sample $A$ and $0 \leq \ell \leq k-1$ items from sample $B$ are set on the $\ell + j - 1$ lower ranks, 1 item from sample $A$ is set on the $k$-th rank, and the $n-j$ remaining items from sample $A$ are set on the $2n - j - \ell$ remaining ranks. The number of distinct ways of obtaining this combination, out of the $\binom{2n}{n}$ possible cases, is therefore equal to

$$\sum_{\ell=0}^{k-1} \binom{\ell + j - 1}{\ell} \binom{2n - j - \ell}{n - j}.$$

Therefore, we obtain the formula:

$$P(N_k \geq j) = \begin{cases} \frac{1}{\binom{2n}{n}} \sum_{\ell=0}^{k-1} \binom{\ell+j-1}{\ell}\binom{2n-j-\ell}{n-j} & \text{for } \ 1 \leq j \leq n, \\ 1 & \text{for } \ j = 0 \end{cases} \qquad [10.9]$$

which, when combined with [10.8], yields [10.7].

COMMENT 10.2.–

1) Balakrishnan and Frattina [BAL 00] investigate the power of the precedence for various choices of $k$, with respect to location-shift (slippage) alternatives. They comment that *for small sample sizes such as* $n = 10$, $15$ *or* $20$, *the precedence test with* $k = 1$ *gives higher power than the precedence tests with higher values of* $k$. Moreover, they observe that, in several cases, *the precedence test is more powerful than the Wilcoxon rank-sum test.* For small values of $k$, the first formula in [10.7] is useful for numerical purposes, since it requires only the summation of $k - 1$ terms. The following particular cases are of significance.

2) For $k = 1$, [10.9] reduces to

$$\mathbb{P}(N_1 \geq j) = \frac{\binom{2n-j}{n-j}}{\binom{2n}{n}} = \prod_{i=0}^{j-1} \left(\frac{n-i}{2n-i}\right) \quad \text{for } \ j = 0, \ldots, n, \qquad [10.10]$$

where we use the convention that $\prod_{\emptyset}(\cdot) = 1$.

3) For $k = 2$, [10.9] reduces to

$$\mathbb{P}(N_2 \geq j) = \left\{1 + \frac{jn}{2n-j}\right\} \frac{\binom{2n-j}{n-j}}{\binom{2n}{n}} \qquad [10.11]$$

$$= \left\{1 + \frac{jn}{2n-j}\right\} \prod_{i=0}^{j-1} \left(\frac{n-i}{2n-i}\right), \qquad [10.12]$$

for $j = 1, \ldots, n$.

4) For $k = 3$, [10.9] reduces to

$$\mathbb{P}(N_3 \geq j) = \left\{1 + \frac{jn}{2n-j} + \frac{j(j+1)n(n-1)}{2(2n-j)(2n-j-1)}\right\} \frac{\binom{2n-j}{n-j}}{\binom{2n}{n}}$$

$$= \left\{1 + \frac{jn}{2n-j} + \frac{j(j+1)n(n-1)}{2(2n-j)(2n-j-1)}\right\} \prod_{i=0}^{j-1} \left(\frac{n-i}{2n-i}\right), \qquad [10.13]$$

for $j = 1, \ldots, n$.

EXAMPLE 10.2.– For $n = 10$, we obtain, in agreement with Table 3.1, p. 358, of [BAL 00], the following tabulation.

| $j$ | $\mathbb{P}(N_1 \geq j)$ | $\mathbb{P}(N_2 \geq j)$ | $\mathbb{P}(N_3 \geq j)$ |
|---|---|---|---|
| 0 | 100.000% | 100.000% | 100.000% |
| 1 | 50.0000% | 76.3158% | 89.4727% |
| 2 | 23.6842% | 50.0000% | 70.8978% |
| 3 | 10.5263% | 29.1022% | 50.0000% |
| 4 | 4.33437% | 15.1703% | 31.4291% |
| 5 | 1.62539% | 7.04334% | 17.4293% |
| 6 | 0.54180% | 2.86378% | 8.49012% |
| 7 | 0.15480% | 0.98833% | 3.48893% |
| 8 | 0.03572% | 0.27387% | 1.15071% |
| 9 | 0.00595% | 0.05467% | 0.27387% |
| 10 | 0.00054% | 0.00595% | 0.03572% |

**Table 10.2.**

For the data of Table 10.1, we obtain the statistics $N_1^* = 6, N_2^* = 6, N_3^* = 7$, which are significative, respectively, at the levels $\alpha = 0.54\%, 2.86\%, 3.49\%$. We see, therefore, that the statistic based on $N_1^*$ is the most significative among all the tests we have considered up to now. On the other hand, significance is less clear with respect to $N_1 = 1, N_2 = 4, N_3 = 6$, which have respective levels of significance given by $\alpha = 50\%, 15.17\%, 8.49\%$. Even though $N_3$ is weakly significant, we conclude that, in the present setup, exceedance tests are much more sensitive than precedence tests.

## 10.3. Negative binomial limit laws

In this section, we start with a derivation of the negative binomial distribution, which is of interest in and of itself. For $i = 1, 2$, let $\{N_i(t) : t \geq 0\}$ denote two independent right-continuous Poisson processes, such that $\mathbb{E}(N_i(t)) = \lambda_i t$ for $t \geq 0$, with $\lambda_i > 0$. Denote, respectively, the arrival times of these two processes, by $0 < \tau_i(1) < \tau_i(2) < \ldots$, $i = 1, 2$, and observe that, for $i = 1, 2$ and each $k \geq 1$, the distribution of $\tau_i(k)$ is $\Gamma(k, \lambda_i)$, with density

$$f_{i;k}(t) = \frac{\lambda_i^k t^{k-1}}{(k-1)!} \exp(-\lambda_i t) \quad \text{for} \quad t > 0.$$

For $k \geq 1$, set now $\nu_k = \nu_k(\lambda_1, \lambda_2) := N_1(\tau_2(k))$, for the number of arrival times of $N_1(\cdot)$ falling below the $k$th arrival times $\tau_2(k)$ of $N_2(\cdot)$. The following proposition gives the distribution of $\nu_k$.

PROPOSITION 10.4.– We have, for each integer $k \geq 1$ and $j = 0, 1, \ldots$,

$$\mathbb{P}(\nu_k = j) = \binom{k+j-1}{k-1}\Big\{\frac{\lambda_1}{\lambda_1+\lambda_2}\Big\}^j\Big(1-\Big\{\frac{\lambda_1}{\lambda_1+\lambda_2}\Big\}\Big)^k, \qquad [10.14]$$

and

$$\begin{aligned}\mathbb{P}(\nu_k \geq j) &= \sum_{\ell=0}^{k-1}\binom{j+\ell-1}{j-1}\Big\{\frac{\lambda_1}{\lambda_1+\lambda_2}\Big\}^j\Big(1-\Big\{\frac{\lambda_1}{\lambda_1+\lambda_2}\Big\}\Big)^\ell \\ &= \sum_{\ell=j}^{\infty}\binom{k+\ell-1}{k-1}\Big\{\frac{\lambda_1}{\lambda_1+\lambda_2}\Big\}^\ell\Big(1-\Big\{\frac{\lambda_1}{\lambda_1+\lambda_2}\Big\}\Big)^k. \qquad [10.15]\end{aligned}$$

PROOF.– Since, for $t \geq 0$, $N_1(t)$ follows a Poisson distribution with expectation $\lambda_1 t$, we have, for $j = 0, 1, \ldots$,

$$\begin{aligned}\mathbb{P}(\nu_k = j) &= \int_0^\infty \mathbb{P}\left(N_1(t) = j \mid \tau_2(k) = t\right) \mathrm{d}\mathbb{P}(\tau_2(k) \leq t) \\ &= \int_0^\infty \Big\{\frac{\lambda_1^j t^j}{j!}\exp(-\lambda_1 t)\Big\}\Big\{\frac{\lambda_2^k t^{k-1}}{(k-1)!}\exp(-\lambda_2 t)\Big\}\mathrm{d}t \\ &= \Big\{\frac{(k+j-1)!}{j!(k-1)!}\Big\}\frac{\lambda_1^j\lambda_2^k}{(\lambda_1+\lambda_2)^{j+k}}, \qquad [10.16]\end{aligned}$$

from which [10.14] is straightforward. The second half of [10.15] is then readily inferred from [10.14] by writing that $\{\nu_k \geq j\} = \bigcup_{\ell=j}^{\infty}\{\nu_k = \ell\}$. Next, we infer from [10.16], after integrating by parts that

$$\begin{aligned}\mathbb{P}(\nu_k \geq j) &= \int_0^\infty \Big\{\sum_{i=j}^\infty \frac{\lambda_1^i t^i}{i!}\exp(-\lambda_1 t)\Big\}\Big\{\frac{\lambda_2^k t^{k-1}}{(k-1)!}\exp(-\lambda_2 t)\Big\}\mathrm{d}t \\ &= \int_0^\infty \Big\{1 - \int_{\lambda_1 t}^\infty \frac{x^{j-1}e^{-x}}{(j-1)!}\mathrm{d}x\Big\}\frac{\mathrm{d}}{\mathrm{d}t}\Big\{-\int_{\lambda_2 t}^\infty \frac{x^{k-1}e^{-x}}{(k-1)!}\mathrm{d}x\Big\} \\ &= \int_0^\infty \frac{\lambda_1^j t^{j-1}}{(j-1)!}\exp(-\lambda_1 t)\Big\{\sum_{\ell=0}^{k-1}\frac{\lambda_2^\ell t^\ell}{\ell!}\exp(-\lambda_2 t)\Big\}\mathrm{d}t \\ &= \sum_{\ell=0}^{k-1}\Big\{\frac{(j+\ell-1)!}{(j-1)!\ell!}\Big\}\frac{\lambda_1^j\lambda_2^\ell}{(\lambda_1+\lambda_2)^{j+\ell}},\end{aligned}$$

which yields the first half of [10.15].

COMMENT 10.3.–

1) The random variable $\nu_k$ follows a *negative binomial distribution* $\mathrm{NB}(k,p)$ (see, e.g., [JOH 92, p.199]), with parameters $k \geq 1$ and $p := \lambda_1/(\lambda_1+\lambda_2)$. The distribution of a random variable $Z$ with a negative binomial distribution $\mathrm{NB}(k,p)$, with $0 < p < 1$, is given by

$$\mathbb{P}(Z=j) = \binom{k+j-1}{k-1} p^j (1-p)^k.$$

By a straightforward extension of [10.16], we obtain that the $\mathrm{NB}(k,p)$ distribution fulfills the following relation, which is new to our best knowledge (see, e.g., the discussion in [JOH 92, pp. 209–210]).

$$\mathbb{P}(Z \geq j) = \sum_{\ell=0}^{k-1} \binom{j+\ell-1}{j-1} p^j (1-p)^\ell \quad \text{for} \quad j = 0,1,\dots . \qquad [10.17]$$

2) The distribution of $\nu_k(\lambda_1,\lambda_2)$ is unchanged if $(\lambda_1,\lambda_2)$ is replaced by $(\rho\lambda_1,\rho\lambda_2)$.

In the next proposition, we show that, under $H_0$, the large sample limiting distribution of $N_k$ (or, equivalently, of $N_k^*$) is *negative binomial*. This result will be established with samples of unequal sizes. Namely, we consider i.i.d. observations $X_1,\dots,X_{n_A}$ and $Y_1,\dots,Y_{n_B}$ from the same continuous distribution. As in the introduction, for $1 \leq k \leq n_B$, we define $N_k = N_k(n_A,n_B)$ as the number of observations from $X_1,\dots,X_{n_A}$ falling below the $k$-th observation from $Y_1,\dots,Y_{n_B}$.

THEOREM 10.1.– Let $n_A \geq 1$ and $n_B \geq 1$ vary in such a way that $n_A \wedge n_B \to \infty$ and $n_A/(n_A+n_B) \to p \in (0,1)$. Then, for each $k \geq 1$, $N_k(n_A,n_B)$ converges to a limiting negative binomial $\mathrm{NB}(k,p)$ distribution, and fulfills

$$\mathbb{P}(N_k = j) \quad = \quad \binom{k+j-1}{k-1} p^j (1-p)^k \quad \text{for} \quad j = 0,1,\dots , \qquad [10.18]$$

and

$$\mathbb{P}(N_k \geq j) \quad = \quad \sum_{\ell=0}^{k-1} \binom{j+\ell-1}{j-1} p^j (1-p)^\ell . \qquad [10.19]$$

PROOF.– Let $N_A$ (respectively, $N_B$) follow a Poisson distribution with expectation $n_A$ (respectively, $n_B$). Denote by $Y \stackrel{d}{=} Z$ the property that $X, Y$ are identically distributed. In view of Comment 10.3, it is straightforward that

$$N_k(N_A,N_B) \stackrel{d}{=} \nu_k(N_A,N_B) \stackrel{d}{=} \nu_k(N_A/(N_A+N_B), N_B/(N_A+N_B)).$$

Now, the conclusion [10.18] follows directly from the observation that, under the assumptions of the theorem, $N_A/(N_A+N_B) \to p$ in probability.

COMMENT 10.4.–

1) For samples with equal sizes, with $n_A = n_B = n$, we readily infer from [10.10]–[10.13] that, for $j = 0, 1, \ldots$, as $n \to \infty$:

$$\mathbb{P}(N_1 \geq j) \to \frac{1}{2^j}, \quad \mathbb{P}(N_2 \geq j) \to \left(1 + \frac{j}{2}\right)\frac{1}{2^j},$$

$$\mathbb{P}(N_3 \geq j) \to \left(1 + \frac{j}{2} + \frac{j(j+1)}{8}\right)\frac{1}{2^j}. \qquad [10.20]$$

It is readily checked that these results are a special case of [10.18].

2) The fact that $\mathbb{P}(N_1 \geq j) \to \frac{1}{2^j}$ as $n \to \infty$ has been observed by Rosenbaum [ROS 54].

COMMENT 10.5.– We return to the data of Seralini *et al.* [SER 12] given in Table 10.1. We have implicitly assumed in the arguments above that the probability that a specified animal develops multiple tumors should be neglected. We will now show that we obtain similar conclusions if we omit this simplifying assumption. The appropriate model is, in this case, to consider the successive order statistics of sample $A$ as the arrival times of a Poisson process $\{N_A(t) : 0 \leq t \leq T\}$, and, likewise, the successive order statistics of sample $B$ as the arrival times of the Poisson process $\{N_B(t) : 0 \leq t \leq T\}$. Here, $T = 721$ is the duration of the experiment. The hypothesis $H_0$ corresponds in this case to the condition that $\mathbb{E}(N_A(t)) = \mathbb{E}(N_B(t)) = \Lambda(t)$ for all $0 \leq t \leq T$, where $\Lambda(\cdot)$ denotes a continuous increasing function on $[0, T]$. Now, if we denote by $\mathcal{N}_5$ the number of arrival times from $N_A(\cdot)$ below the fifth observed arrival time from $N_B(\cdot)$, we obtain from Table 10.1 the value $\mathcal{N}_5 = 10$. A straightforward computation based on [10.18] shows that

$$\begin{aligned}\mathbb{P}(\mathcal{N}_5 \geq 10) &= \frac{1}{2^{10}}\left[1 + \frac{10}{2} + \frac{11 \times 10}{2 \times 2^2} + \frac{12 \times 11 \times 10}{6 \times 2^3}\right. \\ &\quad \left. + \frac{13 \times 12 \times 11 \times 10}{24 \times 2^4}\right] \simeq 8.97827\%.\end{aligned}$$

It turns out that the limit significance levels obtained under this limiting Poisson model are less stringent than those which follow from the finite sample case. A more detailed analysis of this data set under either model will be discussed elsewhere.

## 10.4. Conclusion

The application of the testing methods we have considered for the data of [SER 12], in Table 10.1, clearly shows that *the assumption $H_0$ is rejected, with varying levels of significance, by most of the tests we have considered.* Whereas the classical tests of Kolmogorov–Smirnov (one-sided) and Wilcoxon–Mann–Whitney reach $\alpha$ values of significance of 1.40% and 2.62%, respectively, the most sensitive test turns out to be the Westlake [WES 71] exceedance test, based on the statistic $N_1^*$, which yields $\alpha \simeq 0.54\%$. This last result clearly demonstrates the practical interest of exceedance (or precedence) tests, which turn out, at times, to be much more sensitive than the classical Mann–Whitney–Wilcoxon test. We refer to [BAL 06] for further details on these methods.

## 10.5. Bibliography

[BAL 00] BALAKRISHNAN N., FRATTINA R., "Precedence test and maximal precedence test", in LIMNIOS N., NIKULIN M. (eds), *Recent Advances in Reliability Theory*, Birkhäuser, Boston, pp. 355–378, 2000.

[BAL 06] BALAKRISHNAN N., TONY NG H.K., *Precedence-Type Tests and Applications*, Wiley, New York, 2006.

[BUC 69] BUCKLE N., KRAFT C., VAN EEDEN C., "An approximation to the Wilcoxon–Mann–Whitney distribution", *Journal of the American Statistical Association*, vol. 64, pp. 591–599, 1969.

[BUC 06] BUCKLE N., KRAFT C., VAN EEDEN C., *Extended Tables of the Wilcoxon–Mann–Whitney Distribution*, University Press of Montreal, Montreal, Canada, 2006.

[CON 69] CONOVER W.J., "Two $k$-sample slippage tests", *Journal of the American Statistical Association*, vol. 63, pp. 614–626, 1969.

[DUR 73] DURBIN J., *Distribution Theory for Tests Based on the Sample Distribution Function*, SIAM, Philadelphia, 1973.

[GNE 51] GNEDENKO B.V., KOROLYUK V.S., "On the maximum discrepancy between two empirical distributions", *Doklady Akademya Nauk*, vol. 80, pp. 525–528, 1951. [In Russian]

[HAG 59] HAGA T., "A two-sample rank test for location", *Annals of the Institute of Statistical Mathematics*, vol. 11, pp. 211–219, 1959.

[HOD 58] HODGES J.L., "The significance probability of the Smirnov two-sample test", *Arkiv für Mathematik*, vol. 3, pp. 469–486, 1958.

[HOL 73] HOLLANDER M., WOLFE D.A., *Nonparametric Statistical Methods*, Wiley, New York, 1973.

[JOH 92] JOHNSON N.L., KOTZ S., KEMP A.W., *Univariate Discrete Distributions*, Wiley, New York, 1992.

[KAM 56] KAMAT A.R., "A two-sample distribution-free test", *Biometrika*, vol. 43, pp. 377–387, 1956.

[KRA 68] KRAFT C.H., VAN EEEDEN C., *A Nonparametric Introduction to Statistics*, Macmillan, New York, 1968.

[MAN 47] MANN H.B., WHITNEY D.R., "On a test of whether one of the two random variables is stochastically larger than the other", *Annals of Mathematical Statistics*, vol. 18, pp. 50–60, 1947.

[MOS 48] MOSTELLER F., "A $k$-sample slippage test for an extreme population", *Annals of the Institute of Statistical Mathematics*, vol. 19, pp. 58–65, 1948.

[NEA 72] NEAVE H.R., "Some quick tests for slippage", *The Statistician*, vol. 21, pp. 197–208, 1972.

[NEL 63] NELSON L.S., "Tables of a precedence life test", *Technometrics*, vol. 5, pp. 491–499, 1963.

[NEL 93] NELSON L.S., "Tests on early failures - the precedence life test", *Journal of Quality Technology*, vol. 25, pp. 140–142, 1993.

[ROS 53] ROSENBAUM S., "Tables of a nonparametric test of dispersion", *Annals of the Institute of Statistical Mathematics*, vol. 24, pp. 663–668, 1953.

[ROS 54] ROSENBAUM S., "Tables of a nonparametric test of dispersion", *Annals of the Institute of Statistical Mathematics*, vol. 25, pp. 146–150, 1954.

[SAK 82] SAKS L., *Applied Statistics*, Springer, New York, 1982.

[SER 12] SERALINI G.E., CLAIR E., MESNAGE R., *et al.*, "Long term toxicity of a Roundup herbicide and a Roundup-tolerant genetically modified maize", *Food and Chemical Toxicology*, vol. 50, pp. 4221–4231, 2012.

[TUK 59] TUKEY J.W., "A quick, compact, two-sample test to Duckworth's specifications", *Technometrics*, vol. 1, pp. 31–48, 1959.

[WES 71] WESTLAKE W.J., "A one-sided version of the Tukey–Duckworth test", *Technometrics*, vol. 13, pp. 901–903, 1971.

[WIL 42] WILKS S.S., "Statistical prediction with special reference to the problem of tolerance limits", *Annals of the Institute of Statistical Mathematics*, vol. 13, pp. 400–409, 1942.

[WIL 45] WILCOXON F., "Individual comparison by ranking methods", *Biometrics*, vol. 1, pp. 80–83, 1945.

# Chapter 11

# Asymptotically Parameter-Free Tests for Ergodic Diffusion Processes

In this chapter, we present a review of the several asymptotically parameter-free (APF) goodness-of-fit (GoF) tests in the case of observations of ergodic diffusion processes. The basic hypothesis is supposed to be composite parametric, i.e. the trend coefficient under hypothesis depends on an unknown parameter. The tests studied are similar to the well-known Cramérvon Mises (C-vM) and Kolmogorov Smirnov (K-S) tests of classical (independent and identically distributed (i.i.d.)) statistics. We show that the limit distributions of the test statistics of all presented tests do not depend on unknown parameter.

## 11.1. Introduction

In this chapter, we will consider the following problem. Suppose that we observe (in continuous time) a solution $X^T = (X_t, 0 \leq t \leq T)$ of the stochastic differential equation

$$\mathrm{d}X_t = S\left(X_t\right)\,\mathrm{d}t + \sigma\left(X_t\right)\,\mathrm{d}W_t, \quad X_0, \quad 0 \leq t \leq T, \qquad [11.1]$$

where $W_t$, $t \geq 0$ is a standard Wiener process, $X_0$ is the initial value and $\sigma\left(x\right)^2 > 0$ is the known diffusion coefficient. We suppose that the functions $S\left(\cdot,\cdot\right)$ and $\sigma\left(\cdot\right)$ are such that this equation has a unique solution and this solution has ergodic properties. We have to check a composite basic hypothesis $\mathcal{H}_0$ that the trend coefficient $S\left(\cdot\right)$

Chapter written by Yury A. KUTOYANTS and Li ZHOU.

belongs to some given parametric family $\mathcal{F} = \{S(\vartheta, x), \vartheta \in \Theta\}$, where $S(\cdot,\cdot)$ is some known function and $\vartheta$ is a parameter unknown to us. Our goal is to construct GoF tests $\hat{\psi}_T$, which have asymptotic $(T \to \infty)$ size $\varepsilon$, where $\varepsilon \in (0,1)$ is some given number. The value $\hat{\psi}_T$ as usual is the probability of rejecting the hypothesis $\mathcal{H}_0$. Therefore, we need the following property: for all $\vartheta \in \Theta$

$$\mathbf{E}_\vartheta \hat{\psi}_T = \varepsilon + o(1). \qquad [11.2]$$

Here and in the following, $\mathbf{E}_\vartheta$ means the mathematical expectation under hypothesis $\mathcal{H}_0$ with the value $\vartheta$ of the parameter.

We present a review of several GoF tests based on statistics, whose limit distributions do not depend on an unknown parameter. We call such tests APF by analogy with the well-known asymptotically distribution free (ADF) tests [LEH 05].

First, we recall similar well-known GoF problems for the model of i.i.d. observations $X^n = (X_1, \ldots, X_n)$ with the distribution function (d.f.) $F(x)$ in the cases of simple and composite basic hypotheses. If the hypothesis is simple, i.e.

$$\mathcal{H}_0 : F(x) = F_0(x),$$

where $F_0(x)$ is some known continuous d.f., then it is possible to construct ADF tests satisfying $(n \to \infty)$

$$\mathbf{E}_0 \hat{\psi}_n = \varepsilon + o(1).$$

Here, $\mathbf{E}_0$ means mathematical expectation under hypothesis $\mathcal{H}_0$. These tests can be defined as follows. Let us introduce the empirical distribution function (EDF) $\hat{F}_n(x) = \frac{1}{n}\sum_{j=1}^{n} 1\!\!1_{\{X_j < x\}}$.

The C-vM $\hat{\Psi}_n = 1\!\!1_{\{\Delta_n > c_\varepsilon\}}$ and K-S $\hat{\psi}_n = 1\!\!1_{\{\Gamma_n > d_\varepsilon\}}$ tests are based on the following two statistics

$$\Delta_n = n\int_{-\infty}^{\infty}\left[\hat{F}_n(x) - F_0(x)\right]^2 \mathrm{d}F_0(x), \qquad \Gamma_n = \sup_x \sqrt{n}\left|\hat{F}_n(x) - F_0(x)\right|.$$

It is known that the normalized difference $B_n(x) = \sqrt{n}\left(\hat{F}_n(x) - F_0(x)\right)$ (under hypothesis $\mathcal{H}_0$) converges in distribution to the stochastic process $B(F_0(x)), x \in R$, where $B(s), 0 \le s \le 1$ is the Brownian bridge, i.e. Gaussian process with $\mathbf{E}B(s) = 0$ and $\mathbf{E}B(s_1)B(s_2) = s_1 \wedge s_2 - s_1 s_2$. This convergence provides the following remarkable properties of the underlying statistics (we change the variables $s = F_0(x)$)

$$\Delta_n \Longrightarrow \Delta = \int_0^1 B(s)^2 \, \mathrm{d}s, \qquad \Gamma_n \Longrightarrow \Gamma = \sup_{0\leq s\leq 1} |B(s),|$$

which make the tests $\hat{\Psi}_n$ and $\hat{\psi}_n$ ADF, because the distributions of $\Delta$ and $\Gamma$ do not depend on the model $F_0(\cdot)$. Because of this convergence, the constants $c_\varepsilon$ and $d_\varepsilon$ can be defined as solutions of the equations

$$\mathbf{P}(\Delta > c_\varepsilon) = \varepsilon, \qquad \mathbf{P}(\Gamma > d_\varepsilon) = \varepsilon.$$

If the basic hypothesis is composite parametric, then the situation essentially changes. Suppose now that we have to check the hypothesis

$$\mathcal{H}_0 : F(x) = F_0(\vartheta, x), \quad \vartheta \in \Theta,$$

where $\Theta = (\alpha, \beta)$, i.e. the d.f. $F(x)$ belongs to a parametric family $\mathcal{F} = \{F_0(\vartheta, x), \quad \vartheta \in \Theta\}$. Let us introduce the statistics similar to $\Delta_n$ and $\Gamma_n$

$$\begin{aligned}\hat{\Delta}_n &= n \int_{-\infty}^{\infty} \left[\hat{F}_n(x) - F_0\left(\hat{\vartheta}_n, x\right)\right]^2 \mathrm{d}F_0\left(\hat{\vartheta}_n, x\right), \\ \hat{\Gamma}_n &= \sup_x \sqrt{n} \left|\hat{F}_n(x) - F_0\left(\hat{\vartheta}_n, x\right)\right|,\end{aligned}$$

where $\hat{\vartheta}_n$ is the maximum likelihood estimator (MLE) of the parameter $\vartheta$. Under conditions of regularity, the MLE is consistent and asymptotically normal. Moreover, it admits the representation

$$\sqrt{n}\left(\hat{\vartheta}_n - \vartheta\right) = \frac{1}{\mathrm{I}(\vartheta)\sqrt{n}} \sum_{j=1}^{n} \dot{\ell}(\vartheta, X_j) + o(1), \quad \mathrm{I}(\vartheta) = \int_{-\infty}^{\infty} \frac{\dot{f}(\vartheta, x)^2}{f(\vartheta, x)} \mathrm{d}x.$$

Here, $\mathrm{I}(\vartheta)$ is the Fisher information, $f(\vartheta, x)$ is the density function, $\ell(\vartheta, x) = \ln f(\vartheta, x)$ and the dot means the derivation with respect to (w.r.t.) $\vartheta$.

It can be shown that $B_n(x) = \sqrt{n}\left(\hat{F}_n(x) - F_0\left(\hat{\vartheta}_n, x\right)\right)$ can be written as

$$\begin{aligned}\hat{B}_n(x) &= \sqrt{n}\left(\hat{F}_n(x) - F_0(\vartheta, x)\right) + \sqrt{n}\left(F(\vartheta, x) - F_0\left(\hat{\vartheta}_n, x\right)\right) \\ &= B_n(x) - \sqrt{n}\left(\hat{\vartheta}_n - \vartheta\right) \dot{F}_0(\vartheta, x) + o(1).\end{aligned}$$

This representation is equivalent to the following

$$\hat{B}_n(x) = B_n(x) - \int_{-\infty}^{\infty} \frac{\dot{\ell}(\vartheta, y)}{\sqrt{\mathrm{I}(\vartheta)}} \, \mathrm{d}B_n(y) \int_{-\infty}^{x} \frac{\dot{\ell}(\vartheta, y)}{\sqrt{\mathrm{I}(\vartheta)}} \, \mathrm{d}F_0(\vartheta, y) + o(1),$$

which provides the limit (in distribution) $\hat{B}_n(x) \Rightarrow \hat{B}(x)$, where

$$\hat{B}(x) = B(F_0(\vartheta, x)) - \int_{-\infty}^{\infty} \frac{\dot{\ell}(\vartheta, y)}{\sqrt{\mathrm{I}(\vartheta)}} \, \mathrm{d}B(F_0(\vartheta, y)) \int_{-\infty}^{x} \frac{\dot{\ell}(\vartheta, y)}{\sqrt{\mathrm{I}(\vartheta)}} \, \mathrm{d}F_0(\vartheta, y).$$

Hence, the statistics $\hat{\Delta}_n$ and $\hat{\Gamma}_n$ have the following limits (under hypothesis $\mathcal{H}_0$)

$$\hat{\Delta}_n \Longrightarrow \hat{\Delta} = \int_{-\infty}^{\infty} \left[\hat{B}(x)\right]^2 \mathrm{d}F_0(\vartheta, x), \qquad \hat{\Gamma}_n \Longrightarrow \hat{\Gamma} = \sup_x \left|\hat{B}(x)\right|.$$

Here, $\hat{B}(x) = \hat{B}(F_0, \vartheta)$. It is easy to see that both limits depend strongly on the model $F_0(\cdot, \cdot)$ and on the unknown parameter $\vartheta$, i.e. $\hat{\Delta} = \hat{\Delta}(F_0, \vartheta)$ and $\hat{\Gamma} = \hat{\Gamma}(F_0, \vartheta)$. Therefore, the corresponding tests $\hat{\Psi}_n = 1\!\!\mathrm{I}_{\{\hat{\Delta}_n > c_\varepsilon\}}$ and $\hat{\psi}_n = 1\!\!\mathrm{I}_{\{\hat{\Gamma}_n > d_\varepsilon\}}$ are no longer ADF and the thresholds $c_\varepsilon$ and $d_\varepsilon$ defined as solutions of the equations

$$\mathbf{P}\left(\hat{\Delta} > c_\varepsilon\right) = \varepsilon, \qquad \mathbf{P}\left(\hat{\Gamma} > d_\varepsilon\right) = \varepsilon$$

depend on $F_0(\cdot, \cdot)$ (known) and $\vartheta$ (unknown), i.e. $c_\varepsilon = c_\varepsilon(F_0, \vartheta)$ and $d_\varepsilon = d_\varepsilon(F_0, \vartheta)$. One way to treat this problem is to show that both functions $c_\varepsilon$ and $d_\varepsilon$ are continuous w.r.t. $\vartheta$ and to use the thresholds $c_\varepsilon(F_0, \hat{\vartheta}_n)$ and $d_\varepsilon(F_0, \hat{\vartheta}_n)$.

For some models of observations, it is possible to show that the limits $\hat{\Delta}$ and $\hat{\Gamma}$ do not depend on $\vartheta$, then the thresholds $c_\varepsilon(F_0)$ and $d_\varepsilon(F_0)$ can be calculated before the experiment, say, by Monte Carlo simulations of the limit distributions.

For example, suppose that the unknown parameter $\vartheta$ is one-dimensional shift, i.e. $f(\vartheta, x) = f_*(x - \theta)$, where $f_*(x)$ is some known smooth function. Then, the d.f. $F(\vartheta, x) = F_*(x - \vartheta)$, the Fisher information $\mathrm{I}(\vartheta) = \mathrm{I}_* > 0$ and so on. In this case, the distributions of the limit statistics $\hat{\Delta}$ and $\hat{\Gamma}$ do not depend on $\vartheta$ and the C-vM and K-S tests are APF. This problem has attracted the attention of statisticians for a long time (see, e.g., theorem 7.1 in [DAR 55] and [DUR 73], [KAC 55] and [LEH 05]).

A similar problem exists, of course, for many other statistical models. For inhomogeneous Poisson process, the corresponding problem was studied by Dabye [DAB 13]. In this chapter, we treat this problem in the case of ergodic diffusion processes and present a review of the several recent results devoted to the construction of APF tests in the cases of shift parameter in the trend coefficient and the shift and scale parameters for a special class of ergodic diffusion processes. We show that the limit distributions of the proposed statistics do not depend on the unknown parameter and therefore the thresholds of the corresponding GoF tests can be chosen before the experiment.

## 11.2. Ergodic diffusion process and some limits

We observe a trajectory $X^T = (X_t, 0 \leq t \leq T)$ of diffusion process

$$\mathrm{d}X_t = S(X_t)\,\mathrm{d}t + \sigma(X_t)\,\mathrm{d}W_t, \quad X_0, \quad 0 \leq t \leq T. \qquad [11.3]$$

We suppose that the functions $S(\cdot)$ and $\sigma(\cdot)^2 > 0$ are locally bounded and satisfy the condition

$$xS(x) + \sigma(x)^2 \leq C\left(1 + x^2\right).$$

By this condition, there exists a unique solution of this equation (see, e.g., [DUR 73]). Moreover, we suppose that $S(\cdot), \sigma(\cdot)^{\pm 1} \in \mathcal{P}$, where $\mathcal{P}$ is a class of functions having polynomial majorants, and

$$\lim_{|x|\to\infty} \operatorname{sgn}(x)\, \frac{S(x)}{\sigma(x)^2} < 0. \qquad [11.4]$$

If the functions $S(\cdot)$ and $\sigma(\cdot)$ satisfy these conditions, then equation [11.3] has a unique solution and this solution has ergodic properties. This means that the invariant density has the following form

$$f(x) = \frac{1}{G\sigma(x)^2} \exp\left\{2\int_0^x \frac{S(y)}{\sigma(y)^2}\,\mathrm{d}y\right\},$$

where $G > 0$ is the normalizing constant, and we have the law of large numbers.

We consider the construction of the GoF tests based on the EDF $\hat{F}_T(x)$ and on the empirical density $\hat{f}_T(x)$. Recall that these two statistics are unbiased, consistent, asymptotically normal and asymptotically efficient estimators of the invariant d.f. $F(x)$ and of the invariant density (see propositions 1.51 and 1.57 in [KUT 04]). These two estimators are defined as follows

$$\hat{F}_T(x) = \frac{1}{T}\int_0^T \mathbb{1}_{\{X_t < x\}}\,\mathrm{d}t, \qquad \hat{f}_T(x) = \frac{\Lambda_T(x)}{T\sigma(x)^2}.$$

Here, $\Lambda_T(x)$ is the local time of the diffusion process. Note that empirical density is also called local time estimator.

Our main problem is to check the following parametric hypothesis

$$\mathcal{H}_0 : S(x) = S(\vartheta, x), \quad \vartheta \in \Theta.$$

Therefore, we suppose that under this hypothesis, the observed diffusion process satisfies the equation

$$\mathrm{d}X_t = S(\vartheta, X_t)\, dt + \sigma(X_t)\, \mathrm{d}W_t, \quad X_0, \quad 0 \le t \le T, \tag{11.5}$$

where $S(\vartheta, x)$ is some known function and $\vartheta \in \Theta = (\alpha, \beta)$ is an unknown parameter. The invariant d.f. and density are $F(\vartheta, x)$ and $f(\vartheta, x)$, respectively.

The corresponding C-vM-type statistics are

$$\begin{aligned}\Delta_T\left(X^T\right) &= T\int_{-\infty}^{\infty}\left[\hat{F}_T(x) - F(\hat{\vartheta}_T, x)\right]^2 \mathrm{d}F(\hat{\vartheta}_T, x),\\ \delta_T\left(X^T\right) &= T\int_{-\infty}^{\infty}\left[\hat{f}_T(x) - f(\hat{\vartheta}_T, x)\right]^2 \mathrm{d}F(\hat{\vartheta}_T, x).\end{aligned}$$

Here, $\hat{\vartheta}_T$ is the MLE of $\vartheta$. Such statistics were introduced in [DAC 07]. It was noted in [KUT 12] that these statistics can be written as follows

$$\begin{aligned}\Delta_T\left(X^T\right) &= \int_0^T\left[\hat{F}_T(X_t) - F(\hat{\vartheta}_T, X_t)\right]^2 \mathrm{d}t,\\ \delta_T\left(X^T\right) &= \int_0^T\left[\hat{f}_T(X_t) - f(\hat{\vartheta}_T, X_t)\right]^2 \mathrm{d}t + o(1).\end{aligned}$$

Indeed, for any integrable function $h(x)$, we have the equality

$$\frac{1}{T}\int_0^T h(X_t)\, \mathrm{d}t = \int_{-\infty}^{\infty} h(x)\, \hat{f}_T(x)\, \mathrm{d}x = \int_{-\infty}^{\infty} h(x)\, \mathrm{d}\hat{F}_T(x).$$

Therefore, if we put $\eta_T\left(\hat{\vartheta}_T, x\right) = \sqrt{T}\left(\hat{F}_T(x) - F(\hat{\vartheta}_T, x)\right)$, then

$$\begin{aligned}\Delta_T\left(X^T\right) &= \frac{1}{\sqrt{T}}\int_{-\infty}^{\infty}\eta_T\left(\hat{\vartheta}_T, x\right)^2 \mathrm{d}\eta_T\left(\hat{\vartheta}_T, x\right)\\ &\quad + T\int_{-\infty}^{\infty}\left[\hat{F}_T(x) - F(\hat{\vartheta}_T, x)\right]^2 \mathrm{d}\hat{F}_T(x) \qquad [11.6]\\ &= \int_0^T\left[\hat{F}_T(X_t) - F(\hat{\vartheta}_T, X_t)\right]^2 \mathrm{d}t.\end{aligned}$$

Similar arguments can be applied to $\delta_T\left(X^T\right)$ but special attention has to be paid to the definition of $\hat{f}_T\left(X_t\right)$ because it uses the local time of the diffusion process. For example, the estimator $\hat{f}_T\left(\cdot\right)$ can be replaced by a kernel-type estimator, which has similar asymptotic properties [KUT 04].

The K-S-type statistics are

$$
\begin{aligned}
\Gamma_T\left(X^T\right) &= \sup_x \sqrt{T}\left|\hat{F}_T\left(x\right)-F(\hat{\vartheta}_T,x)\right|,\\
\gamma_T\left(X^T\right) &= \sup_x \sqrt{T}\left|\hat{f}_T\left(x\right)-f(\hat{\vartheta}_T,x)\right|.
\end{aligned}
$$

The limit behavior of these statistics is based on the following representations. We can write

$$
\begin{aligned}
\zeta_T(\hat{\vartheta}_T,x) &= \sqrt{T}\left(\hat{f}_T\left(x\right)-f(\hat{\vartheta}_T,x)\right)\\
&= \sqrt{T}\left(\hat{f}_T\left(x\right)-f(\vartheta,x)\right)-\sqrt{T}\left(\hat{\vartheta}_T-\vartheta\right)\,\dot{f}\left(\vartheta,x\right)+o\left(1\right)
\end{aligned}
$$

and the similar expansion for the process $\eta_T(\hat{\vartheta}_T,x)$. Recall that

$$
\zeta_T(\vartheta,x)\Longrightarrow\zeta\left(\vartheta,x\right)=2f(\vartheta,x)\int_{-\infty}^{\infty}\frac{F\left(\vartheta,y\right)-1\!\!1_{\{y>x\}}}{\sigma\left(y\right)\sqrt{f\left(\vartheta,y\right)}}\,\mathrm{d}W\left(y\right),
$$

$$
\eta_T(\vartheta,x)\Longrightarrow\eta\left(\vartheta,x\right)=2\int_{-\infty}^{\infty}\frac{F\left(\vartheta,y\right)F\left(\vartheta,x\right)-F\left(\vartheta,y\wedge x\right)}{\sigma\left(y\right)\sqrt{f\left(\vartheta,y\right)}}\,\mathrm{d}W\left(y\right),
$$

where $W\left(\cdot\right)$ is double-sided Wiener process [KUT 04]. Note that the stochastic process $\zeta\left(\vartheta,x\right)$ is equal to $\eta'\left(\vartheta,x\right)$ (almost sure (a.s.) derivative w.r.t. $x$). For the MLE under the regularity conditions, we have

$$
\sqrt{T}\left(\hat{\vartheta}_T-\vartheta\right)\Longrightarrow\frac{1}{\mathrm{I}\left(\vartheta\right)}\int_{-\infty}^{\infty}\frac{\dot{S}\left(\vartheta,y\right)}{\sigma\left(y\right)}\sqrt{f\left(\vartheta,y\right)}\,\mathrm{d}W\left(y\right),
$$

with the same double-sided Wiener process. To verify that this is indeed the same Wiener process, we have to show the joint asymptotic normality of $\eta_T\left(\vartheta,x\right)$, $\zeta_T\left(\vartheta,x\right)$ and the normalized likelihood ratio

$$
\frac{\mathrm{d}\mathbf{P}^{(T)}_{\vartheta+u/\sqrt{T}}}{\mathrm{d}\mathbf{P}^{(T)}_{\vartheta}}=\exp\left\{\frac{u}{\sqrt{T}}\int_0^T\frac{\dot{S}\left(\vartheta,X_t\right)}{\sigma\left(X_t\right)}\mathrm{d}W_t-\frac{u^2}{2T}\int_0^T\frac{\dot{S}\left(\vartheta,X_t\right)^2}{\sigma\left(X_t\right)^2}\mathrm{d}t+o\left(1\right)\right\}
$$

(see details in [KUT 04]).

Hence,

$$\zeta_T\left(\hat{\vartheta}_T, x\right) \Longrightarrow \hat{\zeta}(\vartheta, x) = \zeta(\vartheta, x) - \dot{f}(\vartheta, x) \int_{-\infty}^{\infty} \frac{\dot{S}(\vartheta, y)\sqrt{f(\vartheta, y)}}{\mathrm{I}(\vartheta)\,\sigma(y)}\,\mathrm{d}W(y)$$

and

$$\eta_T\left(\hat{\theta}_T, x\right) \Longrightarrow \hat{\eta}(\vartheta, x) = \eta(\vartheta, x) - \dot{F}(\vartheta, x) \int_{-\infty}^{\infty} \frac{\dot{S}(\vartheta, y)\sqrt{f(\vartheta, y)}}{\mathrm{I}(\vartheta)\,\sigma(y)}\,\mathrm{d}W(y).$$

These limits will be applied below in the description of tests.

## 11.3. Shift parameter

Suppose that the observed diffusion process

$$\mathrm{d}X_t = S(X_t)\,\mathrm{d}t + \sigma\,\mathrm{d}W_t, \quad X_0, \quad 0 \le t \le T$$

under hypothesis $\mathcal{H}_0$ has the following stochastic differential

$$\mathrm{d}X_t = S_*(X_t - \vartheta)\,\mathrm{d}t + \sigma\,\mathrm{d}W_t, \quad X_0, \quad 0 \le t \le T,$$

where $S_*(\cdot)$ is some known function, $\sigma$ is a known positive constant and $\vartheta \in \Theta = (\alpha, \beta)$ is an unknown shift parameter.

Our goal is to construct APF tests belonging to the class $\mathcal{K}_\varepsilon$ of tests of asymptotical size $\varepsilon$, i.e.

$$\mathcal{K}_\varepsilon = \{\psi_T \quad : \quad \mathbf{E}_\vartheta \psi_T = \varepsilon + o(1)\},$$

for all $\vartheta \in \Theta$.

The invariant density $f(\vartheta, x)$ (under hypothesis) can be written as

$$f_*(x - \vartheta) = \frac{1}{G}\exp\left\{\frac{2}{\sigma^2}\int_0^{x-\vartheta} S_*(y)\,\mathrm{d}y\right\},$$

where $f_*(x) = f(0, x)$, the normalizing constant $G$ does not depend on $\vartheta$ and the Fisher information after the change of variables is

$$\mathrm{I}(\vartheta) = \int_{-\infty}^{\infty} \frac{S'_*(y)^2}{\sigma^2} f_*(y)\,\mathrm{d}y = \mathrm{I}_*.$$

Our tests are based on the following C-vM- and K-S-type statistics:

$$\begin{aligned}
\Delta_T\left(X^T\right) &= T\int_{-\infty}^{\infty}\left[\hat{F}_T\left(x\right)-F_*(x-\hat{\vartheta}_T)\right]^2 \mathrm{d}F_*(x-\hat{\vartheta}_T),\\
\delta_T\left(X^T\right) &= T\int_{-\infty}^{\infty}\left[\hat{f}_T\left(x\right)-f_*(x-\hat{\vartheta}_T)\right]^2 \mathrm{d}F_*(x-\hat{\vartheta}_T),\\
\Gamma_T\left(X^T\right) &= \sup_x \sqrt{T}\left|\hat{F}_T\left(x\right)-F_*(x-\hat{\vartheta}_T)\right|,\\
\gamma_T\left(X^T\right) &= \sup_x \sqrt{T}\left|\hat{f}_T\left(x\right)-f_*(x-\hat{\vartheta}_T)\right|.
\end{aligned}$$

Let us denote

$$\begin{aligned}
\eta_*\left(x\right) &= 2\int_{-\infty}^{\infty}\left[\frac{F_*\left(y\right)F_*\left(x\right)-F_*\left(y\wedge x\right)}{\sigma\, f_*\left(y\right)}-\frac{f_*\left(x\right)S_*'\left(y\right)}{2\mathrm{I}_*\,\sigma^2}\right]\sqrt{f_*\left(y\right)}\,\mathrm{d}W\left(y\right),\\
\zeta_*\left(x\right) &= 2f_*(x)\int_{-\infty}^{\infty}\left[\frac{F_*\left(y\right)-1\!\!1_{\{y>x\}}}{\sigma\, f_*\left(y\right)}-\frac{S_*\left(x\right)S_*'\left(y\right)}{\mathrm{I}_*\,\sigma^2}\right]\sqrt{f_*\left(y\right)}\,\mathrm{d}W\left(y\right).
\end{aligned}$$

Define the constants $a_\varepsilon, b_\varepsilon, c_\varepsilon$ and $d_\varepsilon$ as solutions of the following equations

$$\begin{aligned}
&\mathbf{P}\left\{\int\eta_*\left(x\right)^2\mathrm{d}F_*(x)>a_\varepsilon\right\}=\varepsilon, \qquad \mathbf{P}\left\{\int\zeta_*\left(x\right)^2\mathrm{d}F_*(x)>b_\varepsilon\right\}=\varepsilon;\\
&\mathbf{P}\left\{\sup_x\left|\eta_*\left(x\right)\right|>c_\varepsilon\right\}=\varepsilon, \qquad \mathbf{P}\left\{\sup_x\left|\zeta_*\left(x\right)\right|>d_\varepsilon\right\}=\varepsilon.
\end{aligned}$$

Note that these constants do not depend on $\vartheta$. The corresponding GoF tests are

$$\begin{aligned}
\hat{\Psi}_T &= 1\!\!1_{\{\Delta_T(X^T)>a_\varepsilon\}}, \quad \hat{\psi}_T=1\!\!1_{\{\delta_T(X^T)>b_\varepsilon\}},\\
\hat{\Phi}_T &= 1\!\!1_{\{\Gamma_T(X^T)>c_\varepsilon\}}, \quad \hat{\varphi}_T=1\!\!1_{\{\gamma_T(X^T)>d_\varepsilon\}}.
\end{aligned}$$

The properties of these tests are given in the following theorem.

THEOREM 11.1.– Suppose that the function $S_*\left(x\right)$ is twice continuously differentiable and the derivatives belong to $\mathcal{P}$, then the tests $\hat{\Psi}_T$, $\hat{\psi}_T$, $\hat{\Phi}_T$ and $\hat{\varphi}_T$ are APF, belong to $\mathcal{K}_\varepsilon$ and are consistent.

PROOF.– For the tests $\hat{\Psi}_T$ and $\hat{\psi}_T$, the proof is given in [NEG 12] and for $\hat{\Phi}_T$ and $\hat{\varphi}_T$ in [ZHO 12]. Note that in [NEG 12], the statistics are defined in slightly different way

$$\Delta_T\left(X^T\right) = T\int_{-\infty}^{\infty}\left[\hat{F}_T\left(x\right) - F_*(x-\hat{\vartheta}_T)\right]^2 \mathrm{d}x,$$
$$\delta_T\left(X^T\right) = T\int_{-\infty}^{\infty}\left[\hat{f}_T\left(x\right) - f_*(x-\hat{\vartheta}_T)\right]^2 \mathrm{d}x,$$

but the proofs given there are valid in this case too.

Of course, this is based on the representations given above for $\eta_T(\hat{\vartheta}_T, x)$ and $\zeta_T(\hat{\vartheta}_T, x)$. Note that the regularity conditions of theorem 2.8 in [KUT 04] in the case of a shift parameter are fulfilled if the function $S_*\left(x\right)$ is twice differentiable. For example, the identifiability condition is: for any $\nu > 0$

$$\inf_{|\vartheta-\vartheta_0|>\nu} \mathbf{E}_{\vartheta_0}\left|S_*\left(\xi-\vartheta\right) - S_*\left(\xi-\vartheta_0\right)\right|^2$$
$$= \inf_{|h|>\nu} \mathbf{E}_*\left|S_*\left(\xi-h\right) - S_*\left(\xi\right)\right|^2 > 0.$$

If this condition is not fulfilled, then there exists a value $h_*$ such that the function $S_*\left(x-h_*\right) = S_*\left(x\right)$ for all $x$, i.e. $S_*\left(x\right)$ is $h_*$-periodic function, but in this case the diffusion process is not ergodic.

For example, we have to verify the convergence of the integrals (under hypothesis $\mathcal{H}_0$)

$$\Delta_T\left(X^T\right) = \int_{-\infty}^{\infty}\hat{\eta}_T\left(\vartheta, x\right)^2 \mathrm{d}F_*(x-\hat{\vartheta}_T) \Longrightarrow \int_{-\infty}^{\infty}\hat{\eta}\left(\vartheta, x\right)^2 \mathrm{d}F_*(x-\vartheta)$$
$$= \int_{-\infty}^{\infty}\hat{\eta}\left(0, x-\vartheta\right)^2 \mathrm{d}F_*(x-\vartheta) = \int_{-\infty}^{\infty}\eta_*\left(x\right)^2 \mathrm{d}F_*(x).$$

The usage of the MLE is motivated by its asymptotic efficiency, which means that the contribution of the estimator in the limit variance in this case is probably less important than if we use other estimators. Note that any other consistent and asymptotically normal estimator will give a similar result.

We can illustrate this by the minimum distance estimator

$$\theta_T^* = \arg\inf_{\vartheta\in\Theta}\int_{-\infty}^{\infty}\left[\hat F_T(x) - F_*(x-\vartheta)\right]^2 \mathrm{d}x.$$

Recall that for this estimator, we have the minimum distance equation

$$\int_{-\infty}^{\infty}\left[\hat F_T(x) - F_*(x-\vartheta_T^*)\right] f_*(x-\vartheta_T^*)\,\mathrm{d}x = 0,$$

which allows us to obtain the following representation and limit

$$\begin{aligned}\sqrt{T}\left(\vartheta_T^* - \vartheta\right) &= -\frac{\int_{-\infty}^{\infty}\eta_T(\vartheta,x) f_*(x-\vartheta)\,\mathrm{d}x}{\int_{-\infty}^{\infty} f_*(x-\vartheta)^2\,\mathrm{d}x} + o(1)\\ &\Longrightarrow -\frac{\int_{-\infty}^{\infty}\eta(x) f_*(x)\,\mathrm{d}x}{\int_{-\infty}^{\infty} f_*(x)^2\,\mathrm{d}x},\end{aligned}$$

where

$$\eta(x) = 2\int_{-\infty}^{\infty}\frac{F_*(y)F_*(x) - F_*(y\wedge x)}{\sigma\sqrt{f_*(y)}}\,\mathrm{d}W(y).$$

Hence, the test $\Psi_T^* = 1\!\!1_{\{\Delta_T^*(X^T)>e_\varepsilon\}}$ with

$$\Delta_T^*\left(X^T\right) = T\int_{-\infty}^{\infty}\left[\hat F_T(x) - F_*(x-\vartheta_T^*)\right]^2 \mathrm{d}F_*(x-\vartheta_T^*),$$

and corresponding threshold $e_\varepsilon$ is APF [ZHO 13].

If the problem of parameter estimation is singular and the rate of convergence of an estimator is better than $\sqrt{T}$, then there is no contribution of the estimator in the limit of test statistics. This property of test statistics was mentioned by Darling [DAR 55]. For example, if the observed process under hypothesis is simple switching

$$\mathrm{d}X_t = -\mathrm{sgn}\left(X_t - \vartheta\right)\,\mathrm{d}t + \sigma\,\mathrm{d}W_t,\quad X_0,\quad 0\le t\le T,$$

then the MLE has non-degenerate distribution with another normalization: $T\left(\hat\vartheta_T - \vartheta\right) \Rightarrow \xi$ and for the statistic $\Delta_T\left(X^T\right)$, we have the following limit

$$\Delta_T\left(X^T\right) \Longrightarrow \int_{-\infty}^{\infty}\eta(x)^2 f_*(x)\,\mathrm{d}x.$$

The consistency of all studied tests follows from standard arguments.

## 11.4. Shift and scale parameters

Let us consider the similar GoF problem in the case of two unknown parameters: shift and scale. The diffusion process under hypothesis can be

$$\mathrm{d}X_t = S\left(\frac{X_t - \alpha}{\beta}\right)\mathrm{d}t + \sigma\,\mathrm{d}W_t, \quad X_0, \quad 0 \le t \le T.$$

To construct an APF test in such generality is a more complicated problem. Note that it is possible to give even ADF test for this model and this problem was treated in our another work [KLE 12]. Here, our goal is to obtain the APF tests by simple methods. That is why, we consider a special class of ergodic diffusion processes and show how for this statistical model the APF tests can be constructed.

Let us suppose that the observed diffusion process

$$\mathrm{d}X_t = S\left(X_t\right)\mathrm{d}t + \sigma\,\mathrm{d}W_t, \quad X_0, \quad 0 \le t \le T$$

under hypothesis $\mathcal{H}_0$ has stochastic differential

$$\mathrm{d}X_t = -\beta\,\mathrm{sgn}\left(X_t - \alpha\right)\left|X_t - \alpha\right|^{\gamma}\,\mathrm{d}t + \sigma\,\mathrm{d}W_t, \qquad X_0, \quad 0 \le t \le T, \qquad [11.7]$$

where $\vartheta = (\alpha, \beta)$ is the unknown parameter, $\vartheta \in \Theta = (a_1, a_2) \times (b_1, b_2), b_1 > 0$. The parameters $\gamma \ge 0$ and $\sigma > 0$ are assumed to be known.

Note that if $\gamma = 1$, then we obtain the well-known Ornstein Uhlenbeck process

$$\mathrm{d}X_t = -\beta\,\left(X_t - \alpha\right)\,\mathrm{d}t + \sigma\,\mathrm{d}W_t, \qquad X_0, \quad 0 \le t \le T,$$

and if $\gamma = 0$, then the solution of [11.7] is the simple switching process

$$\mathrm{d}X_t = -\beta\,\mathrm{sgn}\left(X_t - \alpha\right)\,\mathrm{d}t + \sigma\,\mathrm{d}W_t, \qquad X_0, \quad 0 \le t \le T$$

studied in [KUT 04, section 3.4].

It is easy to verify that this process is positive recurrent with the invariant density

$$f\left(\vartheta, x\right) = \frac{\beta^{\frac{1}{\gamma+1}}}{G_\gamma\,\sigma^{\frac{2}{\gamma+1}}}\,\exp\left\{-\frac{2\beta\left|x-\alpha\right|^{\gamma+1}}{\left(\gamma+1\right)\sigma^2}\right\}. \qquad [11.8]$$

We denote this density for the value $\vartheta = (0,1)$ as $f_*(x)$. The corresponding d.f. and expectation will be $F_*(x)$ and $\mathbf{E}_*$, respectively. Let us introduce two statistics

$$\begin{aligned}\Delta_T\left(X^T\right) &= \hat\beta_T^{\frac{2}{\gamma+1}}\sigma^{\frac{2(\gamma-1)}{\gamma+1}}T\int_{-\infty}^{\infty}\left[\hat F_T(x)-F\left(\hat\vartheta_T,x\right)\right]^2 \mathrm{d}F\left(\hat\vartheta_T,x\right)\\ \delta_T\left(X^T\right) &= \sigma^2 T\int_{-\infty}^{\infty}\left[\hat f_T(x)-f\left(\hat\vartheta_T,x\right)\right]^2 \mathrm{d}F\left(\hat\vartheta_T,x\right),\end{aligned}$$

where $\hat\vartheta_T$ is the MLE. We need the following random variables and random functions

$$\begin{aligned}&\Phi(y) = 2\int_{-\infty}^{\infty}\frac{F_*(z)\,F_*(y)-F_*(z\wedge y)}{\sqrt{f_*(y)}}\,\mathrm{d}W(z),\\ &\tilde\Phi(y) = 2\int_{-\infty}^{\infty}\frac{F_*(z)-1\!\!1_{\{z>y\}}}{\sqrt{f_*(z)}}\,\mathrm{d}W(z),\\ &\Pi = \int_{-\infty}^{\infty}|z|^{\gamma-1}\sqrt{f_*(z)}\,\mathrm{d}W(z), \qquad A = \mathbf{E}_*\,|\xi|^{2\gamma-2},\\ &\Psi = \int_{-\infty}^{\infty}\mathrm{sgn}(z)\,|z|^{\gamma}\sqrt{f_*(z)}\,\mathrm{d}W(z), \quad B = \mathbf{E}_*\,|\xi|^{2\gamma},\\ &\zeta_0(y) = \left[\tilde\Phi(y)+2\mathrm{sgn}(y)\,|y|^{\gamma}\frac{\Pi}{\gamma A}+\left[1-2\,|y|^{\gamma+1}\right]\frac{\Psi}{(\gamma+1)\,B}\right]f_*(y).\end{aligned}$$

Here, $\xi$ is the random variable with the density $f_*(x)$. Let us put

$$\begin{aligned}\Delta_* &= \int_{-\infty}^{\infty}\left[\Phi(y)+\frac{\Pi}{\gamma A}\,f_*(y)+\frac{y\,\Psi}{(\gamma+1)\,B}\,f_*(y)\right]^2 f_*(y)\,\mathrm{d}y,\\ \delta_* &= \int_{-\infty}^{\infty}\zeta_0(y)^2\,f_*(y)\,\mathrm{d}y,\end{aligned}$$

and define the constants $a_\varepsilon$ and $b_\varepsilon$ as solutions of the following equations

$$\mathbf{P}\left(\Delta_* > a_\varepsilon\right) = \varepsilon, \qquad \mathbf{P}\left(\delta_* > b_\varepsilon\right) = \varepsilon.$$

The tests are

$$\Psi_T = 1\!\!1_{\{\Delta_T(X^T)>a_\varepsilon\}}, \qquad \psi_T = 1\!\!1_{\{\delta_T(X^T)>b_\varepsilon\}}$$

and their properties are given in the following theorem.

THEOREM 11.2.– Let $\gamma \geq 1$, then the tests $\Psi_T$ and $\psi_T$ are APF, belong to $\mathcal{K}_\varepsilon$ and are consistent.

PROOF.– The proof can be found in [KUT 12]. It is based on the similar expansion of the normalized estimators. Note that, in this case, the information matrix is diagonal and therefore the estimators $\hat{\alpha}_T$ and $\hat{\beta}_T$ are asymptotically independent. To obtain a representation free of parameters for test statistics we introduce the stochastic process

$$Z_t = \frac{\beta^{\frac{1}{\gamma+1}}}{\sigma^{\frac{2}{\gamma+1}}} \left(X_t - \alpha\right).$$

The equation for this is

$$\mathrm{d}Z_t = -\mathrm{sgn}\left(Z_t\right) \left|Z_t\right|^\gamma \mathrm{d}\left(t\beta^{\frac{2}{\gamma+1}}\sigma^{\frac{2(\gamma-1)}{\gamma+1}}\right) + \beta^{\frac{1}{\gamma+1}}\sigma^{\frac{\gamma-1}{\gamma+1}}\mathrm{d}W_t, \quad 0 \leq t \leq T.$$

Then, we define a new process

$$Y_s = Z_{s\beta^{-\frac{2}{\gamma+1}}\sigma^{-\frac{2(\gamma-1)}{\gamma+1}}}, \qquad 0 \leq s = t\beta^{\frac{2}{\gamma+1}}\sigma^{\frac{2(\gamma-1)}{\gamma+1}} \leq T_* = T\beta^{\frac{2}{\gamma+1}}\sigma^{\frac{2(\gamma-1)}{\gamma+1}}.$$

The process $Y_s$ satisfies the equation

$$\mathrm{d}Y_s = -\mathrm{sgn}\left(Y_s\right) \left|Y_s\right|^\gamma \mathrm{d}s + \mathrm{d}w_s, \qquad Y_0, \quad 0 \leq s \leq T_*,$$

where $w_s = \beta^{\frac{1}{\gamma+1}}\sigma^{\frac{\gamma-1}{\gamma+1}}W_t$ (here $t = s\beta^{-\frac{2}{\gamma+1}}\sigma^{-\frac{2(\gamma-1)}{\gamma+1}}$) is another Wiener process. Now, the unknown parameter $\beta$ is in the "time of observations" $T_*$. The process $Y_s$ is ergodic with the invariant density $f_*\left(x\right)$. Then, we express the test statistics $\Delta_T\left(X^T\right)$ and $\delta_T\left(X^T\right)$ in terms of $Y^{T_*} = \left(Y_s, 0 \leq s \leq T_*\right)$. As a result, we obtain the desired limits

$$\Delta_T\left(X^T\right) \Longrightarrow \Delta_*, \qquad \delta_T\left(X^T\right) \Longrightarrow \delta_*.$$

If $\gamma \in [0, 1/2)$, then the problem of parameter $\alpha$ estimation is singular and the rate of convergence is faster than $\sqrt{T}$. Therefore, the contribution of this estimator in the limit distribution is negligible. The case $\gamma \in (1/2, 1)$ was not treated directly in [KUT 12], but it was noted that it can be shown that the MLE is $\sqrt{T}$-asymptotically normal and is similar to the case $\gamma \geq 1$. Therefore, the limit distribution of the test statistics is the same as described above.

## 11.5. Bibliography

[DAB 13] DABYE A., "On the Cramér-von Mises test with parametric hypothesis for Poisson processes", *Stat. Inference Stoch. Process.*, vol. 16, no. 1, pp. 1–13, 2013.

[DAC 07] DACHIAN S., KUTOYANTS Y., "On the goodness-of-fit tests for some continuous time processes", *Statistical Models and Methods for Biomedical and Technical Systems*. VONTA F. *et al.* (eds), Birkhäuser, Boston, pp. 395–413, 2007.

[DAR 55] DARLING D.A., "The Cramér-Smirnov test in the parametric case", *Ann. Math. Statist.*, vol. 1, pp. 279–290, 1955.

[DUR 73] DURBIN, "Weak convergence of the sample distribution function when parameters are estimated", *Ann. Statist.*, vol. 1, pp. 279–290, 1973.

[KAC 55] KAC M., KIEFER J., WOLFOWITZ J., "On tests of normality and other tests of goodness of fit based on distance methodes", *Ann. Statist.*, vol. 26, pp. 189–211, 1955.

[KLE 12] KLEPTSYNA M., KUTOYANTS Y., "On asymptotically distribution free tests with parametric hypothesis for ergodic diffusion processes", submitted, 2012.

[KUT 04] KUTOYANTS Y., *Statistical Inference for Ergodic Diffusion Processes*, Springer, London, 2004.

[KUT 12] KUTOYANTS Y.A., "On asymptotic distribution of parameter free tests for ergodic diffusion processes", submitted, 2012.

[LEH 05] LEHMANN E., ROMANO J., *Testing Statistical Hypotheses (3rd ed.)*, Springer, N.Y, 2005.

[NEG 12] NEGRI I., ZHOU L., "On goodness-of-fit testing for ergodic diffusion process with shift parameter", *Stat. Inference Stoch. Process.*, 2012.

[ZHO 12] ZHOU L., "On asymptotically parameter free tests for ergodic diffusion process", sumitted, 2012.

[ZHO 13] ZHOU L., Problèmes Statistiques pour des EDS et les EDS Rétrogrades, PhD Thesis, March 2013, University of Maine, Le Mans, 2013.

# Chapter 12

# A Comparison of Homogeneity Tests for Different Alternative Hypotheses

A comparison of two survival curves is not a new problem today, as it has been known in biomedical research and reliability theory for years. This can be carried out using homogeneity tests. Usually, classical homogeneity tests are two-sample modifications of goodness-of-fit tests. For example, the two-sample Kolmogorov–Smirnov test [SMI 48] corresponds to the Kolmogorov–Smirnov goodness-of-fit test, the two-sample Lehmann–Rosenblatt test [LEH 51, ROS 52, FIS 60] corresponds to the Cramer–von Mises goodness-of-fit test, the two-sample Anderson–Darling–Pettitt test [PET 76] corresponds to the Anderson–Darling goodness-of-fit test. For some alternatives, the Wilcoxon rank test can be used for homogeneity testing [WIL 45]. Another approach to the homogeneity test is based on distance between non-parametric density function estimators in metrics $L_1$ and $L_2$ [ALL 97].

In survival data analysis, it is often impossible to observe every subject under study until the end, thus we deal with right-censored data. For such cases, special statistical methods are required for comparing two survival curves: Gehan's generalized Wilcoxon test, the Cox–Mantel test, etc. [GEH 65, MAN 67, PET 72].

Recently, a number of new homogeneity tests have been developed. First, Zhang suggested more powerful modifications of the Kolmogorov–Smirnov test, the Cramer–von Mises test and the Anderson–Darling test [ZHA 06]. Furthermore, Bagdonavičius and Nikulin developed tests for homogeneity against crossing

Chapter written by Sergey POSTOVALOV and Petr PHILONENKO.

survival function alternatives in the case of randomly censored data [BAG 04, BAG 06, BAG 08, BAG 12]. Finally, Martinez and Naranjo suggested a pretest to choose a more powerful test [MAR 10].

The aim of this chapter is to compare the power of homogeneity tests for different alternative hypotheses, especially with a different number of intersections and for censored and uncensored data. For randomly right-censored data, we consider two different distribution laws for censoring time and then we compare how this affects the test power.

## 12.1. Homogeneity tests

Suppose that we have two samples of continuos random variables $\xi_1$ and $\xi_2$, respectively, $X_1 = \{t_{11}, t_{12}, ..., t_{1n_1}\}$ and $X_2 = \{t_{21}, t_{22}, ..., t_{2n_2}\}$, with survival distributions $S_1(t)$ and $S_2(t)$, and

$$t_{ij} = \min\left(T_{ij}, C_{ij}\right),$$

where $T_{ij}$ and $C_{ij}$ are the failure and censoring times for the $j$th object of the $i$th group. $T_{ij}$ and $C_{ij}$ are independent and identically distributed (i.i.d.) with cumulative distribution function (CDF) $F_i(t)$ and $F_i^C(t)$, respectively. Survival curve $S_i(t)$ means the probability of survival in the time interval $[0, t]$

$$S_i\left(t\right) = P\{\xi_i > t\} = 1 - F_i\left(t\right).$$

The main hypothesis is

$$H_0 : S_1(t) = S_2(t).$$

Furthermore, we will suppose that the elements of samples are ordered: $t_{11} < ... < t_{1n_1}$ and $t_{21} < ... < t_{2n_2}$. Also we will pool these samples and sort their elements in ascending order $T = X_1 \cup X_2 = \{t_1,\ t_2,\ \ldots,\ t_n\}$, where $t_1 < t_2 < ... < t_n$, $n = n_1 + n_2$. Denote the sample indicator $v_i$ and the censoring indicators $c_{ij}$ and $c_i$ as

$$v_i = \begin{cases} 0,\ if\ t_i \in X_1, \\ 1,\ if\ t_i \in X_2, \end{cases}$$

$$c_{ij} = \begin{cases} 1,\ if\ t_{ij} \text{ is censoring time,} \\ 0, if\ t_{ij} \text{ is failure time,} \end{cases} \qquad c_i = \begin{cases} 1,\ if\ t_i \text{ is censoring time,} \\ 0, if\ t_i \text{ is failure time.} \end{cases}$$

### 12.1.1. *Tests for data without censoring*

1) Kolmogorov–Smirnov test

The test was developed by Smirnov [SMI 48] and it was based on Kolmogorov's distance. The test statistic is

$$S_{KS} = \sqrt{\frac{n_1 n_2}{n_1 + n_2}} D_{n_1,n_2}, \qquad [12.1]$$

where $D_{n_1,n_2} = \max(D^+_{n_1,n_2}, D^-_{n_1,n_2})$, $D^+_{n_1,n_2} = \max\limits_{1 \le r \le n_1} \left[\frac{r}{n_1} - \hat{F}_2(t_{1r})\right]$, $D^-_{n_1,n_2} = \max\limits_{1 \le r \le n_1} \left[\hat{F}_2(t_{1r}) - \frac{r-1}{n_1}\right]$ and $\hat{F}_2(t_{1r})$ is empirical CDF by second sample. Statistic $S_{KS}$ is distributed as $\lim\limits_{\substack{n_1 \to \infty \\ n_2 \to \infty}} P\{S_{KS} < t\} = K(t)$. The large value of $S_{KS}$ rejects the hypothesis $H_0$.

2) Lehmann–Rosenblatt test [LEH 51, ROS 52]

The test statistic is

$$S_{LR} = \frac{n_2 \sum\limits_{i=1}^{n_2} (r_{2i} - i)^2 + n_1 \sum\limits_{j=1}^{n_1} (r_{1j} - j)^2}{n_1 n_2 (n_1 + n_2)} - \frac{4 n_1 n_2 - 1}{6(n_1 + n_2)}, \qquad [12.2]$$

where $r_{2i}$ is a rank of $t_{2i}$ in the pooled sample $T$ and $r_{1j}$ is a rank of $t_{1j}$ in the pooled sample $T$. Statistic $S_{LR}$ is distributed as $\lim\limits_{\substack{n_1 \to \infty \\ n_2 \to \infty}} P\{S_{LR} < t\} = a_1(t)$. The large value of $S_{LR}$ rejects the hypothesis $H_0$.

3) Anderson–Darling–Pettitt test

This test was proposed by Darling [DAR 57] and studied in detail by Pettitt [PET 76]. Scholz and Stephens generalized this test to a $k$-sample version [SCH 87]. The test statistic is

$$S_{ADP} = \frac{1}{n} \sum_{i=1}^{2} \frac{1}{n_i} \sum_{j=1}^{n-1} \frac{(n m_{ij} - j n_i)^2}{j(n-j)}, \qquad [12.3]$$

where $m_{ij}$ is the number of observations in the $i$th sample, which are not greater than $t_j \in T$. Statistic $S_{ADP}$ is distributed as $\lim\limits_{\substack{n_1 \to \infty \\ n_2 \to \infty}} P\{S_{ADP} < t\} = a_2(t)$. The large value of $S_{ADP}$ rejects the hypothesis $H_0$.

4) Zhang's tests

The following three tests were developed by Zhang in 2001. Let us consider $F_k = \frac{k-0.5}{n}$ and

$$F_{ik} = \begin{cases} \frac{j-0.5}{n_i}, \text{ if } k = r_{ij} \ for\ some\ j, \\ \frac{j}{n_i}, \text{ if } r_{ij} < k < r_{i,j+1}, \ for\ some\ j, \end{cases}$$

where $r_{ij}$ is a rank of $t_{ij}$ in the pooled sample $T$, and $r_{i0} = 1, r_{i,n_i+1} = n + 1$.

Statistics $S_{Z_K}$, $S_{Z_C}$, $S_{Z_A}$ are analogs of the traditional two-sample $S_{KS}$, $S_{LR}$ and $S_{ADP}$. Test statistic $S_{Z_A}$ is [ZHA 06]

$$S_{Z_A} = -\sum_{k=1}^{n}\sum_{i=1}^{2} n_i \frac{F_{ik}\log(F_{ik}) + (1-F_{ik})\log(1-F_{ik})}{(k-0.5)(n-k+0.5)}. \quad [12.4]$$

The small value of $S_{Z_A}$ rejects the hypothesis $H_0$. Test statistic $S_{Z_C}$ is

$$S_{Z_C} = \frac{1}{n}\sum_{i=1}^{2}\sum_{j=1}^{n_i} \log\left(\frac{n_i}{j-0.5} - 1\right)\log\left(\frac{n}{r_{ij}-0.5} - 1\right). \quad [12.5]$$

The small value of $S_{Z_C}$ rejects the hypothesis $H_0$. Test statistic $S_{Z_K}$ is

$$S_{Z_K} = \max_{1\le k\le n}\sum_{i=1}^{2} n_i \left[F_{ik}\log\left(\frac{F_{ik}}{F_k}\right) + (1-F_{ik})\log\left(\frac{1-F_{ik}}{1-F_k}\right)\right]. \quad [12.6]$$

The large value of $S_{Z_K}$ rejects the hypothesis $H_0$.

### 12.1.2. *Tests for data with censoring*

Here, we consider tests that are used for randomly right-censored data. The following two criteria are based on the Wilcoxon test [WIL 45].

– Gehan's generalized Wilcoxon test

The test statistic is [LEE 03, p. 107]

$$S_G = \frac{\sum\limits_{i=1}^{n}(1-v_i)h_i}{\sqrt{\frac{n_1 n_2}{n(n-1)}\sum\limits_{i=1}^{n}(1-v_i)h_i^2}}, \quad [12.7]$$

where

$$h_i = \sum_{j=1}^{n} v_j h_{ij}, \; h_{ij} = \begin{cases} +1, \; if \; t_i > t_j \; \& \; c_j = 0 \; \& \; v_i = 0 \; \& \; v_j = 1, \\ -1, \; if \; t_i < t_j \; \& \; c_i = 0 \; \& \; v_i = 0 \; \& \; v_j = 1 \,, \\ 0, \quad otherwise \,. \end{cases}$$

The null hypothesis is rejected with the significance level $\alpha$, if $|S_G| > z_{1-\frac{\alpha}{2}}$, where $z_{1-\frac{\alpha}{2}}$ is the $(1 - \frac{\alpha}{2})$-quantile of the standard normal distribution.

– Peto and Peto's generalized Wilcoxon test

The test statistic is [LEE 03, p. 116]

$$S_P = \frac{n(n-1)\sum\limits_{i=1}^{n} u_i(1-v_i)}{n_1 n_2 \sum\limits_{i=1}^{n} u_i^2}, \; u_i = \begin{cases} \hat{s}(t_i) + \hat{s}(t_{k_i}) - 1, \; if \; c_i = 0, \\ \hat{s}(t_{k_i}) - 1, \; if \; c_i = 1, \end{cases} \qquad [12.8]$$

where $k_i = \max\{j | j \in \{0, ..., i-1\}, c_j = 0\}, c_0 = 0$ and

$$\hat{S}(t) = \prod_{i:t_i<t, c_i=0} \frac{n-i}{n-i+1}$$

is the Kaplan–Meier estimator of $S(t)$ by the pooled sample $T$. The null hypothesis is rejected with the significance level $\alpha$, if $|S_P| > z_{1-\frac{\alpha}{2}}$.

The following two criteria are based on the logrank test [MAN 67].

– Logrank test

The test statistic is [LEE 03, p. 111]

$$S_L = \frac{\sum\limits_{i=1}^{n}(1-c_i) - \sum\limits_{j=1}^{i} \frac{1}{n-j+1}}{\sqrt{\left[\sum\limits_{i=1}^{n}(1-c_i)\frac{n-i}{n-i+1}\right] \frac{n_1 n_2}{n(n-1)}}}, \qquad [12.9]$$

The null hypothesis is rejected with the significance level $\alpha$, if $|S_L| > z_{1-\frac{\alpha}{2}}$.

– Cox–Mantel test

The test statistic is [LEE 03, p. 109]

$$S_{CM} = \frac{r_2 - \sum_{i=1}^{n} (1 - c_i) A_{(i)}}{\sqrt{\sum_{i=1}^{n} (1 - c_i) A_{(i)} \left(1 - A_{(i)}\right)}}, \qquad [12.10]$$

where

$$r_2 = \sum_{i=1}^{n} v_i (1 - c_i), \; A_{(i)} = \frac{1}{n - i} \sum_{j=i}^{n} v_j.$$

The null hypothesis is rejected with the significance level $\alpha$, if $|S_C M| > z_{1-\frac{\alpha}{2}}$.

– Q-test

Martnez and Naranjo suggested a pretest to choose the better test that should be used: Wilcoxon type or logrank type [MAR 10]. The Q-test statistic is

$$S_Q = \begin{cases} S_L \text{ or } S_{CM}, \text{ if } Q < 0 \\ S_P \text{ or } S_G, \text{ otherwise.} \end{cases} \qquad [12.11]$$

where $Q = \left[\hat{S}_2\left(q_{0.6}\right) - \hat{S}_1\left(q_{0.6}\right)\right] - \left[\hat{S}_2\left(q_{0.2}\right) - \hat{S}_1\left(q_{0.2}\right)\right]$, $q_p - \hat{S}_1^{-1}(p)$ and

$$\hat{S}_i(t) = \prod_{j: t_{ij} < t, c_{ij} = 0} \frac{n - j}{n - j + 1}$$

is the Kaplan–Meier estimator of $S_i(t)$. For unambiguity, we will use the pair $S_L$ and $S_P$ further.

The following two tests are developed against alternative with intersections of survival functions.

– Bagdonavičius–Nikulin test 1 (single crossing)

The test statistic is [BAG 06]

$$S_{BN1} = (U_1, U_2)\,\Sigma^{-1}(U_1, U_2)^T, \qquad [12.12]$$

where the elements of the matrix $\Sigma$ are

$$\sigma_{11} = \sum_{i=1}^{2} \sum_{j:c_{ij}=0} \frac{Y_1(t_{ij})Y_2(t_{ij})}{Y^2(t_{ij})}, \qquad [12.13]$$

$$\sigma_{12} = \sigma_{21} = \sum_{i=1}^{2} \sum_{j:c_{ij}=0} \frac{Y_1(t_{ij})Y_2(t_{ij})}{Y^2(t_{ij})} \ln(1 + \Lambda(t_{ij})), \qquad [12.14]$$

$$\sigma_{22} = \sum_{i=1}^{2} \sum_{j:c_{ij}=0} \frac{Y_1(t_{ij})Y_2(t_{ij})}{Y^2(t_{ij})} \ln^2(1 + \Lambda(t_{ij})). \qquad [12.15]$$

The $H_0$ is rejected with the significance level $\alpha$ if $S_{BN1} > \chi^2_{1-\alpha}(2)$, where $\chi^2_{1-\alpha}(2)$ is the $(1-\alpha)$-quantile of the $\chi^2$-distribution with two degrees of freedom.

– Bagdonavičius–Nikulin test 2 (multiple crossing)

The test statistic is [BAG 06]

$$S_{BN2} = (U_1, U_2, U_3)\,\Sigma^{-1}(U_1, U_2, U_3)^T, \qquad [12.16]$$

where

$$U_1 = \sum_{j:c_{1j}=0} \frac{Y_2(t_{1j})}{Y(t_{1j})} - \sum_{j:c_{2j}=0} \frac{Y_1(t_{2j})}{Y(t_{2j})}, \qquad [12.17]$$

$$U_2 = \sum_{j:c_{1j}=0} \frac{Y_2(t_{1j})}{Y(t_{1j})}\Lambda(t_{1j}) - \sum_{j:c_{2j}=0} \frac{Y_1(t_{2j})}{Y(t_{2j})}\Lambda(t_{2j}), \qquad [12.18]$$

$$U_3 = \sum_{j:c_{1j}=0} \frac{Y_2(t_{1j})}{Y(t_{1j})}\Lambda^2(t_{1j}) - \sum_{j:c_{2j}=0} \frac{Y_1(t_{2j})}{Y(t_{2j})}\Lambda^2(t_{2j}). \qquad [12.19]$$

The elements of the matrix $\Sigma$ are

$$\sigma_{11} = \sum_{i=1}^{2} \sum_{j:c_{ij}=0} \frac{Y_1(t_{ij})Y_2(t_{ij})}{Y^2(t_{ij})}, \qquad [12.20]$$

$$\sigma_{12} = \sigma_{21} = \sum_{i=1}^{2} \sum_{j:c_{ij}=0} \frac{Y_1(t_{ij})Y_2(t_{ij})}{Y^2(t_{ij})} \Lambda(t_{ij}), \qquad [12.21]$$

$$\sigma_{13} = \sigma_{31} = \sigma_{22} = \sum_{i=1}^{2} \sum_{j:c_{ij}=0} \frac{Y_1(t_{ij})Y_2(t_{ij})}{Y^2(t_{ij})} \Lambda^2(t_{ij}), \qquad [12.22]$$

$$\sigma_{23} = \sigma_{32} = \sum_{i=1}^{2} \sum_{j:c_{ij}=0} \frac{Y_1(t_{ij})Y_2(t_{ij})}{Y^2(t_{ij})} \Lambda^3(t_{ij}), \qquad [12.23]$$

$$\sigma_{33} = \sum_{i=1}^{2} \sum_{j:c_{ij}=0} \frac{Y_1(t_{ij})Y_2(t_{ij})}{Y^2(t_{ij})} \Lambda^4(t_{ij}). \qquad [12.24]$$

The $H_0$ is rejected with the significance level $\alpha$ if $S_{BN2} > \chi^2_{1-\alpha}(3)$, where $\chi^2_{1-\alpha}(3)$ is the $(1-\alpha)$-quantile of the $\chi^2$-distribution with three degrees of freedom.

## 12.2. Alternative hypotheses

Since there are a lot of homogeneity tests, the best test should be identified by the researcher, which could be used in practice. Unfortunately, the most powerful test is not present in general case. Looking at empirical survival functions (or Kaplan–Meier estimators for randomly censored data), we could choose what kind of alternative we will have. In general, there are two main situations of survival curve differences: whether they have intersections or not, and where the difference is larger, for early or late times. Therefore, we construct 10 different alternative hypotheses.

The alternative hypotheses are shown in Table 12.2 and in Figure 12.1, which are based on exponential [12.25], gamma [12.26], lognormal [12.27], Weibull [12.28] and generalized Weibull [12.29] families of distribution laws.

$$f_{Exp}(t;\theta_1) = \theta_1 e^{-\theta_1 t},\ t \geq 0, \quad [12.25]$$

$$f_{\Gamma}(t;\theta_1,\theta_2) = \frac{\left(\frac{t}{\theta_1}\right)^{\theta_2-1} e^{-\frac{t}{\theta_1}}}{\theta_1\Gamma(\theta_2)},\ t \geq 0, \quad [12.26]$$

$$f_{LgN}(t;\theta_1,\theta_2) = \frac{1}{t\sqrt{2\pi\theta_2^2}} e^{-\frac{(\ln t-\theta_1)^2}{2\theta_2^2}},\ t \geq 0, \quad [12.27]$$

$$f_{We}(t;\theta_1,\theta_2) = \frac{\theta_2 t^{\theta_2-1} e^{-\left(\frac{t}{\theta_1}\right)^{\theta_2}}}{{\theta_1}^{\theta_2}},\ t \geq 0, \quad [12.28]$$

$$f_{GWe}(t;\theta_1,\theta_2) = \frac{\frac{\theta_1}{\theta_2}\left(1+t^{\theta_1}\right)^{\frac{1}{\theta_2}-1} e^{1-(1+t^{\theta_1})^{\frac{1}{\theta_2}}}}{t^{1-\theta_1}},\ t \geq 0. \quad [12.29]$$

| $H_i$ | First sample | Second sample | Number of intersections | Intersections |
|---|---|---|---|---|
| $H_1$ | We(2;2) | Γ(0.56; 3.12) | 2 | ≈0.7 and ≈2.8 |
| $H_2$ | We(2;2) | LgN(0.41; 0.62) | 2 | ≈0.75 and ≈2.5 |
| $H_3$ | Exp(1) | Exp(1.1) | 0 | - |
| $H_4$ | LgN(0;1) | LgN(0.1; 1.0) | 0 | - |
| $H_5$ | GWe(0.95;0.85) | Exp(1) | 0 | - |
| $H_6$ | We(1.1; 1.4) | Γ(0.9; 1.0) | 1 | ≈1.81 |
| $H_7$ | We(1.0; 1.1) | We(1.1; 0.9) | 1 | ≈0.65 |
| $H_8$ | LgN(0.1;1.0) | LgN(0; 1.2) | 1 | ≈1.82 |
| $H_9$ | We(1.0; 1.1) | We(0.7; 0.9) | 1 | ≈4.97 |
| $H_{10}$ | Γ(1; 1) | LgN (0; 0.65) | 2 | ≈2.28 and ≈5.1 |

**Table 12.1.** *Alternative hypotheses*

## 12.3. Power simulation

The test power was calculated by 83,000 repetitions (five series by 16,600 repetitions to calculate confidence interval). For all tests, both samples $X_1$ and $X_2$ had an equal size, except the Kolmogorov–Smirnov test, because it has a strong discrete distribution for equal sample sizes [LEM 05]. Therefore, we use $n_1' = n_1 - 1$ and $n_2' = n_2 + 1$ only for this test. The significance level is 0.05.

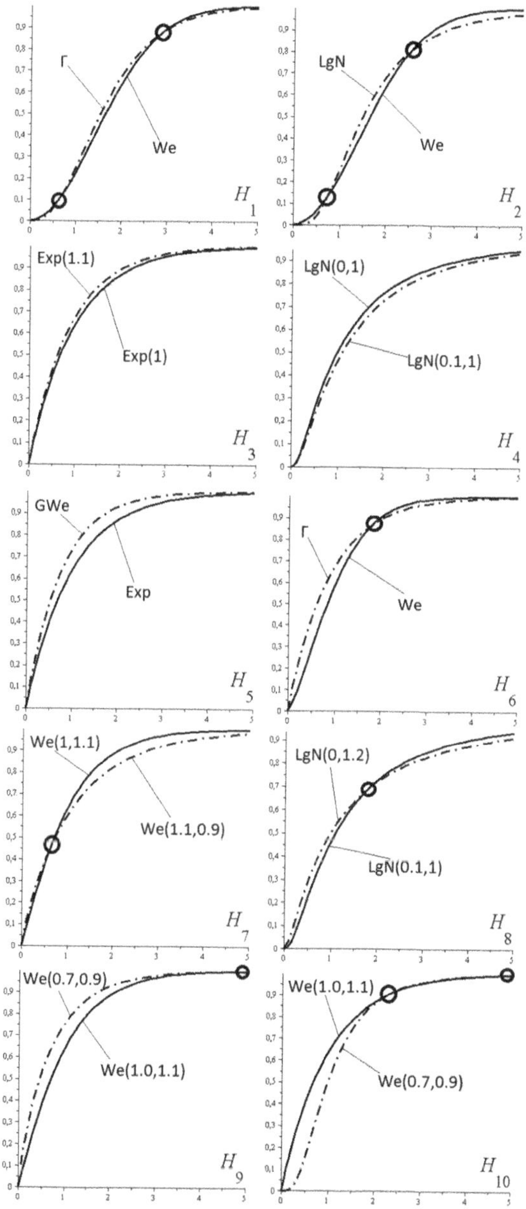


**Figure 12.1.** *CDFs of alternative hypotheses $H_1 - H_{10}$. The circle means the CDFs crossing*

### 12.3.1. *Power of tests without censoring*

In Tables 12.2–12.5, the power of 11 tests, including the Wilcoxon test ($W$), the logrank test ($Lg$), the Q-test ($Q$), the Anderson–Darling–Pettitt test ($ADP$), the Lehmann–Rosenblatt test ($LR$), the Kolmogorov–Smirnov test ($KS$), the Bagdonavičius–Nikulin tests (single crossing – $BN_1$, multiple crossing – $BN_2$) and Zhang's tests ($Z_A$,$Z_C$,$Z_K$), is presented for different sample sizes $n_1 = n_2 =$ 20, 50, 200, 500 without censored observations.

| Tests | $H_1$ | $H_2$ | $H_3$ | $H_4$ | $H_5$ | $H_6$ | $H_7$ | $H_8$ | $H_9$ | $H_{10}$ |
|---|---|---|---|---|---|---|---|---|---|---|
| $W$ | 0.055 | 0.060 | 0.056 | 0.061 | 0.094 | 0.141 | 0.051 | 0.060 | 0.195 | 0.287 |
| $Lg$ | 0.054 | 0.051 | 0.059 | 0.061 | 0.101 | 0.078 | 0.067 | 0.049 | 0.148 | 0.126 |
| $Q$ | 0.054 | 0.049 | 0.061 | 0.060 | 0.100 | 0.081 | 0.062 | 0.048 | 0.157 | 0.246 |
| $ADP$ | 0.056 | 0.064 | 0.058 | 0.060 | 0.095 | 0.155 | 0.058 | 0.066 | 0.193 | 0.367 |
| $LR$ | 0.058 | 0.063 | 0.060 | 0.062 | 0.092 | 0.150 | 0.057 | 0.065 | 0.190 | 0.321 |
| $KS$ | 0.059 | 0.071 | 0.059 | 0.061 | 0.085 | 0.142 | 0.058 | 0.066 | 0.176 | 0.331 |
| $BN_1$ | 0.056 | 0.078 | 0.055 | 0.054 | 0.083 | 0.149 | 0.099 | 0.076 | 0.132 | 0.275 |
| $BN_2$ | 0.054 | 0.078 | 0.056 | 0.056 | 0.082 | 0.121 | 0.081 | 0.070 | 0.125 | 0.280 |
| $Z_A$ | 0.054 | 0.060 | 0.058 | 0.062 | 0.094 | 0.156 | 0.070 | 0.074 | 0.183 | 0.445 |
| $Z_C$ | 0.055 | 0.060 | 0.058 | 0.058 | 0.093 | 0.155 | 0.070 | 0.071 | 0.188 | 0.441 |
| $Z_K$ | 0.060 | 0.071 | 0.061 | 0.063 | 0.094 | 0.155 | 0.087 | 0.082 | 0.165 | 0.508 |

**Table 12.2.** *Tests power in case $n_1 = n_2 =$ 20 without censoring*

| Tests | $H_1$ | $H_2$ | $H_3$ | $H_4$ | $H_5$ | $H_6$ | $H_7$ | $H_8$ | $H_9$ | $H_{10}$ |
|---|---|---|---|---|---|---|---|---|---|---|
| $W$ | 0.066 | 0.068 | 0.070 | 0.081 | 0.167 | 0.282 | 0.054 | 0.076 | 0.414 | 0.600 |
| $Lg$ | 0.051 | 0.048 | 0.076 | 0.072 | 0.197 | 0.093 | 0.112 | 0.051 | 0.289 | 0.211 |
| $Q$ | 0.053 | 0.044 | 0.073 | 0.076 | 0.192 | 0.111 | 0.092 | 0.050 | 0.314 | 0.565 |
| $ADP$ | 0.069 | 0.081 | 0.067 | 0.074 | 0.162 | 0.345 | 0.085 | 0.098 | 0.412 | 0.831 |
| $LR$ | 0.070 | 0.091 | 0.067 | 0.074 | 0.154 | 0.320 | 0.070 | 0.086 | 0.408 | 0.721 |
| $KS$ | 0.071 | 0.096 | 0.067 | 0.073 | 0.137 | 0.299 | 0.073 | 0.088 | 0.353 | 0.744 |
| $BN_1$ | 0.075 | 0.170 | 0.066 | 0.066 | 0.146 | 0.369 | 0.212 | 0.133 | 0.305 | 0.596 |
| $BN_2$ | 0.072 | 0.180 | 0.060 | 0.065 | 0.130 | 0.277 | 0.157 | 0.103 | 0.258 | 0.623 |
| $Z_A$ | 0.059 | 0.102 | 0.068 | 0.073 | 0.172 | 0.377 | 0.149 | 0.137 | 0.373 | 0.938 |
| $Z_C$ | 0.059 | 0.097 | 0.067 | 0.073 | 0.173 | 0.382 | 0.144 | 0.133 | 0.377 | 0.936 |
| $Z_K$ | 0.066 | 0.113 | 0.064 | 0.069 | 0.137 | 0.320 | 0.144 | 0.127 | 0.313 | 0.933 |

**Table 12.3.** *Tests power in case $n_1 = n_2 =$ 50 without censoring*

The Bagdonavičius–Nikulin tests $BN_1$ and $BN_2$ are the best tests against alternatives $H_1$ and $H_2$ with two intersections. The test $BN_1$ is also the best test against alternative $H_7$ with one intersection. The best tests against alternatives $H_6$, $H_8$ and $H_{10}$ are Zhang's tests $Z_A$ and $Z_C$. The Wilkoxon test is the best test against alternatives $H_4$ and $H_9$, and the logrank test is the best test against alternatives $H_3$ and $H_5$.

| Tests | $H_1$ | $H_2$ | $H_3$ | $H_4$ | $H_5$ | $H_6$ | $H_7$ | $H_8$ | $H_9$ | $H_{10}$ |
|---|---|---|---|---|---|---|---|---|---|---|
| $W$ | 0.109 | 0.121 | 0.129 | 0.166 | 0.513 | 0.791 | 0.065 | 0.144 | 0.939 | 0.993 |
| $Lg$ | 0.052 | 0.065 | 0.158 | 0.148 | 0.635 | 0.169 | 0.385 | 0.052 | 0.769 | 0.586 |
| $Q$ | 0.058 | 0.053 | 0.147 | 0.160 | 0.581 | 0.205 | 0.330 | 0.051 | 0.808 | 0.991 |
| $ADP$ | 0.125 | 0.270 | 0.132 | 0.162 | 0.521 | 0.928 | 0.336 | 0.341 | 0.942 | 1.000 |
| $LR$ | 0.140 | 0.234 | 0.127 | 0.157 | 0.479 | 0.887 | 0.173 | 0.231 | 0.934 | 1.000 |
| $KS$ | 0.142 | 0.250 | 0.115 | 0.139 | 0.414 | 0.863 | 0.183 | 0.235 | 0.898 | 1.000 |
| $BN_1$ | 0.194 | 0.729 | 0.121 | 0.127 | 0.521 | 0.943 | 0.726 | 0.434 | 0.887 | 0.996 |
| $BN_2$ | 0.182 | 0.751 | 0.098 | 0.107 | 0.437 | 0.892 | 0.604 | 0.342 | 0.829 | 0.998 |
| $Z_A$ | 0.137 | 0.754 | 0.121 | 0.138 | 0.545 | 0.954 | 0.659 | 0.547 | 0.909 | 1.000 |
| $Z_C$ | 0.132 | 0.733 | 0.123 | 0.141 | 0.552 | 0.956 | 0.656 | 0.549 | 0.916 | 1.000 |
| $Z_K$ | 0.131 | 0.586 | 0.104 | 0.115 | 0.438 | 0.903 | 0.530 | 0.432 | 0.859 | 1.000 |

**Table 12.4.** *Tests power in case* $n_1 = n_2 = 200$ *without censoring*

| Tests | $H_1$ | $H_2$ | $H_3$ | $H_4$ | $H_5$ | $H_6$ | $H_7$ | $H_8$ | $H_9$ | $H_{10}$ |
|---|---|---|---|---|---|---|---|---|---|---|
| $W$ | 0.207 | 0.226 | 0.257 | 0.341 | 0.877 | 0.993 | 0.085 | 0.293 | 1.000 | 1.000 |
| $Lg$ | 0.055 | 0.110 | 0.327 | 0.299 | 0.953 | 0.345 | 0.757 | 0.064 | 0.989 | 0.927 |
| $Q$ | 0.072 | 0.091 | 0.300 | 0.327 | 0.913 | 0.397 | 0.718 | 0.058 | 0.992 | 1.000 |
| $ADP$ | 0.256 | 0.781 | 0.264 | 0.333 | 0.898 | 1.000 | 0.842 | 0.807 | 1.000 | 1.000 |
| $LR$ | 0.275 | 0.554 | 0.251 | 0.324 | 0.863 | 1.000 | 0.532 | 0.607 | 1.000 | 1.000 |
| $KS$ | 0.288 | 0.543 | 0.216 | 0.280 | 0.800 | 0.999 | 0.503 | 0.583 | 0.999 | 1.000 |
| $BN_1$ | 0.452 | 0.992 | 0.248 | 0.261 | 0.913 | 1.000 | 0.988 | 0.831 | 0.999 | 1.000 |
| $BN_2$ | 0.447 | 0.995 | 0.198 | 0.211 | 0.868 | 1.000 | 0.970 | 0.759 | 0.998 | 1.000 |
| $Z_A$ | 0.417 | 0.999 | 0.238 | 0.269 | 0.914 | 1.000 | 0.983 | 0.942 | 0.999 | 1.000 |
| $Z_C$ | 0.406 | 0.999 | 0.244 | 0.277 | 0.918 | 1.000 | 0.983 | 0.944 | 1.000 | 1.000 |
| $Z_K$ | 0.331 | 0.982 | 0.200 | 0.225 | 0.838 | 1.000 | 0.940 | 0.860 | 0.999 | 1.000 |

**Table 12.5.** *Tests power in case* $n_1 = n_2 = 500$ *without censoring*

As would be expected, the Wilkoxon test has a substantial loss of the power against the alternative $H_7$, because the point of intersection is near the median. On the other hand, the logrank test and Q-test have substantial loss of the power against the alternatives $H_1$, $H_2$ and $H_8$.

In total, the Anderson–Darling–Pettitt, the Lehmann–Rosenblatt (and their Zhang's modifications) and the Bagdonavičius–Nikulin ($BN_1$) tests have high power against all alternatives and they can be used as robust tests while the alternative hypothesis is not known.

### 12.3.2. *Power of tests with censoring*

In Tables 12.8–12.10, the power of seven tests, including Gehan's generalized Wilcoxon test (Gehan), Peto and Peto's generalized Wilcoxon test (Peto), the logrank test ($Lg$), the Cox–Mantel test ($CM$), the Q-test ($Q$), the Bagdonavičius–Nikulin tests (single crossing – $BN_1$, multiple crossing – $BN_2$), is presented for sample sizes $n_1 = n_2 = 200$ with censoring rates 10%, 20%, 30%, 40% and 50%. We investigated how the censoring rate and censoring distribution affect the power of tests.

#### 12.3.2.1. *How does the distribution of censoring time affect the power of the test?*

We consider two ways of modeling censoring times using Weibull distribution law (Table 12.6) or gamma distribution law (Table 12.7). The results are shown in Tables 12.8–12.10 for $n_1 = n_2 = 200$.

| $H$ | $i$ | 10% | 20% | 30% | 40% | 50% |
|---|---|---|---|---|---|---|
| $H_1$ | 1 | We(3.44, 6.88) | We(2.87, 5.74) | We(2.48, 4.96) | We(2.16, 4.32) | We(1.87, 3.74) |
| $H_1$ | 2 | We(4.00, 4.00) | We(4.00, 2) | We(4.00, 1.25) | We(2.25, 2.10) | We(3.74, 0.40) |
| $H_2$ | 1 | We(3.44, 6.88) | We(2.87, 5.74) | We(2.48, 4.96) | We(2.16, 4.32) | We(1.87, 3.74) |
| $H_2$ | 2 | We(4.16, 4.20) | We(3.35, 3.11) | We(3.00, 2.49) | We(2.35, 2.10) | We(3.47, 0.45) |
| $H_3$ | 1 | We(3.4, 2.47) | We(2.00, 3.25) | We(2, 1.25) | We(1.00, 6.50) | We(1.00, 1.00) |
| $H_3$ | 2 | We(3.11, 2.47) | We(2.00, 2.14) | We(2.00, 1.09) | We(1.00, 3.00) | We(1.00, 0.75) |

**Table 12.6.** *First way of modeling censoring time using Weibull distribution as $F_i^C(t)$*

| $H$ | $i$ | 10% | 20% | 30% | 40% | 50% |
|---|---|---|---|---|---|---|
| $H_1$ | 1 | $\Gamma(2.00, 3.555)$ | $\Gamma(2.50, 1.580)$ | $\Gamma(2.50, 1.185)$ | $\Gamma(2.65, 0.869)$ | $\Gamma(2.73, 0.678)$ |
| $H_1$ | 2 | $\Gamma(2.9, 1.870)$ | $\Gamma(2.6, 1.460)$ | $\Gamma(2.5, 1.150)$ | $\Gamma(2.2, 1.050)$ | $\Gamma(2.0, 0.927)$ |
| $H_2$ | 1 | $\Gamma(2.00, 3.555)$ | $\Gamma(2.50, 1.580)$ | $\Gamma(2.50, 1.185)$ | $\Gamma(2.65, 0.869)$ | $\Gamma((2.73, 0.678)$ |
| $H_2$ | 2 | $\Gamma(2.1, 3.35)$ | $\Gamma(3.2, 1.13)$ | $\Gamma(2.0, 1.59)$ | $\Gamma(2.2, 1.06)$ | $\Gamma(4.0, 0.42)$ |
| $H_3$ | 1 | $\Gamma(2.3, 1.71)$ | $\Gamma(3.7, 0.55)$ | $\Gamma(3.5, 0.41)$ | $\Gamma(3.3, 0.32)$ | $\Gamma(3.7, 0.209)$ |
| $H_3$ | 2 | $\Gamma(7.71, 0.316)$ | $\Gamma(7.00, 0.235)$ | $\Gamma(7.20, 0.166)$ | $\Gamma(7.60, 0.117)$ | $\Gamma(7.70, 0.086)$ |

**Table 12.7.** *Second way of modeling censoring time using gamma distribution as $F_i^C(t)$*

When survival curves are not crossed (see Table 12.10), test powers with different $F^C$ are almost equal. The differences between power values can be explained by the simulation error. However, when survival curves are crossed, test powers are significantly different for different distributions of censoring times (see Tables 12.8 and 12.9). The most dependency appears for the Bagdonavičius–Nikulin tests, because these tests use information about point of the intersections and early censoring times can hide this information.

| Test | Censoring times from Weibull | | | | | | Censoring times from Gamma | | | | | |
|---|---|---|---|---|---|---|---|---|---|---|---|---|
| | 0% | 10% | 20% | 30% | 40% | 50% | 0% | 10% | 20% | 30% | 40% | 50% |
| Gehan | 0.111 | 0.121 | 0.121 | 0.110 | 0.105 | 0.080 | 0.111 | 0.117 | 0.116 | 0.103 | 0.092 | 0.080 |
| Peto | 0.106 | 0.116 | 0.116 | 0.111 | 0.113 | 0.088 | 0.106 | 0.110 | 0.108 | 0.104 | 0.102 | 0.093 |
| Lg | 0.053 | 0.072 | 0.083 | 0.090 | 0.106 | 0.082 | 0.053 | 0.062 | 0.069 | 0.074 | 0.085 | 0.086 |
| CM | 0.054 | 0.073 | 0.090 | 0.103 | 0.113 | 0.096 | 0.054 | 0.061 | 0.070 | 0.082 | 0.085 | 0.086 |
| Q-test | 0.062 | 0.082 | 0.094 | 0.098 | 0.102 | 0.082 | 0.062 | 0.071 | 0.077 | 0.085 | 0.091 | 0.088 |
| $BN_1$ | 0.184 | 0.130 | 0.105 | 0.098 | 0.096 | 0.086 | 0.184 | 0.147 | 0.123 | 0.103 | 0.091 | 0.080 |
| $BN_2$ | 0.186 | 0.134 | 0.123 | 0.114 | 0.106 | 0.107 | 0.186 | 0.150 | 0.122 | 0.100 | 0.097 | 0.090 |

**Table 12.8.** *The power of tests against $H_1$*

| Test | Censoring times from Weibull | | | | | | Censoring times from Gamma | | | | | |
|---|---|---|---|---|---|---|---|---|---|---|---|---|
| | 0% | 10% | 20% | 30% | 40% | 50% | 0% | 10% | 20% | 30% | 40% | 50% |
| Gehan | 0.120 | 0.139 | 0.160 | 0.173 | 0.161 | 0.106 | 0.120 | 0.128 | 0.138 | 0.132 | 0.121 | 0.105 |
| Peto | 0.120 | 0.132 | 0.147 | 0.161 | 0.157 | 0.121 | 0.120 | 0.124 | 0.136 | 0.129 | 0.127 | 0.132 |
| Lg | 0.061 | 0.049 | 0.076 | 0.103 | 0.132 | 0.102 | 0.061 | 0.046 | 0.050 | 0.058 | 0.074 | 0.107 |
| CM | 0.070 | 0.049 | 0.076 | 0.108 | 0.138 | 0.126 | 0.070 | 0.050 | 0.050 | 0.058 | 0.080 | 0.111 |
| Q-test | 0.050 | 0.052 | 0.092 | 0.120 | 0.131 | 0.102 | 0.050 | 0.043 | 0.055 | 0.064 | 0.084 | 0.122 |
| $BN_1$ | 0.729 | 0.378 | 0.236 | 0.170 | 0.133 | 0.109 | 0.729 | 0.562 | 0.375 | 0.271 | 0.171 | 0.111 |
| $BN_2$ | 0.749 | 0.412 | 0.292 | 0.238 | 0.199 | 0.192 | 0.749 | 0.604 | 0.405 | 0.323 | 0.223 | 0.130 |

**Table 12.9.** *The power of tests against $H_2$*

| Test | Censoring times from Weibull | | | | | | Censoring times from Gamma | | | | | |
|---|---|---|---|---|---|---|---|---|---|---|---|---|
| | 0% | 10% | 20% | 30% | 40% | 50% | 0% | 10% | 20% | 30% | 40% | 50% |
| Gehan | 0.128 | 0.121 | 0.122 | 0.104 | 0.105 | 0.086 | 0.128 | 0.122 | 0.121 | 0.110 | 0.098 | 0.093 |
| Peto | 0.135 | 0.131 | 0.120 | 0.116 | 0.108 | 0.091 | 0.135 | 0.129 | 0.123 | 0.120 | 0.107 | 0.094 |
| Lg | 0.159 | 0.136 | 0.132 | 0.121 | 0.107 | 0.103 | 0.159 | 0.136 | 0.131 | 0.122 | 0.110 | 0.098 |
| CM | 0.160 | 0.142 | 0.133 | 0.126 | 0.114 | 0.100 | 0.160 | 0.141 | 0.134 | 0.123 | 0.110 | 0.101 |
| Q-test | 0.145 | 0.136 | 0.134 | 0.122 | 0.110 | 0.103 | 0.145 | 0.142 | 0.132 | 0.127 | 0.112 | 0.100 |
| $BN_1$ | 0.119 | 0.117 | 0.107 | 0.100 | 0.095 | 0.085 | 0.119 | 0.113 | 0.110 | 0.100 | 0.091 | 0.082 |
| $BN_2$ | 0.107 | 0.101 | 0.097 | 0.091 | 0.092 | 0.078 | 0.107 | 0.096 | 0.095 | 0.090 | 0.091 | 0.085 |

**Table 12.10.** *The power of tests against $H_3$*

12.3.2.2. *How does the censoring rate affect the power of the test?*

Figure 12.2 shows the power of tests against alternatives $H_1$ (a), $H_2$ (b) and $H_3$ (c) for censoring rates 50%, while the distribution of censoring times $F^C(t)$ is the Weibull distribution. The power of the tests usually decreases while the censoring rate increases. However, the power of ranking tests against alternatives $H_1$ and $H_2$ increases up to a 40% censoring rate.

The Bagdonavičius–Nikulin tests are the most powerful tests against alternative $H_1$ (survival curves have two intersections) while the censoring rate is less than 30%. When the censoring rate is 4050%, the most powerful tests are the tests of Wilcoxon type (Gehan's and Peto and Peto's generalizations).

The Bagdonavičius–Nikulin tests are the most powerful tests against alternative $H_2$ (survival curves also have two intersections) while the censoring rate is less than 50%.

The logrank test and the Cox–Mantel test are the most powerful tests against alternative $H_3$ (survival curves do not have intersections). Gehan's generalized Wilcoxon test and Peto and Peto's generalized Wilcoxon test (Wilcoxon-type tests) are less powerful. The Bagdonavičius–Nikulin tests have the least power in this case.

## 12.4. Statistical inference

Let us summarize the received results.

*In the case of non-censoring data,* the Anderson–Darling–Pettitt, the Lehmann–Rosenblatt (and their Zhang's modifications) and the Bagdonavičius–Nikulin ($BN1$) tests have high power against all considered alternatives and they can be used as robust tests while the alternative hypothesis is not known. If survival curves have early or middle intersections, it is better to use the Bagdonavičius–Nikulin tests. If survival curves have late intersections, it is better to use Zhang's modifications of the Anderson–Darling–Pettitt and the Lehmann–Rosenblatt tests. If survival curves do not have intersections, it is better to use the Wilcoxon test, the logrank test or the more robust Q-test.

*In the case of censoring data,* test power depends on distribution of censoring time, if survival curves are crossed, especially for the Bagdonavičius–Nikulin tests. In contrast, if survival curves are not crossed, the choice of censoring time distribution is not so important. If survival curves have intersections, then the Bagdonavičius–Nikulin tests will be the most powerful, but this decreases sharply when the censoring rate increases. If survival curves do not have intersections, then the logrank test, the Cox–Mantel test, Gehan's generalized Wilcoxon test and Peto and Peto's generalized Wilcoxon test are more powerful than the Bagdonavičius–Nikulin tests.

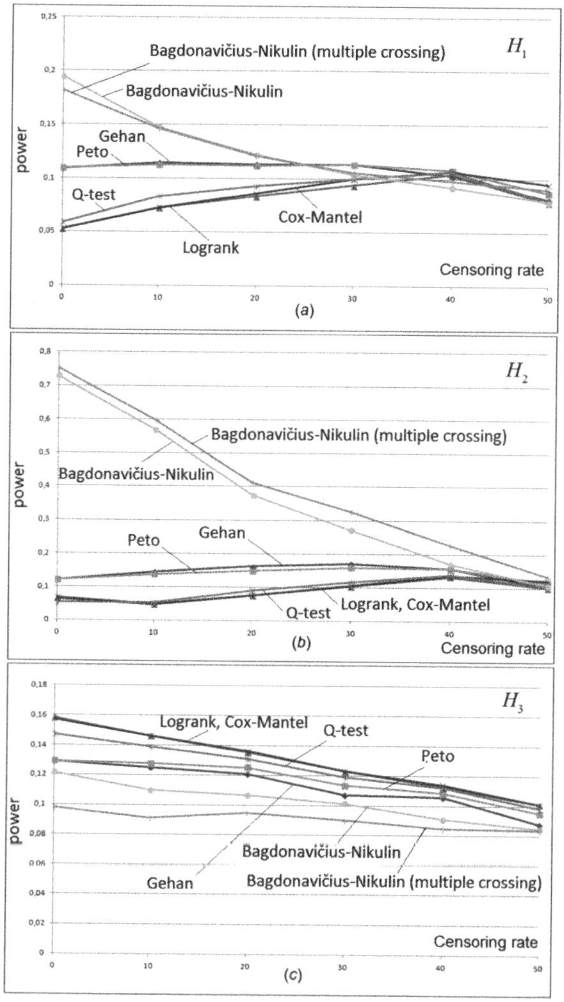


**Figure 12.2.** *The power of tests against alternatives $H_1$ (a), $H_2$ (b) and $H_3$ (c) for censoring rates 0–50%. The distribution of censoring times $F^C(t)$ is the Weibull distribution*

## 12.5. Acknowledgment

This work was supported by the Ministry of Education and Science of the Russian Federation within the state assignment (project 8.1274.2011) and the Federal target programme "Research and Pedagogical Staff for Innovative Russia" (agreement #14.B37.21.0860).

## 12.6. Bibliography

[ALL 97] ALLEN D.L., "Hypothesis testing using an L1-distance bootstrap", *The American Statistician*, vol. 51, pp. 145–150, 1997.

[BAG 04] BAGDONAVIČIUS V.B., LEVULIENE R.J., NIKULIN M.S., *et al.*, "Tests for equality of survival distributions against non-location alternatives", *Lifetime Data Analysis*, vol. 10, no. 4, pp. 445–460, 2004.

[BAG 06] BAGDONAVIČIUS V.B., NIKULIN M., "On goodness-of-fit tests for homogeneity and proportional hazards", *Applied Stochastic Models in Business and Industry*, vol. 22, no. 1, pp. 607–619, 2006.

[BAG 08] BAGDONAVIČIUS V., MALOV S.V., NIKULIN M., "Testing of the homogeneity of marginal distribution in copula models", *Comptes Rendus de l'Academie de Paris*, vol. 346, no. 7–8, pp. 445-450, 2008.

[BAG 12] BAGDONAVIČIUS V., MALOV S.V., NIKULIN M., "Homogeneity tests for related samples under censoring", *Communications in Statistics - Simulation and Computation*, vol. 41, no. 6, pp. 764–775, 2012.

[DAR 57] DARLING D.A., "The Kolmogorov-Smirnov, Cramer-von Mises tests", *Ann. Math. Statist*, vol. 28, pp. 823–838, 1957.

[FIS 60] FISZ M., "On a result by M. Rosenblatt concerning the Mises-Smirnov test", *Annals of Mathematical Statistics*, vol. 31, no. 1, pp. 427–429, 1960.

[GEH 65] GEHAN E.A., "A generalized Wilcoxon test for comparing arbitrarily singly-censored samples", *Biometrika*, vol. 52, no. 1/2, pp. 203–223, 1965.

[LEE 03] LEE E.T., WANG J.W., *Statistical Methods for Survival Data Analysis*, Wiley Series in Probability and Statistics, Wiley, 2003.

[LEH 51] LEHMANN E.L., "Consistency and unbiasedness of certain nonparametric tests", *Ann. Math. Statist.*, vol. 22, no. 1, pp. 165–179, 1951.

[LEM 05] LEMESHKO B.Y., LEMESHKO S.B., "Statistical distribution convergence and homogeneity test power for Smirnov and Lehmann-Rosenblatt tests", *Measurement Techniques*, vol. 48, no. 12, pp. 1159–1166, 2005.

[MAN 67] MANTEL N., "Evaluation of survival data and two new rank order statistics arising in its consideration", *Cancer Chemotherapy Rep.*, vol. 50, pp. 163–170, 1967.

[MAR 10] MARTINEZ RUVIE L.M.C., NARANJO JOSHUA D., "A pretest for choosing between Logrank and Wilcoxon tests in the two-sample problem", *International Journal of Statistics*, vol. LXVIII, no. 2, pp. 111–125, 2010.

[PET 72] PETO R., PETO J., "Asymptotically efficient rank invariant test procedures", *Journal of the Royal Statistical Society, Series A (General)*, vol. 135, no. 2, pp. 185–207, 1972.

[PET 76] PETTITT A.N., "A two-sample Anderson-Darling rank statistic", *Biometrika*, vol. 63, no. 1, pp. 161–168, 1976.

[ROS 52] ROSENBLATT M., "Limit theorems associated with variants of the von Mises statistic", *Ann. Math. Statist.*, vol. 23, no. 1, pp. 617–623, 1952.

[SCH 87] SCHOLZ F.W., STEPHENS M.A., "K-sample Anderson-Darling Tests", *Journal of the American Statistical Association*, vol. 82, no. 399, pp. 918–924, 1987.

[SMI 48] SMIRNOV N.V., "Table for estimating the goodness of fit of empirical distributions", *Ann. Math. Stat.*, vol. 19, pp. 279–281, 1948.

[WIL 45] WILCOXON F., "Individual comparisons by ranking methods", *Biometrics Bulletin*, vol. 1, no. 6, pp. 80–83, International Biometric Society, 1945.

[ZHA 06] ZHANG J., "Powerful two-sample tests based on the likelihood ratio", *Technometrics*, vol. 48, no. 1, pp. 95–103, 2006.

Chapter 13

# Some Asymptotic Results for Exchangeably Weighted Bootstraps of the Empirical Estimator of a Semi-Markov Kernel with Applications

The aim of this chapter is to present a general bootstrap by exchangeable weight random variables for empirical estimators of the semi-Markov kernels and of the conditional transition probabilities for semi-Markov processes with countable state space. We show how to apply our results in statistical problems such as confidence bands.

## 13.1. Introduction

Recall that the bootstrap technique, which is a form of resampling procedure for statistical inference, was introduced in [EFR 79]. The bootstrap may be described briefly as follows. Let $T(F)$ be a functional of an unknown distribution function [df] $F(\cdot)$, $X_1,\ldots,X_n$ a sample from $F(\cdot)$, and $X_1^*,\ldots,X_n^*$ an independent and identically distributed (i.i.d.) sample with common distribution given by the empirical distribution $F_n(\cdot)$ of the original sample. The distribution of $\{T(F_n) - T(F)\}$ is then approximated by that of $\{T(F_n^*) - T(F_n)\}$ conditionally upon $X_1,\ldots,X_n$, with $F_n^*(\cdot)$ being the empirical distribution of $X_1^*,\ldots,X_n^*$. The key idea behind the bootstrap is that if a sample is representative of the underlying

Chapter written by Salim BOUZEBDA and Nikolaos LIMNIOS.

population, then we can make inferences about the population characteristics by resampling from the current sample. Roughly speaking, it is known that the bootstrap works in the i.i.d. case if and only if the central limit theorem holds for the random variable under consideration. For further discussion, we refer the reader to [GIN 90]. Note that the bootstrap, according to Efron's original formulation (see [EFR 79]), presents some drawbacks. For example, some observations may be used more than once while others are not sampled at all. To overcome this problem, a more general formulation of the bootstrap has been introduced, the *weighted* (or *smooth*) bootstrap, which has also been shown to be computationally more efficient in several applications. For a survey of further results on weighted bootstrap, the reader is referred to [BAR 95, SHA 95] and the references therein. A substantial body of literature, reviewed in [BER 03], gives conditions for the bootstrap to be satisfied in order to provide desirable distributional approximations. In [BIC 97], the performance of different kinds of bootstrap procedure is investigated through asymptotic results and small sample simulation studies. A first general resampling scheme was proposed in [RUB 81] and was extensively studied by Bickel and Freedman [BIC 81], who suggested the name "weighted bootstrap", e.g. Bayesian bootstrap when the vector of weights

$$(W_{n1},\ldots,W_{nn}) = (D_{n1},\ldots,D_{nn}),$$

is equal in distribution to the vector of $n$ spacings of $n-1$ ordered uniform $(0,1)$ random variables, that is,

$$(D_{n1},\ldots,D_{nn}) \sim \text{Dirichlet}(n;1,\ldots,1).$$

The reader may refer to [LO 93]. The case

$$(D_{n1},\ldots,D_{nn}) \sim \text{Dirichlet}(n;4,\ldots,4),$$

was considered in [WEN 89, remark 2.3] and [ZHE 88, remark 13.5]. These resampling methods led to the interest of a unified approach, generically designated as general weighted resampling, which was first proposed by Mason and Newton [MAS 92] and among others extended by Præstgaard and Wellner [PRÆ 93] and Del Barrio and Matrán [DEL 00]. Holmes and Reinert [HOL 04] provided new proofs for many known results about the convergence in law of the bootstrap distribution to the true distribution of smooth statistics employing the techniques based on Stein's method for empirical processes. Note that other variations of Efron's bootstrap are studied in [CHA 05] using the term "generalized bootstrap". We may also refer to [BOU 12a, BOU 12b, BOU 12c] and the references therein. There is a huge literature on the application of the bootstrap methodology to non-parametric kernel density and regression estimation, among other statistical procedures, and it is not the purpose of this chapter to survey this extensive literature.

The theory of the semi-Markov processes is given by Pyke [PYK 61a, PYK 61b]. For recent references in this area along with statistical applications, see, e.g., [LIM 01] and [BAR 08]. Nonparametric estimation of semi-Markov kernel and conditional transition probability has been the subject of intense investigation for many years leading to the development of a large variety of methods, see, e.g., [LIM 04] and [BAR 08] and the references therein. Note that the limiting distributions of these estimators, or their functionals, are rather complicated, which does not permit explicit computation in practice. To overcome this difficulty, we will propose a general bootstrap of empirical semi-Markov kernels and the conditional transition probabilities and study some of its asymptotic properties by means of martingale techniques. The importance of considering general bootstrap instead of particular cases lies in the fact that we need, in general, a more flexible modeling to handle the problems in practice. The results presented in this chapter are useful in many statistical problems, in the semi-Markov framework, as will be illustrated in the construction of confidence intervals. Notice that all the results presented here are presented without proof, all details and mathematical developments are given in [BOU 13b, BOU 13a].

This chapter is organized as follows. Section 13.2 contains some basic definitions and notation regarding the semi-Markov process. The bootstrap, for empirical estimators of the semi-Markov kernels and the conditional transition probabilities for semi-Markov processes with countable state space, are introduced, in detail, and their asymptotic properties are given in section 13.3 and the extension to the multivariate case is given in section 13.4. In section 13.5, an application concerning the confidence intervals for semi-Markov kernels is given.

## 13.2. Semi-Markov setting

This section first sets out the basic definitions and assumptions that are used throughout the chapter. To define equivalently Markov renewal processes, it is natural, first, to define semi-Markov kernels (see, e.g., [LIM 01]). All random processes will be supposed to be defined on a complete probability space $(\Omega, \mathcal{F}, \mathbb{P})$. Consider an infinite countable set, say $E$, and an $E$-valued càdlàg time-homogeneous semi-Markov process $Z_t$, $t \in \mathbb{R}_+$, with embedded Markov renewal process $(J_k, S_k)$, for $k \in \mathbb{N}$, where $(J_k)$ is the $E$-valued embedded Markov chain (EMC) of the successive visited states, and

$$0 = S_0 \leq S_1 \leq \cdots \leq S_k \leq S_{k+1} \leq \cdots$$

are the jump times of $(Z_t)$. Define also $X_k := S_k - S_{k-1}, k \geq 1$, the interjump times, and the process $N(t)$, $t \in \mathbb{R}_+$, which counts the number of jumps of $(Z_t)$, in the time interval $(0, t]$, by

$$N(t) := \sup\{k \geq 0 : S_k \leq t\}.$$

Let us also define $N_i(t)$ as the number of visits of $(Z_t)$ to state $i \in E$ up to time $t$, and $N_{ij}(t)$ the number of direct jumps of $(Z_t)$ from state $i$ to state $j$ up to time $t$, (see, e.g., [LIM 01]). More precisely, we have

$$N_i(t) := \sum_{k=1}^{N(t)} \mathbb{1}_{\{J_{k-1}=i\}} \text{ and } N_{ij}(t) := \sum_{k=1}^{N(t)} \mathbb{1}_{\{J_{k-1}=i, J_k=j\}},$$

here and elsewhere, $\mathbb{1}_A$ stands for the indicator function of the event $A$. Consider the renewal process $(S_n^i)_{n\in\mathbb{N}}$ of the successive times of visits to state $i$, $N_i(t)$ is its counting process of jumps. Let us define the following observation in the interval $(0, t]$, for $t > 0$,

$$\mathcal{H}_t := \{Z_u, 0 \leq u \leq t\} = \begin{cases} \{J_0, X_1, \ldots, J_{N(t)}, U_t\} & \text{if } N(t) > 0, \\ \{J_0, U_t = t\} & \text{if } N(t) = 0, \end{cases} \quad [13.1]$$

where $U_t := t - S_{N(t)}$ the backward recurrence time. Let us denote by $Q(t) = \{Q_{ij}(t) : i, j \in E\}, t \geq 0$ the semi-Markov kernel, which is defined by

$$\begin{aligned} Q_{ij}(t) &:= \mathbb{P}(J_{k+1} = j, X_{k+1} \leq t | J_k = i) \\ &= P(i,j) F_{ij}(t), \quad t \geq 0, \quad i, j \in E, \end{aligned} \quad [13.2]$$

where

$$P(i,j) := \mathbb{P}(J_{k+1} = j | J_k = i)$$

is the transition kernel of the EMC $(J_k)$, and

$$F_{ij}(t) := \mathbb{P}(X_{k+1} \leq t | J_k = i, J_{k+1} = j)$$

is the conditional distribution function of the sojourn time in the state $i$, given that the next visited state is $j$, $j \neq i$. The empirical estimator $\widehat{Q}(x,t) := \{\widehat{Q}_{ij}(x,t) : i, j \in E\}$, $x \in (0, t]$, for $t > 0$, of the semi-Markov kernel $Q(x)$ is defined by the following formula

$$\widehat{Q}_{ij}(x,t) := \frac{1}{N_i(t)} \sum_{k=1}^{N(t)} \mathbb{1}_{\{J_{k-1}=i, J_k=j, X_k \leq x\}}, \quad 0 \leq x \leq t, \quad i, j \in E. \quad [13.3]$$

The empirical estimator $\widehat{F}(x,t) := \{\widehat{F}_{ij}(x,t) : i, j \in E\}$, $x \in (0, t]$, for $t > 0$, of the sojourn-time distribution function $F_{ij}(x)$ is given by

$$\widehat{F}_{ij}(x,t) := \frac{1}{N_{ij}(t)} \sum_{k=1}^{N(t)} \mathbb{1}_{\{J_{k-1}=i, J_k=j, X_k \leq x\}}, \quad 0 \leq x \leq t, \quad i, j \in E. \quad [13.4]$$

Let us denote by $\mu_{ii}$ the mean recurrence times of $(S_n^i)$, i.e.,

$$\mu_{ii} := \mathbb{E}[S_2^i - S_1^i],$$

and by $\nu = (\nu)_{i \in E}$ the stationary distribution of the EMC $J$. Let us define the mean sojourn time of the semi-Markov process $Z$ as

$$\overline{m} := \sum_{i \in E} \nu_i m_i$$

with $m_i$ defined to be the mean sojourn time in state $i \in E$, i.e.,

$$m_i := \int_0^\infty [1 - \sum_{i \in E} Q_{ij}(t)]dt.$$

We also recall the property

$$\mu_{ii} = \overline{m}/\nu_i.$$

Empirical estimation of the stationary distribution of the EMC and for the mean sojourn time are presented in [LIM 05]. In the continuous time case, we can define the filtration $\mathbf{F} = (\mathcal{F}_t)_{t \in \mathbb{R}}$ with respect to a non-empty space $\Omega$ contained in $\mathcal{F}$, that is an increasing and right-continuous family of sub-$\sigma$-algebras of $\mathcal{F}$. Moreover, we assume that $\mathcal{F}_0$ contains all $\mathbb{P}$-null sets of $\mathcal{F}$. Consider the continuous-time filtered probability space with respect to filtration $\mathbf{F}$ as the quadruple $\mathcal{B} = (\Omega, \mathcal{F}, \mathbf{F}, \mathbb{P})$. For each $n \in \mathbb{N}$, let $\mathcal{B}^n = (\Omega^n, \mathcal{F}^n, \mathbf{F}^n, \mathbb{P}^n)$, $\mathbf{F}^n = (\mathcal{F}_t^n)_{t \in \mathbb{R}_+}$, be a stochastic basis. In the discrete time case, the filtration is denoted as $\widetilde{\mathbf{F}} = (\mathcal{F}_k)_{k \in \mathbb{N}}$. Therefore, the discrete-time filtered probability space with respect to filtration $\widetilde{\mathbf{F}}$ is accordingly the quadruple $\widetilde{\mathcal{B}} = (\Omega, \mathcal{F}, \widetilde{\mathbf{F}}, \mathbb{P})$. Consider, for all $n \in \mathbb{N}$, a discrete-time stochastic basis $\widetilde{\mathcal{B}}^n = (\Omega^n, \mathcal{F}^n, \widetilde{\mathbf{F}}^n, \mathbb{P}^n)$. In our case, assume that, for any $n \in \mathbb{N}^*$, it holds $\mathcal{F}_k^n = \mathcal{F}_k = \sigma(J_0, J_i, X_i; 0 \leq i \leq k)$ with $X_0 = 0$, and consider the continuous-time filtration $\mathcal{F}_t^n = \mathcal{F}_{N(nt)}$.

To establish our results, the following conditions are tacitly assumed in our analysis:

1) the $(J_n)$ is ergodic with stationary distribution $\nu$;

2) the mean sojourn times $(m_i)_{i \in E}$ are finite;

3) there are no degenerate distributions (i.e. Dirac distributions concentrated at 0).

For any states $i, j \in E$ and any fixed time $x \in \mathbb{R}_+$, we set $Y_k = \{Y_k^{i,j} : i, j \in E\} \in \mathbb{R}^{d^2}$, where $(Y_k^{ij})_{k\in\mathbb{N}}$ are random sequences on $\widetilde{\mathcal{B}}$ defined by

$$Y_k^{ij} := \mathbb{1}_{\{J_{k-1}=i, J_k=j, X_k \leq x\}} - \mathbb{1}_{\{J_{k-1}=i\}} Q_{ij}(x).$$

Let us define the double sequence $(Y_k^n)_{k\in\mathbb{N}^*}$, $n \in \mathbb{N}^*$, on $\widetilde{\mathcal{B}}^n$, by

$$Y_k^n = \frac{Y_k}{\sqrt{n}}. \qquad [13.5]$$

Also, for any $n \in \mathbb{N}^*$, let us define the stochastic processes

$$S_t^{ij,n} := \sum_{k=1}^{N(nt)} Y_k^{ij,n} = \frac{1}{\sqrt{n}} \sum_{k=1}^{N(nt)} Y_k^{ij}, \quad t \in \mathbb{R}_+^*.$$

Making use of the above notation, we have the sequence of stochastic processes on $\mathcal{B}^n$

$$S_t^n = \{S_t^{ij,n} : i, j \in E\}, \quad t \in \mathbb{R}_+^*, n \in \mathbb{N}^*.$$

Let $D[0, \infty)$ be the Skorohod space, and by "$\rightsquigarrow$" we denote the weak convergence in this space and by $\xrightarrow{D}$ we denote the convergence in distribution of a sequence of random variables. The process $W_t := \{W_t^{ij} : i, j \in E\}$ is the $d^2$-dimensional standard Brownian motion on $\mathcal{B}$. For any fixed $x \in \mathbb{R}_+$, the following weak convergence holds, as $n$ tends to infinity

$$\{S_t^n : t \in \mathbb{R}_+^*\} \rightsquigarrow \{\mathcal{C}^{1/2} W(t) : t \in \mathbb{R}_+^*\}, \qquad [13.6]$$

where

$$\mathcal{C} = (\mathcal{C}^{i,j,\ell,r})_{(i,j),(\ell,r)\in E\times E} = \left(\delta_{i\ell} \frac{1}{\mu_{ii}} Q_{ij}(x)(\delta_{jr} - Q_{ir}(x))\right)_{(i,j),(\ell,r)\in E\times E}$$

is the covariance matrix and $\delta_{ij}$ stands for the Kronecker symbol. For more details on the convergence in [13.6], we may refer to [GEO 12].

## 13.3. Main results

In this section, we will establish the asymptotic properties of bootstrapped empirical estimators, under quite general conditions, in the framework of semi-Markov processes. Let $\mathbf{W} \equiv (W_{nj}, j = 1, \ldots, n, n = 1, 2 \ldots)$ be a triangular array of random variables. This array determines a weighted bootstrap of the empirical estimator of the semi-Markov kernel by

$$\widehat{Q}_{ij}^{W}(x,t) := \frac{1}{N_i(t)} \sum_{k=1}^{N(t)} W_{N(t)k} \mathbb{1}_{\{J_{k-1}=i, J_k=j, X_k \le x\}}, \quad 0 \le x \le t, \quad i, j \in E, \quad [13.7]$$

and the weighted bootstrap of the empirical estimator of the conditional transition distribution functions by

$$\widehat{F}_{ij}^{W}(x,t) := \frac{1}{N_{ij}(t)} \sum_{k=1}^{N(t)} W_{N(t)k} \mathbb{1}_{\{J_{k-1}=i, J_k=j, X_k \le x\}}, \quad 0 \le x \le t, \quad i, j \in E. \, [13.8]$$

The bootstrap weights $W_{ni}$'s are assumed to belong to the class of weights investigated in [PRÆ 93]. We will assume the following conditions.

W.1 *The vector* $W_n = (W_{n1}, \ldots, W_{nn})^\top$ *is exchangeable for any* $n = 1, 2, \ldots$, *i.e., for any permutation* $\pi = (\pi_1, \ldots, \pi_n)$ *of* $(1, \ldots, n)$, *the joint distribution of*

$$\pi(W_n) = (W_{n\pi_1}, \ldots, W_{n\pi_n})^\top$$

*is the same as that of* $W_n$.

W.2 $W_{ni} \ge 0$ *for all* $n$, $i$ *and* $\sum_{i=1}^{n} W_{ni} = n$ *for all* $n$.

W.3 $\limsup_{n\to\infty} \|W_{n1}\|_{2,1} \le C < \infty$, *where*

$$\|W_{n1}\|_{2,1} = \int_0^\infty \sqrt{\mathbb{P}(W_{n1} > u)}du.$$

W.4 $\lim_{\lambda\rightsquigarrow\infty} \limsup_{n\to\infty} \sup_{t\ge\lambda} t^2 \mathbb{P}(W_{n1} > t) = 0.$

W.5 $(1/n)\sum_{i=1}^{n}(W_{ni} - 1)^2 \xrightarrow{\mathbb{P}} c^2 > 0.$

It is worth noticing that this corresponds to the choice

$$W_n \sim \text{Multinomial}(n; n^{-1}, \ldots, n^{-1})$$

for which conditions W.1–W.5 are satisfied. In general, conditions W.3–W.5 are satisfied under some moment conditions on $W_{ni}$ (see [PRÆ 93, lemma 3.1]). In addition to Efron's nonparametric bootstrap, the sampling schemes that satisfy conditions W.1–W.5 include *Bayesian bootstrap, Multiplier bootstrap, Double bootstrap and Urn bootstrap*. This list is sufficiently long to indicate that conditions W.1–W.5 are not unduly restrictive. Note that the value of $c$ in W.5 is independent of the sample size and depends on the resampling method, e.g. $c = 1$ for the nonparametric bootstrap and Bayesian bootstrap, whereas $c = \sqrt{2}$ for the double bootstrap. A more precise discussion of this general formulation of the bootstrap and further details can be found in [MAS 92], [PRÆ 93], [VAN 96, section 3.6.2, p. 353], [KOS 08, section 10, p. 179], [CHE 10] and the references therein. The reader may refer to [BIL 68] for excellent general coverage of the theory of exchangeability. Throughout this chapter, we assume that the bootstrap weights $W_{ni}$ are independent from the data $\mathcal{H}_t$s, given in [13.1].

COMMENT 13.1.– Note that an appropriate choice of the bootstrap weights $W_{ni}$ implies a smaller limit variance, that is $c^2$ is smaller than 1. For instance, typical example is the multivariate hypergeometric bootstrap, (see [PRÆ 93, example 3.4]).

For any fixed states $i, j \in E$ and fixed $x \in \mathbb{R}_+$, let us define the random sequences $(Y_\ell)_{\ell\geq 1}$ and $(Y_\ell^*)_{\ell\geq 1}$, respectively, by

$$\begin{aligned} Y_\ell &:= Y_\ell(x) := \mathbb{1}_{\{J_{\ell-1}=i, J_\ell=j, X_\ell\leq x\}} - \mathbb{1}_{\{J_{\ell-1}=i\}} Q_{ij}(x), \\ Y_\ell^* &:= Y_\ell^*(x) := W_{\lfloor nt \rfloor \ell} \left( \mathbb{1}_{\{J_{\ell-1}=i, J_\ell=j, X_\ell\leq x\}} - \mathbb{1}_{\{J_{\ell-1}=i\}} Q_{ij}(x, nt) \right), \end{aligned}$$

with $\lfloor u \rfloor \leq u < \lfloor u \rfloor + 1$ denoting the integer part of $u$, and for $k \geq 1$, define the following sums

$$S_k := \sum_{\ell=1}^{k} Y_\ell \text{ and } S_k^* := \sum_{\ell=1}^{k} Y_\ell^*. \qquad [13.9]$$

We will use the following basic definitions throughout. By a Brownian bridge $\{B(t) : 0 \leq t \leq 1\}$, respectively, by a standard Wiener process $\{W(y) : y \geq 0\}$, is meant a centered Gaussian process with continuous sample paths, and covariance function

$$E(B(s)B(t)) = s \wedge t - st, \quad \text{for} \quad 0 \leq s, t \leq 1,$$

respectively,

$$E(W(x)W(y)) = x \wedge y, \quad \text{for} \quad x, y \geq 0.$$

Limnios [LIM 04] established the following result. For any fixed states $i, j \in E$ and fixed $x \in \mathbb{R}_+$, the following weak convergence holds, as $n$ tends to infinity,

$$n^{-1/2} S_{\lfloor nt \rfloor} \rightsquigarrow \sigma_{ij}(x) W(t), \qquad [13.10]$$

provided that $\sigma_{ij}^2(x) > 0$, where

$$\sigma_{ij}^2(x) := \nu_i Q_{ij}(x)(1 - Q_{ij}(x)).$$

The following theorem gives the bootstrap version of [13.10].

THEOREM 13.1.– Let $\mathbf{W}$ be a triangular array of bootstrap weights satisfying assumptions W.1–W.5. For any fixed states $i, j \in E$ and fixed $x \in \mathbb{R}_+$, the following weak convergence holds, as $n \to \infty$,

$$n^{-1/2} S^*_{\lfloor nt \rfloor} \rightsquigarrow c\sigma_{ij}(x) W(t), \qquad [13.11]$$

provided that $\sigma_{ij}^2(x) > 0$, where

$$\sigma_{ij}^2(x) := \nu_i Q_{ij}(x)(1 - Q_{ij}(x)).$$

THEOREM 13.2.– Let $\mathbf{W}$ be a triangular array of bootstrap weights satisfying assumptions W.1–W.5. For any arbitrary but fixed $i, j \in E$ and fixed $x \in \mathbb{R}_+$, we have the following weak convergence, as $n \to \infty$, for $t > 0$,

$$n^{1/2}(\widehat{Q}_{ij}^W(x, nt) - \widehat{Q}_{ij}(x, nt)) \rightsquigarrow c b_{ij}(x) W(t)/t, \qquad [13.12]$$

provided that $b_{ij}^2(x) > 0$, where

$$b_{ij}^2(x) = \mu_{ii} Q_{ij}(x)(1 - Q_{ij}(x)).$$

THEOREM 13.3.– Let $\mathbf{W}$ be a triangular array of bootstrap weights satisfying assumptions W.1–W.5. For any arbitrary but fixed $i, j \in E$, we have, almost surely, along $\mathcal{H}_t$, the following weak convergence, as $n \to \infty$, for $t > 0$ and $x \in \mathbb{R}_+$,

$$n^{1/2}(\widehat{Q}_{ij}^W(x, nt) - \widehat{Q}_{ij}(x, nt)) \rightsquigarrow \frac{\mu_{ii}^{1/2}}{t^{1/2}} c B(Q_{ij}(x)). \qquad [13.13]$$

COMMENT 13.2.– We can see that theorem 13.3 coincides with theorem 3.1 of [MAS 92], if we replace the $Z_t$ by a sequence of independent random variables.

The following corollary is a straightforward result of theorem 13.2.

COROLLARY 13.1.– Let $\mathbf{W}$ be a triangular array of bootstrap weights satisfying assumptions W.1–W.5. For any arbitrary but fixed $i, j \in E$ and fixed $x \in \mathbb{R}_+$, we have the following convergence in distribution, as $t \to \infty$,

$$t^{1/2}(\widehat{Q}_{ij}^W(x,t) - \widehat{Q}_{ij}(x,t)) \xrightarrow{D} N(0, c^2 b_{ij}^2(x)), \qquad [13.14]$$

where $N(0, c^2 b_{ij}^2(x))$ is a normal random variable with mean 0 and variance $c^2 b_{ij}^2(x)$.

For any fixed states $i, j \in E$ and fixed $x \in \mathbb{R}_+$, let us define the random sequences $(Y_\ell')_{\ell \geq 1}$ and $(Y_\ell'^*)_{\ell \geq 1}$ by

$$\begin{aligned} Y_\ell' &:= Y_\ell'(x) := \mathbb{1}_{\{J_{\ell-1}=i, J_\ell=j, X_\ell \leq x\}} - \mathbb{1}_{\{J_{\ell-1}=i\}} F_{ij}(x), \\ Y_\ell^{*\prime} &:= Y_\ell^{*\prime}(x) := (W_{\lfloor nt \rfloor \ell} - 1)\left(\mathbb{1}_{\{J_{\ell-1}=i, J_\ell=j, X_\ell \leq x\}} - \mathbb{1}_{\{J_{\ell-1}=i\}} F_{ij}(x)\right), \end{aligned}$$

and, for $k \geq 1$, define the following sums

$$S_k' := \sum_{\ell=1}^{k} Y_\ell' \text{ and } S_k^{*\prime} := \sum_{\ell=1}^{k} Y_\ell^{*\prime}.$$

Limnios [LIM 04] established the following result. For any fixed states $i, j \in E$ and fixed $x \in \mathbb{R}_+$, the following weak convergence holds, as $n$ tends to infinity,

$$n^{-1/2} S_{\lfloor nt \rfloor}' \rightsquigarrow \sigma_{ij}'(x) W(t), \qquad [13.15]$$

provided that $\sigma_{ij}'^2(x) > 0$, where

$$\sigma_{ij}'^2(x) := \nu_i Q_{ij}(x)(1 - F_{ij}(x)).$$

We state our result in the following theorem that gives the bootstrap version [13.15].

THEOREM 13.4.– Let $\mathbf{W}$ be a triangular array of bootstrap weights satisfying assumptions W.1–W.5. For any fixed states $i, j \in E$ and fixed $x \in \mathbb{R}_+$, the following weak convergence holds, as $n \to \infty$,

$$n^{1/2} S_{\lfloor nt \rfloor}^{*\prime} \rightsquigarrow c\sigma_{ij}'(x) W(t), \qquad [13.16]$$

provided that $\sigma_{ij}'^2(x) > 0$, where

$$\sigma_{ij}'^2(x) := \nu_i Q_{ij}(x)(1 - F_{ij}(x)).$$

THEOREM 13.5.– Let **W** be a triangular array of bootstrap weights satisfying assumptions W.1–W.5. For any arbitrary but fixed $i, j \in E$ and fixed $x \in \mathbb{R}_+$, we have the following weak convergence, as $n \to \infty$, for $t > 0$,

$$n^{1/2}(\widehat{F}_{ij}^{W}(x, nt) - \widehat{F}_{ij}(x, nt)) \rightsquigarrow c c_{ij}(x) W(t)/t, \qquad [13.17]$$

provided that $c_{ij}^2(x) > 0$, where

$$c_{ij}^2(x) = \frac{\mu_{ii}}{P(i,j)} Q_{ij}(x)(1 - F_{ij}(x)).$$

THEOREM 13.6.– Let **W** be a triangular array of bootstrap weights satisfying assumptions W.1–W.5. For any arbitrary but fixed $i, j \in E$, we have, almost surely, along $\mathcal{H}_t$, the following weak convergence, as $n \to \infty$, for $t > 0$ and $x \in \mathbb{R}_+$,

$$n^{1/2}(\widehat{F}_{ij}^{W}(x, nt) - \widehat{F}_{ij}(x, nt)) \rightsquigarrow \frac{\mu_{ii}^{1/2}}{t^{1/2}} c B(F_{ij}(x)). \qquad [13.18]$$

The following corollary is a direct result of theorem 13.5.

COROLLARY 13.2.– Let **W** be a triangular array of bootstrap weights satisfying assumptions W.1–W.5. For any arbitrary but fixed $i, j \in E$ and fixed $x \in \mathbb{R}_+$, we have the following convergence in distribution, as $t \to \infty$,

$$t^{1/2}(\widehat{F}_{ij}^{W}(x, t) - \widehat{F}_{ij}(x, t)) \xrightarrow{D} N(0, c^2 c_{ij}^2(x)). \qquad [13.19]$$

We conclude this section by the following remark.

COMMENT 13.3.– Note that the choice of weights depends on the problem at hand: the accuracy of the estimation of the entire distribution of the statistic, the accuracy of a confidence interval, the accuracy in the large deviation sense, accuracy for a finite sample size, we may refer to [JAM 97] and the references therein for more details. Barbe and Bertail [BAR 95] point out that the area where the weighted bootstrap clearly performs better than the classical bootstrap is in term of coverage accuracy.

## 13.4. Bootstrap for a multidimensional empirical estimator of a continuous-time semi-Markov kernel

We now provide extensions of the results of section 13.3 to the multidimensional case. In this section, we consider a finite set $E = \{1, \ldots, d\}$. For any states $i, j \in E$ and any fixed time $x \in \mathbb{R}_+$, we set $Y_k^* = \{Y_k^{*ij} : i, j \in E\} \in \mathbb{R}^{d^2}$, where $(Y_k^{*ij})_{k \in \mathbb{N}^*}$ are random sequences on $\widetilde{\mathcal{B}}$ defined by

$$Y_k^{*ij} := (W_{nk} - 1)\left\{\mathbb{1}_{\{J_{k-1}=i, J_k=j, X_k \leq x\}} - \mathbb{1}_{\{J_{k-1}=i\}} Q_{ij}(x)\right\}.$$

Let us define the double sequence $(Y_k^n)_{k\in\mathbb{N}^*}$, $n \in \mathbb{N}^*$, on $\widetilde{\mathcal{B}}^n$, by

$$Y_k^{*n} = \frac{Y_k^*}{\sqrt{n}}. \quad [13.20]$$

Also, for any $n \in \mathbb{N}^*$, let us define the stochastic processes

$$S_t^{ij*,n} := \sum_{k=1}^{N(nt)} Y_k^{ij*,n} = \frac{1}{\sqrt{n}} \sum_{k=1}^{N(nt)} Y_k^{*ij}, \quad t \in \mathbb{R}_+^*. \quad [13.21]$$

Making use of the notation above, we have the sequence of stochastic processes on $\mathcal{B}^n$

$$S_t^{*n} = \{S_t^{ij*,n} : i,j \in E\}, \quad t \in \mathbb{R}_+^*, n \in \mathbb{N}^*. \quad [13.22]$$

THEOREM 13.7.– Let $\mathbf{W}$ be a triangular array of bootstrap weights satisfying assumptions W.1–W.5. For any fixed $x \in \mathbb{R}_+$, the following weak convergence holds, as $n$ tends to infinity

$$\{S_t^{*n} : t \in \mathbb{R}_+^*\} \rightsquigarrow \{c \times \mathcal{C}^{1/2} W(t) : t \in \mathbb{R}_+^*\},$$

where $c$ is the constant given in condition $W.5$. For any $i, j \in E$, we set the processes

$$\Delta Q_{ij}(\cdot, nt) := \widehat{Q}_{ij}^W(\cdot, nt) - \widehat{Q}_{ij}(\cdot, nt).$$

THEOREM 13.8.– Let $\mathbf{W}$ be a triangular array of bootstrap weights satisfying assumptions W.1–W.5. For any fixed $x \in \mathbb{R}_+$, it holds the weak convergence, as $n \to \infty$,

$$\{n^{1/2}\Delta Q_{ij}(x, nt) : i,j \in E, t \in \mathbb{R}_+^*\} \rightsquigarrow \{\mathcal{G}_{ij}(t) : i,j \in E, t \in \mathbb{R}_+^*\}, \quad [13.23]$$

where

$$\mathcal{G}_{ij}(t) = \mu_{ii} c \sqrt{\mathcal{C}^{i,j,\ell,r}} W^{ij}(t)/t$$

is a continuous Gaussian process.

The following corollary is a straightforward result of theorem 13.4.

COROLLARY 13.3.– Let $\mathbf{W}$ be a triangular array of bootstrap weights satisfying assumptions W.1–W.5. For any fixed $x \in \mathbb{R}_+$, we have, as $n$ tends to infinity,

$$\{n^{1/2}\Delta Q_{ij}(x, n) : i,j \in E\} \xrightarrow{D} N(\mathbf{0}, c^2\Sigma_Q), \quad [13.24]$$

where $N(\mathbf{0}, c^2\Sigma_Q)$ is a $d^2$-dimensional normal random variable with mean $\mathbf{0}$ and covariance matrix

$$c\Sigma_Q = c^2 \left(\mu_{ii}^2 C^{i,j,\ell,r}\right)_{(i,j),(\ell,r)\in E\times E}.$$

We proceed now to the multidimensional functional central limit theorem concerning the bootstrapped empirical estimator [13.8] of the conditional sojourn–time distribution function. For any states $i,j \in E$ and any fixed time $x \in \mathbb{R}_+$, we set $V_k^* = \{V_k^{*ij} : i,j \in E\} \in \mathbb{R}^{d^2}$, where $(V_k^{*ij})_{k\in\mathbb{N}^*}$ are random sequences on $\widetilde{\mathcal{B}}$ defined by

$$V_k^* := (W_{nk} - 1)\left\{\mathbb{1}_{\{J_{k-1}=i, J_k=j, X_k\le x\}} - \mathbb{1}_{\{J_{k-1}=i\}}F_{ij}(x)\right\}.$$

Let us define the double sequence $(V_k^n)_{k\in\mathbb{N}^*}$, $n \in \mathbb{N}^*$, on $\widetilde{\mathcal{B}}^n$, by

$$V_k^{*n} = \frac{V_k^*}{\sqrt{n}}. \qquad [13.25]$$

Also, for any $n \in \mathbb{N}^*$, let us define the stochastic processes

$$T_t^{ij*,n} := \sum_{k=1}^{N(nt)} V_k^{ij*,n} = \frac{1}{\sqrt{n}} \sum_{k=1}^{N(nt)} V_k^{*ij}, \quad t \in \mathbb{R}_+^*.$$

Making use of the above notation, we have the sequence of stochastic processes on $\mathcal{B}^n$

$$T_t^{*n} = \{T_t^{ij*,n} : i,j \in E\}, \quad t \in \mathbb{R}_+^*, n \in \mathbb{N}^*.$$

THEOREM 13.9.– Let $\mathbf{W}$ be a triangular array of bootstrap weights satisfying assumptions W.1–W.5. For any fixed $x \in \mathbb{R}_+$, the following weak convergence holds, as $n$ tends to infinity,

$$\{T_t^{*n} : t \in \mathbb{R}_+^*\} \rightsquigarrow \{c \times \mathcal{D}^{1/2} W(t) : t \in \mathbb{R}_+^*\},$$

where

$$\mathcal{D} = (\mathcal{D}^{i,j,\ell,r})_{(i,j),(\ell,r)\in E\times E} = \left(\delta_{i\ell}\delta_{jr}\frac{1}{\mu_{ii}} Q_{ij}(x)(1 - F_{ij}(x))\right)_{(i,j),(\ell,r)\in E\times E}.$$

For any $i,j \in E$, we set the processes

$$\Delta F_{ij}(\cdot, nt) := \widehat{F}_{ij}^{W}(\cdot, nt) - \widehat{F}_{ij}(\cdot, nt).$$

THEOREM 13.10.– Let $\mathbf{W}$ be a triangular array of bootstrap weights satisfying assumptions W.1–W.5. For any fixed $x \in \mathbb{R}_+$, it holds the weak convergence, as $n \to \infty$,

$$\{n^{1/2}\Delta F_{ij}(x, nt) : i, j \in E, t \in \mathbb{R}_+^*\} \rightsquigarrow \{\mathcal{J}_{ij}(t) : i, j \in E, t \in \mathbb{R}_+^*\}, \quad [13.26]$$

where

$$\mathcal{J}_{ij}(t) = \frac{\mu_{ii}}{P(i,j)} c\sqrt{\mathcal{D}^{i,j,\ell,r}} W^{ij}(t)/t$$

is a continuous Gaussian process.

The following corollary is a straightforward result of theorem 13.4.

COROLLARY 13.4.– Let $\mathbf{W}$ be a triangular array of bootstrap weights satisfying assumptions W.1–W.5. For any fixed $x \in \mathbb{R}_+$, we have, as $n$ tends to infinity,

$$\{n^{1/2}\Delta F_{ij}(x, n) : i, j \in E\} \xrightarrow{D} N(\mathbf{0}, c^2\Sigma_F) \quad [13.27]$$

where $N(\mathbf{0}, c^2\Sigma_F)$ is a $d^2$-dimensional normal random variable with mean $\mathbf{0}$ and covariance matrix

$$\Sigma_F = \left(\frac{\mu_{ii}^2}{P(i,j)^2} \mathcal{C}^{i,j,\ell,r}\right)_{(i,j),(\ell,r)\in E\times E}.$$

COMMENT 13.4.– Although the idea to prove the results of this section follows that in [GEO 12], we mention that the tools used in their proof are not applied directly in our setting since we reweight the original empirical estimators by stochastic weights. Their results are not only useful in their own right but also essential to the first part of the proof.

COMMENT 13.5.– The results presented in [GEO 12] are a special case of the theorems of this section, since we may take $W_{ni} = 1$ a.s., in equations [13.7] and [13.8], for $i = 1, \ldots, n$.

COMMENT 13.6.– All our results remain valid in the particular case of Markov process with the generator $A = (a_{ij})_{i,j\in E}$ and the semi-Markov kernel function

$$Q_{ij}(t) = \frac{a_{ij}}{a_i}(1 - \exp(-a_i t)), \quad \text{for } t > 0,$$

where $a_i = -a_{ii}$.

## 13.5. Confidence intervals

Theorem 13.3 turns out to be useful in obtaining confidence bands for the semi-Markov kernel $Q_{ij}(x)$, for any arbitrary $i, j \in E$ and $x \in \mathbb{R}_+$. We formalize this idea as follows: for $0 < \alpha < 1$, from theorem 13.2 in [LIM 04], we have

$$\lim_{n\to\infty} \mathbb{P}\left(\widehat{Q}_{ij}(x,nt) - c(\alpha)n^{-1/2} \leq Q_{ij}(x) \leq \widehat{Q}_{ij}(x,nt) + c(\alpha)n^{-1/2}\right)$$
$$= \mathbb{P}\left(|b_{ij}(x)W(t)/t| \leq c(\alpha)\right),$$

where $c(\alpha)$ is the $\alpha$-quantile of $|b_{ij}(x)W(t)/t|$, for $t > 0$. In practice, of course, $c(\alpha)$ cannot be computed since the covariance of $b_{ij}(x)W(t)/t$ depends on the unknown parameter $b_{ij}(x)$. Instead, suppose

$$(W^{(1)}_{\lfloor nt\rfloor 1},\ldots,W^{(1)}_{\lfloor nt\rfloor \lfloor nt\rfloor}),\ldots,(W^{(N)}_{\lfloor nt\rfloor 1},\ldots,W^{(N)}_{\lfloor nt\rfloor \lfloor nt\rfloor})$$

are $N$ independent vectors, sampled independently from the $Z_t$s. Define the random variables

$$\psi^{(\ell)}_{n,ij}(x) := \left|c^{-1}\alpha^{(\ell)}_{n,ij}(x,t)\right|, \quad \ell = 1,\ldots,N,$$

where

$$\alpha^{(\ell)}_{n,ij}(x,t) = n^{1/2}(Q^{W,\ell}_{ij}(x,nt) - \widehat{Q}_{ij}(x,nt))$$

is computed using the sample $(W^{(\ell)}_{\lfloor nt\rfloor 1},\ldots,W^{(\ell)}_{\lfloor nt\rfloor \lfloor nt\rfloor})$, for $\ell = 1,\ldots,N$. As a result of theorem 13.3, as an estimator of $c(\alpha)$, we can use the smallest $z > 0$ such that

$$\frac{1}{N}\sum_{\ell=1}^{N} \mathbb{1}\left\{\psi^{(\ell)}_{n,ij}(x) \leq z\right\} \geq 1 - \alpha.$$

It is well known that theroem 13.3 can be used to obtain easily, through routine bootstrap sampling, the critical value of the tests, which we describe briefly as follows. Let $N$ be a large integer. Let $W^{(k)}_n = (W^{(k)}_{n1},\ldots,W^{(k)}_{nn})^\top$, for $k = 1,\ldots,N$, be exchangeable vectors of non-negative weights satisfying the conditions W.1–W.5 and are independent of $Z_t$. Moreover, for any $\ell = 1,\ldots,N$, recall

$$\alpha^{(\ell)}_{n,ij}(x,t) = n^{1/2}(Q^{W,(\ell)}_{ij}(x,nt) - \widehat{Q}_{ij}(x,nt))$$

Now, according to theorem 13.3, we readily obtain that,

$$(\alpha^{(1)}_{n,ij}(x,nt),\ldots,\alpha^{(N)}_{n,ij}(x,nt))$$
$$\rightsquigarrow \left(\frac{\mu_{ii}^{1/2}}{t^{1/2}}cB^{(1)}(Q_{ij}(x)),\ldots,\frac{\mu_{ii}^{1/2}}{t^{1/2}}cB^{(N)}(Q_{ij}(x))\right),$$

where $B^{(1)}(\cdot), \ldots, B^{(N)}(\cdot)$ are independent copies of $B(\cdot)$. To approximate the limiting distribution of $n^{1/2}(\widehat{Q}_{ij}(x, nt) - Q_{ij}(x))$, we can use the empirical distribution of $\alpha_{n,ij}^{(1)}(x, nt), \ldots, \alpha_{n,ij}^{(N)}(x, nt)$, for $N$ large enough. The statistics that can be written as a functional of the empirical semi-Markov kernels and then their asymptotic behavior can be deduced from the weak convergence properties of the bootstrapped semi-Makov process. To be more precise, if we are interested in performing a statistical test based on a *smooth* functional

$$S_n := \varphi(n^{1/2}(\widehat{Q}_{ij}(x, nt) - Q_{ij}(x))),$$

with the convention that large values of $S_n$ lead to the rejection of the null hypothesis, $\mathcal{H}_0$ say, under some regularity conditions, a valid approximation to the $p$-value of the test based on $S_n$, for $N$ large enough, is given by

$$\frac{1}{N} \sum_{k=1}^{N} \mathbb{1}\{S_n^{(k)} \geq S_n\},$$

where

$$S_n^{(k)} := \varphi(\alpha_{n,ij}^{(k)}(x, nt)).$$

## 13.6. Bibliography

[BAR 95] BARBE P., BERTAIL P., *The Weighted Bootstrap*, vol. 98 of Lecture Notes in Statistics, Springer-Verlag, New York, 1995.

[BAR 08] BARBU V.S., LIMNIOS N., Semi-Markov chains and hidden semi-Markov models toward applications. Their use in reliability and DNA analysis, Lecture Notes in Statistics, vol. 191, Springer, New York, 2008.

[BER 03] BERAN R., "The impact of the bootstrap on statistical algorithms and theory", *Statist. Sci.*, vol. 18, no. 2, pp. 175–184, 2003,

[BIC 81] BICKEL P.J., FREEDMAN D.A., "Some asymptotic theory for the bootstrap", *Ann. Statist.*, vol. 9, no. 6, pp. 1196–1217, 1981.

[BIC 97] BICKEL P.J., GÖTZE F., VAN ZWET W.R., "Resampling fewer than $n$ observations: gains, losses, and remedies for losses", *Statist. Sinica*, vol. 7, no. 1, pp. 1–31, 1997.

[BIL 68] BILLINGSLEY P., *Convergence of Probability Measures*, John Wiley & Sons Inc., New York, 1968.

[BOU 12a] BOUZEBDA S., "On the strong approximation of bootstrapped empirical copula processes with applications", *Math. Methods Statist.*, vol. 21, no. 3, pp. 153–188, 2012.

[BOU 12b] BOUZEBDA S., CHERFI M., "General bootstrap for dual $\phi$-divergence estimates", *J. Probab. Stat.*, Art. ID 834107, vol. 33, 2012.

[BOU 12c] BOUZEBDA S., CHERFI M., "Test of symmetry based on copula function", submitted paper, *J. Statist. Plann. Inference*, vol. 142, no. 5, pp. 1262–1271, 2012.

[BOU 13a] BOUZEBDA S., LIMNIOS N., "On a multidimensional general bootstrap for empirical estimator of continuous-time semi-Markov kernels with applications", 2013, in press.

[BOU 13b] BOUZEBDA S., LIMNIOS N., "On general bootstrap of empirical estimator of a semi-Markov kernel with applications", *Journal of Multivariate Analysis*, vol. 116, pp. 52–62, 2013.

[CHA 05] CHATTERJEE S., BOSE A., "Generalized bootstrap for estimating equations", *Ann. Statist.*, vol. 33, no. 1, pp. 414–436, 2005.

[CHE 10] CHENG G., HUANG J.Z., "Bootstrap consistency for general semiparametric $M$-estimation", *Ann. Statist.*, vol. 38, no. 5, pp. 2884–2915, 2010.

[DEL 00] DEL BARRIO E., MATRÁN C., "The weighted bootstrap mean for heavy-tailed distributions", *J. Theoret. Probab.*, vol. 13, no. 2, pp. 547–569, 2000.

[EFR 79] EFRON B., "Bootstrap methods: another look at the jackknife", *Ann. Statist.*, vol. 7, no. 1, pp. 1–26, 1979.

[GEO 12] GEORGIADIS S., LIMNIOS N., "A multidimensional functional central limit theorem for an empirical estimator of a continuous-time semi-markov kernel", *J. Nonparametr. Stat.*, vol. 24, no. 4, pp. 1007–1017, 2012.

[GIN 90] GINÉ E., ZINN J., "Bootstrapping general empirical measures", *Ann. Probab.*, vol. 18, no. 2, pp. 851–869, 1990.

[HOL 04] HOLMES S., REINERT G., "Stein's method for the bootstrap", *Stein's Method: Expository Lectures and Applications*, vol. 46 of IMS Lecture Notes in Monograph Series, Institute of Mathematical Statistics, Beachwood, OH, pp. 95–136, 2004.

[JAM 97] JAMES L.F., "A study of a class of weighted bootstraps for censored data", *Ann. Statist.*, vol. 25, no. 4, pp. 1595–1621, 1997.

[KOS 08] KOSOROK M.R., *Introduction to Empirical Processes and Semiparametric Inference*, Springer Series in Statistics, Springer, New York, 2008.

[LIM 01] LIMNIOS N., OPRIŞAN G., *Semi-Markov Processes and Reliability*, Statistics for Industry and Technology, Birkhäuser Boston Inc., Boston, MA, 2001.

[LIM 04] LIMNIOS N., "A functional central limit theorem for the empirical estimator of a semi-Markov kernel", *J. Nonparametr. Stat.*, vol. 16, no. 1–2, pp. 13–18, 2004.

[LIM 05] LIMNIOS N., OUHBI B., SADEK A., "Empirical estimator of stationary distribution for semi-Markov processes", *Comm. Statist. Theory Methods*, vol. 34, no. 4, pp. 987–995, 2005.

[LO 93] LO A.Y., "A Bayesian method for weighted sampling", *Ann. Statist.*, vol. 21, no. 4, pp. 2138–2148, 1993.

[MAS 92] MASON D.M., NEWTON M.A., "A rank statistics approach to the consistency of a general bootstrap", *Ann. Statist.*, vol. 20, no. 3, pp. 1611–1624, 1992.

[PRÆ 93] PRÆSTGAARD J., WELLNER J.A., "Exchangeably weighted bootstraps of the general empirical process", *Ann. Probab.*, vol. 21, no. 4, pp. 2053–2086, 1993.

[PYK 61a] PYKE R., "Markov renewal processes: definitions and preliminary properties", *Ann. Math. Statist.*, vol. 32, pp. 1231–1242, 1961.

[PYK 61b] PYKE R., "Markov renewal processes with finitely many states", *Ann. Math. Statist.*, vol. 32, pp. 1243–1259, 1961.

[RUB 81] RUBIN D.B., "The Bayesian bootstrap", *Ann. Statist.*, vol. 9, no. 1, pp. 130–134, 1981.

[SHA 95] SHAO J., TU D.S., *The Jackknife and Bootstrap*, Springer Series in Statistics, Springer-Verlag, New York, 1995.

[VAN 96] VAN DER VAART A.W., WELLNER J.A., *Weak Convergence and Empirical Processes. With Applications to Statistics*, pp. xvi–508, Springer-Verlag, New York, 1996.

[WEN 89] WENG C.-S., "On a second-order asymptotic property of the Bayesian bootstrap mean", *Ann. Statist.*, vol. 17, no. 2, pp. 705–710, 1989.

[ZHE 88] ZHENG Z.G., TU D.S., "Random weighting methods in regression models", *Sci. Sinica Ser. A*, vol. 31, no. 12, pp. 1442–1459, 1988.

## Chapter 14

# On Chi-Squared Goodness-of-Fit Test for Normality

In this chapter, we consider the chi-squared goodness-of-fit test based on the chi-square statistic $\mathbf{Y}_n^2$ of Nikulin–Rao–Robson (NRR), on the efficient estimator (such as the maximum likelihood estimation). For testing normality, the optimal choice of boundaries cells is considered. A simulation study for power test $\mathbf{Y}_n^2$ was carried out to evaluate its performance and stability based on the classes of Neyman–Pearson and to compare it with the power of the Shapiro–Wilk test for normality against composite hypotheses. The results of our simulation study are also discussed.

### 14.1. Chi–squared test for normality

In 1973, Nikulin [NIK 73b] proposed one modification of the famous Pearson's chi-squared test for testing the composite hypothesis $H_0$ of normality when data are not censored, i.e. when we observe a complete simple sample. Chernoff and Lehmann [CHE 54], Watson [WAT 57], Moore [MOO 71] and Nikulin [NIK 73b] have shown that if the parameters of the normal distribution are unknown and are estimated by the maximum likelihood method, then the statistic of Pearson will never have the chi-squared distribution as the limit distribution, and hence we cannot use Pearson's test (see also recent publications of Greenwood and Nikulin [GRE 96], Bagdonavičius *et al.* [BAG 11b] and Voinov *et al.* [VOI 13]). Let us consider this problem one more time. Let $X = (X_1, X_2, \cdots, X_n)^T$ be a simple sample, i.e.

---

Chapter written by Mikhail NIKULIN, Léo GERVILLE-RÉACHE and Xuan Quang TRAN.

$X_1, X_2, \cdots, X_n$ are independent identically distributed random variables, and we wish to test the hypothesis $H_0$, according to which:

$$H_0 : \mathbf{P}_\theta(X_1 \leq x) = \Phi\left(\frac{x-\mu}{\sigma}\right), \ (\mu, \sigma^2)^T \in \mathbf{\Theta} = \{|\mu| < \infty, \sigma > 0\}, \ x \in R^1,$$

where $\Phi(x)$ is the distribution function of the standard normal law. The probability density function of $X_i$:

$$\frac{1}{\sigma}\varphi\left(\frac{x-\mu}{\sigma}\right) = \frac{1}{\sigma\sqrt{2\pi}} exp\left(-\frac{(x-\mu)^2}{2\sigma^2}\right), \qquad x \in R^1.$$

It is well known that under $H_0$

$$\mathbf{E}_\theta X_i = \mu, \quad \text{and} \quad \mathbf{Var}_\theta X_i = \sigma^2, \quad i = 1, 2, ..., n.$$

If parameter $\theta = (\mu, \sigma^2)^T$ is unknown, then we can estimate it using the data by the maximum likelihood estimator $\hat{\theta}_n$. If the $H_0$ is true, then the likelihood function $L(\cdot)$ on $\mathbf{\Theta}$ is given by the formula

$$\begin{aligned} L(\theta) &= \frac{(2\pi)^{-n/2}}{\sigma^n} \exp\left\{-\frac{1}{2\sigma^2}\sum_{i=1}^n (X_i - \mu)^2\right\} \\ &= \frac{(2\pi)^{-n/2}}{\sigma^n} \exp\left\{-\frac{1}{2\sigma^2}\left[\sum_{i-1}^n X_i^2 - 2\mu\sum_{i=1}^n X_i + n\mu^2\right]\right\}, \end{aligned}$$

from where we can see that that a minimal sufficient statistic for $\theta$ is

$$\mathbf{T} = \left\{\sum_{i=1}^n X_i, \ \sum_{i=1}^n X_i^2\right\},$$

and that under $H_0$ the statistics

$$\overline{X}_n = \frac{1}{n}\sum_{i=1}^n X_i, \quad \tilde{s}_n^2 = \frac{1}{n}\sum_{i=1}^n (X_i - \mu)^2, \quad s_n^2 = \frac{1}{n}\sum_{i=1}^n (X_i - \overline{X}_n)^2 \quad [14.1]$$

are the maximum likelihood estimators for the components $\mu$ and $\sigma^2$ of the parameter $\theta$. It is well known that for the regular family of distributions, there exists a sequence of maximum likelihood estimators $\hat{\theta}_n$ such that

$$\hat{\theta}_n = \arg max_{\theta \in \mathbf{\Theta}} L(\theta),$$

where $L(\theta)$ is the likelihood functions of $X_1, X_2, \cdots, X_n$, and

$$\sqrt{n}(\hat{\theta}_n - \theta) \xrightarrow{d} N_2(\mathbf{0}, \mathbf{i}^{-1}(\theta)),$$

where $\mathbf{i}(\theta)$ is Fisher's information matrix of one observation $X_1$

$$\mathbf{i}(\theta) = \mathbf{i}(\mu, \sigma^2) = \frac{1}{\sigma^2}\begin{pmatrix} 1 & 0 \\ 0 & 2 \end{pmatrix}, \quad \theta = (\mu, \sigma^2)^T.$$

Set

$$\eta_i = \begin{cases} \frac{1}{\sigma}(X_i - \mu), & \text{if } \mu \text{ and } \sigma \text{ are known,} \\ \frac{1}{\sigma}\sqrt{\frac{n}{n-1}}(X_i - \bar{X}_n), & \text{if } \mu \text{ is unknown }, \sigma \text{ is known,} \\ \frac{1}{\bar{s}_n}(X_i - \mu), & \text{if } \mu \text{ is known }, \sigma \text{ is unknown }, \\ \frac{1}{s_n}(X_i - \bar{X}_n), & \text{if } \mu \text{ and } \sigma \text{ are known,} \end{cases}$$

where $i = 1, 2, ..., n$.

Let $\mathbf{p} = (p_1, p_2, \cdots, p_r)^T$ be the vector of positive probabilities, $1 > p_i > 0$, such that:

$$\mathbf{p}^T\mathbf{1}_r = \sum_{j=1}^{r} p_j = 1, \quad \text{where} \quad \mathbf{1}_r = (1, ..., 1)^T \in R^r,$$

and let us define the quantiles of the standard normal distribution

$$-\infty = x_0 < x_1 < x_2 < \cdots x_{r-1} < x_r = +\infty,$$

with

$$x_j = \Phi^{-1}\left(\sum_{i=1}^{j} p_i\right), j = 1, 2, \cdots, r-1.$$

Furthermore, let $\nu = (\nu_1, \nu_2, \cdots, \nu_r)^T$ be a frequency vector obtained by grouping the statistics $\eta_1, ..., \eta_n$ over the $r$ intervals of the partition

$$(x_0, x_1], (x_1, x_2], ..., (x_{r-1}, x_r).$$

It is evident that the same statistic is obtained by grouping the initial random variables $X_1, X_2, \cdots, X_n$ over the random intervals $J_i$, $(i = 1, ..., r)$

$$J_i = \begin{cases} (x_{i-1}\sigma + \mu\,;\, x_i\sigma + \mu], & \text{if } \mu \text{ and } \sigma \text{ are known,} \\ (x_{i-1}\sigma + \overline{X}_n\,;\, x_i\sigma + \overline{X}_n], & \text{if } \mu \text{ is unknown, } \sigma \text{ is known,} \\ (x_{i-1}\tilde{s}_n + \mu\,;\, x_i\tilde{s}_n + \mu], & \text{if } \mu \text{ is known, } \sigma \text{ is unknown,} \\ (x_{i-1}s_n + \overline{X}_n\,;\, x_i s_n + \overline{X}_n], & \text{if } \mu \text{ and } \sigma \text{ are unknown.} \end{cases}$$

It is clear that if $H_0$ is true, then for any $j = 1, ..., n$

$$\mathbf{P}\{\eta_j \in (x_{i-1}, x_i]\} = \mathbf{P}\{X_j \in J_i\} = p_i, \quad i = 1, ..., r.$$

REMARK 14.1.– Note that under $H_0$, the statistic $\nu$ follows the multinomial distribution $M_r(\mathbf{p}, n)$ with parameters $r, n, \mathbf{p}$. It is easy to show that in our case

$$\mathbf{E}\nu = n\mathbf{p} \quad \text{and} \quad \mathbf{E}(\nu - n\mathbf{p})(\nu - n\mathbf{p})^T = n\mathbf{B},$$

the matrix $\mathbf{B} = \mathbf{P} - \mathbf{p}\mathbf{p}^T$ is degenerated since $\mathbf{1}_r^T\nu = n$ and hence $rank(\mathbf{B}) = r - 1$, $\mathbf{P}$ is a non-degenerated diagonal matrix with $p_1, ..., p_r$ on the main diagonal. In the general case, the statistic $\nu = (\nu_1, \nu_2, \cdots, \nu_r)^T$ is not sufficient and it cannot replace the initial data $X_1, ..., X_n$, which means that some information is lost if we use only grouped data. Moreover, the statistic $\nu$ is obtained using the so-called censoring mechanism of grouping data. For example, Fisher's chi-square test, based on the minimum chi-square estimator

$$\tilde{\theta}_n = \arg min_{\theta\in\mathbf{\Theta}} \mathbf{X}_n^2(\mathit{0}),$$

is not the best if we have information that the statistic $\nu = (\nu_1, \nu_2, \cdots, \nu_r)^T$ is not sufficient. From the properties of the multinomial distribution, we get the following theorem.

THEOREM 14.1.– If $n \to \infty$, $r = const \geq 2$, then from the central limit theorem it follows that under the $H_0$ the statistic $\nu$ is asymptotically normal with parameters

$$\mathbf{E}\nu = n\mathbf{p} + O(\mathbf{1}_r) \quad \text{and} \quad \mathbf{E}(\nu - n\mathbf{p})(\nu - n\mathbf{p})^T = n\mathbf{B} + O(\mathbf{1}_{r\times r}), \qquad [14.2]$$

where

$$\mathbf{B} = \begin{cases} \mathbf{P} - \mathbf{p}\mathbf{p}^T, & \text{if } \mu \text{ and } \sigma \text{ are known ,} \\ \mathbf{P} - \mathbf{p}\mathbf{p}^T - \mathbf{a}\mathbf{a}^T, & \text{if } \mu \text{ is unknown , } \sigma \text{ is known ,} \\ \mathbf{P} - \mathbf{p}\mathbf{p}^T - \frac{1}{2}\mathbf{b}\mathbf{b}^T, & \text{if } \mu \text{ is known , } \sigma \text{ is unknown ,} \\ \mathbf{P} - \mathbf{p}\mathbf{p}^T - \mathbf{a}\mathbf{a}^T - \frac{1}{2}\mathbf{b}\mathbf{b}^T, & \text{if } \mu \text{ and } \sigma \text{ are unknown ,} \end{cases}$$

the matrix $\mathbf{B}$ is degenerated, $rank(\mathbf{B}) = r - 1$, in all the considered four cases,

$$\mathbf{a} = (a_1, \cdots, a_r)^T, \quad a_j = \varphi(x_j) - \varphi(x_{j-1}), \qquad [14.3]$$

$$\mathbf{b} = (b_1, \cdots, b_r)^T, \quad b_j = -x_j\varphi(x_j) + x_{j-1}\varphi(x_{j-1}). \qquad [14.4]$$

To prove this theorem, we can apply the methods and techniques of Chernoff and Lehmann [CHE 54], Watson [WAT 57], Moore [MOO 75], Chibisov [CHI 71] and Nikulin [NIK 73a]. We will use one trivial theorem stating one important property of multivariate normal distribution $N_r(\mathbf{a}, \mathbf{\Sigma})$ according to which if we have an $r$-dimensional normal vector $\mathbf{Z} \sim N_r(\mathbf{a}, \mathbf{\Sigma})$, then

$$(\mathbf{Z} - \mathbf{a})^T \mathbf{\Sigma}^- (\mathbf{Z} - \mathbf{a}) \sim \chi^2_m,$$

where $\mathbf{\Sigma}^-$ is a general inverse of the matrix $\mathbf{\Sigma}$, $m$ is the rank of the matrix $\mathbf{\Sigma}$, $m \leq r$. For example, in the first case of Pearson, we have

$$\mathbf{B} = \mathbf{P} - \mathbf{p}\mathbf{p}^T, \quad \text{and} \quad \mathbf{B}^- = \mathbf{P}^{-1} + \frac{1}{p_r}\mathbf{1}_r\mathbf{1}_r^T,$$

since

$$\mathbf{1}_r^T\mathbf{P} = \mathbf{p}^T, \quad \mathbf{1}_r^T\mathbf{p} = 1, \quad \mathbf{P}\mathbf{1}_r = \mathbf{p}.$$

Note that by definition, a general inverse matrix $\mathbf{B}^-$ verifies the relation: $\mathbf{B}\mathbf{B}^-\mathbf{B} = \mathbf{B}$. Let us consider now the famous Pearson's [PEA 00] statistics

$$\mathbf{X}_n^2 = X_n^T X_n = \sum_{j=1}^{r} \frac{(\nu_j - np_j)^2}{np_j}$$
$$= \frac{1}{n}[\mathbf{P}^{-1/2}(\nu - n\mathbf{p})]^T \mathbf{P}^{-1/2}(\nu - n\mathbf{p}) = \frac{1}{n}(\nu - n\mathbf{p})^T \mathbf{P}^{-1}(\nu - n\mathbf{p}),$$

which represent the square of the length of the vector

$$X_n = \left(\frac{\nu_1 - np_1}{\sqrt{np_1}}, \frac{\nu_2 - np_2}{\sqrt{np_2}}, \ldots, \frac{\nu_r - np_r}{\sqrt{np_r}}\right)^T = \frac{1}{\sqrt{n}}\mathbf{P}^{-1/2}(\nu - n\mathbf{p}).$$

From the central limit theorem, it follows that the probability distribution of the statistic $\nu$ can be approximated by the chi-squared distribution with $r - 1$ degrees of freedom. Thus, the probability distribution of $X_n(\theta)$ can be approximated by the multivariate normal distribution, that is

$$X_n(\theta) \xrightarrow{d} \mathbf{X} \sim N_r(\mathbf{0}, \mathbf{\Sigma}),$$

where $\mathbf{\Sigma}$ is the covariance matrix of $X_n(\theta)$, $rank(\mathbf{\Sigma}) = r - 1$

$$\mathbf{\Sigma} = \mathbf{P}^{-1/2}\mathbf{B}\mathbf{P}^{-1/2} =$$

$$= \begin{cases} \mathbf{E}_r - \mathbf{q}\mathbf{q}^T, & \text{if } \mu \text{ and } \sigma \text{ are known,} \\ \mathbf{E}_r - \mathbf{q}\mathbf{q}^T - \mathbf{P}^{-1/2}\mathbf{a}\mathbf{a}^T\mathbf{P}^{-1/2}, & \text{if } \mu \text{ is unknown, } \sigma \text{ is known,} \\ \mathbf{E}_r - \mathbf{q}\mathbf{q}^T - \frac{1}{2}\mathbf{P}^{-1/2}\mathbf{b}\mathbf{b}^T\mathbf{P}^{-1/2}, & \text{if } \mu \text{ is known, } \sigma \text{ is unknown,} \\ \mathbf{E}_r - \mathbf{q}\mathbf{q}^T - \mathbf{P}^{-1/2}[\mathbf{a}\mathbf{a}^T + \frac{1}{2}\mathbf{b}\mathbf{b}^T]\mathbf{P}^{-1/2}, & \text{if } \mu \text{ and } \sigma \text{ are unknown,} \end{cases}$$

where $\mathbf{E}_r$ is $r \times r$ unit matrix, $\mathbf{q} = (\sqrt{p_1}, \sqrt{p_2}, \cdots, \sqrt{p_r})^T$, $\mathbf{a}$ and $\mathbf{b}$ are defined in [14.4]–[14.4].

So, we obtain

$$\mathbf{Y}_n^2 = X_n^T(\hat{\theta}_n)\mathbf{\Sigma}^- X_n(\hat{\theta}_n) \xrightarrow{d} \mathbf{Y}^2 \sim \chi^2_{r-1}.$$

Thus, the test statistic $\mathbf{Y}_n^2$ proposed by Nikulin [NIK 73a] is

$$\mathbf{Y}_n^2 = X_n^T(\hat{\theta}_n)\mathbf{\Sigma}^- X_n(\hat{\theta}_n) =$$

$$= \mathbf{X}_n^2 + \mathbf{Q} = \begin{cases} \mathbf{X}_n^2, & \text{if } \mu \text{ and } \sigma \text{ are known,} \\ \mathbf{X}_n^2 + \frac{1}{n\lambda_1}\alpha^2, & \text{if } \mu \text{ is unknown, } \sigma \text{ is known,} \\ \mathbf{X}_n^2 + \frac{1}{n\lambda_2}\beta^2, & \text{if } \mu \text{ is known, } \sigma \text{ is unknown,} \\ \mathbf{X}_n^2 + \frac{\lambda_2\alpha^2 - 2\lambda_3\alpha\,\beta + \lambda_1\beta^2}{n(\lambda_1\lambda_2 - \lambda_3^2)}, & \text{if } \mu \text{ and } \sigma \text{ are unknown,} \end{cases}$$

where

$$\mathbf{\Sigma}^- = \begin{cases} \mathbf{E}_r + \mathbf{q}\mathbf{q}^T, & \text{if } \mu \text{ and } \sigma \text{ are known,} \\ \mathbf{E}_r + \mathbf{q}\mathbf{q}^T + \frac{1}{\lambda_1}\mathbf{a}\mathbf{a}^T, & \text{if } \mu \text{ is unknown, } \sigma \text{ is known,} \\ \mathbf{E}_r + \mathbf{q}\mathbf{q}^T + \frac{1}{\lambda_2}\mathbf{b}\mathbf{b}^T, & \text{if } \mu \text{ is known, } \sigma \text{ is unknown,} \\ \mathbf{E}_r + \mathbf{q}\mathbf{q}^T + \frac{(\lambda_2\mathbf{a}\mathbf{a}^T - \lambda_3(\mathbf{a}\mathbf{b}^T + \mathbf{b}\mathbf{a}^T) + \lambda_1\mathbf{b}\mathbf{b}^T)}{\lambda_1\lambda_2 - \lambda_3^2}, & \text{if } \mu \text{ and } \sigma \text{ are unknown} \end{cases}$$

is the generalized inverse matrix of $\mathbf{\Sigma}$

$$\mathbf{X}_n^2 = X_n^T X_n = \frac{1}{n}\sum_{j=1}^{r}\frac{\nu_j^2}{p_j} - n, \quad \alpha = \sum_{j=1}^{r}\frac{a_j\nu_j}{p_j} = \mathbf{a}^T\mathbf{P}^{-1}\nu,$$

$$\beta = \sum_{j=1}^{r}\frac{b_j\nu_j}{p_j} = \mathbf{b}^T\mathbf{P}^{-1}\nu, \quad \lambda_1 = 1 - \sum_{j=1}^{r}\frac{a_j^2}{p_j} = 1 - \mathbf{a}^T\mathbf{P}^{-1}\mathbf{a},$$

$$\lambda_2 = 2 - \sum_{j=1}^{r} \frac{b_j^2}{p_j} = 2 - \mathbf{b}^T \mathbf{P}^{-1} \mathbf{b}, \quad \lambda_3 = - \sum_{j=1}^{r} \frac{a_j b_j}{p_j} = -\mathbf{a}^T \mathbf{P}^{-1} \mathbf{b}.$$

Under the hypothesis $H_0$, the statistic $\mathbf{Y}_n^2$ has an asymptotically chi-squared distribution with $(r-1)$ degrees of freedom.

*Statistical inference for the hypothesis $H_0$:* the hypothesis is rejected with approximate significance level $\alpha$ if $\mathbf{Y}_n^2 > \chi_\alpha^2(r-1)$.

REMARK 14.2.– In the case of application of the minimum chi-square $\tilde{\theta}_n = \arg min_{\theta \in \mathbf{\Theta}} \mathbf{X}_n^2(\theta)$, and Pearson's statistic $\mathbf{X}_n^2(\tilde{\theta}_n)$, we have Fisher's theorem [FIS 28]: *if the hypothesis of normality $H_0$ is true then*

$$\lim_{n \to \infty} \mathbf{P}\{\mathbf{X}_n^2(\tilde{\theta}_n) > x\} = \mathbf{P}\{\chi_{r-3}^2 > x\}, \quad \textit{for any } x > 0.$$

Now we present some remarks on the choice of the intervals for grouping data.

REMARK 14.3.– The choice of random grouping intervals $a_j$ depends on the choice of vector $\mathbf{p} = (p_1, p_2, \cdots, p_r)^T$ such that $\mathbf{p}^T \mathbf{1}_r = 1$ and all $p_i > 0$. If there is no alternative, we would choose $\mathbf{p}$ to maximize power, or choose $p_1 = p_2 = \cdots = p_r = \frac{1}{r}$, then $\lambda_3 = 0$.

REMARK 14.4.– If the vector $\mathbf{p} = (p_1, p_2, \cdots, p_r)^T$ is chosen with symmetric probabilities (SP), so that $p_1 = p_r$, $p_2 = p_{r-1}$, $p_3 = p_{r-2}$ and so on, then $\lambda_3 = 0$. Ogawa [OGA 74] found that this symmetric choice is optimal for testing normality.

REMARK 14.5.– In the case of a goodness-of-fit test for normality against composite hypothesis, to construct grouping cells $a_j$ we use two Neyman–Pearson-type classes.

$$\begin{aligned} H_0 : \mathbf{P}_\theta\{T_1 \leq x\} &= \Phi\left(\frac{x-\mu}{\sigma}\right), \ |x| < \infty, \\ \theta &= (\mu, \sigma^2)^T \in \mathbf{\Theta} = \{|\mu| < \infty, \sigma > 0\}, \end{aligned}$$

against the alternative hypothesis

$$H_1 : \mathbf{P}_\gamma\{T_1 \leq x\} = G(x, \gamma), \gamma \in \mathbf{G} \subset \mathbf{R}^m.$$

Let us consider $\varphi(x, \theta) = \frac{1}{\sigma}\varphi\left(\frac{x-\mu}{\sigma}\right)$ and $g(x, \gamma)$ are the density probability functions of null and alternative hypotheses, respectively.

To use the Neyman–Pearson classes, it is necessary to find the intersection points $a_1, a_2, \cdots, a_r$ of two density probability by solving the following equation

$$\frac{1}{\hat{\sigma}}\varphi\left(\frac{x-\hat{\mu}}{\hat{\sigma}}\right) = g(x, \hat{\gamma}_n),$$

where $\hat{\theta}_n = (\hat{\mu}, \hat{\sigma}^2)^T$ and $\hat{\gamma}_n$ are the maximum likelihood estimators of parameters $\theta = (\mu, \sigma^2)^T$ and $\gamma$, respectively.

We obtain $r$ Neyman–Pearson-type classes (rNPG)

$$I_1 = (-\infty, a_1] \cup [a_r, +\infty),\ I_2 = (a_1, a_2], \cdots, I_r = (a_{r-1}, a_r], \qquad [14.5]$$

and Greenwood and Nikulin [GRE 96] proposed two Neyman–Pearson-type classes (2NPG) that are

$$\mathbf{I}_1 = \{t : \varphi(t, \hat{\theta}_n) \geq g(t, \hat{\gamma}_n)\}, \quad \mathbf{I}_2 = \{t : \varphi(t, \hat{\theta}_n) < g(t, \hat{\gamma}_n)\}. \qquad [14.6]$$

For two Neyman–Pearson classes $\mathbf{I}_1, \mathbf{I}_2$, we obtain the frequency vector $\nu = (\nu_1, \nu_2)^T$, and positive probability vector $\mathbf{p} = (p_1, p_2)^T$, which are given as

$$\nu_j = n\int_{\mathbf{I}_j} dF_n(t), \quad j = 1, 2,$$

$$p_1 := p_1(\theta) = \int_{\mathbf{I}_1} \varphi(u, \hat{\theta}_n)du, \quad p_2 := p_2(\theta) = \int_{\mathbf{I}_2} \varphi(u, \hat{\theta}_n)du, \quad \theta - (\mu, \sigma)^T,$$

where $F_n$ is the empirical distribution function based on our data. We note that

$$\nu_1 + \nu_2 = n, \quad p_1 + p_2 = 1,$$

and the first derivative of probability vector

$$a_1 = \frac{\partial p_1(\theta)}{\partial \mu}, \quad a_2 = \frac{\partial p_2(\theta)}{\partial \mu}, \quad b_1 = \frac{\partial p_1(\theta)}{\partial \sigma}, \quad b_2 = \frac{\partial p_2(\theta)}{\partial \sigma},$$

satisfies $a_1 + a_2 = b_1 + b_2 = 0$. We construct the statistic $\mathbf{Y}_n^2$ as

$$\mathbf{Y}_n^2 = \mathbf{X}_n^2 + \frac{\lambda_2\alpha^2 - 2\lambda_3\alpha\beta + \lambda_1\beta^2}{n(\lambda_1\lambda_2 - \lambda_3^2)},$$

where

$$\mathbf{X}_n^2 = \frac{1}{n}\left(\frac{\nu_1^2}{p_1} + \frac{\nu_2^2}{p_2}\right), \ \lambda_1 = 1 - \frac{a_1^2}{p_1p_2}, \quad \lambda_2 = 2 - \frac{b_1^2}{p_1p_2}, \quad \lambda_3 = -\frac{a_1b_1}{p_1p_2},$$

$$\alpha = \frac{a_1\nu_1}{p_1} + \frac{a_2\nu_2}{p_2}, \quad \beta = \frac{b_1\nu_1}{p_1} + \frac{b_2\nu_2}{p_2}.$$

To construct a goodness-of-fit NRR test based on the statistic $\mathbf{Y}_n^2$ against composite hypothesis, we can use two Neyman–Pearson grouping classes $\mathbf{I}_1$, $\mathbf{I}_2$ in [14.6] so that the distribution limiting of $\mathbf{Y}_n^2$ belongs to chi-squared distribution with one degree of freedom as $n$ to $\infty$. We also refer to calculating $\mathbf{Y}_n^2$ by using $r$ Neyman–Pearson $I_j$, which are defined in [14.5], then the limit of distribution of $\mathbf{Y}_n^2$ has chi-squared distribution with $r-1$ degrees of freedom as $n$ to $\infty$.

REMARK 14.6.– We note here that in [BAG 11a], we can find the chi-square test for normality for right-censored data.

## 14.2. Simulation study

In this section, we analyze the power of the NRR's test based on the statistic $\mathbf{Y}_n^2$ and compare it with the powers of the Shapiro–Wilk test based on the statistics $\mathbf{W}$ and the Shapiro–Francia test based on the statistic $\mathbf{W}'$ in the cases of complete samples for the normal null hypothesis $H_0$

$$H_0 \, : X \simeq \Phi\left(\frac{x-\mu}{\sigma}\right), \quad \theta = (\mu, \sigma)^T \in \mathbf{\Theta} = \{|\mu| < \infty, \sigma > 0\}, \quad x \in \mathbf{R}^1,$$

against alternative hypotheses given below:

– Logistic distribution:

$$H_1 \, : X \sim F(x, \alpha, \beta) = \frac{1}{1 + \exp\left\{-\frac{x-\alpha}{\beta}\right\}}, \quad |x| < \infty, \quad |\alpha| < \infty, \beta > 0.$$

– Laplace distribution:

$$H_1 \, : X \sim F(x, \alpha, \beta) = \frac{1}{2} + \frac{1}{2} sgn(x-\alpha)\left(1 - \exp\left(-\frac{|x-\alpha|}{\beta}\right)\right),$$

$$|x| < \infty, \ |\alpha| < \infty, \ \beta > 0.$$

– Weibull distribution:

$$H_1 \; : \; X \sim F(x, \theta, \nu) = \exp\left\{ -\left( \frac{x}{\theta} \right)^{\nu} \right\}, \qquad \theta > 0, \, \nu > 0.$$

To construct NRR's statistic, we use two grouping interval methods two Neyman–Pearson classes (2NPG) and equiprobable (EQPG) grouping cells. The numbers of intervals $r$ were used for NRR's test based on the equiprobable cells. To take two Neyman–Pearson-type classes for NRR's statistic with respect to the alternative close to each other, we selected parameters of the alternative in such a manner that graphs of the alternative would be visually as close to each other as possible. Figure 14.1 shows the curve of probability density of normal $N(1, \, 0.2^2)$, logistic $Log(1, \, 0.2)$, Laplace $L(1, \, 0.25)$ and Weibull $W(1, \, 5)$ distributions.

Following Figure 14.1, we remark that there are four points of intersection among normal and alternative references. Thus, two Neyman–Pearson-type classes can be obtained on the basis of these points. For example, under logistic distribution alternative $Log(1, \, 0.2)$, to construct $\mathbf{Y}_n^2$ statistic based on two Neyman–Pearson classes it is necessary to find the roots of the following equation

$$\frac{1}{\sigma\sqrt{2\pi}} \exp\left\{ -\frac{(x-\mu)^2}{2\sigma^2} \right\} = \frac{5\exp\left\{-\frac{x-1}{0.2}\right\}}{\left(1+\exp\left\{-\frac{x-1}{0.2}\right\}\right)^2}. \qquad [14.7]$$

To do this, we use the dichotomy method based on a simulated population from a logistic distribution $Log(1, \, 0.2)$ with sample size $n = 10,000$. We take the mean and standard deviation of population to obtain the following four solutions of equation [14.7]

$$a_1 = 0.1686993, \quad a_2 = 0.7501097, \quad a_3 = 1.237212, \quad a_4 = 1.8538186,$$

and the following two Neyman–Pearson classes can then be considered

$$\mathbf{I}_1 = (0.1686993, \, 0.7501097] \cup (1.237212, \, 12.0632832],$$

$$\text{and} \quad \mathbf{I}_2 = (-\infty, \, 0.1686993] \cup (0.7501097, \, 1.237212] \cup [1.8538186, \, +\infty).$$

Under the Laplace $L(1, \, 0.25)$ alternative, based on the size of 10,000 simulated samples of Laplace $L(1, \, 0.25)$ distribution, we take its mean and standard deviation and obtain the points of intersection of two density functions

$$a_1 = 0.1761564, \quad a_2 = 0.8260756, \quad a_3 = 1.1723921, \, a_4 = 1.8313972.$$

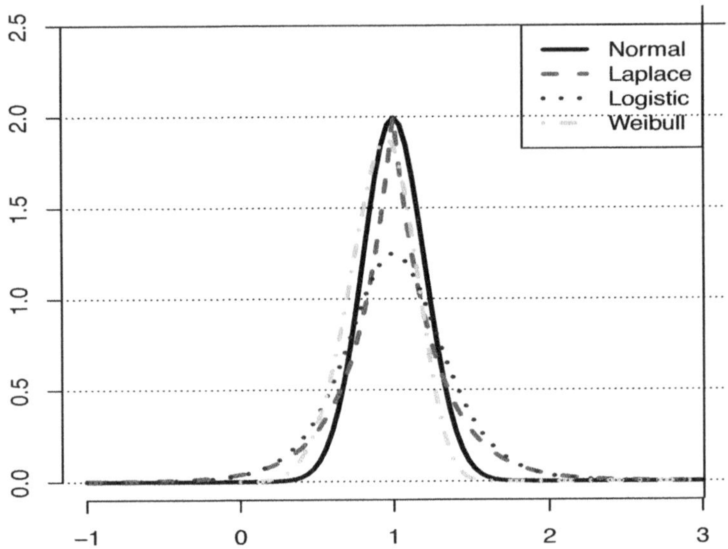


**Figure 14.1.** *Probability density functions of* $N(1,\,0.2^2)$, $Log(1,\,0.2)$, $L(1,\,0.25)$ *and* $W(1,\,5)$ *distributions*

Thus, under the Laplace $L(1,\,0.25)$ alternative, two Neyman–Pearson classes are

$$\mathbf{I}_1 = (0.1761564,\, 0.8260756] \cup (1.1723921,\, 1.8313972],$$

$$\text{and}\quad \mathbf{I}_2 = (-\infty,\, 0.1761564] \cup (0.8260756,\, 1.1723921] \cup (1.8313972,\, +\infty).$$

By doing so, we obtain the following solution and two Neyman–Pearson classes under Weibull $W(1,\,5)$ alternative

$$a_1 = 0.14823, \quad a_2 = 0.58350, \quad a_3 = 0.94616, \quad a_4 = 1.28586.$$

Thus, the following Neyman–Pearson-type classes for normal against Weibull $W(1,\,5)$ distribution,

$$\mathbf{I}_1 = (-\infty,\, 0.14823] \cup (0.58350,\, 0.94616] \cup [1.28586,\, \infty)$$

$$\text{and} \quad \mathbf{I}_2 = (0.14823,\ 0.58350] \cup (0.94616,\ 1.28586)$$

are considered.

The Shapiro–Wilk [SHA 72b] statistic is defined as

$$\mathbf{W} = \frac{\left(\sum\limits_{i=1}^{n} a_i X_{(i)}\right)^2}{\sum\limits_{i=1}^{n} (X_i - \overline{X}_n)^2}, \qquad [14.8]$$

where $\mathbf{X}^{(\cdot)} = (X_{(1)}, ..., X_{(n)})^T$ is the vector of order statistics associated with the simple sample $X_1, ..., X_n$ and $\mathbf{a} = (a_1,\ a_2,\ \cdots,a_n)^T$ are the coefficients having the forms

$$\mathbf{a} = \frac{\mathbf{m}^T \Sigma^{-1}}{(\mathbf{m}^T \Sigma^{-1} \Sigma^{-1} \mathbf{m})^{\frac{1}{2}}},$$

$$\mathbf{m} = (m_1,\ m_2,\ \cdots,\ m_n)^T; \quad \Sigma = [\sigma_{ij}]_{n \times n},$$

the expected values of normal order statistics and the covariance matrix of order statistics from the standard normal distribution, which do not depend on unknown parameters $\mu$ and $\sigma$ under the normality hypothesis.

In 1972, Shapiro–Francia [SHA 72a] proposed a modification of the Shapiro–Wilk statistic, which can be used for large samples and is defined as

$$\mathbf{W}' = \frac{\left(\sum\limits_{i=1}^{n} \mathbf{b}_i X_{(i)}\right)^2}{\sum\limits_{i=1}^{n} (X_i - \overline{X}_n)^2}, \qquad [14.9]$$

where, $\mathbf{b} = (b_1,\ b_2,\ \cdots,b_n)^T$ are the coefficients having the form

$$\mathbf{b} - \frac{\mathbf{m}^T}{(\mathbf{m}^T \mathbf{m})^{\frac{1}{2}}},$$

where $\mathbf{m}$ is the expected value of normal order statistics $\mathbf{X}^{(\cdot)}$.

The Shapiro–Wilk and Shapiro–Francia statistics are widely used methods for evaluating the condition of normality. However, the distribution of these test statistics is unknown under the null hypothesis and must be approximated.

Thus, to estimate the power of Shapiro–Wilk (**W**), Shapiro–Francia ( **W**′), using the Monte–Carlo method for 50,000 simulated samples from a standard normal distribution with sample size $n$, we obtained its estimated critical values corresponding at two levels of significances $\alpha = 5\%$ and $10\%$. In Table 14.1, we present the approximated critical values for each test used for reference.

For each sample size $n$, simulations are repeated 10,000 times ($N = 10,000$) for the sample size $n$. For each operation, we compute the values of each test, then we calculate the powers of each test, i.e. the difference between $N$ and the number of times where each statistic is less than its critical value divided by $N$ by subtraction, the significance level $\alpha = 0.05$ is used. The results of power tests for each alternative are given in Tables 14.2–14.4.

| $\alpha$ | Tests | Sample size $n$ | | | | | | | | | |
|---|---|---|---|---|---|---|---|---|---|---|---|
| | | 40 | 50 | 80 | 100 | 150 | 200 | 250 | 300 | 400 | 500 |
| 0.05 | **W** | 0.9444 | 0.9539 | 0.9690 | 0.9745 | 0.9823 | 0.9862 | 0.9888 | 0.9905 | 0.9927 | 0.9940 |
| | **W**′ | 0.9439 | 0.9537 | 0.9688 | 0.9744 | 0.9822 | 0.9862 | 0.9887 | 0.9905 | 0.9927 | 0.9940 |
| 0.10 | **W** | 0.9531 | 0.9612 | 0.9738 | 0.9784 | 0.9850 | 0.9883 | 0.9904 | 0.9919 | 0.9937 | 0.9949 |
| | **W**′ | 0.9542 | 0.9619 | 0.9743 | 0.9788 | 0.9851 | 0.9884 | 0.9906 | 0.9920 | 0.9938 | 0.9950 |

**Table 14.1.** *The approximated critical values for corresponding tests*

| Tests | Sample size $n$ | | | | | | | | | |
|---|---|---|---|---|---|---|---|---|---|---|
| | 40 | 50 | 80 | 100 | 150 | 200 | 250 | 300 | 400 | 500 |
| **W** | 0.8300 | 0.8027 | 0.7360 | 0.6972 | 0.5911 | 0.5147 | 0.4286 | 0.3543 | 0.2494 | 0.1709 |
| **W**′ | 0.7913 | 0.7555 | 0.6812 | 0.6329 | 0.5242 | 0.4449 | 0.3669 | 0.2916 | 0.2016 | 0.1310 |
| $\mathbf{Y}_n^2$*(EQPG)* | 0.1024 | 0.1235 | 0.1603 | 0.1801 | 0.2441 | 0.2960 | 0.3389 | 0.3903 | 0.4677 | 0.5503 |
| $\mathbf{Y}_n^2$ *(2NPG)* | 0.0833 | 0.1002 | 0.1438 | 0.1737 | 0.2503 | 0.3243 | 0.3932 | 0.4653 | 0.5798 | 0.6783 |
| $r$ | 4 | 5 | 8 | 10 | 15 | 20 | 25 | 30 | 35 | 40 |

**Table 14.2.** *Power for normality against logistic* $Log(1,\ 0.2)$ *distribution*

From a practical point of view, the coefficients a and b of the Shapiro–Wilk and Shapiro–Francia tests are given only for $20 < n \leq 50$ by Shapiro, Wilk and Francia. They are not available for $n > 50$. However, in the case of the small sample size, these tests then may not have enough power to detect non-normality in the population. Furthermore, the results of the simulations show that the Shapiro–Wilk and Shapiro–Francia tests are less powerful when the sample size increases. Thus, the Shapiro–Wilk and Shapiro–Francia tests have disadvantages.

| Tests | Sample size $n$ | | | | | | | | | |
|---|---|---|---|---|---|---|---|---|---|---|
| | 40 | 50 | 80 | 100 | 150 | 200 | 250 | 300 | 400 | 500 |
| **W** | 0.5609 | 0.486 | 0.2968 | 0.2025 | 0.076 | 0.024 | 0.0074 | 0.0016 | 0 | 0 |
| **W**′ | 0.4913 | 0.416 | 0.2399 | 0.1584 | 0.0555 | 0.017 | 0.0053 | $9.10^{-4}$ | 0 | 0 |
| $\mathbf{Y}^2_n$*(EQPG)* | 0.3406 | 0.4227 | 0.6136 | 0.7008 | 0.8449 | 0.9217 | 0.9580 | 0.9793 | 0.9962 | 0.9994 |
| $\mathbf{Y}^2_n$ *(2NPG)* | 0.3902 | 0.4737 | 0.6797 | 0.7809 | 0.9216 | 0.9764 | 0.9920 | 0.9972 | 0.9998 | 1 |
| $r$ | 4 | 5 | 8 | 10 | 15 | 20 | 25 | 30 | 35 | 40 |

**Table 14.3.** *Power for normality against Laplace* $L(1,\ 0.25)$ *distribution*

| Tests | Sample size $n$ | | | | | | | | | |
|---|---|---|---|---|---|---|---|---|---|---|
| | 40 | 50 | 80 | 100 | 150 | 200 | 250 | 300 | 400 | 500 |
| **W** | 0.9263 | 0.9158 | 0.8869 | 0.8606 | 0.7993 | 0.7405 | 0.6626 | 0.6105 | 0.4757 | 0.3741 |
| **W**′ | 0.9354 | 0.9293 | 0.9099 | 0.8860 | 0.8409 | 0.7890 | 0.7303 | 0.6802 | 0.5525 | 0.4431 |
| $\mathbf{Y}^2_n$*(EQPG)* | 0.0594 | 0.0647 | 0.0669 | 0.0751 | 0.088 | 0.097 | 0.1066 | 0.1112 | 0.1450 | 0.1660 |
| $\mathbf{Y}^2_n$ *(2NPG)* | 0.0733 | 0.0832 | 0.1184 | 0.1421 | 0.1908 | 0.2438 | 0.2965 | 0.3377 | 0.4448 | 0.5361 |
| $r$ | 4 | 5 | 8 | 10 | 15 | 20 | 25 | 30 | 35 | 40 |

**Table 14.4.** *Power for normality against Weibull* $W(1,\ 5)$ *distribution*

Following the simulated results in Tables 14.2–14.4, the power of NRR's statistic $\mathbf{Y}^2_n$ based on two Neyman–Pearson classes is the highest and has the best stability among all.

## 14.3. Bibliography

[BAG 11a] BAGDONAVIČIUS V., KRUOPIS J., NIKULIN M.S., *Non-Parametric Tests for Censored Data*, ISTE, London, John Wiley and Sons, New York, 2011.

[BAG 11b] BAGDONAVIČIUS V., KRUOPIS J., NIKULIN M.S., *Non-Parametric Tests for Complete Data*, ISTE, London, John Wiley and Sons, New York, 2011.

[CHE 54] CHERNOFF H., LEHMANN E.L., "The use of maximum likelihood estimates in $\chi^2$ tests for goodness of fit", *The Annals of Mathematical Statistics*, vol. 25, no. 3, pp. 579–586, 1954.

[CHI 71] CHIBISOV D.M., "Certain chi-square type tests for continuous distributions", *Theory of Probability & Its Applications*, vol. 16, no. 1, pp. 1–22, 1971.

[FIS 28] FISHER R.A., "On a property connecting the $\chi^2$ measure of discrepancy with the method of maximum likelihood", *Journal of the Royal Statistical Society*, vol. 87, pp. 442–449, 1928.

[GRE 96] GREENWOOD P.E., NIKULIN M.S., *A Guide to Chi-Squared Testing*, John Wiley and Sons, New York, 1996.

[NIK 73a] NIKULIN M.S., "Chi-square test for continuous distributions with location and scale parameters", *Teoriya Veroyatnostei i ee Primeneniya*, vol. 18, no. 3, pp. 583–591, 1973.

[NIK 73b] NIKULIN M.S., "Chi-square test for normality", *Proceedings of International Vilnius Conference on Probability Theory and Mathematical Statistics*, vol. 2, pp. 119–122, 1973.

[MOO 71] MOORE D.S., "A chi-square statistic with random cell boundaries", *The Annals of Mathematical Statistics*, vol. 42, no. 1, pp. 147–156, 1971.

[MOO 75] MOORE D.S., SPRUILL M.C., "Unified large-sample theory of general chi-squared statistics for tests of fit", *The Annals of Statistics*, vol. 3, pp. 599–616, 1975.

[OGA 74] OGAWA J., *A Note on the Optimal Spacing of the Systematic Statistics Normal Distribution*, Department of Mathematics, Statistics and Computing Science, University of Calgary, 1974.

[PEA 00] PEARSON K., "On the criterion that a given system of deviations from the probable in the case of a correlated system of variables is such that it can be reasonably supposed to have arisen from random sampling", *The London, Edinburgh, and Dublin Philosophical Magazine and Journal of Science*, vol. 50, no. 302, pp. 157–175, 1900.

[SHA 72a] SHAPIRO S.S., FRANCIA R.S., "An approximate analysis of variance test for normality", *Journal of the American Statistical Association*, vol. 67, no. 337, pp. 215–216, 1972.

[SHA 72b] SHAPIRO S.S., WILK M.B., "An analysis of variance test for the exponential distribution (complete samples)", *Technometrics*, vol. 14, no. 2, pp. 355–370, 1972.

[VOI 13] VOINOV V.G., NIKULIN M.S., *Chi-squared Goodness of Fit Tests with Applications*, Elsevier, 2013.

[WAT 57] WATSON G.S., "The Chi-square goodness-of-fit test for normal distributions", *Biometrika*, vol. 44, pp. 336–348, 1957.

# PART 2

# Statistical Models and Methods in Survival Analysis

Chapter 15

# Estimation/Imputation Strategies for Missing Data in Survival Analysis

## 15.1. Introduction

We consider the problem of estimation from right-censored data, when the censoring indicator is possibly missing. Let $T$ be a random variable representing the time to death from the cause of interest. Let $C$ denote a right-censoring random time. Under usual random censorship, the observation is $Y = T \wedge C$ and $\delta = \mathbf{1}(T \leq C)$, and it is assumed that $T$ and $C$ admit densities, respectively, denoted by $f_T$ and $g$. In addition, $C$ is assumed to be independent of $T$.

When the cause of death is not recorded, the censoring indicator is missing: this is the missing censoring indicator (MCI) model (see [SUB 06]), which is defined as follows. Let $\xi$ be the missingness indicator, that is $\xi = 1$ if $\delta$ is observed and $\xi = 0$ otherwise. The observed data are then given for individual $i \in \{1, \ldots, n\}$:

$$(Y_i, \delta_i, \xi_i = 1) \quad \text{or} \quad (Y_i, \xi_i = 0).$$

We consider the so-called MAR context, which means that we assume that the indicators are missing at random, i.e. $\xi$ and $\delta$ are independent conditionally to $Y$ (contrary to the MCAR case where it is assumed that the indicator is missing completely at random, i.e. $\xi$ is independent of $T$ and $C$). In this chapter, we mainly concentrate on different strategies for dealing with missing indicators.

---

Chapter written by Elodie BRUNEL, Fabienne COMTE and Agathe GUILLOUX.

This model has been considered by several authors in the last decade. For a complete overview on missing data, we refer to [LIT 02]. Most papers are interested in survival function and cumulative hazard rate estimation. In particular, van der Laan and McKeague [VAN 98] build a sieved non-parametric maximum likelihood (M.L.) estimator of the survival function in the MAR case and prove its efficiency. Their estimator is a generalization of the Kaplan–Meier estimator in this context and is the first proposal to reach the efficiency bound. Subramanian [SUB 04] also proposes an estimator of the survival function in the MAR case; he proves his estimate to be efficient as well.

Kernel methods have also been used to build different estimators in the MAR context. Subramanian [SUB 06] estimates the cumulative hazard rate with a ratio of kernel estimators. He provides an almost sure representation and a central limit theorem (CLT). He deduces results of the same type for the survival function. A study in a similar context is also provided by Wang and Ng [WAN 08]. Recently, Wang *et al.* [WAN 09] proposed density estimators based on kernels and Kaplan–Meier-type corrections of censoring. They prove a CLT and suggest a bandwidth selection strategy. Extensions of these works to conditional functions (both cumulative hazard and survival functions) in the presence of covariates are developed by Wang and Shen [WAN 08].

Our aim here is to study the non-parametric hazard rate estimation and, more specifically, regression and model selection estimation of the function defined, for all $t > 0$, by:

$$\lambda(t) = \frac{f_T(t)}{1 - F_T(t)}, \tag{15.1}$$

where $f_T$ and $F_T$ are, respectively, the probability density function (PDF) and the cumulative distribution function (CDF) of the survival time $T$. Two contrasts and estimators are proposed, and both involve the estimation of a specific nuisance functional parameter, $\pi$ or $\zeta$:

$$\pi(y) = \mathbb{E}(\xi|Y = y) = \mathbb{P}(\xi = 1|Y = y), \tag{15.2}$$

$$\zeta(y) = \mathbb{E}(\delta|Y = y) = \mathbb{P}(\delta = 1|Y = y). \tag{15.3}$$

We compare different estimation–imputation strategies for recovering $\pi$ and $\zeta$. More precisely, we propose either a parametric strategy following a logistic model – standard for $\pi$, but involving imputation steps for $\zeta$ – or a pure non-parametric regression strategy. We provide theoretical properties and numerical comparisons for these procedures.

## 15.2. Model and strategies

### 15.2.1. *Model assumptions*

The unknown function $\lambda$ to be estimated is the hazard rate of the random variable $T$, as defined in [15.1]. We will denote by $G_C$ the CDF of $C$. We define the conditional expectations of $\xi$ and $\delta$ by [15.2] and [15.3].

The global model assumption is denoted by (A15) and has several parts, specified as follows:

A15.1) The random vectors $(Y_i, C_i)$ are independent copies, for $i = 1, \ldots, n$, of $(Y, C)$.

A15.2) For $i = 1, \ldots, n$, we observe $Y_i = T_i \wedge C_i$, $\xi_i \in \{0, 1\}$, and $\delta_i = \mathbf{I}(T_i \leq C_i)$ if $\xi_i = 1$, otherwise $\xi_i = 0$.

A15.3) $C$ is independent of $T$.

A15.4) $\xi$ and $\delta$ are independent given $Y$.

Our estimation strategies are based on penalized contrasts. Namely, we obtain different collections of estimator by minimization of empirical criteria on functional spaces spanned by orthonormal bases defined on an interval $[0, \tau]$. We consider in the following two specific and classical examples of bases and model collections:

1) *Trigonometric bases.* These are defined by $\varphi_0(x) = (1/\sqrt{\tau})\mathbf{I}_{[0,\tau]}(x)$,

$$\varphi_{2j+1}(x) = \sqrt{2/\tau}\sin(2\pi jx/\tau)\mathbf{I}_{[0,\tau]}(x), \varphi_{2j}(x) = \sqrt{2/\tau}\cos(2\pi jx/\tau)\mathbf{I}_{[0,\tau]}(x).$$

Considering $(\varphi_j)_{0\leq j\leq m-1}$ yields spaces $S_m$ spanned by them, with odd dimension $D_m = m$. We denote by $\mathcal{S}_n$ the nesting space of the collection, i.e. the space corresponding to the maximal dimension for $S_m$ and we set $\mathcal{D}_n := \dim(\mathcal{S}_n) \leq \sqrt{n/\log(n)}$.

2) *Histogram bases.* These are defined by $\varphi_j(x) = \sqrt{2^m/\tau}\mathbf{I}_{[(j-1)\tau/2^m, j\tau/2^m[}(x)$, for $j = 1, \ldots, 2^m$, so that $S_m = \text{span}(\varphi_j, j = 1, \ldots, 2^m)$ and $D_m := \dim(S_m) = 2^m$. We will take $m \leq [\log_2(n)/2]$, where $[z]$ denotes the integer part of $z$ and $\log_2(x) = \log(x)/\log(2)$. We denote by $\mathcal{S}_n$ the nesting space of the collection and we set $\mathcal{D}_n := \dim(\mathcal{S}_n) \leq n/\log(n)$.

These bases are representative examples of localized bases for the second one (as piecewise polynomials, wavelets) or bounded non-localized bases for the first one.

### 15.2.2. *Strategy involving knowledge of* $\zeta$

The crucial property for the construction of an estimation procedure is the following: for any integrable function $h$, we have:

$$\begin{aligned} \mathbb{E}(\zeta(Y)h(Y)) &= \mathbb{E}[\mathbb{E}(\delta|Y)h(Y)] = \mathbb{E}(\delta h(Y)) \\ &= \mathbb{E}[\mathbf{1}(T \leq C)h(T)] \\ &= \int h(t)(1 - G_C)(t) f_T(t) dt. \end{aligned}$$

This yields the equality:

$$\mathbb{E}(\zeta(Y)h(Y)) = \mathbb{E}(\delta h(Y)) = \int h(y)\lambda(y)(1 - L(y))dy, \qquad [15.4]$$

with $1 - L(y) = \bar{L}(y) := \mathbb{P}(Y \geq y) = (1 - F_T(y))(1 - G_C(y))$.

Next, we note that, with assumption A15.4 and the definition of $\zeta$, we have:

$$\begin{aligned} \mathbb{E}(\delta_i\xi_i + (1 - \xi_i)\zeta(Y_i)|Y_i) &= \mathbb{E}(\delta_i|Y_i)\mathbb{E}(\xi_i|Y_i) + \mathbb{E}[(1 - \xi_i)\mathbb{E}(\delta_i|Y_i)|Y_i] \\ &= \mathbb{E}(\mathbb{E}(\delta_i|Y_i)(\xi_i + (1 - \xi_i))|Y_i) \\ &= \mathbb{E}(\delta_i|Y_i). \end{aligned} \qquad [15.5]$$

If $\zeta$ was known, we would consider the contrast:

$$\Gamma^{th}_{n,1}(h) = \frac{1}{n}\sum_{i=1}^{n}\int_0^1 h^2(y)\mathbf{1}(Y_i \geq y)dy - \frac{2}{n}\sum_{i=1}^{n}(\xi_i\delta_i + (1 - \xi_i)\zeta(Y_i))\, h(Y_i),$$

which is a particular case of the contrast introduced in [BRU 13]. Indeed, if we compute the expectation of this theoretical contrast, we obtain, under the MAR assumption and using [15.4] and [15.5],

$$\mathbb{E}(\Gamma^{th}_{n,1}(h)) = \|h\|^2_{\mu_1} - 2\int h(y)\lambda(y)d\mu_1(y) = \|h - \lambda\|^2_{\mu_1} - \|\lambda\|^2_{\mu_1},$$

with $d\mu_1(y) = (1 - L(y))dy$.

Clearly, the above quantity is small if $h$ is near $\lambda$, and the measure denoted by $\mu_1$ plays the role of a reference weighting norm. This explains why minimizing $\Gamma^{th}_{n,1}$ over an appropriate set of functions would be a relevant strategy to estimate $\lambda$.

As $\zeta$ is unknown, we must substitute another quantity in the contrast. The first natural strategy is to use an estimator $\tilde{\zeta}$ of $\zeta$. Consequently, we consider:

$$\Gamma^{\text{EST}}_{n,1}(h) = \frac{1}{n}\sum_{i=1}^{n}\int_0^1 h^2(y)\mathbf{1}(Y_i \geq y)dy - \frac{2}{n}\sum_{i=1}^{n}\left(\xi_i\delta_i + (1-\xi_i)\tilde{\zeta}(Y_i)\right)h(Y_i). \quad [15.6]$$

Estimators of $\zeta$ are constructed below. This strategy of estimation of the unknown hazard rate $\lambda$, via an estimation of $\zeta$, is also considered in [WAN 09]. Such a regression strategy was also considered in [BRU 05] without a missing indicator.

### 15.2.3. *Strategy involving knowledge of* $\pi$

A similar reasoning can be applied in case $\pi$ is known. Analogously, we consider the contrast:

$$\Gamma^{th}_{n,2}(h) = \frac{1}{n}\sum_{i=1}^{n}\int_0^1 h^2(y)\pi(y)\mathbf{1}(Y_i \geq y)dy - \frac{2}{n}\sum_{i=1}^{n}\delta_i\xi_i h(Y_i).$$

Indeed, we can compute the expectation of $\Gamma^{th}_{n,2}(h)$. First, we have:

$$\mathbb{E}(\delta_i\xi_i h(Y_i)) = \mathbb{E}(\mathbb{E}(\delta_i\xi_i|Y_i)h(Y_i)),$$

and with (A15.4), we get $\mathbb{E}(\delta_i\xi_i h(Y_i)) = \mathbb{E}(\zeta(Y_i)\pi(Y_i)h(Y_i))$. Now, applying [15.4] to $h\pi$ (instead of $h$) yields

$$\mathbb{E}(\delta_i\xi_i h(Y_i)) = \int h(y)\pi(y)\lambda(y)(1 - L(y))dy.$$

Consequently:

$$\mathbb{E}(\Gamma^{th}_{n,2}(h)) = \int h^2(y)d\mu_2(y) - 2\int h(y)\lambda(y)d\mu_2(y) = \|h-\lambda\|^2_{\mu_2} - \|\lambda\|^2_{\mu_2},$$

with $d\mu_2(y) = \pi(y)(1-L_Y(y))dy$. As previously, this explains why minimizing $\Gamma^{th}_{n,2}$ should allow us to estimate $\lambda$.

As $\pi$ is unknown, we replace it by an estimator $\tilde{\pi}$ in the contrast. Thus, we consider:

$$\Gamma^{\text{EST}}_{n,2}(h) = \frac{1}{n}\sum_{i=1}^{n}\int_0^1 h^2(y)\tilde{\pi}(y)\mathbf{1}(Y_i \geq y)dy - \frac{2}{n}\sum_{i=1}^{n}\delta_i\xi_i h(Y_i). \quad [15.7]$$

### 15.2.4. *Estimation of $\zeta$ or $\pi$: logit or non-parametric regression*

In both [15.6] and [15.7], a step has to be completed: we must make proposals for $\tilde{\zeta}$ and $\tilde{\pi}$. Two strategies can naturally be considered. First, coherently with the non-parametric setting of the first step, we can use a non-parametric regression method.

*Non-parametric regression* for estimation of $\zeta(y)$ or $\pi(y)$. The definition of $\zeta$ and $\pi$ as conditional expectations given $Y$ (see [15.3] and [15.2]) allows us to estimate these functions by non-parametric mean square contrast minimization. More precisely, we obtain a collection of estimators of $\zeta$ or $\pi$ on an interval $A$ by minimizing for $T$ varying over spaces $S_m$ the contrasts:

$$\tilde{\gamma}_{n,1}(T) = \frac{1}{n}\sum_{i=1}^{n}[\xi_i T^2(Y_i) - 2\xi_i\delta_i T(Y_i)] \text{ or } \tilde{\gamma}_{n,2}(T) = \frac{1}{n}\sum_{i=1}^{n}[T^2(Y_i) - 2\xi_i T(Y_i)],$$

respectively.

On the other hand, a natural strategy to estimate conditional indicators is the parametric logistic regression.

*Logistic regression*. If we add a logistic assumption:

$$\zeta(y) = \frac{\exp(\alpha_0 + \alpha_1 y)}{1 + \exp(\alpha_0 + \alpha_1 y)}, \quad \pi(y) = \frac{\exp(\beta_0 + \beta_1 y)}{1 + \exp(\beta_0 + \beta_1 y)},$$

then we may use an ML procedure. Note that, nevertheless, $\zeta$ and $\pi$ are different from this point of view. Indeed, concerning the estimation of $\pi$, $\xi$ and $Y$ are observed and, therefore, $\beta_0$ and $\beta_1$ correspond to standard logit estimation. However, this is not true for $\zeta$: we cannot directly compute the ML estimators for $(\alpha_0, \alpha_1)$ since there are censoring indicators $\delta_i$ missing. Thus, we follow a procedure proposed by Rubin [RUB 87]. For a binary classification variable, a new logistic regression model is simulated from the posterior predictive distribution of the parameters and is used to impute the missing values for each variable.

### 15.2.5. *Computing the hazard estimators*

We consider that we estimate the hazard rate on a compact set:

$$A = [0, \tau].$$

We set standard assumptions about boundedness from above and below.

A15.5) $\forall y \in A, \lambda(y) \leq \|\lambda\|_{A,\infty} < +\infty$.

A15.6) $\forall y \in A, 1 - L_Y(y) \geq \bar{L}_0 = \min_{y\in A} 1 - L_Y(y) > a > 0$ for fixed positive constant $a$.

A15.7) $\forall y \in A, \pi(y) \geq \pi_0 = \inf_{y\in A} \pi(y) > b > 0$ for fixed positive constant $b$.

A15.8) $\forall y \in A, 0 < f_0^* \leq f_Y(y) \leq f_1^* < +\infty$.

Assumptions A15.5 and A15.6 are common natural assumptions, and assumption A15.7 is specific to the strategy involving $\pi$ in the reference measure. Assumption A15.8 is a common technical assumption.

First, we define two types of estimator $\hat{\lambda}_m^{(i)}$ for $i = 1, 2$ on the space $S_m$ by:

$$\hat{\lambda}_m^{(i)} = \arg\min_{h\in S_m} \Gamma_{n,i}^{\text{EST}}(h), \quad \text{where } S_m = \text{span}\{\varphi_j, j = 1, \ldots, D_m\}.$$

The $\varphi_j$'s constitute an $\mathbb{L}^2$-orthonormal basis, and the function $h$ is of the form $h = \sum_j a_j \varphi_j$.

The estimators are obtained by minimization of the contrasts [15.6] and [15.7] given in sections 15.2.2 and 15.2.3. For the histogram basis, we can give their explicit expression in the MAR setting: $\hat{\lambda}_{\hat{m}}^{(i),[\text{H}]}(y) = \sum_{j=1}^{D_{\hat{m}}} \hat{a}_j^{(i),[\text{H}]} \varphi_j(y)$, with the coefficient $\hat{a}_j^{(i),[\text{H}]}$ having the following form, provided the denominators are non-zero:

$$\hat{a}_j^{(1),[\text{H}]} = \frac{\displaystyle\sum_{i=1}^{n} [\delta_i \xi_i + (1 - \xi_i)\zeta(Y_i)]\varphi_j(Y_i)}{\displaystyle\sum_{i=1}^{n} \int_0^1 \varphi_j^2(y) \mathbf{1}_{(Y_i \geq y)} dy},$$

$$\hat{a}_j^{(2),[\text{H}]} = \frac{\displaystyle\sum_{i=1}^{n} \delta_i \xi_i \varphi_j(Y_i)}{\displaystyle\sum_{i=1}^{n} \int_0^1 \varphi_j^2(y) \bar{\pi}(y) \mathbf{1}(Y_i > y) dy}.$$

More generally, let us define the matrices and vectors: $\overrightarrow{\hat{a}}^{(i)} = \left(\hat{a}_j^{(i)}\right)_{1\leq j\leq D_m}$,

$$\Theta_m^{(1)} := \left(\frac{1}{n} \sum_{i=1}^{n} \int \varphi_j(y) \varphi_{j'}(y) \mathbf{1}_{\{Y_i \geq y\}} dy\right)_{1\leq j,j'\leq D_m},$$

$$\Delta_m^{(1)} = \text{vec}\left(\frac{1}{n}\sum_{i=1}^{n}\left(\delta_i\xi_i + (1-\xi_i)\tilde{\zeta}(Y_i)\right)\varphi_j(Y_i))_{1\le j\le D_m}\right),$$

$$\Theta_m^{(2)} := \left(\frac{1}{n}\sum_{i=1}^{n}\int \varphi_j(y)\varphi_{j'}(y)\tilde{\pi}(y)\mathbf{1}_{\{Y_i\ge y\}}dy\right)_{1\le j,j'\le D_m}$$

$$\Delta_m^{(2)} = \left(\frac{1}{n}\sum_{i=1}^{n}\delta_i\xi_i\varphi_j(Y_i)\right)_{1\le j\le D_m}.$$

Then the coefficients of the estimators must fulfill the matrix constraint:

$$\Theta_m^{(i)}\overrightarrow{\hat{a}}_m^{(i)} = \Delta_m^{(i)}.$$

It follows that the estimator is well defined if $\Theta_m^{(i)}$ is invertible. We define $\rho(M)$ as the spectral radius of a matrix $M$, i.e. the largest eigenvalue in modulus of $M$. We set:

$$\overrightarrow{\hat{a}}_m^{(i)} = (\Theta_m^{(i)})^{-1}\Delta_m^{(1)} \text{ if } \rho(\Theta_m^{(i)}) \ge \max(\hat{c}_i, n^{-1/2}) \qquad [15.8]$$

and $\overrightarrow{\hat{a}}_m^{(i)} = 0$ otherwise, with $\hat{c}_1 = \hat{\bar{L}}_0/3$ and $\hat{c}_2 = 4\tilde{\pi}_0\hat{\bar{L}}_0/9$.

The quantities $\hat{\bar{L}}_0$ and $\tilde{\pi}_0$ are estimators of $\bar{L}_0 = \min_{y\in A} 1 - L_Y(y) > 0$ under assumption A15.6 and $\pi_0 = \min_{y\in A}\pi(y) > 0$ under assumption A15.7. An estimator $\hat{\bar{L}}_0$ is defined in [COM 09], and proved to have the required properties, namely to satisfy:

A15.9) For any integer $k \ge 1$, there exists a constant $C_k^{(L_0)} > 0$ such that $\mathbb{P}(|\hat{\bar{L}}_0 - \bar{L}_0| > \bar{L}_0/2) \le C_k^{(L_0)}/n^k$.

For $\tilde{\pi}_0$, we could study $\min\tilde{\pi}$ but we state results in terms of the conditions $\tilde{\pi}$ must fulfill:

A15.10) $\quad \mathbb{P}\left(\left\|\frac{\tilde{\pi}-\pi}{\pi}\right\|_{\infty,A} > 1/4\right) := \mathbb{P}(\Omega_\pi^c) \le \frac{C}{n^k}, \quad$ for $\quad k \ge 4 \quad$ and $C$ a given constant.

After this stage, the coefficients $\hat{a}_j^{(i)}$, for $i = 1, 2$, provide the development of an estimator:

$$\hat{\lambda}_m^{(i)}(y) = \sum_j \hat{a}_j^{(i)}\varphi_j(y). \qquad [15.9]$$

The model selection device is now based on the following criterion: for $i = 1, 2$,

$$\hat{m}^{(i)} = \arg\min_{m \in \mathcal{M}_n} (\Gamma^{\text{EST}}_{n,i}(\hat{\lambda}_m) + \widehat{\text{pen}}^{(i)}(m)), \quad [15.10]$$

where

$$\mathcal{M}_n = \{m \in \mathbb{N}, \dim(S_m) \leq N_n\}.$$

Moreover:

$$\widehat{\text{pen}}^{(i)}(m) = \kappa_i \|\hat{\lambda}\|_{\infty,A} \frac{\dim(S_m)}{n}, \quad [15.11]$$

where $\Phi_0$ is such that $\forall t \in S_m$, $\|t\|_\infty \leq \Phi_0 \|t\|$, and $\hat{\lambda} = \hat{\lambda}_{m_0}$ is an estimator in the collection, on a space $S_{m_0}$, with dimension $D_{m_0}$ such that $\log(n) \leq D_{m_0} \leq n^{1/4}$. Moreover, the dimension $D_m$ of the spaces $S_m$ has to be larger than $\log(n)$.

The computation of $\tilde{\zeta}$ and $\tilde{\pi}$ is similar and the vector of their coefficients in the basis $(\varphi_j)_{1 \leq j \leq d_m}$ is defined by the inverse of matrices:

$$\Upsilon^{(1)}_m = \left(\frac{1}{n}\sum_{i=1}^n \xi_i \varphi_j(Y_i)\varphi_{j'}(Y_i)\right)_{1 \leq j,j' \leq D_m},$$

$$\Upsilon^{(2)}_m = \left(\frac{1}{n}\sum_{i=1}^n \varphi_j(Y_i)\varphi_{j'}(Y_i)\right)_{1 \leq j,j' \leq D_m},$$

multiplied, respectively, by vectors:

$$\Xi^{(1)}_m = \Delta^{(2)}_m, \quad \Xi^{(2)}_m = \left(\frac{1}{n}\sum_{i=1}^n \xi_i \varphi_j(Y_i)\right)_{1 \leq j \leq D_m}$$

with adequate thresholds.

### 15.2.6. *Theoretical results*

We consider that we estimate the hazard rate on a compact set $A = [0, \tau]$.

The empirical reference norms are defined by:

$$\|h\|^2_{n,1} = \frac{1}{n}\sum_{i=1}^n \int_0^1 h^2(y)\mathbf{1}(Y_i \geq y)dy,$$

$$\|h\|_{n,2}^2 = \frac{1}{n}\sum_{i=1}^{n}\int_0^1 h^2(y)\pi(y)\mathbf{1}(Y_i \geq y)dy,$$

associated with the natural scalar product $\langle h_1, h_2\rangle_{n,i}$ obtained by polarization identity.

Let us also define another integral norm with respect to $d\varrho(y) = f_Y(y)dy$, where $f_Y$ is the PDF of $Y_1$, that is:

$$\|\psi\|_\varrho^2 = \int \psi^2(y)d\varrho(y) = \int \psi^2(y)f_Y(y)dy. \quad [15.12]$$

When considering estimators $\tilde{\pi}$ and $\tilde{\zeta}$ computed on independent samples of observations, we obtain the following result:

THEOREM 15.1.– Let $\hat{\lambda}^{(i)}_{\hat{m}^{(i)}}$ be the estimator defined by [15.8–15.11] for $i = 1, 2$, under assumptions A15.1–A15.4. Assume that the collection of models is nested and the estimators $\tilde{\zeta}$ and $\tilde{\pi}$ are computed on independent samples.

1) If assumptions A15.5, A15.6, A15.8 and A15.9 hold, there exists a choice of $\kappa_1$ such that, for $n$ large enough,

$$\begin{aligned}\mathbb{E}(\|\lambda\mathbf{I}_A - \hat{\lambda}^{(1)}_{\hat{m}^{(1)}}\|_{n,1}^2) \;&\leq\; C_1 \inf_{m\in\mathcal{M}_n}(\|\lambda\mathbf{I}_A - \lambda_m\|^2 \\ &+\|\lambda\|_{A,\infty}\frac{D_m}{n}) + C_1'\mathbb{E}(\|\tilde{\zeta} - \zeta\|_\varrho^2) + \frac{C_1''}{n},\end{aligned} \quad [15.13]$$

where $C_1$ is a numerical constant and $C_2'$, $C_2''$ are constants depending on the constants of the problem.

2) If assumptions A15.5–A15.8 and A15.9–A15.10 hold, there exists a choice of $\kappa_2$ such that, for $n$ large enough,

$$\begin{aligned}\mathbb{E}(\|\lambda\mathbf{I}_A - \hat{\lambda}^{(2)}_{\hat{m}^{(2)}}\|_{n,2}^2) \;&\leq\; C_2 \inf_{m\in\mathcal{M}_n}(\|\lambda\mathbf{I}_A - \lambda_m\|^2 + \|\lambda\|_{A,\infty}\frac{D_m}{n}) \\ &+C_2'\mathbb{E}(\|\tilde{\pi} - \pi\|_\varrho^2) + \frac{C_2''}{n},\end{aligned} \quad [15.14]$$

where $C_2$ is a numerical constant and $C_2'$, $C_2''$ are constants depending on the constants of the problem.

Inequality [15.13] is proved in [BRU 13] in a more general setting, and a sketch of the proof of [15.14] is given in the appendix.

Note that the larger $\kappa_i$, the larger $\text{pen}^{(i)}(m)$ in the upper bound.

The results stated in [15.13] and [15.14] involve three terms:

– The first terms $\inf_{m\in\mathcal{M}_n}(\|\lambda\mathbf{I}_A - \lambda_m\|^2 + \|\lambda\|_{A,\infty}D_m/n$ are common and correspond to the squared-bias ($\|\lambda\mathbf{I}_A - \lambda_m\|^2$)/variance ($\|\lambda\|_{A,\infty}D_m/n$) compromise.

– The second term is $\mathbb{E}(\|\tilde{\zeta} - \zeta\|_\varrho^2)$ or $\mathbb{E}(\|\tilde{\pi} - \pi\|_\varrho^2)$, that is the mean-square risk of the estimator of $\zeta$ or $\pi$ on $A$. These quantities can have different orders, depending on the second step strategy.

– The last terms $C_1''/n$ or $C_2''/n$ are negligible.

We can obtain adaptive non-parametric rate with the same model selection principle for $\zeta$ and $\pi$, if we apply the non-parametric strategy (see theorem 2 for $\zeta$ in [BRU 13]).

On the other hand, with the logit strategy, we have:

$$\tilde{\zeta}(y) = \frac{e^{\hat{\alpha}_0 + y\hat{\alpha}_1}}{1 + e^{\hat{\alpha}_0 + y\hat{\alpha}_1}}, \quad \tilde{\pi}(y) = \frac{e^{\hat{\beta}_0 + y\hat{\beta}_1}}{1 + e^{\hat{\beta}_0 + y\hat{\beta}_1}},$$

and we get:

$$\mathbb{E}(\|\tilde{\zeta} - \zeta\|_\varrho^2) \leq 2\left(\mathbb{E}((\hat{\alpha}_0 - \alpha_0)^2) + \mathbb{E}(Y_1^2)\mathbb{E}((\hat{\alpha}_1 - \alpha_1)^2)\right),$$

$$\mathbb{E}(\|\tilde{\pi} - \pi\|_\varrho^2) \leq 2\left(\mathbb{E}((\hat{\beta}_0 - \beta_0)^2) + \mathbb{E}(Y_1^2)\mathbb{E}((\hat{\beta}_1 - \beta_1)^2)\right).$$

All these terms are of order $1/n$ as soon as all parameters are estimated with parametric rate.

The questions that we have to study empirically are the following:

1) If the logit assumptions are fulfilled, is the non-parametric strategy much worse than the parametric one?

2) If the logit model is not satisfied, can the logit approximation correctly estimate $\pi$ or $\zeta$?

3) How much does the quality of the estimator of these quantities influence that of $\lambda$?

## 15.3. Imputation-based strategy

The second approach we aim to explore is inspired from stochastic regression imputation. The idea is no longer to substitute an estimator of the unknown quantity $\zeta$ but rather to substitute a random number $\gamma_i$ to $\zeta(Y_i)$ that has the same expectation.

Let $\gamma_i$ be generated from a Bernoulli distribution with parameter $\zeta(Y_i)$ given $\xi_i = 0$. Here, the quantity $\zeta(Y_i)$ also has to be estimated as in section 15.2.4, and we consider again both non-parametric and logit strategies. Then, we are in a position to define the following contrast:

$$\Gamma_n^{\text{IMP}}(h) = \frac{1}{n}\sum_{i=1}^{n}\int_0^1 h^2(y)\mathbf{1}(Y_i \geq y)dy - \frac{2}{n}\sum_{i=1}^{n}(\xi_i\delta_i + (1-\xi_i)\gamma_i)\, h(Y_i), \quad [15.15]$$

and the imputation strategy gives the following estimator:

$$\hat{\lambda}_m^{(3)} = \arg\min_{h\in S_m} \Gamma_n^{\text{IMP}}(h), \quad \text{where } S_m = \text{span}\{\varphi_j, j = 1, \cdots, D_m\}.$$

We do not provide the theoretical study of the estimator defined by this contrast since $\mathbb{E}(\Gamma_n^{\text{IMP}}(h)) = \mathbb{E}(\Gamma_{n,1}^{th}(h))$. We propose a numerical comparison of this strategy with the other strategies developed in section 15.2.

## 15.4. Numerical comparison

To illustrate the behaviors of the different estimators, we consider three models to simulate data from the MCI, i.e. $n$ independent replications of:

$$(Y_i = T_i \wedge C_i, \delta_i, \xi_i = 1) \quad \text{or} \quad (Y_i, \xi_i = 0).$$

In each model, the parameters are set in order to obtain around 30% of censoring and 85% (respectively 55%) of non-missing indicators.

MODEL 15.1.– (Weibull-logistic) Each $T_i$ is drawn from a Weibull distribution with scale $a = 10$ and shape $b = 4$, each $C_i$ from an exponential distribution with mean $\mu = 25$. For each individual $i$, the missing indicator is simulated from a Bernoulli distribution with parameter $\pi(y_i)$ with:

$$\pi(y) = \frac{1}{1 + \exp(-\beta_1 - \beta_2 y)},$$

where we set $\beta_1 = 0.1$ and $\beta_2 = 0.25$ (respectively 0.005) to obtain 85% (respectively 55%) non-missing indicators. In this model:

$$\zeta(y) = (1 + \frac{\mu}{ba^{-b}y^{b-1}})^{-1}.$$

MODEL 15.2.– (Monotone) This is based on an example from Subramanian [SUB 09]. For each $i$, $Y_i$ is drawn from an exponential distribution with parameter 1.

The censoring and missing indicators are simulated from Bernoulli distributions with parameters $\zeta(y_i)$ and $\pi(y_i)$, respectively, with:

$$\pi(y) = 1 - \exp(-\theta \exp(y)) \text{ and } \zeta(y) = \frac{1}{1 + \exp(-\alpha_1 - \alpha_2 y)},$$

with $\theta = 1$ (respectively 0.32) to obtain 85% (respectively 55%) non-missing indicators, and $\alpha_1 = -2$, $\alpha_2 = 5.2$. Note that the hazard rate $h(y)$ coincides with the auxiliary function $\zeta(y)$ in this model.

MODEL 15.3.– (Non-monotone) Consider the same schema as model 15.2, with

$$\zeta(y) = 0.9|\sin(\pi/2 - 1.5y)| + 0.1.$$

For $K = 500$ replications over different paths, we compute the (empirical) average MISE of the penalized estimators $\tilde{\lambda}$ over a grid of size 100:

$$\text{MISE} = \frac{1}{K} \sum_{k=1}^{K} \left( \frac{\tau_k}{100} \sum_{i=1}^{100} \left( \lambda(t_i) - \tilde{\lambda}^{(k)}(t_i) \right)^2 \right),$$

where $\tau_k$ is the inter-quantile interval length associated with the 10% and 90% empirical quantiles of the $Y_i$s . The value of the constant $\kappa$ appearing in the penalty has been calibrated over the three models and fixed to 2 (respectively 3.5) for the contrast $\Gamma^{\text{EST}}_{n,1}$ (respectively $\Gamma^{\text{EST}}_{n,2}$). We also give in parenthesis the standard deviation value of the MISE evaluated over the 500 samples. The results are summarized in Tables 15.1–15.4. For each model, we give the MISEs of our penalized estimators $\hat{\lambda}^{(1)}_{\hat{m}^{(1)}}$ for the estimation strategy involving $\zeta$ in columns 1 and 2, $\hat{\lambda}^{(2)}_{\hat{m}^{(2)}}$ for the estimation strategy involving $\pi$ in columns 3 and 4, and $\hat{\lambda}^{(3)}_{\hat{m}^{(3)}}$ for the imputation strategy in columns 5 and 6.

For each strategy and each model, we consider various sample sizes, $n = 200, 1{,}000$ and 5,000, and two levels of missing indicators, namely 15% and 45%. The abbreviations "NP" and "logit" stand for the way of estimating the auxiliary functions $\zeta$ or $\pi$ according to the strategy under study (estimation of $\zeta$ with $\Gamma^{\text{EST}}_{n,1}$ and $\Gamma^{\text{IMP}}_{n}$ or estimation of $\pi$ with $\Gamma^{\text{EST}}_{n,2}$). First, whatever the model we considered, we observe that the strategy involving $\pi$ gives very poor performances compared to the other ones. The difference between the strategies involving $\pi$ or $\zeta$ is of great importance for a numerical point of view. In fact, as we can see by examining the sequence of coefficients $\hat{a}_j^{(1),[\text{H}]}$ and $\hat{a}_j^{(2),[\text{H}]}$, the estimator $\tilde{\pi}$ appears in the denominator (while the estimator $\tilde{\zeta}$ does not) and it results that a small estimation error for $\pi$ can strongly increase the final error on the hazard estimator. We can see in Figure 15.1 that despite a quite good estimation of $\pi$, the MISE in Tables 15.1–15.3

is always the worst for the estimation strategy involving $\pi$. Consequently, we focus on the other strategies. Both imputation and estimation strategies involving the function $\zeta$ have very similar behavior and performance. We also notice that the logit estimator nearly always outperforms the non-parametric one, at least for sample sizes of $n = 200$ and 1,000. Of course, when the logit assumption is violated, for large sample ($n = 5{,}000$) and important missing rate (45%), the results are deteriorating (see Table 15.3).

| Weibull/ Logistic | Estimation strategy with $\Gamma^{\text{EST}}_{n,1}$ | | Estimation strategy with $\Gamma^{\text{EST}}_{n,2}$ | | Imputation strategy $\Gamma^{\text{IMP}}_{n}$ | |
|---|---|---|---|---|---|---|
| Missing rate | 15% | 45% | 15% | 45% | 15% | 45% |
| $n = 200$ | | | | | | |
| NP | 0.055 | 0.061 | 0.120 | 0.269 | 0.055 | 0.067 |
| | (0.032) | (0.043) | (0.081) | (0.162) | (0.032) | (0.051) |
| logit | 0.055 | 0.053 | 0.151 | 0.286 | 0.053 | 0.055 |
| | (0.029) | (0.032) | (0.123) | (0.160) | (0.031) | (0.032) |
| $n = 1{,}000$ | | | | | | |
| NP | 0.017 | 0.021 | 0.035 | 0.086 | 0.018 | 0.024 |
| | (0.010) | (0.014) | (0.018) | (0.037) | (0.010) | (0.015) |
| logit | 0.016 | 0.016 | 0.062 | 0.092 | 0.016 | 0.019 |
| | (0.008) | (0.009) | (0.036) | (0.026) | (0.008) | (0.012) |
| $n = 5{,}000$ | | | | | | |
| NP | 0.008 | 0.009 | 0.011 | 0.025 | 0.008 | 0.009 |
| | (0.007) | (0.008) | (0.010) | (0.023) | (0.006) | (0.009) |
| logit | 0.007 | 0.008 | 0.023 | 0.024 | 0.007 | 0.008 |
| | (0.005) | (0.006) | (0.022) | (0.023) | (0.006) | (0.006) |

**Table 15.1.** *Average and standard deviation (in parenthesis) of the MISE over 500 replicated samples for hazard rate estimators of model 15.1 (Weibull-Logistic), for estimation strategies with $\Gamma^{\text{EST}}_{n,1}$ and $\Gamma^{\text{EST}}_{n,2}$ and imputation strategy $\Gamma^{\text{IMP}}_{n}$. Censoring rate $\simeq 30\%$*

## 15.5. Proofs

PROOF OF THEOREM 15.1.–

The first inequality is obtained as a particular case of the bound proved in [BRU 13] (see also [COM 11]). Below, we give steps for obtaining the second inequality.

The reference centered empirical process is

$$\nu_n(h) = \frac{1}{n}\sum_{i=1}^{n}\left(\delta_i\xi_i h(Y_i) - \int h(y)\lambda(y)\pi(y)\mathbf{1}_{\{Y_i \geq y\}}dy\right).$$

| Subramanian monotone | Estimation strategy with $\Gamma^{\text{EST}}_{n,1}$ | | Estimation strategy with $\Gamma^{\text{EST}}_{n,2}$ | | Imputation strategy $\Gamma^{\text{IMP}}_{n}$ | |
|---|---|---|---|---|---|---|
| Missing rate | 15% | 45% | 15% | 45% | 15% | 45% |
| $n = 200$ | | | | | | |
| NP | 0.102 | 0.125 | 0.201 | 0.410 | 0.102 | 0.124 |
| | (0.051) | (0.052) | (0.143) | (0.445) | (0.057) | (0.051) |
| logit | 0.105 | 0.101 | 0.177 | 0.182 | 0.102 | 0.120 |
| | (0.062) | (0.060) | (0.083) | (0.071) | (0.0592) | (0.0576) |
| $n = 1{,}000$ | | | | | | |
| NP | 0.037 | 0.037 | 0.073 | 0.181 | 0.037 | 0.038 |
| | (0.016) | (0.014) | (0.043) | (0.131) | (0.017) | (0.014) |
| logit | 0.036 | 0.035 | 0.063 | 0.084 | 0.037 | 0.039 |
| | (0.018) | (0.018) | (0.035) | (0.035) | (0.017) | (0.016) |
| $n = 5{,}000$ | | | | | | |
| NP | 0.012 | 0.012 | 0.028 | 0.080 | 0.012 | 0.012 |
| | (0.011) | (0.011) | (0.025) | (0.068) | (0.012) | (0.011) |
| logit | 0.013 | 0.013 | 0.019 | 0.023 | 0.013 | 0.012 |
| | (0.012) | (0.011) | (0.018) | (0.022) | (0.012) | (0.011) |

**Table 15.2.** *Average and standard deviation (in parenthesis) of the MISE over 500 replicated samples for hazard rate estimators of the Model 15.2, for estimation strategies with $\Gamma^{\text{EST}}_{n,1}$ and $\Gamma^{\text{EST}}_{n,2}$ and imputation strategy $\Gamma^{\text{IMP}}_{n}$. Censoring rate $\simeq$ 30%*

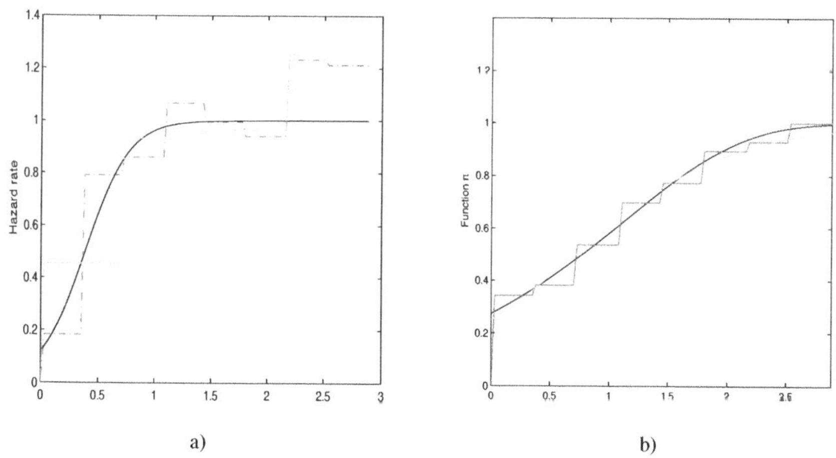


**Figure 15.1.** *Strategy involving estimation of $\pi$: a) true hazard rate (black curved line) for model 15.2 and its estimators $\hat{\lambda}^{(2)}_{\hat{m}^{(2)}}$ with both non-parametric (dotted line) and logit (gray line) estimators of $\pi$ and b) true auxiliary function $\pi$ (black curved line) and its non-parametric (black plain line) and logit (gray line) estimators. $n = 1{,}000$ with $\simeq$ 30% censoring and 45% missing indicators*

| Subramanian non-monotone | Estimation strategy with $\Gamma^{\text{EST}}_{n,1}$ | | Estimation strategy with $\Gamma^{\text{EST}}_{n,2}$ | | Imputation strategy $\Gamma^{\text{IMP}}_{n}$ | |
|---|---|---|---|---|---|---|
| Missing rate | 15% | 45% | 15% | 45% | 15% | 45% |
| $n = 200$ | | | | | | |
| NP | 0.132 (0.061) | 0.163 (0.060) | 0.218 (0.088) | 0.269 (0.248) | 0.132 (0.061) | 0.163 (0.060) |
| logit | 0.138 (0.060) | 0.180 (0.058) | 0.234 (0.081) | 0.257 (0.073) | 0.137 (0.062) | 0.170 (0.060) |
| $n = 1{,}000$ | | | | | | |
| NP | 0.064 (0.019) | 0.076 (0.016) | 0.105 (0.029) | 0.125 (0.055) | 0.065 (0.019) | 0.076 (0.016) |
| logit | 0.066 (0.019) | 0.096 (0.018) | 0.113 (0.026) | 0.128 (0.024) | 0.065 (0.019) | 0.077 (0.016) |
| $n = 5{,}000$ | | | | | | |
| NP | 0.038 (0.039) | 0.039 (0.038) | 0.055 (0.054) | 0.067 (0.066) | 0.038 (0.040) | 0.040 (0.040) |
| logit | 0.040 (0.040) | 0.065 (0.064) | 0.061 (0.062) | 0.069 (0.069) | 0.040 (0.040) | 0.065 (0.064) |

**Table 15.3.** *Average and standard deviation (in parenthesis) of the MISE over 500 replicated samples for hazard rate estimators of model 15.3, for estimation strategies with $\Gamma^{\text{EST}}_{n,1}$ and $\Gamma^{\text{EST}}_{n,2}$ and imputation strategy $\Gamma^{\text{IMP}}_{n}$. Censoring rate $\simeq 30\%$*

It is composed of two different processes that have different orders, $\nu_n(h) = \nu_{n,1}(h) - \nu_{n,2}(h)$, with

$$\nu_{n,1}(h) = \frac{1}{n}\sum_{i=1}^{n}(\delta_i\xi_i h(Y_i) - \langle h, \lambda\rangle_{\mu_2}), \qquad [15.16]$$

$$\nu_{n,2}(h) = \frac{1}{n}\sum_{i=1}^{n}\left(\int h(y)\lambda(y)\pi(y)\mathbf{1}_{\{Y_i\geq y\}}dy - \langle h, \lambda\rangle_{\mu_2}\right). \qquad [15.17]$$

We define a set $\Omega = \Delta \cap \Omega_{\bar{L}_0} \cap \Omega_\pi$, where $\Omega_\pi$ is defined in assumption A15.10 and

$$\hat{\mathcal{G}} = \bigcap_{m\in\mathcal{M}_n} \hat{\mathcal{G}}_m, \quad \hat{\mathcal{G}}_m = \{\min Sp(\Theta_m^{(2)}) \geq \max(4\tilde{\pi}_0\hat{\bar{L}}_0/9, n^{-1/2})\},$$

$$\Delta = \left\{\forall h \in \mathcal{S}_n, \left|\frac{\|h\|^2_{n,2}}{\|h\|_{\mu_2}} - 1\right| \leq \frac{1}{2}\right\} \quad \text{and} \quad \Omega_{\bar{L}_0} = \left\{|\frac{\hat{\bar{L}}_0}{\bar{L}_0} - 1| \leq \frac{1}{2}\right\}.$$

It is easy to see that (see [BRU 31]):

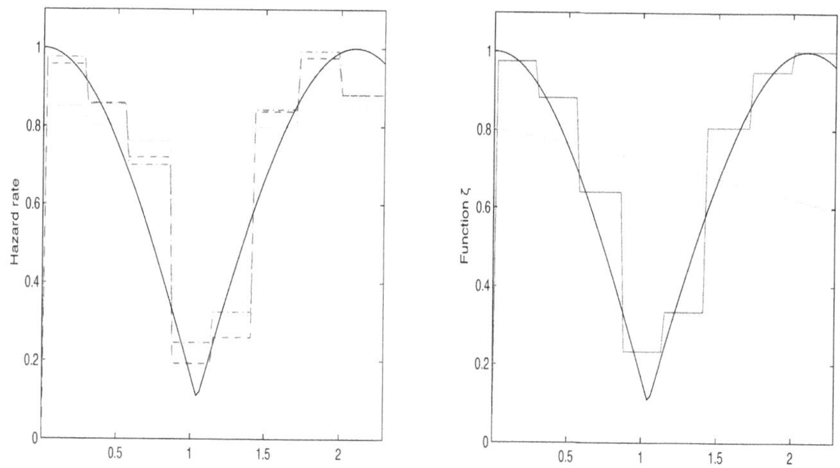


a) b)

**Figure 15.2.** *Strategies involving estimation of $\zeta$: a) true hazard rate (black v-shaped line) for model 15.3 and its estimators $\hat{\lambda}^{(1)}_{\hat{m}(1)}$ with non-parametric (black dotted line) and logit (gray line) estimators of $\zeta$ and $\hat{\lambda}^{(3)}_{\hat{m}(3)}$ with non-parametric estimator of $\zeta$ (gray dotted line), and b) true auxiliary function $\zeta$ (black v-shaped line) and its non-parametric (black plain line) and logit (gray line) estimators. $n = 1,000$ with $\simeq 30\%$ censoring and 45% missing indicators*

LEMMA 15.1.– Under the assumptions of theorem 15.1, for $n$ large enough, the following embedding holds: $\Delta \cap \Omega_{\bar{L}_0} \cap \Omega_\pi \subset \hat{\mathcal{G}} \cap \Omega_{\bar{L}_0} \cap \Omega_\pi$.

As a consequence, for all $m \in \mathcal{M}_n$, the matrices $\Theta^{(2)}_m$ are invertible on $\Omega$.

Let us denote $\lambda_A = \lambda \mathbf{1}_A$ and $\pi_A = \pi \mathbf{1}_A$. We start with the decomposition

$$\Gamma^{\text{EST}}_{n,2}(h) - \Gamma^{\text{EST}}_{n,2}(\ell) = \|h - \lambda_A\|^2_{n,2} - \|\ell - \lambda_A\|^2_{n,2} - 2\nu_n(h - \ell)$$

$$+\frac{1}{n}\sum_{i=1}^{n} \int (h - \lambda_A)^2(y)\mathbf{1}_{\{Y_i \geq y\}}(\tilde{\pi} - \pi_A)(y)dy$$

$$-\frac{1}{n}\sum_{i=1}^{n} \int (\ell - \lambda_A)^2(y)\mathbf{1}_{\{Y_i \geq y\}}(\tilde{\pi} - \pi_A)(y)dy$$

$$+\frac{2}{n}\sum_{i=1}^{n} \int (h - \ell)(y)\lambda(y)\mathbf{1}_{\{Y_i \geq y\}}(\tilde{\pi} - \pi_A)(y)dy.$$

Then, let us denote by $\lambda_m$ the orthogonal projection of $\lambda_A$ on $S_m$ and use the definition of $\hat{\lambda}_{\hat{m}}$ on $\Omega$ (see lemma 15.1). We assume first that $\|\lambda\|_{A,\infty}$ is known and denote $\text{pen}(m) = \kappa\|\lambda\|_{A,\infty}D_m/n$, and write $\Gamma^{\text{EST}}_{n,2}(\hat{\lambda}_{\hat{m}}) + \text{pen}(\hat{m}) \leq \Gamma^{\text{EST}}_{n,2}(\lambda_m) + \text{pen}(m)$.

Then we get, on $\Omega$,

$$\begin{aligned}\|\hat{\lambda}_{\hat{m}} - \lambda_A\|^2_{n,2} \leq\ & \|\lambda_m - \lambda_A\|^2_{n,2} + \text{pen}(m) + 2\nu_n(\hat{\lambda}_{\hat{m}} - \lambda_m) - \text{pen}(\hat{m}) \\ & -\frac{1}{n}\sum_{i=1}^{n}\int(\hat{\lambda}_{\hat{m}} - \lambda_A)^2(y)\mathbf{1}_{\{Y_i\geq y\}}(\tilde{\pi} - \pi_A)(y)dy \\ & +\frac{1}{n}\sum_{i=1}^{n}\int(\lambda_m - \lambda_A)^2(y)\mathbf{1}_{\{Y_i\geq y\}}(\tilde{\pi} - \pi_A)(y)dy \\ -\frac{2}{n}\sum_{i=1}^{n}\int(\hat{\lambda}_{\hat{m}} - \lambda_m)(y)\lambda(y)1_{\{Y_i\geq y\}}(\tilde{\pi} - \pi_A)(y)dy.\end{aligned} \quad [15.18]$$

First, we write

$$2|\nu_n(\hat{\lambda}_{\hat{m}} - \lambda_m)| \leq \frac{1}{8}\|\hat{\lambda}_{\hat{m}} - \lambda_m\|^2_{\mu_2} + 8 \sup_{h\in B_{m,\hat{m}}(0,1)} \nu_n^2(h),$$

where $B_{m,m'}(0,1) = \{h \in S_m + S_{m'}, \|h\|_{\mu_2} \leq 1\}$. It follows that

$$\begin{aligned}2|\nu_n(\hat{\lambda}_{\hat{m}} - \lambda_m)|\mathbf{1}_\Omega \leq\ & \frac{1}{4}\|\hat{\lambda}_{\hat{m}} - \lambda_m\|^2_{n,2}\mathbf{1}_\Omega + 16\left(\sup_{h\in B_{m,\hat{m}}(0,1)} \nu_{n,1}^2(h) - p(m,\hat{m})\right)_+ \\ & +16p(m,\hat{m}) + 16 \sup_{h\in B_{m,\hat{m}}(0,1)} \nu_{n,2}^2(h) \\ & \leq \frac{1}{2}\|\hat{\lambda}_{\hat{m}} - \lambda\|^2_{n,2}\mathbf{1}_\Omega + \frac{1}{2}\|\lambda_m - \lambda\|^2_{n,2} \\ & +16\left(\sup_{h\in B_{m,\hat{m}}(0,1)} \nu_n^2(h) - p(m,\hat{m})\right)_+ \\ & +16p(m,\hat{m}) + 16 \sup_{h\in B_{m,\hat{m}}(0,1)} \nu_{n,2}^2(h),\end{aligned} \quad [15.19]$$

where $p(m, m')$ is defined below.

Next, using the definition of $\Omega_\pi$ and assumption A15.10, we have:

$$\begin{aligned}&\frac{1}{n}\left|\sum_{i=1}^{n}\int(\lambda_m-\lambda_A)^2(y)\mathbf{1}_{\{Y_i\geq y\}}(\tilde{\pi}-\pi_A)(y)dy\right|\mathbf{1}_{\Omega_\pi}\\ \leq\ &\frac{1}{4n}\sum_{i=1}^{n}\int(\lambda_m-\lambda_A)^2(y)\mathbf{1}_{\{Y_i\geq y\}}\pi_A(y)dy\leq\frac{1}{4}\|\lambda_m-\lambda_A\|_{n,2}^2. \quad [15.20]\end{aligned}$$

In the same way

$$\frac{1}{n}\left|\sum_{i=1}^{n}\int(\hat{\lambda}_{\hat{m}}-\lambda)^2(y)\mathbf{1}_{\{Y_i\geq y\}}(\tilde{\pi}-\pi)(y)dy\right|\mathbf{1}_{\Omega_\pi}\leq\frac{1}{4}\|\hat{\lambda}_{\hat{m}}-\lambda\|_{n,2}^2. \quad [15.21]$$

Moreover, using first the Schwarz inequality and then $2xy\leq x^2/a+ay^2$ for $x,y$ and $a>0$ (here $a=16$), we get

$$\begin{aligned}&\frac{2}{n}\left|\sum_{i=1}^{n}\int(\hat{\lambda}_{\hat{m}}-\lambda_m)(y)\lambda(y)\mathbf{1}_{\{Y_i\geq y\}}(\tilde{\pi}-\pi_A)(y)dy\right|\\ &\leq\frac{1}{16n}\sum_{i=1}^{n}\int(\hat{\lambda}_{\hat{m}}-\lambda_m)^2(y)\mathbf{1}_{\{Y_i\geq y\}}\pi(y)dy\\ &+\frac{16}{n}\sum_{i=1}^{n}\int[(\tilde{\pi}-\pi_A)^2(y)/\pi(y)]\lambda^2(y)\mathbf{1}_A(y)dy\\ &\leq\frac{1}{16}\|\hat{\lambda}_{\hat{m}}-\lambda_m\|_{n,2}^2+\frac{16\|\lambda\|_{\infty,A}}{n\pi_0}\sum_{i=1}^{n}\int(\tilde{\pi}-\pi_A)^2(y)\lambda(y)\mathbf{1}_A(y)dy.\end{aligned}$$

Therefore, as $\tilde{\pi}$ is computed on an independent sample of observations, we get

$$\begin{aligned}&\frac{2}{n}\mathbb{E}\left(\left|\sum_{i=1}^{n}\int(\hat{\lambda}_{\hat{m}}-\lambda_m)(y)\lambda(y)\mathbf{1}_{\{Y_i\geq y\}}(\tilde{\pi}-\pi)(y)dy\right|\mathbf{1}_\Omega\right)\\ &<\frac{1}{8}\mathbb{E}(\|\hat{\lambda}_{\hat{m}}-\lambda\|_{n,2}^2\mathbf{1}_\Omega)+\frac{1}{8}\|\lambda-\lambda_m\|_{\mu_2}^2+\frac{16\|\lambda\|_{\infty,A}}{\pi_0}\mathbb{E}\left(\int_A(\tilde{\pi}-\pi)^2(y)\lambda(y)dy\right)\\ &\leq\frac{1}{8}\mathbb{E}(\|\hat{\lambda}_{\hat{m}}-\lambda\|_{n,2}^2\mathbf{1}_\Omega)+\frac{1}{8}\|\lambda-\lambda_m\|_{\mu_2}^2+\frac{16\|\lambda\|_{\infty,A}}{\bar{L}_0\pi_0}\mathbb{E}\left(\int_A(\tilde{\pi}-\pi)^2(y)f_Y(y)dy\right)\\ &\leq\frac{1}{8}\mathbb{E}(\|\hat{\lambda}_{\hat{m}}-\lambda\|_{n,2}^2\mathbf{1}_\Omega)+\frac{1}{8}\|\lambda-\lambda_m\|_{\mu_2}^2+\frac{16\|\lambda\|_{\infty,A}}{\bar{L}_0\pi_0}\mathbb{E}(\|\tilde{\pi}-\pi\|_\varrho^2). \quad [15.22]\end{aligned}$$

Now, gathering [15.18]–[15.22], we get, as $1 - 1/2 - 1/4 - 1/8 = 1/8$ and $1 + 1/2 + 1/4 + 1/8 = 15/8$,

$$\frac{1}{8}\mathbb{E}(\|\hat{\lambda}_{\hat{m}} - \lambda_A\|^2_{n,2}\mathbf{1}_\Omega) \leq \frac{15}{8}\|\lambda_m - \lambda_A\|^2_{\mu_2} + \mathrm{pen}(m) + \frac{16\|\lambda\|_{\infty,A}}{\bar{L}_0\pi_0}\mathbb{E}(\|\tilde{\pi} - \pi\|^2_\varrho)$$

$$+16\mathbb{E}\left(\sup_{h\in B_{m,\hat{m}}(0,1)} \nu^2_{n,2}(h)\right)$$

$$+16\mathbb{E}\left(\left(\sup_{h\in B_{m,\hat{m}}(0,1)} \nu^2_{n,1}(h) - p(m,\hat{m})\right)_+\right) + 16\mathbb{E}(p(m,\hat{m}) - \mathrm{pen}(\hat{m})). \quad [15.23]$$

Now write $\nu_{n,2}(h) = \int_A h(y)\lambda(y)(\hat{\bar{L}}(y) - \mathbb{E}(\hat{\bar{L}}(y))\pi(y)dy$, with $\hat{\bar{L}}(y) = (1/n)\sum_{i=1}^n \mathbf{1}_{Y_i\geq y}$. Clearly

$$\sup_{h\in B_{m,\hat{m}}(0,1)} \nu^2_{n,2}(h) \leq \frac{1}{\bar{L}_0^2}\sup_{h\in B_{m,\hat{m}}(0,1)}\int_A h^2(y)d\mu_2(y)\int_A \lambda^2(y)(\hat{\bar{L}}(y)$$

$$-\mathbb{E}(\hat{\bar{L}}(y))^2 d\mu_2(y)$$

$$= \frac{1}{\bar{L}_0^2}\int_A \lambda^2(y)(\hat{\bar{L}}(y) - \mathbb{E}(\hat{\bar{L}}(y))^2 d\mu_2(y).$$

Therefore, since $\mathbb{E}[(\hat{\bar{L}}(y) - \mathbb{E}(\hat{\bar{L}}(y))^2] = \bar{L}(y)(1-\bar{L}(y))/n \leq 1/(4n)$ and $\pi(y) \leq 1$, we get

$$\mathbb{E}\left(\sup_{h\in B_{m,\hat{m}}(0,1)} \nu^2_{n,2}(h)\right) \leq \frac{\|\lambda\|^2_{A,\infty}}{4n\bar{L}_0^2}\int \bar{L}(y)dy = \frac{\|\lambda\|^2_{A,\infty}\mathbb{E}(Y_1)}{4n\bar{L}_0^2}. \quad [15.24]$$

We can use the Talagrand inequality (see [TAL 96]) to prove the proposition below, and we refer to [BRU 13] for details.

PROPOSITION 15.5.– Under the assumptions of theorem 15.1, there exists a numerical constant $\kappa$ such that, for

$$p(m,m') = (\kappa/16)\left(\|\lambda\|^2_{A,\infty}\frac{D_m + D_{m'}}{n}\right),$$

we have

$$\mathbb{E}\left(\sup_{h\in B_{m,\hat{m}}(0,1)} (\nu^2_{n,1}(h) - p(m,\hat{m}))_+\right) \leq \frac{C}{n},$$

and this, together with [15.23] and [15.24], yields, as $16p(m, m') \leq \text{pen}(m) + \text{pen}(m')$,

$$\frac{1}{8}\mathbb{E}(\|\hat{\lambda}_{\hat{m}} - \lambda_A\|_{n,2}^2 \mathbf{I}_\Omega) \leq \frac{15}{8}\|\lambda_m - \lambda_A\|_{\mu_2}^2 + 2\text{pen}(m) + \frac{C}{n}.$$

Next, we study $\mathbb{E}(\|\hat{\lambda}_{\hat{m}} - \lambda\|_{n,2}^2 \mathbf{I}_{\Omega^c})$. We have first (see [BRU 13]):

LEMMA 15.2.– Under the assumptions of theorem 15.1, we have $\|\hat{\lambda}_{\hat{m}}\|_{n,2}^2 \leq Cn^3$ for a constant $C$ depending on the basis.

Moreover assumption A15.10 ensures that $\mathbb{P}(\Omega_\pi^c) \leq C/n^k$, assumption A15.9 implies that $\mathbb{P}(\Omega_{\hat{L}_0}^c) \leq C_k^{(L_0)}/n^k$ and proposition 4 in [COM 11] can be adapted here to get $\mathbb{P}(\Delta^c) \leq C_k^{(\Delta)}/n^k$, under the condition that ${\mathcal{D}_n}^2 \leq n/\log^2(n)$ if basis 1 is used and $\mathcal{D}_n \leq n/\log^2(n)$ if basis 2 is used. Gathering these elements yields $\mathbb{E}(\|\hat{\lambda}_{\hat{m}} - \lambda\|_n^2 \mathbf{I}_{\Omega^c}) \leq C/n$ by choosing $k = 4$.

For the method allowing us, to replace $\|\lambda\|_{A,\infty}$ by an estimator, we refer to [COM 11].

## 15.6. Bibliography

[BRU 05] BRUNEL E., COMTE F., "Penalized contrast estimation of density and hazard rate with censored data", *Sankhya*, vol. 67, part 3, pp. 441–475, 2005.

[BRU 13] BRUNEL E., COMTE F., GUILLOUX A., "Nonparametric estimation for survival data with censoring indicators missing at random", *Journal of Statistical Planning and Inference*, vol. 143, pp. 1653–1671, 2013.

[COM 11] COMTE F., GAÏFFAS S., GUILLOUX A., "Adaptive estimation of the conditional intensity of marker-dependent counting processes", *Annales de l'Institut Henri Poincaré Probabilités et Statistiques*, vol. 47, pp. 1171–1196, 2011.

[LIT 02] LITTLE R.J.A., RUBIN D.B., *Statistical Analysis with Missing Data,* 2nd ed, Wiley Series in Probability and Statistics, Wiley-Interscience, John Wiley & Sons, Hoboken, NJ, 2002.

[RUB 87] RUBIN D.B., *Multiple Imputation for Nonresponse in Surveys,* Wiley Series in Probability and Mathematical Statistics: Applied Probability and Statistics, John Wilcy & Sons, New York, 1987.

[TAL 96] TALAGRAND M., "New concentration inequalities in product spaces", *Inventiones Mathematicae*, vol. 126, pp. 505–563, 1996.

[VAN 98] VAN DER LAAN M.J., MCKEAGUE I.W., "Efficient estimation from right-censored data when failure indicators are missing at random", *Annals of Statistics,* vol. 26, pp. 164–182, 1998.

[SUB 04] SUBRAMANIAN S., "Asymptotically efficient estimation of a survival function in the missing censoring indicator model", *Journal of Nonparametric Statistics*, vol. 16, pp. 797–817, 2004.

[SUB 06] SUBRAMANIAN S., "Survival analysis for the missing censoring indicator model using kernel density estimation techniques", *Statistical Methodology*, vol. 3, pp. 125–136, 2006.

[SUB 09] SUBRAMANIAN S., "The multiple imputations based Kaplan–Meier estimator", *Statistics & Probability Letters* , vol. 79, pp. 1906–1914, 2009.

[SUB 11] SUBRAMANIAN S., "Multiple imputations and the missing censoring indicator model", *Journal of Multivariate Analysis,* vol. 102, pp. 105–117, 2011.

[WAN 08] WANG Q., SHEN J., "Estimation and confidence bands of a conditional survival function with censoring indicators missing at random", *Journal of Multivariate Analysis,* vol. 99, pp. 928–948, 2008.

[WAN 08] WANG Q., NG K.W., "Asymptotically efficient product-limit estimators with censoring indicators missing at random", *Statistica Sinica,* vol. 18, pp. 749–768, 2008.

[WAN 09] WANG Q., LIU W., LIU C., "Probability density estimation for survival data with censoring indicators missing at random", *Journal of Multivariate Analysis,* vol. 100, pp. 835–850, 2009.

[ZHO 08] ZHOU X., SUN L., "Additive hazards regression with missing censoring information", *Statistica Sinica,* vol. 13, pp. 1237–1257, 2003.

Chapter 16

# Non-Parametric Estimation of Linear Functionals of a Multivariate Distribution Under Multivariate Censoring with Applications

## 16.1. Introduction

The study of dependent durations arises in many fields. For example, in epidemiology, one can be interested in the lifetimes of twins or the times before healing after using two kinds of drugs on eye problems. In actuarial sciences, provision of pension contracts with a reversion clause requires us to model the dependence between the two lifetimes involved in the contract. The reliability of several components of an engine could be studied in industry. Data sets used to perform statistical estimation in such problems often suffer from incompleteness of some observations due to censoring. Several approaches have been proposed to overcome this difficulty [HOU 00]. Among them, the estimation of the cumulative distribution function in a non-parametric way has been studied [HOU 00, DAB 88, WAN 97, LIN 93, PRE 92, TSA 86, PRU 91, VAN 96a, AKR 03]. However, most of the estimators present some drawbacks, for instance Tsai *et al.* [TSA 86] studied a method based on the conditional survival functions but the almost sure consistency is slow and the estimator is not symmetric in the sense that it is not equivariant under reversal of coordinates. Dabrowska [DAB 88] proposed a product limit estimator that can be seen as a bivariate Kaplan–Meier estimator. The multivariate distribution

Chapter written by Olivier LOPEZ and Philippe SAINT-PIERRE.

function is rewritten with quantities that are easily estimated using their empirical counterparts; however, the estimator can assign negative mass to some observations [PRU 91]. Another approach based on slightly modifying the observations is proposed by van der Laan [VAN 96a]. A non-parametric estimator based on copula modeling is introduced by Wang and Wells [WAN 97], whereas Lin and Ying [LIN 93] focus on univariate censoring. These estimators are designed to estimate the cumulative distribution function, but are not adapted to estimate the distribution. They do not define a proper joint distribution as they are not equivariant under reversal coordinates or assign negative mass to some points. Therefore, we cannot use them to estimate functionals of the multivariate distribution [GIL 92, FAN 00]. An estimator of the distribution based on weighted mean of marginal distribution have been introduced by Akritas and Van Keilegom [AKR 03]. This estimator defines a proper distribution, however, the convergence rate does not achieve a $\sqrt{n}$-rate, the computation and the generalization to multivariate case seems difficult. More recently, Lopez and Saint-Pierre [LOP 12] introduced an estimator that does not present these drawbacks. This estimator is asymptotically normal, can easily be extended to multivariate censored lifetimes and defines a proper joint distribution. Therefore, it provides a valuable tool to study functionals of a bivariate distribution.

This chapter deals with multivariate right-censored survival data but the notations can be quite complex. To facilitate the readability, a bivariate framework is considered in this chapter even if the proposed estimation method can easily be adapted to deal with multivariate survival data. We are interested in estimating integrals of the distribution of a bivariate right-censored variable $(T^{(1)}, T^{(2)})$. Let us define $F(t^{(1)}, t^{(2)}) = \mathbb{P}(T^{(1)} \leq t^{(1)}, T^{(2)} \leq t^{(2)})$ as the joint distribution function and $\phi(t^{(1)}, t^{(2)})$ as a measurable function from $\mathbb{R}^2$ to $\mathbb{R}$ such as $\iint |\phi(t^{(1)}, t^{(2)})| dF(t^{(1)}, t^{(2)}) < \infty$. The aim is to estimate integrals of $F$ of the following form

$$\xi(\phi) = \mathbb{E}(\phi(T^{(1)}, T^{(2)})) = \iint \phi(t^{(1)}, t^{(2)}) dF(t^{(1)}, t^{(2)}). \qquad [16.1]$$

Such functionals are involved in several useful quantities when analyzing the relationship between bivariate data. For example, $\mathbb{E}(T^{(1)}T^{(2)})$ allows us to obtain the covariance and the correlation coefficient. The distribution functions $\mathbb{E}(\mathbf{1}_{T^{(1)} \leq t^{(1)}, T^{(2)} \leq t^{(2)}})$, $\mathbb{E}(T^{(1)} - T^{(2)})$, $\mathbb{E}(\min(T^{(1)}, T^{(2)}))$ or $\mathbb{P}(T^{(1)} < T^{(2)})$ are other desirable quantities for the analysis. Kendall's coefficient and other dependance measures [GIL 83, FAN 00] can also be derived and will be discuss later in the chapter.

In the absence of censoring, the estimation of integrals $\xi$ has been studied by many authors. In the case of univariate survival data, among others, Stute [STU 95] has provided several works on the estimation of Kaplan–Meier integrals. In a multivariate framework, only a few papers concern the estimation $\xi$. Indeed, most of the estimators of the bivariate distribution function are not adapted to this question. A class of asymptotically normal estimators designed for univariate censoring has been proposed by Lu and Burke [LU 08]. With regard to bivariate censoring, Gill [GIL 92] and Fan *et al.* [FAN 00] have studied quantities [16.1] by correcting the Dabrowska's estimator [DAB 88] with a view to obtaining a regular distribution estimator. The non-parametric estimator proposed by Lopez and Saint-Pierre [LOP 12] is well adapted as it allows us to obtain a proper joint distribution. Moreover, it presents some desirable properties: classical empirical estimator is obtained in absence of censoring, estimates of marginal distribution lead to Kaplan–Meier estimates and it is equivariant under reversal of coordinates. Several applications useful in multivariate analysis can be derived from such estimator. Estimation of several dependence measures as Kendall's coefficient is discussed. A bootstrap procedure for multivariate survival data is derived. The estimator [LOP 12] that defines a proper distribution (as it does not assign negative mass to some points) can be used to simulate samples. Efron's resampling procedure [EFR 81] can then be adapted to the multivariate framework. Moreover, a regression model where the response and the covariate are both randomly right-censored is also studied.

The remaining chapter is organized as follows. Section 16.2 introduces the bivariate distribution estimator to estimate the quantities [16.1]. Section 16.3 provides an asymptotic independent and indentically distributed (i.i.d.) representation for the estimators. Consistency and asymptotic normality are derived. Section 16.4 is devoted to the estimation of dependence measures, a bootstrap procedure and regression modeling. Finally, these methods are applied to the lifetimes of twins in section 16.5.

## 16.2. Non-parametric estimation of the distribution

The aim of this section is to estimate the distribution of a random vector $T = (T^{(1)}, T^{(2)})$, subject to bivariate censoring. Therefore, $T = (T^{(1)}, T^{(2)})$ is not fully observed and the available data in the case of right-censoring are an i.i.d. sample $(Y_j, \Delta_j)_{1 \leq j \leq n}$ where

$$Y_j = (Y_j^{(1)}, Y_j^{(2)}) = (\inf(T_j^{(1)}, C_j^{(1)}), \inf(T_j^{(2)}, C_j^{(2)})),$$

$$\Delta_j = (\Delta_j^{(1)}, \Delta_j^{(2)}) = (\mathbf{1}_{T_j^{(1)} \leq C_j^{(1)}}, \mathbf{1}_{T_j^{(2)} \leq C_j^{(2)}}),$$

and $C = (C^{(1)}, C^{(2)})$ is the censoring vector. Some identifiability assumptions are needed to derive a consistent estimate of the joint distribution of $T = (T^{(1)}, T^{(2)})$. Multivariate counterparts of traditional assumptions used in univariate survival analysis are considered in the remainder of this chapter.

ASSUMPTION 16.1.– $T$ and $C$ are independent variables,$\mathbb{P}(T^{(i)} = C^{(i)}) = 0$ for $i = 1, 2$.

In absence of censoring, the joint distribution function can easily be estimated using the empirical function $F_{emp}(t^{(1)}, t^{(2)}) = \frac{1}{n}\sum_{j=1}^{n} \mathbf{1}_{T_j^{(1)} \leq t^{(1)}, T_j^{(2)} \leq t^{(2)}}$. Unfortunately, the empirical distribution is unavailable if $T = (T^{(1)}, T^{(2)})$ are censored and the empirical distribution of observed variables is a biased estimator. The well-known Kaplan–Meier product limit estimate is often used in case of univariate right-censored data. This estimator can be rewritten as a weighted empirical sum [STU 95, SAT 01]. Each observation $Y = (Y^{(1)}, Y^{(2)})$ is weighted in order to compensate the censoring. Lopez and Saint-Pierre [LOP 12] have extended this approach to bivariate right-censored lifetimes. The non-parametric estimator of the distribution function is given by

$$\hat{F}(y^{(1)}, y^{(2)}) = \frac{1}{n}\sum_{j=1}^{n} \hat{W}_j \mathbf{1}_{Y_j^{(1)} \leq y^{(1)}, Y_j^{(2)} \leq y^{(2)}}, \qquad [16.2]$$

where the weights $\hat{W}_j$ are converging toward some limit weights $W_j$, which must satisfy the following equality

$$\mathbb{E}[W_j \phi(Y^{(1)}, Y^{(2)})] = E[\phi(T^{(1)}, T^{(2)})].$$

The weights $W_j$ represent the mass assigned to the $j$th observation to correct the effect of bivariate censoring. Assuming the independence between $T$ and $C$, we can easily show that

$$W_j = \frac{\Delta_j^{(1)}\Delta_j^{(2)}}{\mathbb{E}[\Delta_j^{(1)}\Delta_j^{(2)} | T^{(1)} = y^{(1)}, T^{(2)} = y^{(2)}]} \qquad [16.3]$$

$$= \frac{\Delta_j^{(1)}\Delta_j^{(2)}}{\mathbb{P}(C^{(1)} > y^{(1)}, C^{(2)} > y^{(2)})}. \qquad [16.4]$$

We can observe that weights only affect mass to doubly uncensored observations. In the case of independence between $C^{(1)}$ and $C^{(2)}$, the estimation of weights only consists of estimating the marginal distribution functions. In the case of dependence between censoring times, a copula approach can be used to model the cumulative distribution function. In this chapter, we suppose that

ASSUMPTION 16.2.– $\mathbb{P}(C^{(1)} > y^{(1)}, C^{(2)} > y^{(2)}) = \mathcal{C}(G_1(y^{(1)}), G_2(y^{(2)}))$

where $\mathcal{C}$ is the known copula. Indeed, the copula estimation for censored data has already been considered in the literature ([SHI 95]). Moreover, the theoretical results presented in the next section are still valid if the known copula is replaced by a consistent estimator of $\mathcal{C}$ [SHI 95]. In this case, it is sufficient to estimate the marginal distribution functions $G_1$ and $G_2$ to obtain an estimator of the weights

$$\hat{W}_j = \frac{\Delta_j^{(1)}\Delta_j^{(2)}}{\mathcal{C}(\hat{G}_1(Y_j^{(1)}), \hat{G}_2(Y_j^{(2)}))}, \qquad [16.5]$$

where $\hat{G}_i$ is the Kaplan–Meier estimator of the survival function of censoring. Indeed, by reversing the labels of $T^{(i)}$ and $C^{(i)}$, the censoring $C^{(i)}$ can be considered as right-censored variable. Therefore, the survival function of the censoring times can be estimated by the Kaplan–Meier estimator, for $i = 1, 2$,

$$\hat{G}_i(t) = \prod_{k:Y_k^{(i)} \leq t} \left(1 - \frac{d\hat{H}_i^0(Y_k^{(i)})}{\hat{H}_i(Y_k^{(i)})}\right), \qquad [16.6]$$

where $\hat{H}_i(t) = \sum_{j=1}^n \mathbf{1}_{Y_j^{(i)} \geq t}$ and $\hat{H}_i^0(t) = \sum_{j=1}^n (1 - \Delta_j^{(i)})\mathbf{1}_{Y_j^{(i)} \geq t}$.

The estimations of weights are used to define estimators of the cumulative distribution function and integrals of the distribution

$$\hat{F}(y^{(1)}, y^{(2)}) = \frac{1}{n} \sum_{j-1}^{n} \hat{W}_j \mathbf{1}_{Y_j^{(1)} \leq y^{(1)}, Y_j^{(2)} \leq y^{(2)}}, \qquad [16.7]$$

$$\hat{\xi}(\phi) = \iint \phi(y^{(1)}, y^{(2)}) d\hat{F}(y^{(1)}, y^{(2)}) = \frac{1}{n} \sum_{j=1}^{n} \hat{W}_j \phi(Y_j^{(1)}, Y_j^{(2)}). \qquad [16.8]$$

These estimates can be generalized to multivariate observations using a multivariate copula function. Left-censored data or left- and right-censored data can also be considered. Indeed, the left-censored counterpart of the Kaplan–Meier estimator proposed by Gómez *et al.* [GÓM 92] can be used to estimate the marginal survival function in case of left censoring.

## 16.3. Asymptotic properties

The weights used in equations [16.7] and [16.8] depend on the whole sample. Hence, they are not independent and the central limit theorem does not apply. An asymptotic i.i.d. representation for integrals [16.8] can be obtained under some assumptions related to the copula function and to the function $\phi$ belonging to a Donsker class of functions. Some conditions on regularity of the copula function $\mathcal{C}$ are needed.

ASSUMPTION 16.3.– The function $(x_1, x_2) \rightarrow \mathcal{C}(x_1, x_2)$ is twice continuously differentiable. Moreover, we will denote $\partial_1 \mathcal{C}(x_1, x_2)$ (respectively $\partial_2 \mathcal{C}(x_1, x_2)$) the partial derivative of $\mathcal{C}$ with respect to $x_1$ (respectively $x_2$), and we assume that

$$\sup_{x_1,x_2} |\partial_1 \mathcal{C}(x_1, x_2)| + \sup_{x_1,x_2} |\partial_2 \mathcal{C}(x_1, x_2)| + \sum_{i,j \in \{1,2\}} \sup_{x_1,x_2} |\partial_i \partial_j \mathcal{C}(x_1, x_2)| < \infty,$$

and that $\mathcal{C}(x_1, x_2) \neq 0$ for $x_1 \neq 0$ and $x_2 \neq 0$.

ASSUMPTION 16.4.– Assume that

$$\mathcal{C}(x_1, x_2) \geq x_1^{\alpha_1} x_2^{\alpha_2},$$

with $0 \leq \alpha_i \leq 1$ for $i = 1, 2$.

For instance, this assumption holds in the case of independence copula or copula from Frank's family with $\alpha_1 = \alpha_2 = 1$. Consistency of alternative estimators in multivariate analysis have only been studied on a compact set strictly include in the support of the distribution of $Y = (Y^{(1)}, Y^{(2)})$. Such result is restrictive, for instance in a regression framework, since only functions that vanish at the tail of the distribution are considered. Tightness arguments similar to ones used in [GIL 83] to study the consistency of the Kaplan–Meier estimator on the whole line can be used to extend the consistency to functions taking non-zero values on the whole support of $Y$.

ASSUMPTION 16.5.– $\mathcal{F}$ is a Donsker class of locally bounded functions with positive locally bounded envelope function $\Phi$ such that

$$\int \frac{\Phi^2(y) dF(y)}{\mathcal{C}(G_1(y^{(1)}), G_2(y^{(2)}))} < \infty, \qquad [16.9]$$

and, for some $\varepsilon > 0$ arbitrary small,

$$\int \Phi(y) \left[ \frac{G_1^{1-\alpha_1}(y^{(1)}) \mathcal{K}_1^{1/2+\varepsilon}(y^{(1)})}{G_2^{\alpha_2}(y^{(2)})} + \frac{G_2^{1-\alpha_2}(y^{(2)}) \mathcal{K}_2^{1/2+\varepsilon}(y^{(2)})}{G_1^{\alpha_1}(y^{(1)})} \right] dF(y) < \infty, \qquad [16.10]$$

where for $i = 1, 2$,

$$\mathcal{K}_i(u) = -\int_0^u \frac{dG_i(t)}{G_i(t)^2 F_i(t)}.$$

The functions $H_i$, $G_i$ and $F_i$ represent the survival function of the variables $Y^{(i)}$, $C^{(i)}$ and $T^{(i)}$, respectively. The moment condition [16.9] corresponds to a finite asymptotic variance and condition [16.10] is useful to obtain the tightness of the process. This assumption is a bivariate adaptation of assumption 1.6 described in [STU 95]. It is shown in [STU 95] that such assumption holds for a large class of distributions, for instance, when the lifetime distributions are subexponential, the function $\Phi$ is polynomial. Donsker classes of functions [VAN 96b] satisfy a uniform central limit theorem property. Their asymptotic properties is a valuable tool to derive our main theorem. For any function $\phi \in L^1(\xi)$, it is natural to compare $\hat{\xi}(\phi)$ with the following quantity,

$$\tilde{\xi}(\phi) = \frac{1}{n} \sum_{j=1}^{n} W_j \phi(Y_j^{(1)}, Y_j^{(2)}).$$

Indeed, $\tilde{\xi}(\phi)$ is a sum of i.i.d. quantities that tends to $\xi(\phi)$ by the strong law of large numbers. The following i.i.d. representation can be obtained.

THEOREM 16.1.– Let $\mathcal{F}$ be a class of functions satisfying assumption 16.5. Then, under identifiability assumptions 16.1 and 16.2 and under regularity assumptions 16.3 and 16.4,

$$\hat{\xi}(\phi) - \tilde{\xi}(\phi) = \frac{1}{n} \sum_{j=1}^{n} \int \eta(Y_j, \Delta_j, t) \phi(t) \mathcal{C}(G_1(t^{(1)}), G_2(t^{(2)})) dF(t) + R_n(\phi),$$

with $\sup_{\phi \in \mathcal{F}} |R_n(\phi)| = o_P(n^{-1/2})$, and where $\eta$ is

$$\eta(Y_j, \Delta_j; y) = -\sum_{i=1,2} \frac{\partial_i \mathcal{C}(G_1(y^{(1)}), G_2(y^{(2)}))}{\mathcal{C}(G_1(y^{(1)}), G_2(y^{(2)}))^2} \left[ \left\{ \frac{\Delta_j^{(i)} \mathbf{1}_{Y_j^{(i)} > y}}{F_i(Y_j^{(i)})} - G_i(y) \right\} \right.$$
$$\left. + \left\{ \frac{(1 - \Delta_j^{(i)}) G_i(Y_j^{(i)} \vee y)}{H_i(Y_j^{(i)})} - \int \frac{\mathbf{1}_{Y_j^{(i)} \geq u} G_i(u \vee y) dF_i(u)}{H_i(u) F_i(u)} \right\} \right].$$

The proof of this theorem is obtained and detailed in [LOP 12]. Note that the asymptotic normality of integrals [16.8] can easily be deduced from the i.i.d. representation and the central limit theorem. In some applications, the function $\phi$

belongs to a Donsker class of functions with bounded envelope $\Phi$ satisfying $\Phi(y^{(1)}, y^{(2)}) \equiv 0$ if $y^{(1)} > \tau^{(1)}$ or $y^{(2)} > \tau^{(2)}$ for $\tau^{(i)} < \inf\{y : \mathbb{P}(Y^{(i)} > y) = 0\}$, $i = 1, 2$. In such cases, the function $\phi$ are bounded with compact support strictly included in the support of the distribution, assumption 16.5 is already satisfied and is not needed to derive the asymptotic representation

## 16.4. Statistical applications of functionals

### 16.4.1. *Dependence measures*

A popular dependence measure between two variables is the Kendall's coefficient of concordance

$$\tau = 2\mathbb{P}[(T_1^{(1)} - T_2^{(1)})(T_1^{(2)} - T_2^{(2)}) > 0] - 1.$$

Alternatively, Kendall's coefficient $\tau$ can be evaluated by integration of the bivariate distribution function

$$\tau = 4 \iint F(t^{(1)}, t^{(2)}) dF(t^{(1)}, t^{(2)}) - 1.$$

This expression is an integral of the form [16.1]. A non-parametric estimate $\hat{\tau}$ can be obtained using the estimator $\hat{F}$. Theorem 16.1 can be used to obtain an asymptotic representation for $\hat{\tau}$ [GRI 13]. The asymptotic normality of $\hat{\tau}$ can be derived from the central limit theorem.

THEOREM 16.2.– Under identifiability assumptions 16.1 and 16.2 and under regularity assumptions 16.3 and 16.4

$$n^{-1/2}(\hat{\tau} - \tau) \rightsquigarrow \mathcal{N}(0, Q).$$

The analytic expression of the limit covariance matrix $Q$ can be derived from the asymptotic representation and then be estimated from the data by replacing each unknown quantity by its empirical counterpart. Nevertheless, the bootstrap can be recommended to evaluate the law of $\hat{\tau}$. Kendall's tau estimation helps to evaluate the dependence between the variables. It can also be useful when studying survival copula models under bivariate censoring. Indeed, Kendall's coefficient is related to the parameter of association for some copula families, for example Frank or Clayton copulas. The association parameter can be estimated using maximum likelihood estimation [SHI 95]. However, due to censored data, the optimization is very sensitive to the choice of initial value. Our estimate of Kendall's coefficient can be used as an initial value of the association parameter in the maximization algorithm. The integrated hazard correlation, the median concordance [HOU 00] or the dependence measures defined in [GIL 92, FAN 00] can also be estimated using [16.8].

### 16.4.2. *Bootstrap*

Estimator of integrals [16.1] is proven to be asymptotically normal [LOP 12]. The expression of the asymptotic variance can be derived even if the explicit expression is very complex. The variance can be estimated by replacing the unknown functions by their empirical estimates and asymptotic confidence intervals can be obtained. However, such intervals are often larger than those obtained with bootstrap methods. In the case of univariate survival data, a bootstrap procedure introduced by Efron [EFR 81] is consistent and seems to give the best results [AKR 86]. Instead of resampling the pairs of observed variables $(Y, \Delta)$ directly, Efron's methodology consists of simulating lifetimes and censoring times using non-parametric estimators of their distributions. The aim is to adapt this approach to a multivariate framework using the estimator [16.7](which defines a proper distribution). The lifetimes $(T^{(1)}, T^{(2)})$ are simulated using the estimated distribution $\hat{F}$. Assumption 16.2 can be used to estimate the distribution function $G$ of the censoring times $(C^{(1)}, C^{(2)})$

$$\hat{G}(y^{(1)}, y^{(2)}) = \mathcal{C}(\hat{G}_1(y^{(1)}), \hat{G}_2(y^{(2)})), \qquad [16.11]$$

where $\hat{G}_1$ and $\hat{G}_2$ are the Kaplan–Meier estimators defined in [16.6]. Samples of censoring times can be simulated according to the estimator of the distribution $\hat{G}$. We can then construct samples of observed variables and estimate the integral [16.1] on each sample. The bootstrap procedure is as follows:

– Step 0: compute estimates $\hat{F}$ and $\hat{G}$ of the distribution functions of lifetimes $F$ and censoring times $G$ using estimators [16.7] and [16.11].

– Step 1: for $b = 1, ..., B$,

1) simulate independent variables $(T_{j,b}^{(1)}, T_{j,b}^{(2)})_{1 \leq j \leq n}$ according to the distribution $\hat{F}$;

2) simulate independent variables $(C_{j,b}^{(1)}, C_{j,b}^{(2)})_{1 \leq j \leq n}$ according to the distribution $\hat{G}$;

3) compute the $b$th bootstrap sample $(Y_{j,b}^{(1)}, Y_{j,b}^{(2)}, \Delta_{j,b}^{(1)}, \Delta_{j,b}^{(2)})_{1 \leq j \leq n}$ where for $k = 1, 2$,

$$Y_{j,b}^{(k)} = \inf(T_{j,b}^{(k)}, C_{j,b}^{(k)}),$$

$$\Delta_{j,b}^{(k)} = \mathbf{1}_{T_{j,b}^{(k)} \leq C_{j,b}^{(k)}};$$

4) compute the integral [16.8] for the $b$th bootstrap sample $\hat{\xi}_b$.

– Step 2: use $\hat{\xi}_1$,...,$\hat{\xi}_B$ to estimate the distribution of $\hat{\xi}$.

The sample $\hat{\xi}_1, ..., \hat{\xi}_B$ can be used to estimate the distribution, the mean $\hat{\mu}(\hat{\xi})$ or the variance $\hat{\sigma}(\hat{\xi})$ of $\hat{\xi}$ using the empirical estimate. A large sample confidence interval of the mean $\mu(\hat{\xi})$ can be obtained using a normal approximation

$$CI_1(\alpha) = [\hat{\mu}(\hat{\xi}) \pm z_{\alpha/2}\frac{\hat{\sigma}(\hat{\xi})}{\sqrt{n}}].$$

Another confidence interval can be obtained by estimating the quantiles $\hat{q}_{\alpha/2}$ and $\hat{q}_{1-\alpha/2}$ using the empirical distribution. A confidence band is then given by

$$CI_2(\alpha) = [\hat{q}_{\alpha/2}; \hat{q}_{1-\alpha/2}].$$

In practice, the total mass of the distributions defined by $\hat{F}$ and $\hat{G}$ can be lower than 1. A normalization by dividing the distribution by the total mass can be required. Such a bootstrap procedure can be used to estimate the distribution of test statistics [GRI 13]. Bias corrections and weighted bootstrap approaches could also be adapted to our framework.

### 16.4.3. *Linear regression*

The linear regression model has been studied in the particular case where the covariate is observed and the response is censored [STU 93, DEL 08]. We consider the more general case where the response and the covariate are both randomly censored. Assume that

$$T^{(1)} = f(\beta, T^{(2)}) + \epsilon$$

where $\mathbb{E}[\epsilon|T^{(2)}] = 0$ and $f$ is a known family of functions depending on a finite dimensional parameter $\beta$. As the two variables are censored, the non-parametric estimator [16.8] can be used to obtain a consistent least squares estimator of $\beta$

$$\hat{\beta} = \arg\min_{\beta} \iint \phi(t^{(1)}, t^{(2)}) d\hat{F}(t^{(1)}, t^{(2)})$$

with $\phi(t^{(1)}, t^{(2)}) = (t^{(1)} - f(\beta, t^{(2)}))^2$. In the case of linear regression, the function $\phi$ satisfies the assumptions required to obtain the asymptotic representation of the integrals. We can then deduce an asymptotic representation and the asymptotic normality of $\hat{\beta}$ ([LOP 12]).

THEOREM 16.3.– Under assumption 16.5, under identifiability assumptions 16.1 and 16.2 and under regularity assumptions 16.3 and 16.4

$$n^{-1/2}(\hat{\beta} - \beta) \rightsquigarrow \mathcal{N}(0, R).$$

The complex expression of the asymptotic covariance matrix $R$ can be derived but can easily be estimated using the bootstrap procedure. This approach can be seen as a generalization of the results obtained by Stute [STU 93] when only one variable is censored. Note that prediction measures as the coefficient of determination $R^2$ can also be estimated using [16.8]. A generalized linear model with censored variables can be considered using the same approach. The extension to a regression model with two or more covariates is possible as the estimators can be adapted to multivariate censored data.

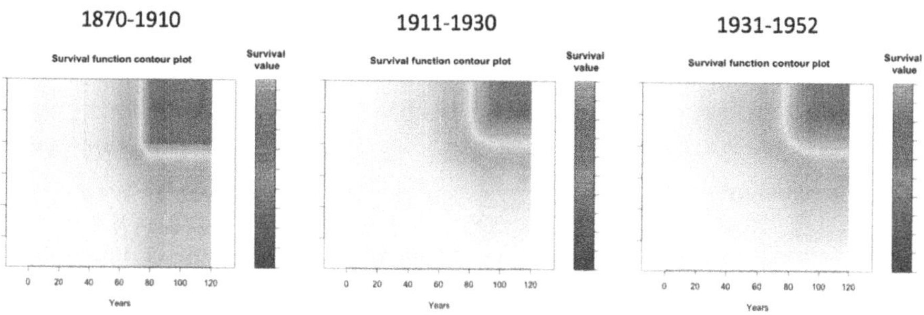


**Figure 16.1.** *Contour plot of the survival function for each birth cohort*

## 16.5. Illustration

The Danish Twin Registry is used to illustrate the analysis of bivariate survival data using integrals of the distribution. The registry includes more than 28,000 pairs born from 1870 to 1952 in Denmark. The ascertainment of twins took place using different methods leading to several birth cohorts (1870–1910; 1911–1930; 1931–1952). One accurate piece of information is the zygosity: monozygotic (MZ), dizygotic (DZ) and unknown zygosity (UZ). We refer to [SKY 02] for detailed descriptions of the registry and the way it was collected. The survival function can be estimated using estimator [16.7]. In this application, the censoring times are supposed to be independent. The estimation of survival function is done for each birth cohort and represented using contour plot (Figure 16.1).

| | Kendall tau | | |
|---|---|---|---|
| | 1870–1910 | 1911–1930 | 1931–1950 |
| MZ | 0.184 | 0.289 | 0.303 |
| DZ | 0.097 | 0.135 | 0.057 |
| UZ | 0.162 | 0.188 | 0.369 |

**Table 16.1.** *Estimation of Kendall tau for each birth cohort*

Kendall's tau coefficient between the twin lifetimes is estimated for each birth cohort as well as the regression of $T_1$ on $T_2$. The estimations of Kendall's tau and regression coefficient displayed in Tables 16.1 and 16.2 lead to the same conclusion. The dependence between lifetimes is stronger for monozygotic and seems to grow with time.

| Regression coefficient | | | |
|---|---|---|---|
| | 1870–1910 | 1911–1930 | 1931–1950 |
| MZ | 0.149 | 0.280 | 0. 354 |
| DZ | 0.088 | 0.115 | 0.037 |
| UZ | 0.161 | 0.212 | 0.279 |

**Table 16.2.** *Estimation of regression coefficients for each birth cohort*

## 16.6. Conclusion

We introduced a non-parametric estimator of the distribution of multivariate failure times. This estimator has several interesting properties, it is asymptotically normal, it generalizes the univariate Kaplan–Meier estimator and the empirical function, it defines a proper distribution and it is fast to compute. Using this estimator, integrals according to the distribution of failure times can be estimated consistently. We proposed several practical applications of the integrals of a distribution. Dependence measures such as the Kendall's coefficient can be estimated. This measure can be used for studying the dependence as well as estimating the association parameter of a copula. A bootstrap procedure according to Efron's methodology is provided to derive confidence intervals. The regression model with censored response and variables is also considered. The obtained results can be generalized to the multivariate case and can be adapted to left censoring.

## 16.7. Acknowledgment

This work is supported by French Agence Nationale de la Recherche (ANR) ANR Grant ANR-09-JCJC-0101-01. The authors wish to thank the Danish Twin Register, University of Southern Denmark for providing the data.

## 16.8. Bibliography

[AKR 86] AKRITAS M.G., "Bootstrapping the Kaplan-Meier estimator", *Journal of the American Statistical Association*, vol. 81, no. 396, pp. 1032–1038, 1986.

[AKR 03] AKRITAS M.G., VAN KEILEGOM I., "Estimation of bivariate and marginal distributions with censored data", *Journal of the Royal Statistical Society: Series B*, vol. 65, no. 2, pp. 457–471, 2003.

[DAB 88] DABROWSKA D.M., "Kaplan-Meier estimate on the plane", *The Annals of Statistics*, vol. 16, no. 4, pp. 1475–1489, 1988.

[DEL 08] DELECROIX M., LOPEZ O., PATILEA V., "Nonlinear censored regression using synthetic data", *Scandinavian Journal of Statistics*, vol. 35, no. 2, pp. 248–265, 2008.

[EFR 81] EFRON B., "Censored data and the bootstrap", *Journal of the American Statistical Association*, vol. 76, no. 374, pp. 312–319, 1981.

[FAN 00] FAN J., HSU L., PRENTICE R.L., "Dependence estimation over a finite bivariate failure time region", *Lifetime Data Analysis*, vol. 6, no. 4, pp. 343–355, 2000.

[GIL 83] GILL R., "Large sample behaviour of the product-limit estimator on the whole line", *The Annals of Statistics*, vol. 11, no. 1, pp. 49–58, 1983.

[GIL 92] GILL R.D., "Multivariate Survival Analysis", *Teoriya Veroyatnostei i ee Primeneniya*, vol. 37, pp. 19–35, 1992.

[GÓM 92] GÓMEZ G., JULIÀ O., UTZET F., "Survival analysis for left censored data", *Survival Analysis: State of the Art*, pp. 269–288, Kluwer Academic Publisher, Dordrecht, 1992,

[GRI 13] GRIBKOVA S., LOPEZ O., SAINT-PIERRE P., "A simplified model for studying bivariate mortality under right-censoring", *Journal of Multivariate Analysis*, vol. 115, pp. 181–192, 2013.

[HOU 00] HOUGAARD P., *Analysis of Multivariate Survival Data*, Statistics for Biology and Health, Springer-Verlag, New York, 2000.

[LIN 93] LIN D.Y., YING Z., "A simple nonparametric estimator of the bivariate survival function under univariate censoring", *Biometrika*, vol. 80, no. 3, pp. 573–581, 1993.

[LOP 12] LOPEZ O., SAINT PIERRE P., "Bivariate censored regression relying on a new estimator of the joint distribution function", *Journal of Statistical Planning and Inference*, vol. 142, no. 8, pp. 2440–2453, 2012.

[LU 08] LU X., BURKE M.D., "Nonparametric estimation of linear functionals of a bivariate distribution under univariate censoring", *Journal of Statistical Planning and Inference*, vol. 138, no. 10, pp. 3238–3256, 2008.

[PRE 92] PRENTICE R.L., CAI J., "Covariance and survivor function estimation using censored multivariate failure time data", *Biometrika*, vol. 79, no. 3, pp. 495–512, 1992.

[PRU 91] PRUITT R.C., "On negative mass assigned by the bivariate Kaplan-Meier estimator", *The Annals of Statistics*, vol. 19, no. 1, pp. 443–453, 1991.

[SAT 01] SATTEN G.A., DATTA S., "The Kaplan-Meier estimator as an inverse-probability-of-censoring weighted average", *The American Statistician*, vol. 55, no. 3, pp. 207–210, 2001.

[SHI 95] SHIH J.H., LOUIS T.A., "Inferences on the association parameter in copula models for bivariate survival data", *Biometrics*, vol. 51, no. 4, pp. 1384–1399, 1995.

[SKY 02] SKYTTHE A., KYVIK K., HOLM N.V., *et al.*, "The danish twin registry: 127 birth cohorts of twins", *Twin Research*, vol. 5, pp. 352–357, 2002.

[STU 93] STUTE W., "Consistent estimation under random censorship when covariables are present", *Journal of Multivariate Analysis*, vol. 45, no. 1, pp. 89–103, 1993.

[STU 95] STUTE W., "The central limit theorem under random censorship", *The Annals of Statistics*, vol. 23, no. 2, pp. 422–439, 1995.

[TSA 86] TSAI W.-Y., LEURGANS S., CROWLEY J., "Nonparametric estimation of a bivariate survival function in the presence of censoring", *The Annals of Statistics*, vol. 14, no. 4, pp. 1351–1365, 1986.

[VAN 96a] VAN DER LAAN M.J., "Efficient estimation in the bivariate censoring model and repairing NPMLE", *The Annals of Statistics*, vol. 24, no. 2, pp. 596–627, 1996.

[VAN 96b] VAN DER VAART A.W., WELLNER J.A., *Weak Convergence and Empirical Processes with Applications to Statistics*, Springer Series in Statistics, Springer-Verlag, New York, 1996.

[WAN 97] WANG W., WELLS M.T., "Nonparametric estimators of the bivariate survival function under simplified censoring conditions", *Biometrika*, vol. 84, no. 4, pp. 863–880, 1997.

# Chapter 17

# Kernel Estimation of Density from Indirect Observation

In this chapter, we consider an observational scheme, as data are interval censored. We suggest a simple non-parametric estimator $\widehat{f}_n$ for unknown density $f$ and under some appropriate conditions prove the consistency of this estimator.

## 17.1. Introduction

### 17.1.1. *Random partition*

Let $\tau$ be a random partition of a real line, that is a countable collection of intervals

$$\tau = \{[a_j, b_j),\ j \in \mathbb{Z}\} \qquad [17.1]$$

so that

$$\mathbb{R} = \bigcup_{j\in\mathbb{Z}} [a_j, b_j), \text{ and } [a_j, b_j) \cap [a_i, b_i) = \emptyset, \text{ as } i \neq j. \qquad [17.2]$$

For a point $x$, we denote by $[U(x), V(x))$ the interval of the partition $\tau$ such that

$$x \in [U(x), V(x)). \qquad [17.3]$$

Chapter written by Valentin SOLEV.

It is clear that for any $x, y \in \mathbb{R}$, there are only two possibilities

$$\begin{aligned} &1)\ [U(x), V(x)) = [U(y), V(y)) \\ &\text{or} \\ &2)\ [U(x), V(x)) \cap [U(x), V(x)) = [U(y), V(y)) = \emptyset. \end{aligned} \qquad [17.4]$$

Therefore, we will identify the random partition $\tau$ with the random process $\Lambda(x) = (U(x), V(x)), x \in \mathbb{R}$, which satisfies conditions [17.3] and [17.4] almost surely (a.s.).

Further, we will assume that the distribution of random vector $\Lambda(x)$ has density $p_x(u, v)$. Under some regularity conditions, there exists the function $p(u, v)$ such that

$$p_x(u, v) = p(u, v)\mathbf{1}_{[u,\, v)}(x),\ \ x \in \mathbb{R}. \qquad [17.5]$$

Here, we denote by $\mathbf{1}_A(x)$ the indicator function of the set $A$. The function $p(u, v)$ is called the basic density of the partition $\tau$. The function $p(u, v)$ is not the density of probability distribution, but

$$\iint\limits_{u \le x < v} p(u, v)\, du\, dv = 1, \ \text{ for all } x \in \mathbb{R}. \qquad [17.6]$$

As an example, we can take the function $p(u, v) = e^{u-v}\mathbf{1}_{u<v}$.

### 17.1.2. *Indirect observation*

Let $X$ be the random variable with density $f$. Denote by $\Delta(X) = [U(X), V(X))$ the interval of the partition $\tau$ that contains $X$. We assume that $X$ and $\tau$ are independent. It is easy to calculate the density $\kappa(u, v)$ of random vector $\Lambda(X) = (U(X), V(X))$

$$\kappa(u, v) = p(u, v)\left(F(v) - F(u)\right),\ u < v, \qquad [17.7]$$

where we denote by $F(x)$ the distribution function of random variable $X$. It is convenient to write

$$\kappa(u, v) = p_*(u, v)\,\frac{1}{v-u}\int\limits_u^v f(s)\, ds, \ \text{where } p_*(u, v) = p(u, v)(v - u). \qquad [17.8]$$

Suppose that instead of $X$ we observe the interval $\Delta(X)$. The main goal in this chapter is to construct an estimator $\widehat{f}_n$ for unknown density $f$ of the distribution of $X$ from indirect observations.

Namely, let $\Delta_1, \ldots, \Delta_n$ be an independent copy of random interval $\Delta(X)$. We need to construct an estimator $\widehat{f}_n$ on observations $W_1 = (U_1, V_1), \ldots, W_n = (U_n, V_n)$. Here, $\Delta_j = [U_j, V_j)$. This problem was investigated by Turnbull [TUR 76] and many other authors. In this chapter we use approaches that were proposed in [HUB 04], [HUB 09], [HUB 10a] and [HUB 10b].

### 17.1.3. *Kernel density estimator*

Let $W_1 = (U_1, V_1), \ldots, W_n = (U_n, V_n)$ be independent identically distributed random vectors with joint density

$$\kappa(u,v) = p(u,v)\{F(v) - F(u)\} = p(u,v)\int_u^v f(s)\,ds, \; u < v,$$

the function $p(u,v)$ is known, and we need to estimate unknown density $f$. At the beginning, we take a kernel $K(w), \; w = (u,v)$, with

$$\iint |K(u,v)|\,du\,dv < \infty \text{ and } \iint K(u,v)\,du\,dv = 1,$$

and construct the kernel density estimator $\widehat{k}_n(u,v)$ for function $k(u,v)$, generated by kernel $K$

$$\widehat{k}_n(u,v) = \frac{1}{nh^2}\sum_{j=1}^{n} K\left(\frac{w - W_j}{h}\right), \; u < v. \qquad [17.9]$$

We further assume that the kernel $K$ is non-negative, symmetric, $K(w) = K(-w)$, and we will use the notation

$$\mathscr{K}_h(w) = K\left(\frac{w}{h}\right), K_h = \frac{1}{h^2}\mathscr{K}_h(w) = \frac{1}{h^2}K\left(\frac{w}{h}\right).$$

Denote by $\mathbf{P}$ the measure with density $k(u,v)$, and by $\mathbf{P}_n$ the empirical version of $\mathbf{P}$

$$\mathbf{P}_n\{A\} = \frac{1}{n}\sum_{j=1}^{n} \mathbf{1}_A(W_j). \qquad [17.10]$$

It is clear that for $w = (u,v)$ and $t = (t_1, t_2)$

$$\iint K_h(w-t)\,\mathbf{P}(dt_1\,dt_2) = \iint K_h(w-t)\,g_*(t_1,t_2) f_{t_1,t_2}\,dt_1\,dt_2,$$

where $f_{t_1,t_2}$ is the average value of function $f$ on the interval $[t_1, t_2]$

$$f_{t_1,t_2} = \frac{1}{t_2 - t_1} \int_{t_1}^{t_2} f(s)\, ds, \; t_1 < t_2.$$

For continuous function $f$, if the length of the interval $[t_1, t_2]$ is small, then

$$f_{t_1,t_2} \approx f\left(\frac{t_2 + t_1}{2}\right).$$

Thus, if the kernel $K(u, v)$ is supported on the finite square $[-a, a] \times [-a, a]$ and the function $f(x)$ is continuous, then, for small bandwidth $h > 0$,

$$\iint K_h\,(w - t)\; \mathbf{P}(dt_1\, dt_2) \approx$$

$$\approx f\left(\tfrac{v+u}{2}\right) \iint K_h\,(w - t)\, g_*(t_1, t_2)\, dt_1\, dt_2.$$

This suggests using as the estimator $\widehat{f}_n(x)$ for value $f(x)$ the ratio

$$\widehat{f}_n(x) = \frac{\iint \mathscr{K}_h\,(w - t)\; \mathbf{P}_n(dt_1\, dt_2)}{\mu(w; h)}, \qquad [17.11]$$

with $w = (x - h, x + h)\; t = (t_1, t_2)$,

where

$$\mu(w; h) = \iint \mathscr{K}_h\,(w - t)\; p_*(t_1, t_2)\, dt_1\, dt_2 = \\ = \iint \mathscr{K}_h\,(w - t)\; p(t_1, t_2)\, (t_2 - t_1)\; dt_1\, dt_2. \qquad [17.12]$$

It should be noted that we have to assume that the function $p(u, v)$ is uniformly bounded away from zero in some neighborhood of the diagonal $u = v$

$$p(u, v) \geq c > 0, \text{ for all } u < v \leq u + \delta, \qquad [17.13]$$

and uniformly continuous on this set.

## 17.2. Density of random vector $\Lambda(X)$

For the partition $\tau$

$$\tau = \{[a_j, b_j),\ j \in \mathbb{Z}\} \tag{17.14}$$

with

$$\mathbb{R} = \bigcup_{j\in\mathbb{Z}} [a_j, b_j),\ \text{and } [a_j, b_j) \cap [a_i, b_i) = \emptyset,\ \text{as } i \neq j, \tag{17.15}$$

and for point $x$ we use notation $[U(x), V(x))$ for the interval of partition that contains $x$, and $\Lambda(x)$ for vector $\Lambda(x) = (U(x), V(x))$. Suppose that random variable $X$ and partition $\tau$ are independent. Denote by $\Lambda(X)$ the random vector $\Lambda(X) = (U(X), V(X))$. Let $F(x)$ be the distribution function of $X$ and $f(x)$ be the density function. First, we plan to prove that the distribution of random vector $\Lambda(x)$ has density under some simple condition on the distribution of $\tau$.

We assume that for each interval $[a_j, b_j)$ of the partition $\tau$, the distribution of random vector $(a_j, b_j)$ has density $p_j(u, v)$. Then, for a non-negative function $\psi(u, v)$ and point $x$, we can calculate the expectation $\mathbf{E}\psi\,(U(x), V(x))$. Since

$$\psi\,(U(x), V(x)) = \sum_{j\in\mathbb{Z}} \psi\,(a_j, b_j)\,\mathbf{1}_{[a_j,b_j)}(x)$$

and

$$\mathbf{E}\psi\,(a_j, b_j)\,\mathbf{1}_{[a_j,b_j)}(x) = \iint\limits_{u\le x<v} \psi(u,v) p_j(u,v)\,du\,dv,$$

then for non-negative function $\psi$

$$\mathbf{E}\psi\,(U(x), V(x)) = \iint\limits_{u\le x<v} \psi(u,v) \sum_{j\in\mathbb{Z}} p_j(u,v)\,du\,dv.$$

Therefore, for fixed $x$, the functional series $\sum_{j\in\mathbb{Z}} p_j(u, v)$ converges a.s. on the set $A_x = \{(u, v):\ u \le x < v\}$ to the non-negative function $p_x(u, v)$. We set $p_x(u, v) = 0$, if $u > v$. It is clear that

$$\iint p_x(u,v)\,du\,dv = 1,\ \text{for all } x \in \mathbb{R} \text{ and}$$

$$p_x(u,v) = p_y(u,v) \text{ a.s. on the set } A_x \cap A_y.$$

So, we can define the function $p(u,v)$ a.s. on the set $A_x \cup A_y$ such that $p(u,v)\mathbf{1}_{[u,\,v)}(z) = p_z(u,v)$, as $z \in \{x,y\}$. In the same way, we can take a countable set $Q \subset \mathbb{R}$ and define the function $p(u,v)$ a.s. on the set $\bigcup\limits_{x\in S} A_x$ such that $p(u,v)\mathbf{1}_{[u,\,v)}(z) = p_z(u,v)$, as $z \in S$.

Thus, there exists such a measurable, non-negative function $p(u,v)$ defined on the set $\{(u,v) : \ u < v\}$ such that

$$p_x(u,v) = p(u,v)\mathbf{1}_{[u,v)}(x) \text{ almost sure on } \{(u,v) : \ u < v\}. \qquad [17.16]$$

If we set $p(u,v) = 0$, if $u > v$, then function $p(u,v)$ will be defined almost surely. Obviously,

$$\iint\limits_{u\leq x<v} p(u,v)\,du\,dv = 1, \text{ for all } x.$$

So, we have proved that there exists non-negative function $p(u,v)$ that is supported on $\{(u,v) : \ u \leq v\}$, such that for any non-negative function $\psi$ and fixed $x$

$$\begin{aligned}\mathbf{E}\psi\left(U(x),V(x)\right) &= \iint\limits_{u\leq x<v} \psi(u,v)\,p(u,v)\,du\,dv = \\ &= \iint \psi(u,v)\,p(u,v)\mathbf{1}_{[u,v)}(x)\,du\,dv.\end{aligned} \qquad [17.17]$$

Suppose that random variable $X$ and partition $\tau$ are independent. If the distribution of $X$ has density $f$, then from [17.17] we deduce that

$$\mathbf{E}\psi\left(U(X),V(X)\right) = \int \left\{\mathbf{E}\psi\left(U(x),V(x)\right)\right\} f(x)\,dx.$$

Since

$$\int \mathbf{1}_{[u,v)}(x)\,f(x)\,dx = F(v) - F(u),$$

where $F(x)$ is the distribution function of $X$, then

$$\mathbf{E}\psi\left(U(X),V(X)\right) = \iint \psi(u,v)\,p(u,v)\left(F(v) - F(u)\right)\,du\,dv.$$

So, we have proved that the distribution of random vector $\Lambda = \{U(X),V(X)\}$ has density $k(u,v)$

$$k(u,v) = p(u,v)\left(F(v) - F(u)\right),\ u < v.$$

## 17.3. Pseudo-kernel density estimator

### 17.3.1. *Pointwise density estimation based on indirect data*

Let $\tau$ be the random partition with the basic density function $p(u,v)$, and $X$ be the random variable with the distribution function $F(x)$ and density $f$. We assume that $X$ is independent on $\tau$, and denote by $\Delta(X) = [U,V)$ the interval of partition $\tau$ that contains $X$ and set $W = (U,V)$. We have proved that random vector $W$ has the distribution with density $k(u,v)$

$$k(u,v) = p(u,v) \int_u^v f(s)\, ds, \ u < v.$$

Let $W_1 = (U_1,V_1),\ldots,W_1 = (U_1,V_1)$ be independent copies of $W = (U,V)$. We need to construct an estimator $\widehat{f}_n(x)$ of the value of the unknown function $f$ in the point $x$ on observations $W_1 = (U_1,V_1),\ldots,W_n = (U_n,V_n)$.

Denote by $\mathbf{P}$ the measure with density $k(u,v)$, and by $\mathbf{P}_n$ the empirical version of $\mathbf{P}$

$$\mathbf{P}_n\{A\} = \frac{1}{n}\sum_{j=1}^n \mathbf{1}_A(W_j). \qquad [17.18]$$

The kernel estimator $\widehat{k}_n(u,v)$ of $k(u,v)$ (generated by kernel $K(w), \ w = (u,v)$) is defined by

$$\widehat{k}_n(w) = \iint K_h(w-t)\,\mathbf{P}_n(dt_1\,dt_2) = \frac{1}{n}\sum_{j=1}^n K_h(w-W_j). \qquad [17.19]$$

Here, the bandwidth $h > 0$, and $K_h(w) = \frac{1}{h^2}K\left(\frac{w}{h}\right), t = (t_1,t_2)$. We assume that the kernel $K$ is non-negative and symmetric: $K(w) = K(-w)$, and we denote

$$\mu(w;h) = \iint \mathscr{K}_h(w-t)\,g(t_1,t_2)(t_2-t_1)\,dt_1\,dt_2, \ \mathscr{K}_h(w) = K\left(\frac{w}{h}\right),$$

and set

$$\widetilde{k}_n(w) = \iint \mathscr{K}_h(w-t)\,\mathbf{P}_n(dt_1\,dt_2) = \frac{1}{n}\sum_{j=1}^n \mathscr{K}_h(w-W_j). \qquad [17.20]$$

Clearly

$$\widetilde{k}_n(w) = h^2 \widehat{k}_n(w).$$

As an estimate of the unknown density $f(x)$, we will use the pseudo-kernel density estimator $\widehat{f}_n(x)$

$$\widehat{f}_n(x) = \frac{\widetilde{k}_n(u,v)}{\mu(u,v;h)} \quad \text{with } u = x - \tfrac{h}{2},\ v = x + \tfrac{h}{2}. \qquad [17.21]$$

Regarding the kernel, we will assume that $K(w)$ is a non-negative, symmetric function, supported on the square $[-1/4; 1/4] \times [-1/4; 1/4]$.

### 17.3.2. *Bias of the kernel estimator*

Denote

$$f_h(x) = \frac{1}{\mu(w;h)} \iint \mathscr{K}_h(w-t)\, g(t)(t_2 - t_1) f_{t_1,t_2}\, dt_1\, dt_2. \qquad [17.22]$$

Here and further

$$u = x - \frac{h}{2},\ v = x + \frac{h}{2},\ \mu(w;h) = \iint \mathscr{K}_h(w-t)\, g(t_1,t_2)(t_2 - t_1)\, dt_1\, dt_2,$$

$$f_{t_1,t_2} = \frac{1}{t_2 - t_1} \int_{t_1}^{t_2} f(s)\, ds.$$

Clearly

$$f_h(x) = \mathbf{E}\widehat{f}_n(x).$$

We assume that the function $f(x)$ satisfies the condition

$$\sup_{x \neq y} \frac{|f(x) - f(y)|}{|x - y|} \leq C, \qquad [17.23]$$

and

$$\mu(w;h) > 0. \qquad [17.24]$$

LEMMA 17.1.– Under conditions [17.23] and [17.24]

$$|f_h(x) - f(x)| \leq \frac{3Ch}{2}. \qquad [17.25]$$

PROOF.– At the beginning, we will estimate the value

$$\left| \frac{F(t_2) - F(t_1)}{t_2 - t_1} - f(x) \right|,$$

as $t_1 \leq x < t_2$. It is clear that

$$\left| \frac{F(t_2) - F(t_1)}{t_2 - t_1} - f(x) \right| \leq \frac{1}{t_2 - t_1} \int_{t_1}^{t_2} |f(y) - f(x)| \, dy.$$

Hence, by [17.23] we obtain

$$\left| \frac{F(t_2) - F(t_1)}{t_2 - t_1} - f(x) \right| \leq \frac{C}{t_2 - t_1} \int_{t_1}^{t_2} |y - x| \, dy \leq C(t_2 - t_1).$$

Again

$$|f_h(x) - f(x)| \leq \frac{1}{\mu(w;h)} \iint \mathscr{K}_h(w - t) p_*(t_1, t_2) \left| \frac{F(t_2) - F(t_1)}{t_2 - t_1} - f(x) \right| dt_1 dt_2,$$

where $w = (u, v), u = x - h/2, \ v = x + h/2$ and $t = (t_1, t_2)$. Since the kernel $\mathscr{K}_h(w)$ is supported on the square $[-h/4; h/4] \times [-h/4; h/4]$, then

$$\iint \mathscr{K}_h(w - t) \, p_*(t_1, t_2) \left| \frac{F(t_2) - F(t_1)}{t_2 - t_1} - f(x) \right| dt_1 \, dt_2 =$$

$$= \iint_G \mathscr{K}_h(w - t) \, p_*(t_1, t_2) \left| \frac{F(t_2) - F(t_1)}{t_2 - t_1} - f(x) \right| dt_1 \, dt_2,$$

where

$$G = \{(t_1, t_2) : \ |x - h/2 - t_1| \leq h/4, \ |x + h/2 - t_2| \leq h/4\}.$$

It is evident that for $(t_1, t_2) \in G$

$$t_1 < x < t_2 \ \text{ and } \ t_2 - t_1 \leq 3h/2.$$

Therefore, for $(t_1, t_2) \in G$

$$\left| \frac{F(t_2) - F(t_1)}{t_2 - t_1} - f(x) \right| \leq \frac{3Ch}{2},$$

and

$$|f_h(x) - f(x)| \leq \frac{Ch}{4} \frac{1}{\mu(w; h)} \iint \mathscr{K}_h(w - t)\, p_*(t_1, t_2)\, dt_1\, dt_2 = \frac{3Ch}{2}.$$

So, we obtain [17.25].

### 17.3.3. *Estimate of variance*

Now, we assume that the function $p(u, v)$ is uniformly bounded away from zero in some neighborhood of the diagonal $u = v$:

$$p(u, v) \geq c > 0, \text{ for all } u < v \leq u + \delta, \qquad [17.26]$$

and uniformly continuous on this set. Also assume

$$\iint K(u, v)\, du\, dv = L > 0. \qquad [17.27]$$

We denote

$$\mu_*(x; h) = \frac{1}{h^2} \mu(w; h), \text{ with } w = (u, v), u = x - h/2,\ v = x + h/2.$$

In this case

$$\mu_*(x; h) = \frac{1}{h^2} \iint K\left(\frac{w - t}{h}\right) g(t_1, t_2)(t_2 - t_1)\, dt_1\, dt_2,$$

$$\text{with } w = (u, v), u = x - h/2,\ v = x + h/2,\ t = (t_1, t_2).$$

LEMMA 17.2.– Under conditions [17.26] and [17.27], there exists the constant $c_* > 0$ such that

$$\mu_*(x; h) \geq c_* h, \text{ if } h \text{ is small enough.} \qquad [17.28]$$

PROOF.– It is clear that

$$\frac{1}{h^2}\iint K\left(\frac{w-t}{h}\right) g(t_1,t_2)(t_2-t_1)\,dt_1\,dt_2 =$$

$$= \iint K(s)\; g(w+hs)((v+hs_2)-(u+hs_1))\,ds_1\,ds_2,$$

where $s = (s_1,s_2), w = (u,v)$. Since the kernel $K(s)$ is supported on the square $\mathscr{D} = [-1/4;1/4]\times[-1/4;1/4]$, then

$$\iint K(s)\; g(w+hs)((v+hs_2)-(u+hs_1))\,ds_1\,ds_2 =$$

$$= \iint\limits_{\mathscr{D}} K(s)\; g(w+hs)((v+hs_2)-(u+hs_1))\,ds_1\,ds_2.$$

For $u = x - h/2,\; v = x + h/2,\; s \in \mathscr{D}$, if $h$ is small enough,

$((v+hs_2)-(u+hs_1)) \geq h/2$ and according to condition [17.26], $g(w+hs) \geq c$.

Therefore

$$\iint K(s)\; g(w+hs)((v+hs_2)-(u+hs_1))\,ds_1\,ds_2$$
$$\geq \frac{ch}{2}\iint K(s)\,ds_1\,ds_2 \geq \frac{chL}{2}.$$

So, we obtain [17.28] with $c_* = cL/2$.

Now, we can estimate the variance $\mathbf{D}\widehat{f}_n(x)$. We will write estimate $\widehat{f}_n$ in the form

$$\widehat{f}_n(x) = \frac{1}{\mu_*(x;h)}\frac{1}{nh^2}\sum_{j=1}^{n} K\left(\frac{w-W_j}{h}\right),$$

where $w = (u,v),\; u = x - h/2$ and $v = x + h/2$.

Denote

$$\Lambda = \frac{1}{nh^2}\sum_{j=1}^{n} K\left(\frac{w-W_j}{h}\right).$$

It is well known from the theory of kernel estimators (see [DEV 01]) that

$$\mathbf{D}\Lambda \leq \frac{1}{nh^2}\left\{\frac{1}{h^2}\iint K^2\left(\frac{w-t}{h}\right) g(t)\,(F(t_2)-F(t_1))\,dt_1\,dt_2\right\}. \quad [17.29]$$

Since the kernel $K(s)$ is supported on the square $\mathscr{D} = [-1/4; 1/4] \times [-1/4; 1/4]$, then

$$\begin{aligned} &\tfrac{1}{h^2}\textstyle\iint K^2\left(\frac{w-t}{h}\right) g(t)\,(F(t_2)-F(t_1))\,dt_1\,dt_2 = \\ &= \tfrac{1}{h^2}\textstyle\iint\limits_{\mathscr{D}} K^2\left(\frac{w-t}{h}\right) g(t)\,(F(t_2)-F(t_1))\,dt_1\,dt_2. \end{aligned} \quad [17.30]$$

Now, we assume that, if $\delta$ is small enough,

$$p(u,v) \leq C_1, \text{ for all } u < v \leq u+\delta, \quad [17.31]$$

and

$$\iint K^2(u,v)\,du\,dv = L_2 < \infty. \quad [17.32]$$

In the same way as we used in lemma 17.2, it is easy to prove that under conditions [17.31] and [17.32]

$$\frac{1}{h^2}\iint\limits_{\mathscr{D}} K^2\left(\frac{w-t}{h}\right) g(t)\,(F(t_2)-F(t_1))\,dt_1\,dt_2 \leq \frac{3C_1Lh}{2}.$$

Therefore, from [17.29] and [17.30], we obtain that

$$\mathbf{D}\Lambda \leq \frac{3C_1L}{2nh}.$$

Thus, for some constant $C_2$

$$\mathbf{D}\widehat{f}_n(x) \leq \frac{C_2}{nh\mu_*^2(x;h)},$$

and we can deduce from lemma 17.2 the following lemma.

LEMMA 17.3.– Under conditions [17.26], [17.27], [17.31] and [17.32], there exists the constant $C_*$ such that

$$\mathbf{D}\widehat{f}_n(x) \leq \frac{C_*}{nh^3}, \text{ if } h \text{ is small enough.} \quad [17.33]$$

If we chose $h = n^{-1/5}$, then for some constant $C$

$$\mathbf{E}\left(\widehat{f}_n(x)\right)^2 \leq Cn^{-2/5} \text{ if } h \text{ is small enough.}$$

## 17.4. Bibliography

[DEV 01] DEVROYE L., GABOR L., *Combinatorial Methods in Density Estimation*, Springer, New York, 2001.

[HUB 04] HUBER C., VONTA F., "Frailty models for arbitrarily censored and truncated data", *Lifetime Data Analysis*, vol. 10, pp. 369–388, 2004.

[HUB 09] HUBER C., VALENTIN S., VONTA F., "Interval censored and truncated data: rate of convergence of NPMLE of the density", *Journal of Statistical Planning and Inference*, vol. 139, pp. 1734–1749, 2009.

[HUB 10a] HUBER C., VONTA F., "On the estimation of structural parameters in frailty models for interval censored and truncated data", *Computer Modelling and New Technologies*, vol. 14, no. 4, pp. 40–49, 2010.

[HUB 10b] HUBER C., VONTA F., "A semiparametric model for interval censored and truncated data", *Journal of Mathematical Sciences*, vol. 163, no. 3, pp. 283–289, 2010.

[TUR 76] TURNBULL B.W., "The empirical distribution function with arbitrary grouped, censored and truncated data", *Journal of the Royal Statistical Society*, vol. 38, pp. 290–295, 1976.

# Chapter 18

# A Comparative Analysis of Some Chi-Square Goodness-of-Fit Tests for Censored Data

In this chapter, we consider some chi-square-type goodness-of-fit tests for right censored samples, viz. the Nikulin–Rao–Robson (NRR) test and generalized Pearson–Fisher (GPF) test. We compare these tests in terms of power in the case of testing composite hypotheses by randomly censored and Type II censored samples. Various grouping methods are considered, including equifrequent grouping, grouping into intervals with equal expected numbers of failures as well as optimal grouping for some given pairs of competing hypotheses. In the case of Type II censored data, we compare by power chi-square goodness-of-fit tests with the Kolmogorov, Cramer–von Mises–Smirnov and Anderson–Darling tests.

## 18.1. Introduction

In this chapter, we consider the problem of testing the composite goodness-of-fit hypotheses $H_0 : F(t) \in \{F_0(t;\theta), \theta \in \Theta\}$, meaning that the distribution $F$ of failure time $T$ belongs to a certain parametric family $F_0(\cdot;\theta)$, where $\theta = (\theta_1, ..., \theta_q)^T \in \Theta \subset \mathbf{R}^q$ is unknown $q$-dimensional parameter. Let us denote by $\lambda_0(t;\theta)$ and $\Lambda_0(t;\theta)$ the hazard rate and cumulative hazard rate functions corresponding to the tested model, respectively. In reliability or survival studies, the observed data are usually of the form

$$\mathrm{X} = (X_1, \delta_1), (X_2, \delta_2), ..., (X_n, \delta_n),$$

Chapter written by Ekaterina CHIMITOVA and Boris LEMESHKO.

where $X_i = \min(T_i, C_i)$ is the observation value, $T_i$ is the lifetime and $C_i$ is a censoring time. The censoring indicator $\delta_i = I\{T_i \leq C_i\}$.

There are various forms of right censoring (for details, see [COH 91]):

– If individuals are observed up to some predetermined time $c$, then censoring is referred to as Type I censoring and in the case when $\delta_i = 0 : X_i = c$.

– If a life test is terminated whenever a specified number of failures (say, $m$) have occurred, it is referred to as Type II censoring and in this case when $\delta_i = 0$: $X_i = X_{(m)}$, where $X_{(m)}$ is the last observed failure time.

– Let the lifetime $T$ and the censoring time $C$ be independent random variables from the distribution functions $F(t)$ and $F^C(t)$, respectively. All lifetimes and censoring times are assumed mutually independent. Then, this form of censoring is referred to as independent random censoring.

In the case of complete data (without censored observations), a well-known modification of the classical chi-square tests is the NRR statistic, which was proposed for the first time in [NIK 73c] and further developed in [NIK 73b], [NIK 73a] and [RAO 74]. Later, it was studied in [LEM 01], [VOI 06] and [VOI 13]. The NRR statistic is based on the differences between two estimators of the probabilities to fall into grouping intervals: one estimator is based on the empirical distribution function and the other on the maximum likelihood estimators of unknown parameters of the tested model using initial non-grouped data. Habib and Thomas [HAB 86] and Hollander and Peña [HOL 92] considered natural modifications of the NRR statistic for the case of censored data. The idea of comparing observed and expected numbers of failures into grouping intervals in the case of randomly censored data was proposed by Akritas [AKR 88] and was developed independently by Hjort [HJO 90]. Bagdonavičius *et al.* [BAG 10] have developed this direction, considering the choice of random grouping intervals as data functions and writing simple formulas of test statistics for most applied families of lifetime distributions. The NRR chi-square goodness-of-fit test has been also developed for parametric accelerated failure time models [BAG 10, BAG 13].

Li and Doss [LI 93] have developed the generalization of the classical Pearson–Fisher chi-square goodness-of-fit test for any situation, for which there is available a non-parametric estimator $\hat{F}$ of $F$, such that $\sqrt{n}\left(\hat{F} - F\right) \xrightarrow{d} W$, where $W$ is a continuous zero mean Gaussian process satisfying a mild regularity condition. The GPF statistic for censored data is based on the differences between two estimators of the probabilities to fall into grouping intervals: one estimator is based on the Kaplan–Meier estimator of lifetime distribution and the other on the minimum chi-square estimator of unknown parameters. This test was also developed for randomly censored data by Kim [KIM 93].

In this chapter, we present an empirical analysis of the NRR and GPF chi-square tests for right censored data. We compare these tests by means of power in the case of testing composite goodness-of-fit hypotheses by randomly censored and Type II censored samples. Various grouping methods are considered. The investigation is carried out through Monte Carlo simulations. From the obtained results, we draw some conclusions.

## 18.2. Chi-square goodness-of-fit tests for censored data

Chi-square-type tests require dividing an observed interval $[0, \tau]$ into $k$ smaller intervals $I_j = (a_{j-1}, a_j]$, $0 = a_0 < a_1 < \ldots < a_{k-1} < a_k = \tau$.

### 18.2.1. *NRR* $\chi^2$ *test*

For any $t > 0$, set

$$N(t) = \sum_{i=1}^{n} I\{X_i \leq t, \, \delta_i = 1\},$$

which counts the number of observed failures in $[0, \, t]$, and

$$Y(t) = \sum_{i=1}^{n} I\{X_i \geq t\},$$

which is the number of objects at risk just before time $t$.

Denote by $U_j = N(a_j) - N(a_{j-1})$ the number of observed failures and by

$$e_j = \int_{a_{j-1}}^{a_j} \lambda_0 \left(u; \hat{\theta}_n\right) Y(u) du$$

the "expected" number of failures in the interval $I_j$, $j = 1, ..., k$.

The NRR $\chi^2$ test statistic can be written in the form [BAG 10]

$${Y_n}^2 = Z^T \hat{V}^- Z,$$

where $Z = (Z_1, ..., Z_k)^T$, $Z_j = \frac{1}{\sqrt{n}} (U_j - e_j)$, $\hat{V}^-$ is the general inverse of the matrix

$$\hat{V} = \hat{A} - \hat{C}^T \hat{\imath}^{-1} \hat{C},$$

$\hat{A}$ is the diagonal $k \times k$ matrix with diagonal elements $\hat{A}_j = U_j/n$

$$\hat{i} = \frac{1}{n}\int_0^T \frac{\partial}{\partial\theta}\ln\lambda_0\left(u;\hat{\theta}_n\right)\left(\frac{\partial}{\partial\theta}\ln\lambda_0\left(u;\hat{\theta}_n\right)\right)^T dN(u),$$

$$\hat{C}_j = \frac{1}{n}\int_{I_j} \frac{\partial}{\partial\theta}\ln\lambda_0\left(u;\hat{\theta}_n\right) dN(u),$$

where $\hat{\theta}_n$ is the maximum likelihood estimate of unknown parameters of the tested model using initial non-grouped censored data. The limiting distribution of the test statistic is $\chi^2_r$, $r = rank(V^-)$. So, the hypothesis is rejected with approximate significance level $\alpha$ if $Y_n^2 > \chi^2_\alpha(r)$.

In [CHI 11a] and [CHI 11b], we have studied the distributions of statistic $Y_n^2$ and the power of NRR test, depending on the form and degree of censoring by means of Monte Carlo simulations.

### 18.2.2. *GPF $\chi^2$ test*

Let $\hat{F}_n(t)$ be the Kaplan–Meier estimator of the lifetime distribution $F$. Define the empirical probabilities to fall into grouping intervals as

$$\hat{p}_j = \hat{F}_n(a_j) - \hat{F}_n(a_{j-1}), j = 1, ..., k,$$

where $\hat{p}_1$ and $\hat{p}_k$ must be defined as $\hat{p}_1 = \hat{F}_n(a_1)$ and $\hat{p}_k = 1 - \hat{F}_n(a_{k-1})$. Denote by

$$p_j(\theta) = F_0(a_j, \theta) - F_0(a_{j-1}, \theta)$$

the theoretical probabilities to fall into grouping intervals. Unknown parameters of the tested distribution are estimated as follows

$$\tilde{\theta}_n = Arg\min_\theta \xi_n^T(\theta)\xi_n(\theta),$$

where $\xi_n(\theta) = \left(\frac{n\hat{p}_1 - np_1(\theta)}{\sqrt{np_1(\theta)}}, \ldots, \frac{n\hat{p}_k - np_k(\theta)}{\sqrt{np_k(\theta)}}\right)^T.$

It is well known [LI 93] that $\xi_n(\tilde{\theta}_n) \xrightarrow{d} N(0, \Sigma)$ , where

$$\Sigma = PDJ\Sigma^{(1)}J^T DP,$$

$$P = I - C\left(C^T C\right)^{-1} C^T, C = D\left(\tfrac{\partial}{\partial\theta}p(\theta)\right)^T, p(\theta) = \left(p_1(\theta), ..., p_k(\theta)\right)^T,$$

$$D = \operatorname{diag}\left(\left(p_1(\theta)\right)^{-1/2}, ..., \left(p_k(\theta)\right)^{-1/2}\right),$$

$$J = \begin{pmatrix} 1 & 0 & 0 & \dots & 0 \\ -1 & 1 & 0 & \dots & 0 \\ 0 & -1 & 1 & \dots & 0 \\ \vdots & \vdots & \vdots & \ddots & \vdots \\ 0 & 0 & 0 & \dots & 1 \\ 0 & 0 & 0 & \dots & -1 \end{pmatrix}_{k\times(k-1)}.$$

The elements of matrix $\Sigma^{(1)}$ are calculated by the formula

$$\sigma_{ij}^{(1)} = (1 - F_0(a_i, \theta)) \cdot (1 - F_0(a_j, \theta)) \cdot v\left(\min(a_i, a_j)\right),$$

where the variance function $v(t) = \int\limits_0^t \frac{dF_0(u;\theta)}{(1-F(u;\theta))^2(1-F^C(u))}$. Here, the Kaplan–Meier estimator, obtained by the inverted sample, in which $\delta_i$ is replaced by $1 - \delta_i$, can be used as the distribution of censoring times $F^C(t)$. Let $\hat{\Sigma}$ be obtained by using the minimum chi-square estimator $\tilde{\theta}_n$ as $\theta$ in formulas for the matrix $\Sigma$, and $\hat{\Sigma}^+$ denotes the Moore–Penrose inverse of $\hat{\Sigma}$. Then, under $H_0$, the statistic of GPF test

$$X_n^2 = \xi_n^T\left(\tilde{\theta}_n\right)\hat{\Sigma}^+\xi_n\left(\tilde{\theta}_n\right)$$

has the $\chi^2$-distribution with $k - q - 1$ degrees of freedom as $n \to \infty$.

## 18.3. The choice of grouping intervals

When using chi-square goodness-of-fit tests, the problem of choosing boundary points and the number of grouping intervals is always important, as the power of these tests considerably depends on the grouping method used. In the case of complete samples (without censored observations), this problem was investigated in [VOI 09], [DEN 89], [LEM 98], [LEM 00], [LEM 01] and [DEN 04]. In particular, in [LEM 00], the investigation of the power of the Pearson and NRR tests for complete samples has been carried out for various numbers of intervals and grouping methods. The partition of the real line into equiprobable intervals (EPG) is not an optimal grouping method, as a rule. In [DEN 79], it was shown for the first time that asymptotically optimal grouping, for which the loss of the Fisher information from grouping is minimized, enables us to maximize the power of the Pearson test against close competing hypotheses. For example, it is possible to maximize the determinant of the Fisher information matrix for grouped data

$$\mathbf{J}_G(\theta) = \sum_{j=1}^{k} \frac{\frac{\partial}{\partial\theta} p_j(\theta) \left(\frac{\partial}{\partial\theta} p_j(\theta)\right)^T}{p_j(\theta)},$$

i.e. to solve the problem of D-optimal grouping

$$\max_{a_1<\ldots<a_{k-1}} \det\left(\mathbf{J}_G(\theta)\right). \qquad [18.1]$$

In the case of an A-optimality criterion, the trace of the information matrix $\mathbf{J}_G(\theta)$ is maximized by the boundary points

$$\max_{a_1<\ldots<a_{k-1}} \mathrm{Tr}\left(\mathbf{J}_G(\theta)\right), \qquad [18.2]$$

and an E-optimality criterion maximizes the minimum eigenvalue of the information matrix

$$\max_{a_1<\ldots<a_{k-1}} \min_{l=\overline{1,q}} \lambda_l\left(\mathbf{J}_G(\theta)\right). \qquad [18.3]$$

The problem of asymptotically optimal grouping by the A- and E-optimality criteria has been solved for certain distribution families, and the tables of A-optimal grouping are given in [LEM 11]. The versions of asymptotically optimal grouping maximize the test power relative to a set of close competing hypotheses, but they do not ensure the highest power against some given competing hypothesis. For the given competing hypothesis $H_1$, it is possible to construct the $\chi^2$ test, which has the highest power for testing hypothesis $H_0$ against $H_1$. For example, in the case of $\chi^2$ Pearson test, it is possible to maximize the non-centrality parameter for the given number of intervals $k$:

$$\max_{a_1<a_2<\ldots.<a_{k-1}} \left( n \sum_{j=1}^{k} \frac{\left(p_j^1\left(\theta^1\right) - p_j^0\left(\theta^0\right)\right)^2}{p_j^0\left(\theta^0\right)} \right), \qquad [18.4]$$

where

$$p_j^0\left(\theta^0\right) = \int_{a_{j-1}}^{a_j} f_0\left(u;\theta^0\right) du \quad \text{and} \quad p_j^1\left(\theta^1\right) = \int_{a_{j-1}}^{a_j} f_1\left(u;\theta^1\right) du$$

are the probabilities to fall into $j$th interval according to the hypotheses $H_0$ and $H_1$, respectively. Let us refer to this grouping method as optimal grouping.

Asymptotically optimal boundary points, corresponding to different optimality criteria, as well as the optimal points, corresponding to [18.4], are considerably different from each other. For example, the boundary points maximizing criteria [18.1]–[18.4] for the following pair of competing hypotheses are given in Table 18.1. The null hypothesis $H_0$ is the normal distribution with density function $f(t) = \frac{1}{\theta_2\sqrt{2\pi}} \exp\left\{-\frac{(t-\theta_1)^2}{2\theta_2^2}\right\}$, and parameters $\theta_1 = 0$, $\theta_2 = 1$, and the competing hypothesis $H_1$ is the logistic distribution with density function $f(t) = \frac{\pi}{\theta_2\sqrt{3}} \exp\left\{-\frac{\pi(t-\theta_1)}{\theta_2\sqrt{3}}\right\} \Big/ \left[1 + \exp\left\{-\frac{\pi(t-\theta_1)}{\theta_2\sqrt{3}}\right\}\right]^2$, and parameters $\theta_1 = 0$, $\theta_2 = 1$.

| Optimality criterion | $a_1$ | $a_2$ | $a_3$ | $a_4$ | $a_5$ | $a_6$ | $a_7$ | $a_8$ |
|---|---|---|---|---|---|---|---|---|
| A-optimum | −2.3758 | −1.6915 | −1.1047 | −0.4667 | 0.4667 | 1.1047 | 1.6915 | 2.3758 |
| D-optimum | −2.3188 | −1.6218 | −1.0223 | −0.3828 | 0.3828 | 1.0223 | 1.6218 | 2.3188 |
| E-optimum | −1.8638 | −1.1965 | −0.6805 | −0.2216 | 0.2216 | 0.6805 | 1.1965 | 1.8638 |
| Optimal grouping | −3.1616 | −2.0856 | −1.2676 | −0.4601 | 0.4601 | 1.2676 | 2.0856 | 3.1616 |

**Table 18.1.** *Optimal boundary points for* $k = 9$

Moreover, in the case of the given competing hypothesis, we can use the so-called Neyman–Pearson classes [GRE 96], for which the random variable domain is partitioned into intervals of two types, according to the inequalities $f_0(t) < f_1(t)$ and $f_0(t) > f_1(t)$, where $f_0(t)$ and $f_1(t)$ are the density functions, corresponding to the competing hypotheses. For $H_0$ and $H_1$ from our example, we have the first-type intervals

$$(-\infty; -2.3747], (-0.6828; 0.6828], (2.3747; \infty),$$

and the second-type intervals

$$(-2.3747; -0.6828], (0.6828; 2.3747].$$

Figures 18.1 and 18.2 illustrate the power of the Pearson $\chi^2$ test for the hypotheses $H_0$ and $H_1$ of our example in the case of different grouping methods, depending on the number of intervals ($\alpha = 0.1$, $n = 500$). The powers of the well-known non-parametric Kolmogorov, Cramer-von Mises-Smirnov and Anderson–Darling goodness-of-fit tests are given for the comparison.

It is obvious that the power of the $\chi^2$ tests for censored samples is also influenced by some other factors, associated with the form of censoring (the type and degree of censoring, as well as the distribution of censoring times). The usage of grouping methods, mentioned above, for censored samples may result in some difficulties, such as the fact that not all boundary points may belong to the observed interval

$[0, \tau]$, or some intervals may not contain complete observations, which does not allow calculation of the NRR statistic. In such situations, it is necessary to combine intervals or change boundary points of intervals. In any case, the boundary points of grouping intervals should be chosen, taking into account the form of censoring. In this chapter, we consider three such grouping methods.

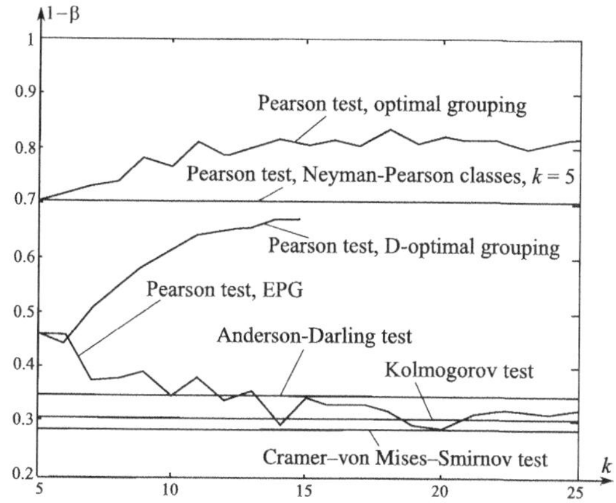


**Figure 18.1.** *The power of the $\chi^2$ Pearson test when testing simple hypothesis*

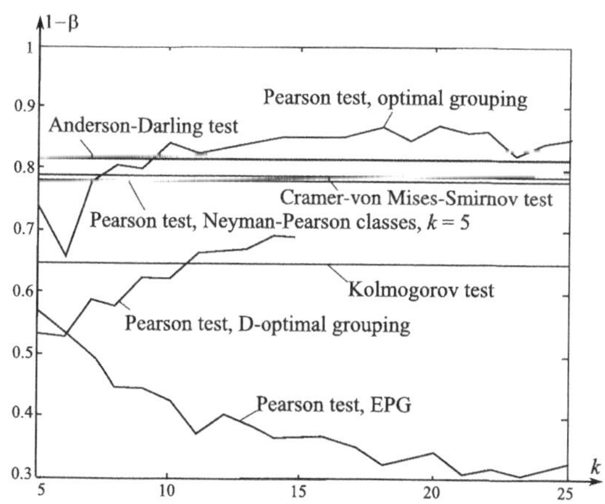


**Figure 18.2.** *The power of the $\chi^2$ Pearson test when testing composite hypothesis*

### 18.3.1. *Equifrequent grouping (EFG)*

In this grouping method, the boundary points are chosen in the interval $[0, \tau]$ from the condition of equal (or almost equal) numbers of complete observations, falling into grouping intervals.

### 18.3.2. *Intervals with equal expected numbers of failures (EENFG)*

In [BAG 10], it is recommended to take $a_j$ as random data functions, dividing the interval $[0, \tau]$ into $k$ intervals with equal expected numbers of complete observations (which are not necessary integer). In this case, $a_j$ are calculated as follows. Define

$$E_k = \int_0^\tau \lambda_0 \left(u; \hat{\theta}_n\right) Y(u) du,$$

$E_j = \frac{j}{k} E_k$ , $j = 1, ..., k$. Denote by $X_{(1)} \leq X_{(2)} \leq ... \leq X_{(n)}$ the ordered sample from $X_1, ..., X_n$, $X_{(0)} = 0$. Set

$$b_i = (n - i) \Lambda_0 \left(X_{(i)}; \hat{\theta}_n\right) + \sum_{l=1}^{i} \Lambda_0 \left(X_{(l)}; \hat{\theta}_n\right).$$

If $i$ is the smallest natural number, verifying $E_j \in [b_{i-1}, b_i]$, $j = 1, ..., k - 1$, then

$$\hat{a}_j = \Lambda_0^{-1} \left( \left( E_j - \sum_{l=1}^{i-1} \Lambda_0 \left(X_{(l)}; \hat{\theta}_n\right) \right) \Big/ (n - i + 1); \hat{\theta}_n \right),$$

$\hat{a}_k = X_{(n)}$, where $\Lambda_0^{-1}(\cdot)$ is the inverse of the cumulative hazard rate function.

### 18.3.3. *Optimal grouping (OptG)*

In the case of testing hypothesis $H_0$ against a given competing hypothesis $H_1$ by randomly censored sample, the optimal boundary points of the grouping intervals can be obtained by solving the maximization problem, which is similar to [18.4]

$$\max_{a_1 < a_2 < .... < a_{k-1}} \left( n \sum_{j=1}^{k} \frac{\left(p_j^1 \left(\theta^1\right) - p_j^0 \left(\theta^0\right)\right)^2}{p_j^0 \left(\theta^0\right)} \right), \qquad [18.5]$$

where

$$p_j^0\left(\theta^0\right) = \int\limits_{a_{j-1}}^{a_j} \left(1 - F^C\left(u\right)\right) f_0\left(u;\theta^0\right) du$$

and

$$p_j^1\left(\theta^1\right) = \int\limits_{a_{j-1}}^{a_j} \left(1 - F^C\left(u\right)\right) f_1\left(u;\theta^1\right) du$$

are the probabilities to fall into $j$th interval according to hypotheses $H_0$ and $H_1$, respectively. If the distribution of censoring times $F^C(t)$ is unknown (which is a typical situation in practice), then we can use the Kaplan–Meier estimator as $F^C(t)$, which is obtained by the inverted sample, in which $\delta_i$ is replaced by $1 - \delta_i$.

REMARK 18.1.– The maximized function in [18.5] is multiextremal, so the global optimization methods are required to solve this problem.

REMARK 18.2.– Strictly speaking, the usage of EENFG and OptG does not guarantee that each grouping interval will contain complete observations, especially in the case of large degrees of censoring. In our simulation study, when such a situation occurs, we change boundary points so that each interval would contain at least one complete observation.

## 18.4. Empirical power study

In this chapter, we study the power of the considered $\chi^2$ goodness-of-fit tests for two pairs of close competing hypotheses through Monte Carlo simulations. Three grouping methods: EFG, EENFG and OptG are used, and the number of intervals $k$ is taken as equal to 3, 4, 5 and 6. Let us emphasize that all results presented below have been obtained in the case of testing composite hypotheses. The estimates of the power are calculated from the distributions of the test statistics $G\left(S|H_0\right)$ and $G\left(S|H_1\right)$, which are simulated on the basis of censored samples of size $n = 200$. The number of simulations used is $N = 10^5$. The values of the power of the tests are calculated with the nominal significance level $\alpha = 0.1$. Let us consider first the null hypothesis $H_0$ is the Weibull distribution with the density function $f\left(t;\theta\right) = \frac{\theta_2}{t}\left(\frac{t}{\theta_1}\right)^{\theta_2} \exp\left(-\left(\frac{t}{\theta_1}\right)^{\theta_2}\right)$ and the parameters $\theta_1 = 2$, $\theta_2 = 2$. The competing hypothesis $H_1$ is the gamma distribution with the density function $f\left(t;\theta\right) = \frac{1}{\theta_1 \cdot \Gamma(\theta_2)}\left(\frac{t}{\theta_1}\right)^{\theta_2 - 1} \exp\left(-\frac{t}{\theta_1}\right)$ and parameters $\theta_1 = 0.5577$, $\theta_2 = 3.1215$.

In the case of random censoring, we need to specify the distribution of censoring times. We have chosen two different families of distributions for censoring times: Type I beta-distribution family with the density function

$$f^C(t;\theta) = B_I(\theta_1, \theta_2, \theta_3) = \frac{1}{\theta_3 \mathrm{B}(\theta_1, \theta_2)} \left(\frac{t}{\theta_3}\right)^{\theta_1 - 1} \left(1 - \frac{t}{\theta_3}\right)^{\theta_1 - 1},$$

where $\mathrm{B}(a, b)$ is the beta function; and the Weibull distribution family. Censoring distributions are given in Table 18.2. The distribution parameters were adjusted so that the average degree of censoring under considered null hypothesis is equal to some given value. In the case of Type I beta distributions, censored observations appear in the variational series of a censored sample approximately uniformly, as opposed to the Weibull distributions, for which censored observations appear generally at the end of the variational series.

| Average degree of censoring | Distribution of censoring times | |
|---|---|---|
| 10% | $B_I(1.81, 1, 7)$ | W(3.44, 6.88) |
| 20% | $B_I(1.19, 1, 7)$ | W(2.87, 5.74) |
| 30% | $B_I(1, 1.24, 7)$ | W(2.48, 4.96) |
| 40% | $B_I(1, 1.83, 7)$ | W(2.16, 4.32) |
| 50% | $B_I(1, 2.58, 7)$ | W(1.87, 3.74) |

**Table 18.2.** *Distributions of censoring times in the case of the lifetime distribution W(2, 2)*

In Table 18.3, the powers of the considered $\chi^2$ tests are presented for various degrees of censoring in the case of randomly censored samples with censoring times from the Weibull distributions with parameter values, given in Table 18.1. In parentheses, there is the number of intervals for which the corresponding estimate of the power was obtained. In Table 18.4, the powers of the considered $\chi^2$ tests are presented in the case of Type I beta distributions of censoring times.

| Test | Grouping method | Average degree of censoring | | | | | |
|---|---|---|---|---|---|---|---|
| | | 0% | 10% | 20% | 30% | 40% | 50% |
| NRR | EFG | 0.51 (5) | 0.38 (5) | 0.32 (5) | 0.28 (5) | 0.25 (5) | 0.22 (5) |
| | EENFG | 0.51 (5) | 0.41 (5) | 0.34 (5) | 0.30 (5) | 0.27 (5) | 0.23 (5) |
| | OptG | 0.61 (6) | 0.46 (4) | 0.41 (4) | 0.37 (4) | 0.32 (3) | 0.29 (3) |
| GPF | EFG | 0.20 (6) | 0.17 (6) | 0.15 (6) | 0.15 (6) | 0.14 (6) | 0.13 (6) |
| | EENFG | 0.26 (6) | 0.22 (6) | 0.19 (6) | 0.18 (6) | 0.16 (6) | 0.16 (6) |
| | OptG | 0.56 (4) | 0.51 (4) | 0.44 (4) | 0.37 (4) | 0.35 (4) | 0.29 (4) |

**Table 18.3.** *Power of the tests for the Weibull-gamma pair in the case of randomly censored data with Weibull distribution of censoring times*

As can be seen from Tables 18.3 and 18.4, the power of the considered tests depends on the distribution of censoring times: in the case of Type I beta distribution, the power of the tests is higher, than for the Weibull distribution of censoring times. The GPF $\chi^2$ test is considerably worse in terms of the power than the NRR test, when using EFG and EENFG. But in the case of optimal grouping, it is impossible to give preference to one test. The power of both tests in the case of optimal grouping is much higher than in the case of equifrequent grouping and grouping into intervals with equal expected numbers of failures.

| Test | Grouping method | Average degree of censoring | | | | | |
|---|---|---|---|---|---|---|---|
| | | 0% | 10% | 20% | 30% | 40% | 50% |
| NRR | EFG | 0.51 (5) | 0.43 (5) | 0.40 (5) | 0.36 (5) | 0.33 (5) | 0.28 (5) |
| | EENFG | 0.51 (6) | 0.46 (6) | 0.42 (6) | 0.38 (6) | 0.34 (6) | 0.30 (6) |
| | OptG | 0.61 (6) | 0.54 (6) | 0.51 (4) | 0.48 (4) | 0.44 (4) | 0.40 (4) |
| GPF | EFG | 0.20 (6) | 0.18 (6) | 0.17 (6) | 0.17 (6) | 0.15 (6) | 0.15 (6) |
| | EENFG | 0.26 (6) | 0.24 (6) | 0.23 (6) | 0.22 (6) | 0.21 (6) | 0.19 (6) |
| | OptG | 0.56 (4) | 0.53 (4) | 0.49 (4) | 0.45 (4) | 0.41 (4) | 0.37 (4) |

**Table 18.4.** *Power of the tests for the Weibull-gamma pair in the case of randomly censored data with Type I beta distribution of censoring times*

It is significant to compare the considered $\chi^2$ tests with the modified Kolmogorov, Cramer–von Mises–Smirnov and Anderson–Darling tests for censored samples, which have been studied in detail through Monte Carlo simulations in [LEM 10a] for Type I and Type II censored data and in [LEM 13] for randomly censored data. In Table 18.5, the powers of the considered $\chi^2$ tests are presented for the Weibull-gamma pair of competing hypotheses in the case of Type II censored samples, when using the optimal grouping. In parentheses, there is the number of intervals, for which the corresponding estimate of the power was obtained. The powers of modified Kolmogorov, Cramer-von Mises-Smirnov and Anderson–Darling tests for censored samples are given for comparison in the same table.

As can be seen from Table 18.5, the NRR $\chi^2$ test has an advantage in power comparing not only with GPF test, but also with modified non-parametric goodness-of-fit tests.

In Table 18.6, the powers of the considered $\chi^2$ tests, as well as the powers of modified non-parametric tests are presented in the case of Type II censored samples for the following pair of competing hypotheses. The null hypothesis $H_0$ is the exponential distribution with the density function $f(t;\theta) = \frac{1}{\theta_1}\exp\left(-\frac{t}{\theta_1}\right)$ and the parameter $\theta_1 = 1$. The competing hypothesis $H_1$ is the Weibull distribution with parameters $\theta_1 = 1$ , $\theta_2 = 1.2$. The powers of $\chi^2$ tests in Table 18.6 are given in the case of optimal grouping.

| Test | Degree of censoring | | | | | |
|---|---|---|---|---|---|---|
| | 0% | 10% | 20% | 30% | 40% | 50% |
| NRR | 0.61 (6) | 0.53 (5) | 0.44 (5) | 0.34 (5) | 0.25 (3) | 0.20 (3) |
| GPF | 0.56 (4) | 0.27 (5) | 0.25 (5) | 0.17 (5) | 0.14 (5) | 0.13 (5) |
| Kolmogorov | 0.51 | 0.40 | 0.33 | 0.26 | 0.22 | 0.18 |
| Cramer-von Mises-Smirnov | 0.58 | 0.44 | 0.35 | 0.27 | 0.22 | 0.18 |
| Anderson–Darling | 0.41 | 0.34 | 0.28 | 0.23 | 0.20 | 0.17 |

**Table 18.5.** *Power of the tests for the Weibull-gamma pair in the case of Type II censored data*

| Test | Degree of censoring | | | | | |
|---|---|---|---|---|---|---|
| | 0% | 10% | 20% | 30% | 40% | 50% |
| NRR | 0.83 (3) | 0.78 (3) | 0.72 (3) | 0.66 (3) | 0.59 (3) | 0.52 (3) |
| GPF | 0.76 (3) | 0.68 (5) | 0.28 (4) | 0.27 (4) | 0.26 (4) | 0.22 (4) |
| Kolmogorov | 0.84 | 0.76 | 0.69 | 0.62 | 0.55 | 0.47 |
| Cramer-von Mises-Smirnov | 0.90 | 0.83 | 0.76 | 0.69 | 0.61 | 0.53 |
| Anderson–Darling | 0.91 | 0.85 | 0.79 | 0.72 | 0.65 | 0.56 |

**Table 18.6.** *Power of the tests for the exponential-Weibull pair in the case of Type II censored data*

As can be seen from Table 18.6, the NRR test here has a higher power than the GPF test. Nevertheless, the power of the NRR test turned out to be slightly less than the power of the Cramer–von Mises–Smirnov and Anderson–Darling tests.

## 18.5. Conclusions

The main advantage of the chi-square goodness-of-fit tests for censored data is that the limiting distribution of these statistics is the well-known $\chi^2$-distribution. As has been shown in [LEM 09], in the case of testing simple hypotheses by complete samples (without censoring), the Pearson $\chi^2$ test has a considerable advantage in power due to choosing optimal boundary points in the comparison with the non-parametric Kolmogorov, Cramer–von Mises–Smirnov and Anderson–Darling tests. As to the case of testing composite hypotheses, the non-parametric tests, as a rule, have higher power, as compared with the chi-square tests [LEM 10b]. The best properties of the chi-square tests, which enable them to compete with the Cramer–von Mises–Smirnov and Anderson–Darling tests, are shown up when using optimal grouping and choosing optimal number of intervals. As has been shown in this study, this result is also confirmed in the case of censored data.

It has been shown that in the case of a given pair of competing hypotheses, it is possible to increase essentially the power of the $\chi^2$ tests due to the choice of optimal boundary points of grouping intervals. The power of both considered tests in the case of optimal grouping is much higher than in the case of equifrequent grouping and grouping into intervals with equal expected numbers of failures. The generalized $\chi^2$ Pearson–Fisher test is considerably worse in terms of the power, than the NRR $\chi^2$ test, when EFG and EENFG are used. But in the case of optimal grouping, it is impossible to give preference to one test. The power of the considered tests also depends on the distribution of censoring times: in the case when censored observations are "uniformly" distributed in the variational series, the power of the tests is higher as compared to when censored observations appear generally at the end of the variational series.

### 18.6. Acknowledgment

This work was supported by the Ministry of Education and Science of the Russian Federation within the state assignment (project 8.1274.2011) and the Federal target program "Research and Educational Personnel for Innovative Russia" (agreement # 14.B37.21.0860).

### 18.7. Bibliography

[AKR 88] AKRITAS M., "Pearson-type goodness-of-fit tests: the univariate case", *Journal of the American Statistical Association*, vol. 83, pp. 222–230, 1988.

[BAG 10] BAGDONAVIČIUS V., KRUOPIS J., NIKULIN M., *Nonparametric Tests for Censored Data*, John Wiley & Sons, Inc., New York, 2010.

[BAG 13] BAGDONAVIČIUS V., LEVULIENE R., NIKULIN M., "Chi-squared goodness-of-fit tests for parametric accelerated failure time models", *Communications in Statistics – Theory and Methods*, vol. 42, pp. 2768–2785, 2013.

[CHI 11a] CHIMITOVA E., NIKULIN M., LEMESHKO B., *et al.*, "Nonparametric goodness-of-fit tests for censored data", *Proceedings of the 7th International Conference on Mathematical Methods in Reliability: Theory, Methods and Applications*, Beijing, China, pp. 817–823, 2011.

[CHI 11b] CHIMITOVA E., TSIVINSKAYA A., "Simulation study for the NRR chi-square test of goodness-of-fit for censored data", *Proceedings of the International Workshop on Applied Methods of Statistical Analysis*, Novosibirsk, Russia, pp. 44–52, 2011.

[COH 91] COHEN A., *Truncated and Censored Samples: Theory and Applications*, Marcel Dekker, New York, 1991.

[DEN 79] DENISOV V., LEMESHKO B., "Optimal grouping in the analysis of experimental data", *Measuring Information Systems*, Novosibirsk, pp. 5–14, 1979. [In Russian]

[DEN 89] DENISOV V., LEMESHKO B., "Optimal grouping in estimation and tests of goodness-of-fit hypotheses", *Wissenschaftliche Schriftenreihe der Technishen universitat Karl-Marx-Stadt.*, vol. 10, pp. 63–81, 1989.

[DEN 04] DENISOV V., EGER K.-H., LEMESHKO B., *et al.*, *Design of Experiments and Statistical Analysis for Grouped Observations*, NSTU Publishing House, Novosibirisk, 2004.

[GRE 96] GREENWOOD P., NIKULIN M., *A Guide to Chi-Squared Testing*, John Wiley & Sons, New York, 1996.

[HAB 86] HABIB M., THOMAS D., "Chi-square goodness-of-fit tests for randomly censored data", *Annals of Statistics*, vol. 14, pp. 759–765, 1986.

[HJO 90] HJORT N., "Goodness of fit tests in models for life history data based on cumulative hazard rates", *The Annals of Statistics*, vol. 18, no. 3, pp. 1221–1258, 1990.

[HOL 92] HOLLANDER M., PEÑA E., "A chi-squared goodness-of-fit test for randomly censored data", *Journal of the American Statistical Association*, vol. 87, pp. 458–463, 1992.

[KIM 93] KIM J., "Chi-square goodness-of-fit tests for randomly censored data", *The Annals of Statistics*, vol. 21, no. 3, pp. 1621–1639, 1993.

[LEM 98] LEMESHKO B., "Asymptotically optimal grouping of observations in goodness-of-fit tests", *Industrial Laboratory*, vol. 64, no. 1, pp. 59–67, 1998 [In Russian].

[LEM 00] LEMESHKO B., CHIMITOVA E., "Maximization of the power of $\chi^2$ tests", *Papers of Siberian Branch of Academy of Science of Higher Education*, vol. 2, pp. 53–61, 2000. [In Russian]

[LEM 01] LEMESHKO B., POSTOVALOV S., CHIMITOVA E., "On statistic distrbutions and the power of the Nikulin $\chi^2$ test", *Industrial Laboratory*, vol. 67, no. 3, pp. 52–58, 2001. [in Russian]

[LEM 09] LEMESHKO B., LEMESHKO S., POSTOVALOV S., "Comparative analysis of the power of goodness-of-fit tests for near competing hypotheses. Part I. Verification of simple hypotheses", *Journal of Applied and Industrial Mathematics*, vol. 3, no. 4, pp. 462-475, 2009.

[LEM 10a] LEMESHKO B., CHIMITOVA E., PLESHKOVA T., "Testing simple and composite goodness-of-fit hypotheses by censored samples", *Nauchniy Vestnik NGTU*, vol. 41, no. 4, pp. 13–28, 2010. [In Russian]

[LEM 10b] LEMESHKO B., LEMESHKO S., POSTOVALOV S., "Comparative analysis of the power of goodness-of-fit tests for near competing hypotheses. Part II. Verification of complex hypotheses", *Journal of Applied and Industrial Mathematics*, vol. 4, no. 1, 2010.

[LEM 11] LEMESHKO B., LEMESHKO S., POSTOVALOV S., *et al.*, *Statistical Data Analysis, Simulation and Research of Probabilistic Regularities. Computer Approach*, NSTU Monographs, NSTU Publishing House, Novosibirsk, 2011. [In Russian]

[LEM 13] LEMESHKO B., CHIMITOVA E., VEDERNIKOVA M., "Modified Kolmogorov, Cramer-von Mises-Smirnov and Anderson-Darling goodness-of-fit tests for randomly censored data", *Nauchniy Vestnik NSTU*, vol. 50, no. 1, pp. 3–16, 2013.

[LI 93] LI G., DOSS H., "Generalized Pearson-Fisher chi-square goodness-of-fit tests, with applications to models with life history data", *The Annals of Statistics*, vol. 21, pp. 772–797, 1993.

[NIK 73a] NIKULIN M., "Chi-square test for continuous distributions", *Theory of Probability and Its Applications*, vol. 18, pp. 638–639, 1973. [In Russian].

[NIK 73b] NIKULIN M., "Chi-square test for continuous distributions with shift and scale parameters", *Theory of Probability and Its Applications*, vol. 18, pp. 559–568, 1973. [In Russian]

[NIK 73c] NIKULIN M., "Chi-square test for normality", *Proceedings of the International Vilnius Conference on Probability Theory and Mathematical Statistics*, vol. 2, Vilnius, pp. 119–122, 1973. [in Russian].

[RAO 74] RAO K., ROBSON D., "A chi-squared statistic for goodness-of-fit tests within the exponential family", *Communications in Statistics*, vol. 3, pp. 1139–1153, 1974.

[VOI 06] VOINOV V., "On optimality of the Rao-Robson-Nikulin test", *Industrial Laboratory*, vol. 72, pp. 65–70, 2006. [In Russian]

[VOI 09] VOINOV V., PYA N., ALLOYAROVA R., "A comparative study of some modified chi-squared tests", *Communications in Statistics-Simulation and Computation*, vol. 38, pp. 355–367, 2009.

[VOI 13] VOINOV V., NIKULIN M., BALAKRISHNAN N., *Chi-Squared Goodness of Fit Tests with Applications*, Academic Press, 2013.

# Chapter 19

# A Non-parametric Test for Comparing Treatments with Missing Data and Dependent Censoring

## 19.1. Introduction

The log-rank test has become a standard tool for comparing survival distributions across several randomized treatment groups. If $\lambda_k(\cdot)$ denotes the hazard rate of failure in the $k$th treatment group ($k = 1, \ldots, K$), the null hypothesis of interest is $\lambda_1(\cdot) = \ldots = \lambda_K(\cdot)$, and the alternative is $\lambda_j(\cdot) \neq \lambda_{j'}(\cdot)$ for some $j \neq j'$. The log-rank test is a non-parametric test. Its construction is based on the simple idea of comparing the Nelson–Aalen estimators of the $\int \lambda_k(t)\, dt$, $k = 1, \ldots, K$, with an estimator of the hypothesized common value $\int \lambda(t)\, dt$ under the null hypothesis. The resulting test statistic is asymptotically $\chi^2$ distributed under the null hypothesis (this can be proved by invoking a martingale central limit theorem; see, for example, [AND 93]).

If we need to control for a covariate that may be predictive of the outcome, and if this covariate is discrete, the log-rank can be generalized to the so-called stratified log-rank test (see, for example, [AND 93], [KLE 97] and [MAR 06]). Consider a clinical trial where $n$ patients are randomly assigned to $K$ different treatment groups, and suppose that we wish to compare survival between groups while adjusting for some discrete factor $S$ with $L$ modalities (also called strata, such as income groups or

---

Chapter written by Amel MEZAOUER, Kamal BOUKHETALA and Jean-François DUPUY.

disease stages for example). If $\lambda_{k,l}(\cdot)$ is the hazard rate of failure in the $k$th treatment group and $l$th stratum, then the test for treatment effect can be formulated as

$$H_0 : \lambda_{1,l}(\cdot) = \ldots = \lambda_{K,l}(\cdot) \quad \text{for every } l = 1, \ldots, L$$

versus

$$H_1 : \text{there exists } j \text{ and } j' \text{ such that } \lambda_{j,l}(\cdot) \neq \lambda_{j',l}(\cdot) \text{ for some } l.$$

Let $T_1^0, \ldots, T_n^0$ be the times from randomization to failure observed in the $K$ pooled groups. Let $C_1, \ldots, C_n$ be right-censoring times (the $C_i$ are assumed to be independent of the $T_i^0$ and non-informative). For each patient $i$, we observe $T_i = \min(T_i^0, C_i)$ and $\Delta_i = 1(T_i^0 \leq C_i)$, where $1(\cdot)$ is the indicator function. Assume that the data consist of $n$ independent and identically distributed quadruplets $(T_i, \Delta_i, G_i, S_i)$, $i = 1, \ldots, n$, where $G_i \in \{1, \ldots, K\}$ and $S_i \in \{1, \ldots, L\}$, respectively, indicate the group and stratum of the $i$th patient. Let $N_i(t) = \Delta_i 1(T_i \leq t)$, $Y_i(t) = 1(T_i \geq t)$ and define

$$E_{k,l}^{(n)}(t) = \frac{\sum_{i=1}^{n} Y_i(t) 1(G_i = k) 1(S_i = l)}{\sum_{i=1}^{n} Y_i(t) 1(S_i = l)}.$$

Then, the stratified log-rank statistic for testing $H_0$ is of the form

$$U = (Z_1, \ldots, Z_{K-1}) \hat{\Theta}^{-1} (Z_1, \ldots, Z_{K-1})',$$

where, for every $k = 1, \ldots, K-1$,

$$Z_k = \sum_{i=1}^{n} \int_0^{\tau} \left\{ 1(G_i = k) \quad \sum_{l=1}^{L} 1(S_i = l) E_{k,l}^{(n)}(t) \right\} dN_i(t), \qquad [19.1]$$

$\tau$ denotes the end of the study period and $\hat{\Theta}$ is the estimated asymptotic covariance matrix of $(Z_1, \ldots, Z_{K-1})'$. Under $H_0$, $U$ is asymptotically distributed as a $\chi^2$ distribution with $K-1$ degrees of freedom (e.g. [MAR 06]).

In some applications, the stratum $S$ may be missing for some patients. For example, consider the case where $S$ represents the histological stage of patients included in a cancer clinical trial. The determination of $S$ may require a biopsy, which due to expensiveness may not be performed on all the study subjects. One simple solution to handle such incomplete stratum information is to perform a complete-case analysis, that is, to discard patients with unobserved stratum. However, generally speaking, complete-case analysis is known to induce a substantial loss of power. Dupuy and Leconte [DUP 08, DUP 09] considered the distinct but related problem of estimation in the stratified proportional hazards model with missing strata. The authors proposed a modified version of the maximum partial likelihood estimator, in which the unobserved stratum indicators are replaced by an estimate of their conditional expectation given available auxiliary covariates (this is the so-called regression calibration idea, see, for example, [CAR 95]). Their results provided evidence that such a replacement substantially improves the complete-case analysis.

In many applications, it also happens that the distribution of the censoring time depends on the treatment group. This arises, for example, when the censoring follows from a study dropout caused by treatment toxicity. The treatment group with the heaviest toxicity will be more likely to have a higher dropout rate, and thus a higher censoring rate than the other groups. Inverse probability of censoring weighted (IPCW) procedures have been proposed as a solution to this problem (see, for example, [ROB 00], [YOS 07], and [CAI 09]).

In this chapter, we propose a new test statistic for comparing randomized treatment groups. This test addresses the situation where the patients' stratum information is only partially available to the investigator, and the distribution of the censoring time depends on the treatment group. The test we propose combines the regression calibration and IPCW principles.

The remaining chapter is organized as follows. In section 19.2, we introduce the new test statistic and we give its asymptotic distribution under $H_0$. The proof is given in section 19.3. Some intermediate technical results are proved in section 19.5.

## 19.2. The proposed test statistic

Assume that $n$ independent patients are randomly assigned to $K$ treatment groups. Assume that the stratum value is missing for some of these patients. Thus, a subsample is available where all variables $(T, \Delta, G, S)$ are observed, while only $(T, \Delta, G)$ are observed for the other patients. We assume that some auxiliary variables $W \in \mathbf{R}^p$ are observed for all patients, and that $W$ provides partial information about $S$ when $S$ is missing. Let $R$ be the indicator variable which is 1 if $S$ is observed and 0 otherwise. Throughout the chapter, we assume that $T^0$ and $C$ are independent given $G, S, W$ and $R$, and that $C$ is independent of $S$ and $W$ given $G$. However, as mentioned above, the distribution of the censoring time depends on the treatment group. We assume that $T^0$ is independent of $W$ given $S$ (that is the auxiliary variables $W$ provide no additional information about failure when the true stratum $S$ is known), and that $G$ is independent of $S$ and $W$, as is the case in randomized clinical trials. We assume that $R$ is independent of $T^0, C$ and $G$, and that $0 < \mathbf{P}(R = 1) < 1$. Finally, we assume that $R$ and $S$ are independent given $W$, which is the so-called missing-at-random assumption. In the following, this set of assumptions will be denoted by **C1**.

As mentioned above, we consider the problem of testing $H_0$ through the stratified log-rank test, based on $n$ independent vectors $(T_i, \Delta_i, G_i, W_i, R_i, R_i S_i)$,

$i = 1, \ldots, n$, of possibly incomplete data when, moreover, the distribution of the censoring time depends on the treatment group. To tackle the missing strata and dependent censoring issues simultaneously, we introduce a modified version of $U$. To obtain this new test statistic, we:

1) replace any missing stratum indicator $1(S_i = l)$ in (19.1) by its conditional expectation given the auxiliary $W$;

2) weight every patient by the inverse of the conditional (given the patient's treatment group) survival function of the censoring time.

Precisely, our test statistic is based on the following modified version of [19.1]

$$\widetilde{Z}_k = \sum_{i=1}^{n} \int_0^\tau \mu(G_i, t) \left\{ G_i^k - \sum_{l=1}^{L} D_i^l \widetilde{E}_{k,l}^{(n)}(t) \right\} dN_i(t), \quad [19.2]$$

where for every $i = 1, \ldots, n$, $k = 1, \ldots, K$, $l = 1, \ldots, L$ and $t \in [0, \tau]$, $G_i^k = 1(G_i = k)$, $D_i^l = R_i 1(S_i = l) + (1 - R_i)\mathbf{E}[1(S_i = l)|W_i]$, $\mu(G_i, t) = 1/\mathbf{P}(C \geq t|G_i)$ (where $\mathbf{P}(C \geq t|G)$ is the survival function of the censoring time in group $G$) and $\widetilde{E}_{k,l}^{(n)}(t) = \widetilde{S}_{k,l}^{(n)}(t)/\widetilde{S}_l^{(n)}(t)$, with

$$\widetilde{S}_{k,l}^{(n)}(t) = \frac{1}{n} \sum_{i=1}^{n} Y_i(t) G_i^k D_i^l \mu(G_i, t) \text{ and } \widetilde{S}_l^{(n)}(t) = \frac{1}{n} \sum_{i=1}^{n} Y_i(t) D_i^l \mu(G_i, t).$$

Before proposing our test statistic, we need to introduce some further notations. For every $i = 1, \ldots, n$ and $k = 1, \ldots, K$, let

$$V_{i,k} = \int_0^\tau \mu(G_i, t) \left\{ G_i^k - \sum_{l=1}^{L} D_i^l \widetilde{E}_{k,l}^{(n)}(t) \right\} dN_i(t) \quad [19.3]$$
$$- \frac{1}{n} \sum_{j=1}^{n} \sum_{l=1}^{L} \int_0^\tau \frac{Y_i(t) D_i^l D_j^l \mu(G_i, t) \mu(G_j, t)}{\widetilde{S}_l^{(n)}(t)} \left\{ G_i^k - \widetilde{E}_{k,l}^{(n)}(t) \right\} dN_j(t)$$

and define $\mathbf{V}_i = (V_{i,1}, \ldots, V_{i,K-1})'$ and $\widehat{\Sigma} = \sum_{i=1}^{n} \mathbf{V}_i \mathbf{V}_i'$. For every $k = 1, \ldots, K$, $l = 1, \ldots, L$ and $t \in [0, \tau]$, define

$$\widetilde{s}_{k,l}(t) = \mathbf{E}[Y(t) G^k D^l \mu(G, t)], \; \widetilde{s}_l(t) = \mathbf{E}[Y(t) D^l \mu(G, t)], \; \widetilde{e}_{k,l}(t) = \widetilde{s}_{k,l}(t)/\widetilde{s}_l(t).$$

The following regularity conditions will also be needed:

C2: There exists a positive constant $c_0$ such that $\mathbf{P}(C \geq \tau | G) > c_0$ for every $G \in \{1, \ldots, K\}$, and the survival function $t \mapsto \mathbf{P}(C \geq t | G)$ is continuous on $[0, \tau]$.

C3: For every $k = 1, \ldots, K$ and $l = 1, \ldots, L$, $\sup_{t \in [0,\tau]} \lambda_{k,l}(t) < c_1$ for some finite positive constant $c_1$.

C4: A constant $c_2 > 0$ exists such that $\inf_{t \in [0,\tau]} \widetilde{s}_l(t) > c_2$ for every $l = 1, \ldots, L$.

We are now in a position to propose the new test statistic

$$\widetilde{U} := (\widetilde{Z}_1, \ldots, \widetilde{Z}_{K-1}) \widehat{\Sigma}^{-1} (\widetilde{Z}_1, \ldots, \widetilde{Z}_{K-1})'$$

and to state its asymptotic distribution under the null hypothesis $H_0$.

THEOREM 19.1.– Assume that the conditions C1–C4 hold. Then, under $H_0$, as $n \to \infty$, $\widetilde{U}$ converges in distribution to a $\chi^2(K-1)$ distribution.

On the basis of this theorem, the proposed test rejects $H_0$ if $\widetilde{U} \geq \chi^2_{1-\alpha}(K-1)$, where $\chi^2_{1-\alpha}(K-1)$ is the quantile of order $1-\alpha$ of the $\chi^2(K-1)$ distribution.

## 19.3. Asymptotic distribution of the proposed test statistic

The key to derive the null asymptotic distribution of the usual stratified log-rank test $U$ is to represent $Z_k$ in [19.1] as a martingale process and to use a central limit theorem for martingales (see, for example, [FLE 91] or [AND 93]). This is not possible in our case, since the sum across strata in $Z_k$ has been replaced, in $\widetilde{Z}_k$, by a sum where each patient $i$ such that $R_i = 0$ contributes to each of the strata. Therefore, in order to obtain the null asymptotic distribution of $\widetilde{U}$, we instead prove that $\widetilde{Z}_k$ is asymptotically linear (lemma 19.1), and we use a central limit theorem for sums of independent and identically distributed (i.i.d.) terms.

Assume that all random variables are defined on a probability space $(\Omega, \mathcal{C}, \mathbf{P})$, and define $\mathcal{F}_{t,i} = \sigma\{N_i(s), (1-\Delta_i)1(T_i \leq s), G_i, S_i, W_i : 0 \leq s \leq t\}$ as the $\sigma$-algebra generated by the event time and censoring histories of the $i$th patient over $[0, t]$, and by the group, stratum and auxiliary informations for this patient. Then, by assumption, the $\mathcal{F}_{t,i}$-intensity of the counting process $N_i(t)$ is given by

$$Y_i(t)\lambda_i(t) = Y_i(t) \sum_{k=1}^{K} \sum_{l=1}^{L} \lambda_{k,l}(t) G_i^k 1(S_i = l).$$

If $S_i$ is missing, the information for the $i$th patient is represented by the smaller $\sigma$-algebra $\mathcal{G}_{t,i} = \sigma\{N_i(s), (1-\Delta_i)1(T_i \leq s), G_i, W_i : 0 \leq s \leq t\} \subseteq \mathcal{F}_{t,i}$. By the innovation theorem, the intensity of $N_i(t)$ with respect to $\mathcal{G}_{t,i}$ is $Y_i(t)\gamma_i(t) := \mathbf{E}[Y_i(t)\lambda_i(t)|\mathcal{G}_{t-,i}]$, where $\gamma_i(t) = \sum_{k=1}^K \sum_{l=1}^L \lambda_{k,l}(t)\times G_i^k \mathbf{E}[1(S_i = l)|\mathcal{G}_{t-,i}]$. Finally, letting $\mathcal{H}_{t,i} = (\mathcal{F}_{t,i})^{R_i}(\mathcal{G}_{t,i})^{1-R_i}$ be the observed filtration, $N_i(t)$ has intensity $Y_i(t)\zeta_i(t) := Y_i(t)[\lambda_i(t)R_i + \gamma_i(t)(1-R_i)]$ with respect to $\mathcal{H}_{t,i}$. It follows that $M_i(t) = N_i(t) - \int_0^t Y_i(s)\zeta_i(s)\,ds$ is a martingale with respect to $(\mathcal{H}_{t,i})_{t\geq 0}$. In the following, we will note $\kappa_l(t) = \mathbf{E}[Y(t)\zeta(t)D^l\mu(G,t)]$, $l = 1,\ldots,L$ (note that under the conditions mentioned above, $\kappa_l(t) < \infty$ for every $t$).

On the basis of these notations, the following preliminary lemma establishes a useful approximation of $n^{-\frac{1}{2}}\widetilde{Z}_k$.

LEMMA 19.1.– For every $i = 1,\ldots,n$ and $k = 1,\ldots,K$, let

$$Q_{i,k} = \int_0^\tau \sum_{l=1}^L D_i^l \mu(G_i,t)\left(G_i^k - \widetilde{e}_{k,l}(t)\right)\left[dN_i(t) - Y_i(t)\frac{\kappa_l(t)}{\widetilde{s}_l(t)}\,dt\right].$$

Then, under C1–C4, $n^{-\frac{1}{2}}\widetilde{Z}_k = n^{-\frac{1}{2}}\sum_{i=1}^n Q_{i,k} + o_p(1)$. Moreover, if $H_0$ holds, $\mathbf{E}[Q_{i,k}] = 0$.

PROOF.– We have

$$\begin{aligned} n^{-\frac{1}{2}}\widetilde{Z}_k &= n^{-\frac{1}{2}}\sum_{i-1}^n \int_0^\tau \mu(G_i,t)G_i^k\,dN_i(t) \\ &\quad - n^{-\frac{1}{2}}\sum_{i=1}^n\sum_{l=1}^L \int_0^\tau \mu(G_i,t)D_i^l \widetilde{E}_{k,l}^{(n)}(t)\,dN_i(t) \\ &:= A_{1,k}^{(n)} - A_{2,k}^{(n)}, \end{aligned}$$

and $A_{2,k}^{(n)}$ can be written as

$$\begin{aligned} A_{2,k}^{(n)} &= n^{\frac{1}{2}}\sum_{l=1}^L \int_0^\tau \widetilde{e}_{k,l}(t)\left[\frac{1}{n}\sum_{i=1}^n \mu(G_i,t)D_i^l dN_i(t) - \kappa_l(t)dt\right] \\ &+ n^{\frac{1}{2}}\sum_{l=1}^L \int_0^\tau v_{k,l}^{(n)}(t)\left[\frac{1}{n}\sum_{i=1}^n \mu(G_i,t)D_i^l dN_i(t) - \kappa_l(t)dt\right] \\ &+ n^{\frac{1}{2}}\sum_{l=1}^L \int_0^\tau \widetilde{E}_{k,l}^{(n)}(t)\kappa_l(t)dt, \end{aligned}$$

where $v_{k,l}^{(n)}(t) := \widetilde{E}_{k,l}^{(n)}(t) - \widetilde{e}_{k,l}(t)$. We leave the first term in $A_{2,k}^{(n)}$ unchanged. The second and third terms satisfy, respectively, the following two technical lemmas, whose proofs are discussed in section 19.5.

LEMMA 19.2.– Under conditions C1–C4,

$$n^{\frac{1}{2}} \sum_{l=1}^{L} \int_0^{\tau} v_{k,l}^{(n)}(t) \left[ \frac{1}{n} \sum_{i=1}^{n} \mu(G_i, t) D_i^l dN_i(t) - \kappa_l(t) dt \right] \xrightarrow{p} 0 \text{ as } n \to \infty.$$

LEMMA 19.3.– Under conditions C1–C4, as $n \to \infty$,

$$n^{\frac{1}{2}} \sum_{l=1}^{L} \int_0^{\tau} \widetilde{E}_{k,l}^{(n)}(t) \kappa_l(t) dt = n^{\frac{1}{2}} \sum_{l=1}^{L} \int_0^{\tau} \left\{ \widetilde{e}_{k,l}(t) + \frac{\widetilde{S}_{k,l}^{(n)}(t)}{\widetilde{s}_l(t)} - \frac{\widetilde{s}_{k,l}(t) \widetilde{S}_l^{(n)}(t)}{\widetilde{s}_l(t)^2} \right\} \times \kappa_l(t) dt + o_p(1).$$

Using these lemmas, we can rewrite $n^{-\frac{1}{2}} \widetilde{Z}_k$ as

$$\begin{aligned} n^{-\frac{1}{2}} \widetilde{Z}_k = {} & \\ & n^{-\frac{1}{2}} \sum_{i=1}^{n} \int_0^{\tau} \mu(G_i, t) G_i^k \, dN_i(t) \\ & - n^{\frac{1}{2}} \sum_{l=1}^{L} \int_0^{\tau} \widetilde{e}_{k,l}(t) \left[ \frac{1}{n} \sum_{i=1}^{n} \mu(G_i, t) D_i^l dN_i(t) - \kappa_l(t) dt \right] \\ & - n^{\frac{1}{2}} \sum_{l=1}^{L} \int_0^{\tau} \frac{1}{\widetilde{s}_l(t)} \left\{ \widetilde{s}_{k,l}(t) + \widetilde{S}_{k,l}^{(n)}(t) - \frac{\widetilde{s}_{k,l}(t) \widetilde{S}_l^{(n)}(t)}{\widetilde{s}_l(t)} \right\} \kappa_l(t) dt + o_p(1). \end{aligned}$$

Rearranging the terms in this expression concludes the proof of the first statement in lemma 19.1. We now focus on the expectation of $Q_{i,k}$ under $H_0$.

For every $i = 1, \ldots, n$, $k = 1, \ldots, K$ and $t \in [0, \tau]$, $\mu(G_i, t) G_i^k$ is $\mathcal{H}_{t,i}$-measurable, hence the process $(\int_0^t \mu(G_i, s) G_i^k \, dM_i(s))_{t \geq 0}$ is a zero-mean martingale. It follows that

$$\mathbf{E} \left[ \int_0^{\tau} \sum_{l=1}^{L} D_i^l \mu(G_i, t) G_i^k \, dN_i(t) \right] = \int_0^{\tau} \mathbf{E} \left[ \mu(G_i, t) G_i^k Y_i(t) \zeta_i(t) \right] dt.$$

Similarly, we have

$$\mathbf{E}\left[\int_0^\tau \sum_{l=1}^L D_i^l \mu(G_i,t)\widetilde{e}_{k,l}(t)\,dN_i(t)\right] = \sum_{l=1}^L \int_0^\tau \widetilde{e}_{k,l}(t)\kappa_l(t)\,dt,$$

$$\mathbf{E}\left[\int_0^\tau \sum_{l=1}^L D_i^l \mu(G_i,t)G_i^k Y_i(t)\frac{\kappa_l(t)}{\widetilde{s}_l(t)}\,dt\right] = \sum_{l=1}^L \int_0^\tau \widetilde{e}_{k,l}(t)\kappa_l(t)\,dt,$$

$$\mathbf{E}\left[\int_0^\tau \sum_{l=1}^L D_i^l \mu(G_i,t)\widetilde{e}_{k,l}(t)Y_i(t)\frac{\kappa_l(t)}{\widetilde{s}_l(t)}\,dt\right] = \sum_{l=1}^L \int_0^\tau \widetilde{e}_{k,l}(t)\kappa_l(t)\,dt.$$

Thus

$$\begin{aligned}\mathbf{E}[Q_{i,k}] &= \int_0^\tau \mathbf{E}\left[\mu(G_i,t)G_i^k Y_i(t)\zeta_i(t)\right]\,dt - \sum_{l=1}^L \int_0^\tau \widetilde{e}_{k,l}(t)\kappa_l(t)\,dt \\ &:= B_{1,k} - B_{2,k}.\end{aligned}$$

We now prove that under $H_0$, $B_{2,k} = \int_0^\tau \mathbf{E}[G^k]\mathbf{E}[Y(t)\zeta(t)\mu(G,t)]\,dt$. First, note that

$$\begin{aligned}\widetilde{s}_{k,l}(t) &= \mathbf{E}\left[Y(t)G^k D^l \mu(G,t)\right] \\ &\overset{H_0}{=} \mathbf{E}\left[G^k D^l \mu(G,t)\mathbf{E}\left[1(T^0 \geq t)|S\right]\mathbf{E}\left[1(C \geq t)|G\right]\right],\end{aligned}$$

where the second line follows from the assumption C1 and the fact that under $H_0$, the distribution of $T^0$ does not depend on $G$. Then, by the independence of $G$ and $(S, W, R)$, we get that $\widetilde{s}_{k,l}(t) = \mathbf{E}\left[G^k\right]\mathbf{E}\left[D^l\mathbf{E}\left[1(T^0 \geq t)|S\right]\right]$. Finally, using the same arguments as above and the properties of conditional expectation, we have

$$\begin{aligned}\widetilde{s}_{k,l}(t) &= \mathbf{E}\left[G^k\right]\mathbf{E}\left[D^l\mathbf{E}\left[1(T^0 \geq t)|S\right]\mu(G,t)\mathbf{E}\left[1(C \geq t)|G\right]\right] \\ &= \mathbf{E}\left[G^k\right]\mathbf{E}\left[D^l\mu(G,t)\mathbf{E}\left[Y(t)|G,S,W,R\right]\right] \\ &= \mathbf{E}\left[G^k\right]\widetilde{s}_l(t).\end{aligned}$$

Thus, under $H_0$,

$$\begin{aligned} B_{2,k} &= \sum_{l=1}^{L} \int_0^{\tau} \mathbf{E}\left[G^k\right] \kappa_l(t)\, dt \\ &= \int_0^{\tau} \mathbf{E}\left[G^k\right] \mathbf{E}[Y(t)\zeta(t)\mu(G,t)]\, dt. \end{aligned}$$

Using similar arguments, we get that $B_{1,k} = \int_0^{\tau} \mathbf{E}\left[G^k\right] \mathbf{E}[Y(t)\zeta(t)\mu(G,t)]\, dt$ under $H_0$ and thus, $\mathbf{E}[Q_{i,k}] = 0$.

We now focus on the proof of theorem 19.1.

PROOF OF THEOREM 19.1.– It follows from lemma 19.1, from the multivariate central limit theorem and from Slutsky's theorem that under $H_0$, $n^{-\frac{1}{2}}(\widetilde{Z}_1, \ldots, \widetilde{Z}_{K-1})'$ converges in distribution, as $n \to \infty$, to a $(K-1)$ dimensional Gaussian vector with mean 0 and covariance matrix $\Sigma = \mathbf{E}[\mathbf{Q}_1\mathbf{Q}_1']$, where $\mathbf{Q}_i = (Q_{i,1}, \ldots, Q_{i,K-1})'$. As a result, under $H_0$, $n^{-1}(\widetilde{Z}_1, \ldots, \widetilde{Z}_{K-1})\Sigma^{-1}(\widetilde{Z}_1, \ldots, \widetilde{Z}_{K-1})' \xrightarrow{d} \chi^2(K-1)$ as $n \to \infty$. $\Sigma$, however, involves several unknown expectations. A consistent estimator of $\Sigma$ is $n^{-1}\widehat{\Sigma} := n^{-1}\sum_{i=1}^{n} \mathbf{V}_i\mathbf{V}_i'$, where $\mathbf{V}_i = (V_{i,1}, \ldots, V_{i,K-1})'$ and $V_{i,k}$ is given by (19.3). To see this, since $n^{-1}\widehat{\Sigma} = n^{-1}\sum_{i=1}^{n}(\mathbf{V}_i\mathbf{V}_i' - \mathbf{Q}_i\mathbf{Q}_i') + n^{-1}\sum_{i=1}^{n}\mathbf{Q}_i\mathbf{Q}_i'$, it is sufficient to prove that $\mathbf{V}_{i,k} - \mathbf{Q}_{i,k} \xrightarrow{p} 0$ as $n \to \infty$. This follows from similar arguments and calculations as in the detailed proof of the lemma 19.2 (see the appendix). The details are, therefore, omitted. Finally, it follows from Slutsky's theorem that $\widetilde{U} := (\widetilde{Z}_1, \ldots, \widetilde{Z}_{K-1})\widehat{\Sigma}^{-1}(\widetilde{Z}_1, \ldots, \widetilde{Z}_{K-1})' \xrightarrow{d} \chi^2(K-1)$ as $n \to \infty$. This concludes the proof of theorem 19.1.

REMARK 19.1.– In practice, the weighting functions $\mu(G_i, \cdot)$ and/or the conditional probabilities $\mathbf{E}[1(S_i = l)|W_i]$ may either be known from previous studies or completely unknown. In the latter case, estimated functions and probabilities have to be substituted in $\widetilde{U}$.

## 19.4. Acknowledgment

A. Mezaouer is funded by a PhD grant from the Progamme National Exceptionnel (PNE, Algeria). Part of this work was conducted while A. Mezaouer was visiting J.-F. Dupuy at La Rochelle University (France), under the PNE program. J.-F. Dupuy acknowlegdes the financial support of the Université des Sciences et Technologie Houari Boumédienne (USTHB, Bab-Ezzouar, Algeria) and INSA of Rennes (France) for his visiting periods at USTHB.

## 19.5. Appendix

PROOF OF LEMMA 19.2.– Let $l \in \{1, \ldots, L\}$, and decompose

$$n^{\frac{1}{2}} \int_0^{\tau} v_{k,l}^{(n)}(t) \left[ \frac{1}{n} \sum_{i=1}^{n} \mu(G_i, t) D_i^l dN_i(t) - \kappa_l(t) dt \right]$$

as $C_{1,k,l}^{(n)} + C_{2,k,l}^{(n)}$, with

$$C_{1,k,l}^{(n)} := n^{\frac{1}{2}} \int_0^{\tau} v_{k,l}^{(n)}(t) \left[ \frac{1}{n} \sum_{i=1}^{n} \mu(G_i, t) D_i^l dN_i(t) - \frac{1}{n} \sum_{i=1}^{n} \mu(G_i, t) D_i^l Y_i(t) \zeta_i(t)\, dt \right]$$

$$C_{2,k,l}^{(n)} := \int_0^{\tau} v_{k,l}^{(n)}(t) \cdot n^{\frac{1}{2}} \left[ \frac{1}{n} \sum_{i=1}^{n} \mu(G_i, t) D_i^l Y_i(t) \zeta_i(t)\, dt - \kappa_l(t) dt \right].$$

In the following development, we first show that $C_{1,k,l}^{(n)} \xrightarrow{P} 0$ as $n \to \infty$. Note that $C_{1,k,l}^{(n)}$ is of the form

$$C_{1,k,l}^{(n)} = n^{-\frac{1}{2}} \sum_{i=1}^{n} \int_0^{\tau} H_{i,k,l}^{(n)}(t) dM_i(t),$$

where $M_i(t) = N_i(t) - \int_0^t Y_i(s)\zeta_i(s)ds$ and $H_{i,k,l}^{(n)}(t) := v_{k,l}^{(n)}(t)\mu(G_i, t) D_i^l$ is a predictable process with respect to $\mathcal{H}_t := \bigvee_{i=1}^n \mathcal{H}_{t,i}$. Moreover, $H_{i,k,l}^{(n)}(t)$ is bounded on $[0, \tau]$ since $|H_{i,k,l}^{(n)}(t)| \leq |v_{k,l}^{(n)}(t)| \cdot \frac{1}{c_0} \leq \frac{2}{c_0}$. Define the process $(C_{1,k,l}^{(n)}(t))_{t \geq 0}$ by

$$C_{1,k,l}^{(n)}(t) = n^{-\frac{1}{2}} \sum_{i=1}^{n} \int_0^{t} H_{i,k,l}^{(n)}(s) dM_i(s).$$

Then, $(C_{1,k,l}^{(n)}(t))_{t \geq 0}$ is an $\mathcal{H}_t$-martingale and $C_{1,k,l}^{(n)} := C_{1,k,l}^{(n)}(\tau)$. Now, the predictable variation process $< C_{1,k,l}^{(n)} > (t)$ of $C_{1,k,l}^{(n)}(t)$ is

$$< C_{1,k,l}^{(n)} > (t) = \int_0^t \frac{1}{n} \sum_{i=1}^{n} \left\{ H_{i,k,l}^{(n)}(s) \right\}^2 Y_i(s) \zeta_i(s) ds := \int_0^t X^{(n)}(s) ds.$$

Under the regularity conditions mentioned in section 19.2, it is not difficult to check that $X^{(n)}(s) \xrightarrow{p} 0$ as $n \to \infty$ for every $s \in [0, \tau]$. Moreover, for any $s \in [0, \tau]$ and $n \geq 1$,

$$\begin{aligned} |X^{(n)}(s)| &= \left| \frac{1}{n} \sum_{i=1}^{n} \left\{ H_{i,k,l}^{(n)}(s) \right\}^2 Y_i(s) \zeta_i(s) \right| \\ &\leq \frac{1}{n} \sum_{i=1}^{n} \left\{ H_{i,k,l}^{(n)}(s) \right\}^2 \zeta_i(s) \\ &\leq \frac{4}{c_0^2} c_1 \end{aligned}$$

hence by proposition II.5.3 of Andersen *et al.* [AND 93], $< C_{1,k,l}^{(n)} > (t) \xrightarrow{p} 0$ as $n \to \infty$. Next, for any $\epsilon > 0$ and $t \in [0, \tau]$, let

$$C_{1,k,l,\epsilon}^{(n)}(t) = n^{-\frac{1}{2}} \sum_{i=1}^{n} \int_0^t H_{i,k,l}^{(n)}(s) 1(n^{-\frac{1}{2}} |H_{i,k,l}^{(n)}(s)| \geq \epsilon) dM_i(s).$$

Then, the non-negative $< C_{1,k,l,\epsilon}^{(n)} > (t)$ satisfies the following

$$\begin{aligned} < C_{1,k,l,\epsilon}^{(n)} > (t) &= \int_0^t \frac{1}{n} \sum_{i=1}^{n} \left\{ H_{i,k,l}^{(n)}(s) \right\}^2 1(n^{-\frac{1}{2}} |H_{i,k,l}^{(n)}(s)| \geq \epsilon) Y_i(s) \zeta_i(s) ds \\ &\leq < C_{1,k,l}^{(n)} > (t), \end{aligned}$$

and therefore $< C_{1,k,l,\epsilon}^{(n)} > (t) \xrightarrow{p} 0$ as $n \to \infty$. It follows (see theorem 5.1.1 in [FLE 91]) that $C_{1,k,l}^{(n)}(t) \xrightarrow{d} 0$ as $n \to \infty$ for every $t \in [0, \tau]$. Therefore, $C_{1,k,l}^{(n)}(t) \xrightarrow{p} 0$ as $n \to \infty$ and, in particular, if $t = \tau$, $C_{1,k,l}^{(n)} \xrightarrow{p} 0$ as $n \to \infty$.

We now prove that $C_{2,k,l}^{(n)} \xrightarrow{p} 0$ as $n \to \infty$. To this end, we show that the integrand of $C_{2,k,l}^{(n)}$ converges to 0 uniformly in probability on $[0, \tau]$. We have

$$\begin{aligned} &\sup_{t \in [0,\tau]} \left| v_{k,l}^{(n)}(t) \cdot n^{\frac{1}{2}} \left( \frac{1}{n} \sum_{i=1}^{n} \mu(G_i, t) D_i^l Y_i(t) \zeta_i(t) - \kappa_l(t) \right) \right| \qquad [19.4] \\ &\quad \leq \sup_{t \in [0,\tau]} \left| v_{k,l}^{(n)}(t) \right| \cdot \sup_{t \in [0,\tau]} \left| n^{\frac{1}{2}} \left( \frac{1}{n} \sum_{i=1}^{n} \mu(G_i, t) D_i^l Y_i(t) \zeta_i(t) - \kappa_l(t) \right) \right|. \end{aligned}$$

Somewhat straightforward Glivenko–Cantelli arguments yield that $\widetilde{E}_{k,l}^{(n)}(t)$ is uniformly consistent (on $[0,\tau]$) for $\widetilde{e}_{k,l}(t)$. It follows that, as $n \to \infty$,

$$\sup_{t\in[0,\tau]} \left| v_{k,l}^{(n)}(t) \right| \xrightarrow{p} 0. \qquad [19.5]$$

Then, the classes $\{Y(t), t \in [0,\tau]\}$ and $\{\mu(G,t), t \in [0,\tau]\}$ are both Donsker (by lemma 4.1 in [KOS 08]), and thus so is $\{\mu(G,t)Y(t), t \in [0,\tau]\}$ since products of bounded Donsker classes are Donsker. Similarly, $\{\zeta(t), t \in [0,\tau]\}$ is Donsker (corollary 9.32 in [KOS 08]) and $\{D^l\}$ is Donsker. Finally, the class $\{\mu(G,t)D^lY(t)\zeta(t), t \in [0,\tau]\}$ is Donsker (as a product of bounded Donsker classes) and, therefore, the process $G_n(\cdot) := n^{\frac{1}{2}}(n^{-1}\sum_{i=1}^{n}\mu(G_i,\cdot)$ $D_i^lY_i(\cdot)\zeta_i(\cdot) - \kappa_l(\cdot))$ converges weakly to some mean zero Gaussian process $G$. By the continuous mapping theorem, $\sup_{t\in[0,\tau]}|G_n(t)|$ converges weakly to $\sup_{t\in[0,\tau]}|G(t)|$ and therefore $\sup_{t\in[0,\tau]}|G_n(t)| = O_p(1)$. Combining this result and result (19.5), in (19.4), yields that

$$\sup_{t\in[0,\tau]} \left| v_{k,l}^{(n)}(t) \cdot n^{\frac{1}{2}} \left( \frac{1}{n}\sum_{i=1}^{n} \mu(G_i,t)D_i^lY_i(t)\zeta_i(t) - \kappa_l(t) \right) \right| \xrightarrow{p} 0$$

as $n \to \infty$. This implies that $C_{2,k,l}^{(n)} \xrightarrow{p} 0$ as $n \to \infty$.

Hence, for any $l \in \{1,\ldots,L\}$, $C_{1,k,l}^{(n)} + C_{2,k,l}^{(n)} \xrightarrow{p} 0$ as $n \to \infty$, and thus $\sum_{l=1}^{L} C_{1,k,l}^{(n)} + C_{2,k,l}^{(n)} \xrightarrow{p} 0$. This concludes the proof of lemma 19.2.

PROOF OF LEMMA 19.3.– Decompose

$$n^{\frac{1}{2}} \sum_{l=1}^{L} \int_0^\tau \widetilde{E}_{k,l}^{(n)}(t)\kappa_l(t)dt$$

as

$$n^{\frac{1}{2}} \sum_{l=1}^{L} \int_0^\tau \left\{ \widetilde{e}_{k,l}(t) + \frac{\widetilde{S}_{k,l}^{(n)}(t)}{\widetilde{s}_l(t)} - \frac{\widetilde{s}_{k,l}(t)\widetilde{S}_l^{(n)}(t)}{\widetilde{s}_l(t)^2} \right\} \kappa_l(t)dt \qquad [19.6]$$

$$+ \sum_{l=1}^{L} \int_0^\tau \left\{ \frac{\widetilde{S}_{k,l}^{(n)}(t)}{\widetilde{s}_l(t)} - \frac{\widetilde{s}_{k,l}(t)\widetilde{S}_l^{(n)}(t)}{\widetilde{s}_l(t)^2} \right\} \cdot n^{\frac{1}{2}} \left( \frac{\widetilde{s}_l(t)}{\widetilde{S}_l^{(n)}(t)} - 1 \right) \kappa_l(t)dt.$$

The second term of [19.6] can be shown to converge to 0 in probability as $n \to \infty$. The arguments are similar to the arguments used in the proof of lemma 19.2 and are therefore omitted.

### 19.6. Bibliography


[AND 93] ANDERSEN P.K., BORGAN Ø., GILL R.D., *et al.*, *Statistical Models Based on Counting Processes*, Springer, New York, 1993.

[CAI 09] CAIN L.E., COLE S.R., "Inverse probability-of-censoring weights for the correction of time-varying noncompliance in the effect of randomized highly active antiretroviral therapy on incident AIDS or death", *Statistics in Medicine*, vol. 28, pp. 1725–1738, 2009.

[CAR 95] CARROLL R.J., RUPPERT D., STEFANSKI L.A., *Measurement Error in Nonlinear Models,* Monographs on Statistics and Applied Probability, vol. 63, Chapman & Hall, London, 1995.

[DIR 01] DIRIENZO A.G., LAGAKOS S.W., "Bias correction for score tests arising from misspecified proportional hazards regression models", *Biometrika,* vol. 88, pp. 421–434, 2001.

[DUP 08] DUPUY J.-F., LECONTE E., "Cox regression with missing values of a covariate having a non-proportional effect on hazard of failure", in HUBER C., LIMNIOS N., MESBAH M., NIKULIN M. (eds), *Mathematical Methods in Survival Analysis, Reliability and Quality of Life,* Applied Stochastic Methods Series, ISTE, London, John Wiley & Sons, New York, pp. 133–150, 2008.

[DUP 09] DUPUY J.-F., LECONTE E., "A study of regression calibration in a partially observed stratified Cox model", *Journal of Statistical Planning and Inference,* vol. 139, pp. 317–328, 2009.

[FLE 91] FLEMING T.R., HARRINGTON D.P., *Counting Processes and Survival Analysis,* John Wiley & Sons, New York, 1991.

[KLE 97] KLEIN J.P., MOESCHBERGER M.L., *Survival Analysis: Techniques for Censored and Truncated Data*, Springer, New York, 1997.

[KOS 08] KOSOROK M.R., *Introduction to Empirical Processes and Semiparametric Inference*, Springer, New York, 2008.

[LOA 99] LOADER C., *Local Regression and Likelihood.*, Springer, New York, 1999.

[MAR 06] MARTINUSSEN T., SCHEIKE T.H., *Dynamic Regression Models for Survival Data*, Springer, New York, 2006.

[ROB 00] ROBINS J.M., FINKELSTEIN D.M., "Correcting for noncompliance and dependent censoring in an AIDS Clinical Trial with inverse probability of censoring weighted (IPCW) log-rank tests", *Biometrics,* vol. 56, pp. 779–788, 2000.

[THU 03] THURSTON S.W., SPIEGELMAN D., RUPPERT D., "Equivalence of regression calibration methods in main study/external validation study designs", *Journal of Statistical Planning and Inference,* vol. 113, pp. 527–539, 2003.

[WEL 07] WELLER E.A., MILTON D.K., EISEN E.A., *et al.*, "Regression calibration for logistic regression with multiple surrogates for one exposure", *Journal of Statistical Planning and Inference,* vol. 137, pp. 449–461, 2007.

[YOS 07] YOSHIDA M., MATSUYAMA Y., OHASHI Y., "Estimation of treatment effect adjusting for dependent censoring using the IPCW method: an application to a large primary prevention study for coronary events (MEGA study)", *Clinical Trials,* vol. 4, pp. 318–328, 2007.

Chapter 20

# Group Sequential Tests for Treatment Effect with Covariates Adjustment through Simple Cross-Effect Models

Group sequential (GS) methods for analyzing censored survival data are reviewed toward parametric or semi-parametric implementation of a simple cross-effect (SCE) model proposed by Bagdonavičius *et al.* [BAG 04]. For a brief review, we sketch the frameworks of GS tests through parametric models, proportional hazards model and accelerated life models. With the SCE model, testing for treatment effect involves two parameters. We calculate two score processes corresponding to the two parameters based on the partial likelihood, and extend the parametric approach of Tsiatis *et al.* [TSI 95] to adapt to a GS test.

## 20.1. Introduction

In many clinical trials, patients are enrolled sequentially. For practical and ethical reasons, GS design is frequently used to compare different treatment arms. When the endpoint is the survival of patient, it is a common practice to test *treatment effect* through the log-rank test implemented at $K$ prespecified different times $t_1, \ldots, t_K$ (called interim analysis). The asymptotic joint distribution of the log-rank statistic calculated at the K analysis times is studied by Harrington *et al.* [HAR 82].

When there are covariates for adjustment, GS log-rank tests based on regression models have been studied extensively. For example, Tsiatis *et al.* [TSI 85] study the

Chapter written by Isaac Wu HONG-DAR.

sequentially computed score test derived from partial likelihood based on the proportional hazards (PHs) model [COX 72]. In their setting, if the treatment variable is independent of the covariates, the sequentially computed test in the presence of other concomitant variables has the same asymptotic joint distribution as the ordinary log-rank test. Specifically, Tsiatis *et al.* [TSI 85] show that a partial score process calculated with the true parameter being replaced by its consistent estimate can be approximated by a sum of independent and identically distributed random variables. By a suitable counting process setup, the martingale central limit theorem and Cramér–Wold device are applied to derive the joint distribution of the estimated score process calculated at the $K$ interim analysis times. This multivariate distribution forms the basis to construct a level-$\alpha$ test using either the methods of Pocock [POC 77], O'Brien and Fleming [O'BR 79] and Slud and Wei [SUD 82], or Lan and DeMet [LAN 83].

Despite the feasibility of assuming independence between treatment $(Z)$ and covariates $(W)$ in clinical trials due to randomization, it happens that $Z$ and $W$ are not independent because the treatment allocation is not made in a randomized manner. With a PH setting, Gu and Ying [GU 95] relax the independence assumption and establish the large sample results. They show when the score process can possess an independent increment property approximately if it is calculated at the parameter space of null hypothesis with suitable plug-in estimates (restricted maximum partial-likelihood estimates) of the parameters associated with $W$.

Although the Cox PH model, which is "semi-parametric", is popular for offering a basis to compare groups possibly adjusting for the confounding effect, Tsiatis *et al.* [TSI 95] consider a class of *parametric* models so that, in view of the Cox regression, the baseline hazard is modeled explicitly. Based on the full likelihood, they show that the sequentially calculated parametric score statistic can have an independent increment structure if suitable *filtration* is defined with respect to the staggered entry data structure. Asymptotic joint distribution is established whenever the evaluated scores are expressed as a linear combination of individual scores, which are martingales, corresponding to each parameter. The asymptotic $K$-dimensional normal distribution can be derived and the subsequent GS tests can also be constructed.

In [JEN 97], GS analysis is studied more generally in terms of different endpoints/responses, and traditional GS survival analysis is also investigated. They present a unified theoretical framework through the counting process model to explain the independent increment structure of the sequentially computed score statistics. The key step is to argue the orthogonality between two individual martingales associated with the observed counting process and the underlying model.

Alternative to the PH model, the accelerated failure time (AFT) model is a choice for modeling non-proportionality. Lin [LIN 92] derives a GS test of "no treatment effect" with covariate adjustment under the AFT model, also with independence assumption between treatment ($Z$) and covariates ($W$). His approach involves (1) to construct the log-rank score estimating functions as the basis of estimating the parameter of interest (corresponding to $Z$) and the nuisance parameters (corresponding to $W$) and (2) to establish the weak convergence of the score process to a K-dimensional zero-mean multivariate normal distribution. This leads to the subsequent construction of GS tests using the method of Slud and Wei [SLU 82] or others.

When there are non-proportional hazards, in two-sample problems, either the survivals separate quickly at an early stage of the study and then get close at a late stage, or the survivals cross each other at some time point (say $t_c$). Pepe and Fleming [PEP 89, PEP 91] propose a class of weighted statistics (denoted as $WKM(t)$), which accumulates the difference of survival estimates of the two groups. Their approach is appealing when the main interest is on the difference of survival probabilities. Li [LI 99a] extends the applications of Pepe and Fleming's $WKM(t)$-statistic to the GS test by deriving the joint asymptotic distribution of the $WKM(t)$-statistic calculated at $(t_1, \ldots, t_K)$. The approach involves approximating the difference between the estimated survival and the true survival for each group by a sum of independent and identically distributed random variables so that the $WKM(t)$-statistic can also be approximated by a sum of martingales, leading to a possibility of applying multivariate central limit theorem. Furthermore, when there is non-proportionality, Sooriyarachchi and Whitehead [SOO 98] proposed GS methods to compare the probabilities of survival for longer than some specified observational time (say $t_h$) between different subpopulations. Houwelingen *et al.* [HOU 05] study the bias issue when proportional hazards are not valid and propose methods to remedy the bias.

This chapter considers a class of non-proportional hazard models, called the simple cross-effect (SCE) model [BAG 04]. We investigate how a score-type test can be adopted in a GS manner to adjust for confounding effects from the covariates. Specifically, we consider the situation when non-proportional hazards are derived from the treatment itself, not from the covariates. Notations, assumptions and model setup are introduced in section 20.2. In section 20.3, the GS test procedure is described. Brief discussions of the proposed procedure are presented in section 20.4.

## 20.2. Notations and models

Consider the time setting of *stagger entry*; let $u_1, \ldots, u_n$ be the calendar times associated with $n$ patients entering the study. For individual $i$, we denote $T_i^*$ and $C_i$ as the true failure time and censoring time, *measured from entry*, such that

$$T_i(t) = \max\{\min(T_i^*, C_i, t - u_i), 0\}$$

is observed at *calendar time* $t$ and $\delta_i(t) = 1_{\{T_i^* < \min(C_i, t-u_i)\}}$. Let $Z_i$ be the treatment indicator and $W(t)$ be a $p$-dimensional *time-dependent* covariate. For ease of exposition, we write $W$ instead of $W(t)$. For individual $i$ at calendar time $t$ and the time from entry $u$, we denote the counting process $N_i(u,t)$ and at-risk process $Y_i(u,t)$ as

$$N_i(u,t) = 1_{\{T_i(t) \leq u, \delta_i(t)=1\}} \text{ and } Y_i(u,t) = 1_{\{T_i(t) \geq u\}}$$

for $0 < u < \tau$, with $\tau$ being a maximal observation time of all individuals. A flexible regression model useful for accommodating non-proportional-hazard specifies the following Cox-type multiplicative structure:

$$\begin{aligned}\lambda(u; Z, W) &= \lambda_0(u)\exp(\beta Z + \theta' W) \\ &\{1 + \Lambda_0(u)\exp(\beta Z + \theta' W + \gamma Z)\}^{-1+\exp(-\gamma Z)}, \end{aligned} \qquad [20.1]$$

where $\Lambda_0(u) = \int_0^u \lambda_0(s)ds$. In [20.1], the power "$-1 + \exp(-\gamma Z)$" models the heterogeneity effect and time-dependent effect simultaneously only through the treatment variable $Z$. The situation of including $W$ in the power is discussed in the last section. With model [20.1], the hypothesis of *no treatment effect* is $H_0 : \beta = \gamma = 0$.

Throughout the following context, we define $\mu_i = \exp(\beta Z_i), \sigma_i = \exp(\gamma Z_i)$, and $P_i(u) = \Lambda_0(u)\exp(\beta Z_i + \theta' W_i + \gamma Z_i)$ for subject $i$. Although the statistical inference can be based on the full likelihood, it is convenient to construct the *partial likelihood* for model [20.1] as

$$L_p(\eta, t) = \prod_{i: u_i \leq t} \left( \frac{g_i\{T_i, \eta, \Lambda_0(T_i)\}}{\sum_j Y_j(T_i, t) g_j\{T_i, \eta, \Lambda_0(T_i)\}} \right)^{\delta_i}, \qquad [20.2]$$

where $g_i\{u, \eta, \Lambda_0(u)\}$, or simply $g_i(u)$, equals $\mu_i \exp(\theta' W_i)\{1 + \Lambda_0(u)\mu_i\sigma_i \exp(\theta' W_i)\}^{-1+1/\sigma_i}$ with $\eta = (\beta, \gamma, \theta')'$. By taking partial derivatives of $\log(L_p)$ with respect to $\beta$, $\gamma$ and $\theta$, we have the following three score functions for estimating the parameters:

$$\begin{aligned}\bar{S}_\beta(t) &= \sum \int_{u=0}^{\tau} \left\{ V_i(u) - \frac{\sum Y_j(u,t) V_j(u) g_j(u)}{\sum Y_j(u,t) g_j(u)} \right\} dN_i(u,t), \\ \bar{S}_\gamma(t) &= \sum \int_{u=0}^{\tau} \left\{ \zeta_i(u) - \frac{\sum Y_j(u,t) \zeta_j(u) g_j(u)}{\sum Y_j(u,t) g_j(u)} \right\} dN_i(u,t), \\ \bar{S}_\theta(t) &= \sum \int_{u=0}^{\tau} \left\{ U_i(u) - \frac{\sum Y_j(u,t) U_j(u) g_j(u)}{\sum Y_j(u,t) g_j(u)} \right\} dN_i(u,t). \end{aligned} \qquad [20.3]$$

In [20.3], $V_i(u) = Z_i\{\sigma_i + P_i(u)\}/\sigma_i\{1 + P_i(u)\}$, $U_i(u) = W_i\{\sigma_i + P_i(u)\}/\sigma_i\{1 + P_i(u)\}$ and $\zeta_i(u) = (Z_i/\sigma_i)\{-\log(1 + P_i(u)) + (1 - \sigma_i)P_i(u)/(1 + P_i(u))\}$. Note that $U(u)$ is a vector because $W$ is a vector. Under $H_0 : \beta = \gamma = 0$, we have $\sigma_i = 1, V_i = Z_i, U_i = W_i, P_i(u) = \Lambda_0(u)\exp(\theta' W_i)$, and $g_i(u)$ reduces to $\exp(\theta' W_i)$. With these simplifications, [20.3] reduces to

$$S_\beta(t) = \sum \int \left\{ V_i(u) - \frac{\sum Y_j(u,t) Z_j e^{\theta' W_j}}{\sum Y_j(u,t) e^{\theta' W_j}} \right\} dN_i(u,t),$$

$$S_\gamma(t) = \sum \int \left\{ \zeta_i^{(0)}(u) - \frac{\sum Y_j(u,t) \zeta_j^{(0)}(u) e^{\theta' W_j}}{\sum Y_j(u,t) e^{\theta' W_j}} \right\} dN_i(u,t),$$

$$S_\theta(t) = \sum \int \left\{ W_i(u) - \frac{\sum Y_j(u,t) W_j(u) e^{\theta' W_j}}{\sum Y_j(u,t) e^{\theta' W_j}} \right\} dN_i(u,t), \qquad [20.4]$$

where $\zeta_i^{(0)}(u) = -Z_i \log\{1 + \Lambda_0(u)\exp(\theta' W_i)\}$.

To construct score tests, we need to calculate the second-order derivatives of the log-likelihood ($\log(L_p)$). Here, let us express $(-1/n)\partial^2 S_\beta(t)/\partial\beta\partial\theta$ for illustration. The other terms can be calculated in a similar manner and we only present the final results. By taking the partial derivative of $S_\beta(t)$ with respect to $\theta$, we have

$$\begin{aligned} \frac{-1}{n}\frac{\partial S_\beta}{\partial\theta} = & \frac{-1}{n}\sum \int \left\{ \dot{V}_i - \frac{\sum Y_j \dot{V}_j g_j}{\sum Y_j g_j} \right\} dN_i(u,t) \\ + & \frac{1}{n}\sum \int \left\{ \frac{\sum Y_j V_j \dot{g}_j}{\sum Y_j g_j} - \frac{\sum Y_j V_j g_j \cdot \sum Y_j \dot{g}_j}{(\sum Y_j g_j)^2} \right\} dN_i(u,t), \end{aligned}$$

where, for a function $h$, $\dot{h} = \partial h/\partial\theta$. The first term on the right-hand side of the equation converges to 0 in probability. So $(-1/n)\partial S_\beta/\partial\theta$ has the approximation:

$$\frac{-1}{n}\frac{\partial S_\beta}{\partial\theta} \approx \frac{1}{n}\sum \int \left\{ \frac{\sum Y_j Z_j W_j e^{\theta' W_j}}{\sum Y_j e^{\theta' W_j}} - \frac{\sum Y_j Z_j e^{\theta' W_j} \cdot \sum Y_j W_j e^{\theta' W_j}}{(\sum Y_j e^{\theta' W_j})^2} \right\} dN_i(u,t).$$

By the law of large numbers, the above expression converges to

$$\int_{u=0}^{t} (e_{wz}(u,t) - \frac{e_w(u,t) e_z(u,t)}{e_1(u,t)}) d\Lambda_0(u) \equiv \imath_{\theta\beta}(t) = \imath'_{\beta\theta}(t),$$

where $e_V(u,t) = \lim(1/n)\sum Y_i(u,t)V_i(u)\exp(\theta' W_i(u))$ for a predictable process $V(u)$ that may depend on covariates [LI 99b]. The other second-derivative terms include $\partial S_\beta/\partial\beta(= \partial^2 \log(L_p)/\partial\beta^2)$, $\partial S_\beta/\partial\gamma(= \partial S_\gamma/\partial\beta)$, $\partial S_\gamma/\partial\gamma$, $\partial S_\gamma/\partial\theta(= \partial S_\theta/\partial\gamma)$, and $\partial S_\theta/\partial\theta$. The observed information matrix $\mathcal{I}(\eta)$ is

$$\mathcal{I}(\eta) = \frac{-1}{n}\begin{pmatrix} \partial S_\beta/\partial\beta & \partial S_\beta/\partial\gamma & \partial S_\beta/\partial\theta \\ \partial S_\gamma/\partial\beta & \partial S_\gamma/\partial\gamma & \partial S_\gamma/\partial\theta \\ \partial S_\theta/\partial\beta & \partial S_\theta/\partial\gamma & \partial S_\theta/\partial\theta \end{pmatrix} \equiv \begin{pmatrix} \mathcal{I}_{\beta\beta} & \mathcal{I}_{\beta\gamma} & \mathcal{I}_{\beta\theta} \\ \mathcal{I}_{\gamma\beta} & \mathcal{I}_{\gamma\gamma} & \mathcal{I}_{\gamma\theta} \\ \mathcal{I}_{\theta\beta} & \mathcal{I}_{\theta\gamma} & \mathcal{I}_{\theta\theta} \end{pmatrix}. \quad [20.5]$$

Under $H_0$, these terms have the following expressions:

$$\mathcal{I}_{\beta\beta} = \frac{1}{n}\sum\int\left\{\frac{\sum Y_j Z_j^2 e^{\theta' W_j}}{\sum Y_j e^{\theta' W_j}} - \left(\frac{\sum Y_j Z_j e^{\theta' W_j}}{\sum Y_j e^{\theta' W_j}}\right)^2\right\} dN_i(u,t),$$

$$\mathcal{I}_{\beta\gamma} = \frac{1}{n}\sum\int\left\{\frac{\sum Y_j Z_j \zeta_j^{(0)} e^{\theta' W_j}}{\sum Y_j e^{\theta' W_j}} - \frac{\sum Y_j Z_j e^{\theta' W_j}\cdot\sum Y_j \zeta_j^{(0)} e^{\theta' W_j}}{(\sum Y_j e^{\theta' W_j})^2}\right\} dN_i(u,t),$$

$$\mathcal{I}_{\gamma\gamma} = \frac{1}{n}\sum\int\left\{\frac{\sum Y_j {\zeta_j^{(0)}}^2 e^{\theta' W_j}}{\sum Y_j e^{\theta' W_j}} - \left(\frac{\sum Y_j \zeta_j^{(0)} e^{\theta' W_j}}{\sum Y_j e^{\theta' W_j}}\right)^2\right\} dN_i(u,t),$$

$$\mathcal{I}_{\gamma\theta} = \frac{1}{n}\sum\int\left\{\frac{\sum Y_j \zeta_j^{(0)} W_j e^{\theta' W_j}}{\sum Y_j e^{\theta' W_j}} - \frac{\sum Y_j \zeta_j^{(0)} e^{\theta' W_j}\cdot\sum Y_j W_j e^{\theta' W_j}}{(\sum Y_j e^{\theta' W_j})^2}\right\} dN_i(u,t),$$

$$\mathcal{I}_{\theta\theta} = \frac{1}{n}\sum\int\left\{\frac{\sum Y_j W_j^2 e^{\theta' W_j}}{\sum Y_j c^{\theta' W_j}} - \left(\frac{\sum Y_j W_j e^{\theta' W_j}}{\sum Y_j e^{\theta' W_j}}\right)^2\right\} dN_i(u,t), \quad [20.6]$$

which, also by the law of large numbers, converge in probability to $\imath_{\beta\beta}(t) = \int_{u=0}^t (e_{zz} - e_z^2/e_1)d\Lambda_0(u)$, $\imath_{\beta\gamma}(t) = \int_{u=0}^t (e_{z\zeta^{(0)}} - e_z e_{\zeta^{(0)}}/e_1)d\Lambda_0(u)$, $\imath_{\gamma\gamma}(t) = \int_{u=0}^t (e_{\zeta^{(0)}\zeta^{(0)}} - e^2_{\zeta^{(0)}}/e_1)d\Lambda_0(u)$, $\imath_{\gamma\theta}(t) = \int_{u=0}^t (e_{\zeta^{(0)}w} - e_{\zeta^{(0)}}e_w/e_1)$ $d\Lambda_0(u)$, and $\imath_{\theta\theta}(t) = \int_{u=0}^t (e_{ww} - e_w^2/e_1)d\Lambda_0(u)$, respectively.

## 20.3. Group sequential test

For the prespecified GS times $t_1, \ldots, t_K$, denote the sequential estimates of $\eta = (\beta, \gamma, \theta')'$ by $\eta_{t_k} = (\beta_{t_k}, \gamma_{t_k}, \theta'_{t_k})'$, $k = 1, \ldots, K$. For $\Lambda_0(u)$ or $\lambda_0(u)$, if it is modeled by a parametric distribution, extra parameters are then added to $\eta$. In general, $\Lambda_0(u)$ is non-parametric and we can use the Breslow-type estimate:

$$\widehat{\Lambda}_0(u) = \sum_i \int_{s=0}^{T} \frac{dN_i(s,u)}{\sum_j Y_j(s,u) g_i(s,\eta,\Lambda_0(s))},$$

which should be solved iteratively. In view of the observed information matrix of [20.5], hereafter let us denote $\mathcal{I}_{\beta\bar{\beta}} = \mathcal{I}'_{\bar{\beta}\beta} = (\mathcal{I}_{\beta\gamma}, \mathcal{I}_{\beta\theta}), \mathcal{I}_{\gamma\bar{\gamma}} = \mathcal{I}'_{\bar{\gamma}\gamma} = (\mathcal{I}_{\gamma\beta}, \mathcal{I}_{\gamma\theta})$,

$$\mathcal{I}_{\bar{\beta}\bar{\beta}} = \begin{pmatrix} \mathcal{I}_{\gamma\gamma} & \mathcal{I}_{\gamma\theta} \\ \mathcal{I}_{\theta\gamma} & \mathcal{I}_{\theta\theta} \end{pmatrix} \text{ and } \mathcal{I}_{\bar{\gamma}\bar{\gamma}} = \begin{pmatrix} \mathcal{I}_{\beta\beta} & \mathcal{I}_{\beta\theta} \\ \mathcal{I}_{\theta\beta} & \mathcal{I}_{\theta\theta} \end{pmatrix}.$$

With respect to a suitable filtration, the score process $(1/\sqrt{n})(S_\beta(t_k), S_\gamma(t_k), S_\theta(t_k)')'$ has a joint multivariate normal distribution with mean zero. To express the covariance, let $(\beta,\gamma,\theta')' \equiv (\eta_1,\eta_2,\ldots,\eta_{p+2})$ since $\beta$ and $\gamma$ are both one-dimensional and dim($\theta$)=p. The covariance between $(1/\sqrt{n})(S_{\eta_i}(t_k)$ and $(1/\sqrt{n})(S_{\eta_j}(t_l)(t_k < t_l)$ is $\mathcal{I}_{ij}(t_k)$ for $i,j = 1,\ldots,p+2; k,l = 1,\ldots,K$, and this $\mathcal{I}_{ij}$ is simply a component of [20.5] evaluated at the null hypothesis. By simple algebraic manipulations, we have

$$U_\beta(t) = S_\beta(t) - \mathcal{I}_{\beta\bar{\beta}}\mathcal{I}^{-1}_{\bar{\beta}\bar{\beta}}\left(S_\gamma(t), S_\theta(t)\right)' \text{ and}$$

$$U_\gamma(t) = S_\gamma(t) - \mathcal{I}_{\gamma\bar{\gamma}}\mathcal{I}^{-1}_{\bar{\gamma}\bar{\gamma}}\left(S_\beta(t), S_\theta(t)\right)'$$

which are sums of independent and identically distributed random variables because they ($U_\beta(t)$ and $U_\gamma(t)$) appear to be simple linear combinations of $S_\beta(t)$, $S_\gamma(t)$, and/or $S_\theta(t)$. Moreover, it can be shown that $(S_\beta(t), S_\gamma(t))'$ and $(U_\beta(t), U_\gamma(t))'$ are asymptotically equivalent [TSI 95]. This means that $S_\beta(t)$ and $S_\gamma(t)$ calculated at $t_1,\ldots,t_K$ have an asymptotic joint multivariate normal distribution with zero means. The covariance of $((1/\sqrt{n})S_\beta(t_i)$ and $(1/\sqrt{n})(S_\beta(t_j)(t_i < t_j)$ is $\mathcal{I}_{\beta\beta}(t_i) - \mathcal{I}_{\beta\bar{\beta}}(t_i)\mathcal{I}^{-1}_{\bar{\beta}\bar{\beta}}(t_i)\mathcal{I}_{\bar{\beta}\beta}(t_i) \equiv \Sigma_{\beta,i}$, and the covariance of $((1/\sqrt{n})S_\gamma(t_i)$ and $(1/\sqrt{n})(S_\gamma(t_j)$ is $\mathcal{I}_{\gamma\gamma}(t_i) - \mathcal{I}_{\gamma\bar{\gamma}}(t_i)\mathcal{I}^{-1}_{\bar{\gamma}\bar{\gamma}}(t_i)\mathcal{I}_{\bar{\gamma}\gamma}(t_i) \equiv \Sigma_{\gamma,i}$. Let

$$Q_{\beta,k} = \frac{1}{\sqrt{n}} S_\beta(t_k; \widehat{\theta}^{(0)}_{t_k})\widehat{\Sigma}^{-1/2}_{\beta,k} \text{ and } Q_{\gamma,k} = \frac{1}{\sqrt{n}} S_\gamma(t_k; \widehat{\theta}^{(0)}_{t_k})\widehat{\Sigma}^{-1/2}_{\gamma,k},$$

where $\widehat{\theta}^{(0)}_{t_k}$ is the *restricted estimate* of $\theta$ at $t_k$ under $H_0$, and $\widehat{\Sigma}_{\beta,k}$ and $\widehat{\Sigma}_{\gamma,k}$ are obtained by substituting corresponding restricted estimates into $\Sigma_{\beta,k}$ and $\Sigma_{\gamma,k}$. The GS test proposed in this chapter is:

$$Pr\{(|Q_{\beta,1}| < c_{\beta,1}, |Q_{\gamma,1}| < c_{\gamma,1}),\ldots,|Q_{\beta,k-1}| < c_{\beta,k-1}, |Q_{\gamma,k-1}| < c_{\gamma,k-1}),$$
$$(|Q_{\beta,k}| > c_{\beta,k} \text{ or } |Q_{\gamma,k}| > c_{\gamma,k})\} = \alpha_k, k = 1,\ldots,K; \sum \alpha_k = \alpha, \quad [20.7]$$

with the boundaries $c_{\beta,k}$ and $c_{\gamma,k}$ being numerically determined by the aforementioned multivariate joint distribution. We may specify $c_{\beta,k} = c_{\gamma,k} = c_k$ to avoid complexity. Numerical calculations of the boundaries $c_{\beta,k}$ and $c_{\gamma,k}$ (or simply $c_k$) need extended computer programming. The choice of $(\alpha_1, \ldots, \alpha_K)$ can be referred to [POC 77], [O'BR 79], [LAN 83] and [JEN 00]. The boundary values can be determined by various methods discussed in [JEN 00].

## 20.4. Discussion

The large sample properties stated in the previous context ignore the effect of the cumulative baseline hazard $\Lambda_0(t)$ when it is replaced by a Breslow-type estimator. It may possibly require some mathematical work to investigate the large sample properties.

The entire framework proposed in this chapter is a straightforward application of Tsiatis *et al.*'s [TSI 95] results, except that the proposed GS test accommodates a two-dimensional extension and that the baseline hazard should be explicitly modeled or substituted by its estimate. Since the GS test involves two parameters simultaneously, explicit computation of the probability of [20.7] depends on the joint distribution of the score processes calculated at the $K$ interim analysis.

The SCE model considered in this study is slightly different from that proposed in [BAG 04]. The reason for this is that, from the viewpoint of hypothesis testing, if the power component appearing in [BAG 04] contains parameter corresponding to $W$, the impact of this "heterogeneity effect" will be absorbed into the component of $\exp(\theta' W)$ appearing in the ordinary "risk function" component. Hence, although there could be bias in the estimation of $\beta$ and $\gamma$, the "bias" of power of the statistical test can be small.

## 20.5. Acknowledgment

This study was supported by the National Science Council of Taiwan under grant no. NSC 100-2118-M-005-003.

## 20.6. Bibliography

[BAG 04] BAGDONAVIČIUS V., HAFDI M.A., NIKULIN M., "Analysis of survival data with cross-effects of survival functions", *Biostatistics*, vol. 5, pp. 415–425, 2004.

[COX 72] COX D.R., "Regression models with life tables (with discussion)", *Journal of The Royal Statistical Society: Series B*, vol. 34, pp. 187–220, 1972.

[GU 95] GU M., YING Z., "Group sequential methods for survival data using partial likelihood score processes with covariate adjustment", *Statistica Sinica*, vol. 5, pp. 793–804, 1995.

[HAR 82] HARRINGTON D.P., FLEMING T.R., GREEN S.J., "Procedures for serial testing in censored survival data", in CROWLEY J., JOHNSON R.A. (eds), *Survival Analysis*, IMS Monograph Series, Hayward, California, pp. 269–286, 1982.

[JEN 97] JENNISON C., TURNBULL B.W., "Group-sequential analysis incorporating covariate information", *Journal of the American Statistical Association,* vol. 92, pp. 1330–1341, 1997.

[JEN 00] JENNISON C., TURNBULL B.W., *Group Sequential Methods with Applications to Clinical Trials*, Chapman and Hall/CRC, 2000.

[LAN 83] LAN K.K.G., DEMET D.L., "Discrete sequential boundaries for clinical trials", *Biometrika*, vol. 70, pp. 165–173, 1983.

[LI 99a] LI Z., "A group sequential test for survival trials: an alternative to rank-based procedures", *Biometrics*, vol. 55, pp. 277–283, 1999.

[LI 99b] LI Z., "Group sequential checking of the proportional hazards assumption based on a time-dependent covariate", *Communications in Statistics – Theory and Methods*, vol. 28, pp. 1931–1943, 1999b.

[LIN 92] LIN D.Y., "Sequential log rank tests adjusting for covariates with the accelerated life model", *Biometrika*, vol. 79, pp. 523–529, 1992.

[O'BR 79] O'BRIEN P.C., FLEMING T.R. "A multiple testing procedure for clinical trials", *Biometrics*, vol. 35, pp. 549–556, 1979.

[PEP 89] PEPE M.S., FLEMING T.R., "Weighted Kaplan-Meier statistics: a class of distance tests for censored survival data", *Biometrics*, vol. 45, pp. 497–507, 1989.

[PEP 91] PEPE M.S., FLEMING T.R., "Weighted Kaplan-Meier statistics: large sample and optimality considerations", *Journal of The Royal Statistical Society: Series B*, vol. 53, pp. 341–352, 1991.

[POC 77] POCOCK S.J., "Group sequential methods in the design and analysis of clinical trials", *Biometrika*, vol. 64, pp. 191–199, 1977.

[SLU 82] SLUD E.V., WEI L.J., "Two-sample repeated significance tests based on the modified Wilcoxon statistic", *Journal of the American Statistical Association*, vol. 77, pp. 862–868, 1982.

[SOO 98] SOORIYARACHCHI M.R., WHITEHEAD J., "A method for sequential analysis of survival data with nonproportional hazards", *Biometrics*, vol. 54, pp. 1072–1084, 1998.

[TSI 85] TSIATIS A.A., ROSNER G.L., TRITCHLER D.L., "Group sequential tests with censored survival data adjusting for covariates", *Biometrika*, vol. 72, pp. 365–373, 1985.

[TSI 95] TSIATIS A.A., BOUCHER H., KIM K., "Sequential methods for parametric survival models", *Biometrika*, vol. 82, pp. 165–173, 1995.

[VAN 05] VAN HOUWELINGEN H.C., VAN DE VELDE C.J.H., STIJNEN T., "Interim analysis on survival data: its potential bias and how to repair it", *Statistics in Medicine*, vol. 24, pp. 2823–2835, 2005.

# Part 3

# Reliability and Maintenance

# Chapter 21

# Optimal Maintenance in Degradation Processes

In this chapter, we consider a simple degradation model based on a Wiener process with drift. A failure occurs when the degradation reaches a given level $h$ first time. For preventive maintenance, inspections of the degradation are regularly carried out. If at inspection time the degradation is larger than a predefined level $a$, then the damage level is decreased. In this chapter, we consider the problem of defining cost optimal maintenance policies. Some first ideas about sequential techniques and incomplete maintenance are given.

## 21.1. Introduction

For investigating the lifetime of products or biological organisms that are affected by degradation, damage, stress or shocks, it is often necessary to consider the development of a stochastic process that models aging, wearing, damage accumulation and other changes of the state. In the following, we will call such processes *degradation models*.

The applicability of a purely lifetime-based reliability analysis is limited for several reasons. Collecting a sample of lifetime data is often expensive. In applications from engineering, observed lifetimes are frequently very long. Furthermore, there are many products for which failure causes serious difficulties. When designing the mathematical model of a degradation process, the following steps have to be taken:

1) Starting from basic technical considerations, a suitable mathematical model for the degradation measure process $Z(t)$ has to be defined. For instance, the hypothesis of

---

Chapter written by Waltraud KAHLE.

additive degradation accumulation may be used. In this case, we may select the class of homogeneous processes with independent increments as a suitable model class. Another possibility is the description of degradations by shock models.

2) By applying the level exceeding theory, it is possible to find the type of lifetime distribution, i.e. the distribution of the time at which the stochastic process first exceeds the degradation level.

3) The parameters of the selected model have to be estimated by means of samples related to the degradation measure $Z(t)$.

4) Estimations of the parameters of lifetime distribution can be obtained from the estimated parameters of the model $Z(t)$, if the limit level of degradation is known.

As an example, we can model the degradation $Z(t)$ by the linear path model

$$Z(t) = t/A \quad \text{where } A \text{ is a positive random variable}$$

(see [MEE 88]). A generalization and statistical analysis by non-parametric methods of estimation from joint linear degradation and failure time data with partial renewals and competing failure modes is given in [BAG 07].

Another widely used model is the gamma degradation process: a continuous-time stochastic process with independent and gamma distributed increments. A detailed description and many further references are given in [VAN 09] and [BAG 01].

In section 21.2, we describe in more detail continuous degradation models based on a Wiener process with drift.

## 21.2. The degradation model

The choice of the mathematical model for these degradation processes is based on the assumption of an additive accumulation of degradation with constant wear intensity. Regarding every degradation increment as an additive superposition of a large number of small effects, we can assume the degradation process to be normally distributed. Therefore, the degradation measure $Z(t)$ can be described by the following model [KAH 94]:

$$Z(t) = z_0 + \sigma W(t - t_0) + \mu \cdot (t - t_0) \, , \quad t \geq t_0 \qquad [21.1]$$

with
- $z_0$ – constant initial degradation ($z_0 \in \mathbf{R}$);
- $t_0$ – beginning of the degradation ($t_0 \in \mathbf{R}$);
- $\mu$ – drift parameter ($\mu \in \mathbf{R}$);
- $\sigma$ – variance parameter ($\sigma > 0$);
- $W(t)$ – standard Wiener process on $[0, \infty)$.

We suppose that a failure of a product will occur if the degradation process arrives at a certain boundary degradation level. For a given boundary level $h$, the lifetime $T_h$ of the product is then determined as the instant at which the degradation process $Z(t)$ exceeds the level $h$ for the first time:

$$T_h = \inf\{t \geq t_0 : Z(t) \geq h\} \,. \qquad [21.2]$$

It is well known that for $z_0 < h$, the lifetime $T_h$ follows an inverse Gaussian distribution with the Lebesgue density function

$$f_{T_h}(t) = \frac{h - z_0}{\sqrt{2\pi\sigma^2(t - t_0)^3}} \exp\left( -\frac{(h - z_0 - \mu(t - t_0))^2}{2\sigma^2(t - t_0)} \right) I_{\{t>t_0\}}, \qquad [21.3]$$

where $I_{\{\cdot\}}$ denotes an indicator variable, assuming the value 1 if the relation in brackets is true, and the value 0 otherwise. Obviously, $\mathrm{P}^{T_h}$ is the Dirac measure $\varepsilon_{t_0}$ concentrated at $t_0$ if $z_0 \geq h$.

Usually, another parametrization of the inverse Gaussian distribution is used. The connection between the process parameters above and parameters $\lambda$, $\mu_1$, as an example in [SES 99], is

$$\lambda = \frac{(h - z_0)^2}{\sigma^2} \quad \text{and} \quad \mu = \frac{h - z_0}{\mu} \,.$$

Several generalizations of this simple model have been given. It is possible to include measurement errors [WHI 95], or to transform the time scale [WHI 97]. It is also possible to consider bivariate models [WHI 98]. A similar model and its application in medicine is described in [DOK 96].

The advantages of using the Wiener process and its generalizations for describing the degradation process are its simple form (at least for the univariate Wiener process), and second that a statistical analysis can be carried out for observations at any discrete time point. The parameters of this distribution coincide with those of the degradation process. Note that the degradation level $h$ can be an unknown parameter. To estimate the parameters of this model, several observations schemes are possible.

1) The increments of the degradation process are observed: in this case, the likelihood function of the observation is a product of Gaussian densities. It is possible to estimate the parameters of the degradation process $\mu$, $\sigma^2$, $x_0$ and $t_0$, but it is impossible to estimate the degradation level $h$. Note that neither the same distance between observation points nor observation of all realizations at the same time points is required. Parameter estimations can be found if the observations are taken at individual time points in each realization.

2) The failure times are observed: in this case, we get the classical likelihood function from a sample of inverse Gaussian distributed random variables. From this likelihood function, we can estimate the parameters $\mu$, $\sigma^2$ and $t_0$. The degradation reserve $h - z_0$ must be given in advance.

3) The sample includes both process increments and failure times: assume that the sample consists of $N$ failure times and $\sum_{i=1}^{n} m_i$ increments, being mutually independent. Then, the likelihood function is the product of the likelihood functions mentioned above. In the case that failure times and process increments belong to the same realization of the degradation process, the likelihood function is more complicated. In this case, the density of an observed process increment is a conditional density under the assumption that the process does not cross the degradation level in the current time interval. If the sample includes both process increments and failure times, then it is possible to estimate all parameters of the model.

These models and their parameter estimation are described in [KAH 94] in [KAH 98].

In many situations, especially if a failure causes high costs or serious difficulties, a preventive maintenance action can be carried out to reduce the degradation level. In the following sections, we consider some possible optimal maintenance policies.

## 21.3. Optimal replacement after an inspection

First, we consider the following situation:

– The degradation level is observed at equidistant time points $\delta$, $2\delta$, .... We will call such an observation an *inspection*. Each inspection causes costs $c_I$.

– If at inspection the degradation level is larger than a predefined level $a < h$, a preventive renewal is carried out. The costs of such preventive maintenance are $c_M$.

– It is possible that a failure occurs between two inspections, if at some inspection time $k \cot \delta$ the degradation level is smaller than $a$, the process passes the limit level $h$ for first time before the next inspection at time $(k+1) \cot \delta$. We denote the costs of a failure as $c_F$.

The three possible situations are illustrated in Figure 21.1.

Our aim is to find both the optimal distance $\delta$ between inspections and the optimal "maintenance level" $a$. For this, we consider the average costs per time unit:

$$C = \frac{E\,(j * c_I + c_M I_M + c_F(1 - I_M))}{\text{expected length of a cycle}} \qquad [21.4]$$

where:

– $j$ is the number of inspections before either preventive maintenance is carried out or a failure occurs;

– $I_M = 1$ if the cycle ends by maintenance, and $I_M = 0$ if the cycle ends by a failure.

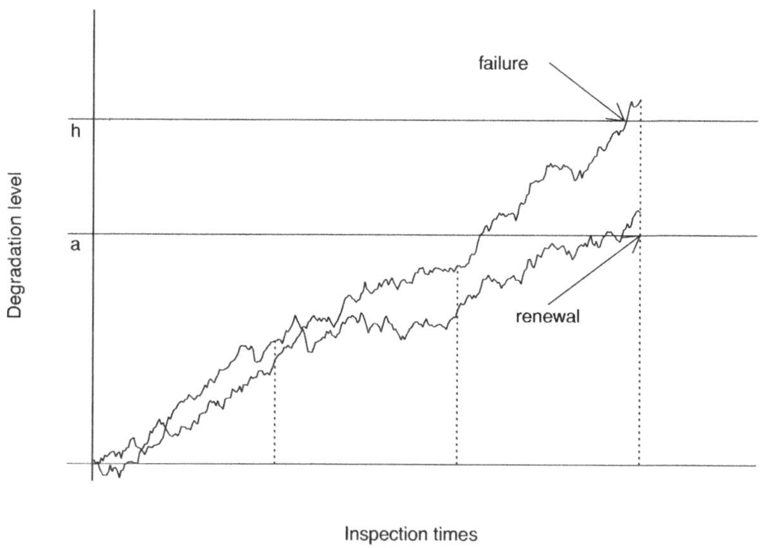


**Figure 21.1.** *Preventive maintenance*

The function $C$ is quite complicated. The following results are based on simulated degradation paths.

## 21.4. The simulation of degradation processes

Let us assume that the process has not exceeded the boundary level before the inspection point $k \cdot \delta$. At this time, the actual degradation level $Z(k \cdot \delta) = z_k$ is observed. There are two possibilities within the next time period of length $\delta$:

1) The system fails within this time interval. The time to failure is inverse Gaussian distributed, with parameters

$$\lambda = \frac{(h - z_k)^2}{\sigma^2} \quad \text{and} \quad \mu_1 = \frac{h - z_k}{\mu}. \qquad [21.5]$$

Therefore, the probability of a failure is given by the distribution function of the inverse Gaussian distribution with above parameters:

$$p_f = P(T_h \leq \delta) = \Phi\left(\sqrt{\frac{\lambda}{\delta}}\,(\frac{\delta}{\mu_1} - 1)\right) + \exp(\frac{2\lambda}{\mu_1})\,\Phi\left(-\sqrt{\frac{\lambda}{\delta}}\,(\frac{\delta}{\mu_1} + 1)\right) \qquad [21.6]$$

where $\Phi(\cdot)$ is the distribution function of the standard normal distribution.

2) With probability $1 - p_f$ the system does not fail within the interval to next inspection. But now, the random increment of the degradation process is not normally distributed. We observe the degradation measure at $z_{k+1}$ at time $(k+1) \cdot \delta$ under the condition that the process has not yet exceeded the level $h$ until the time $k+1) \cdot \delta$. In [KAH 98], it was found that this conditional density is given by

$$f(z) = \frac{1}{\sigma\sqrt{\delta}}\,\phi\left(\frac{z - z_k - \mu \cdot \delta}{\sigma\sqrt{\delta}}\right) \times \left[1 - \exp\left(-\frac{2(h - z_k)(h - z)}{\sigma^2(\delta)}\right)\right] I_{\{z \leq h\}}, \qquad [21.7]$$

where $\phi$ is the density of the standard normal distribution. Some such densities for different values of $z_k$ and $\mu = 5$, $\sigma^2 = 1$, $\delta = 1$, $h = 20$ are shown in Figure 21.2.

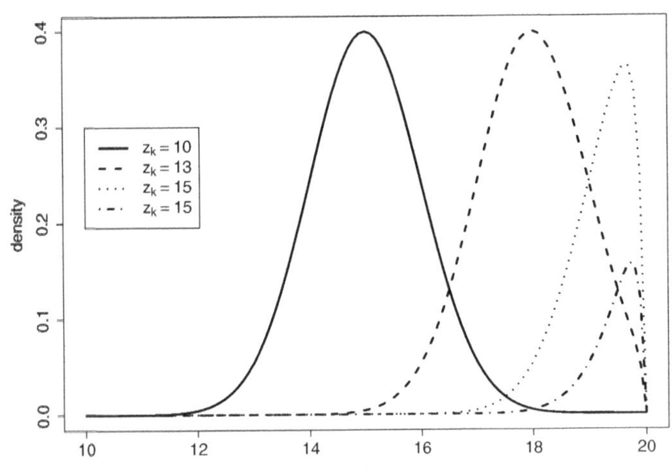


**Figure 21.2.** *Conditional densities of process increments*

If the actual degradation level is far away from the boundary level $h$, the conditional density is quite similar to a normal density. It becomes skewed if $z_k$ is relatively close to $h$. Note that these densities do not integrate to 1. The area under the density function is the probability $1 - p_f$ of not crossing the boundary level.

For simulating one path of a degradation process, we use the following algorithm:

1) The parameters of the process $\mu$, $\sigma^2$, $h$ as well as duration time between two inspections $\delta$ are defined. The initial parameters $x_0$ and $t_0$ do not influence the further maintenance, therefore we let the process begin at time 0 in state 0.

2) A uniformly at [0,1] distributed random number $zz$ is simulated. If $zz > p_p$, we get a failure at time $t_h$ that is the $zz$-quantile of the inverse Gaussian distribution. The process stops at failure time.

3) Otherwise, a process increment using density [21.7] is simulated. Here, the rejection method can be used. After getting a new degradation level, the next interval is considered.

## 21.5. Shape of cost functions and optimal $\delta$ and $a$

Now, we consider the cost function [21.4]. We get the following behavior with respect to $\delta$ and a fixed $a$:

– If $\delta$ tends to 0, then there is an infinite number of inspections per time and $C$ tends to infinity.

– If $\delta$ tends to infinity, then each cycle ends with a failure. Therefore,

$$\lim_{\delta \to \infty} C = c_F \cdot \frac{\mu}{h - x_0}.$$

Regarding $a$, it is easy to see that for a given $\delta$.

– If $a$ is very closed to the boundary level $h$, then with high probability the cycle ends with a failure.

– Otherwise, if $a$ is very small, the cycle ends by a preventive maintenance.

Further, the behavior of the cost function depends on the ratio of maintenance and failure cost. In Figure 21.3, some cost functions in dependence on $\delta$ for different $a$ are plotted. For this, 100 degradation processes with parameters $\mu = 5$, $\sigma^2 = 10$, $h = 40$, $x_0 = t_0 = 0$ were simulated. The average costs from these 100 simulations are plotted for integer values of $\delta$ and for better visibility connected by lines. In Figure 21.3(a), the costs of a failure and the cost of maintenance are the same. In this case, the cost function decreases in $\delta$. Figure 21.3(b) shows cost functions for the same parameters and a ratio of maintenance to failure costs 1/2. It can be seen that for relatively large $a$, the costs are decreasing in $\delta$, too. For smaller values of $a$, the cost function has a unique minimum. In both figures, the straight line is the limit value $\lim_{\delta \to \infty} C$.

Figure 21.4 shows that there exist an optimal $a$. The parameters of the process are the same as before. Here, we get minimal costs for $a = 24$ while the minimal costs with respect to $\delta$ are bigger for $a = 22$ as well as for $a = 26$.

It should be noted that there is an direct connection between the optimal maintenance level $a$ and the probability of a failure before the next inspection time. In the above example, this probability is 0.495 for $\delta = 3$ and $h = 40$.

This leads to the idea that a sequential technique might very usefully:

– define $\alpha$: probability of no failure before the next inspection;

– choose the next inspection point with respect to alpha;

– carry out a preventive maintenance action, if the time to next inspection becomes very small.

This idea needs further research.

## 21.6. Incomplete preventive maintenance

In the last few years, there has been increasing number of papers concerning incomplete repairs and maintenance; see as example [DOY 11] and the literature given in this chapter. In such models, the impact of a maintenance (or a repair) action is not minimal (as bad as old) and not perfect (as good as new) but lies between these boundary cases. The most often used model is the Kijima imperfect repair model after a failure [KIJ 88] and [KIJ 89]. In these models, a new item has the failure rate $\lambda_1(t) = \lambda(t)$. Let $t_1$ be the time of the first failure. After failure, the item will be repaired with the degree $\xi_1$, where $\xi_1 = 1$ in case of minimal repair and $\xi_1 = 0$ in case of perfect repair (renewal). An imperfect repair occurs if $0 < \xi_1 (< 1)$. By repair, the age of the item is decreased to $v_1 = \xi_1 t_1$, which is called the virtual age of the item at time $t_1$. The distribution of the time until the next failure now has the failure rate $\lambda_2(t) := \lambda(t - t_1 + v_1)$.

The process defined by

$$v(t) := t - t_n + v_n, \quad t_n \le t < t_{n+1}\,,\ n \ge 1$$

is called the *virtual age process*.

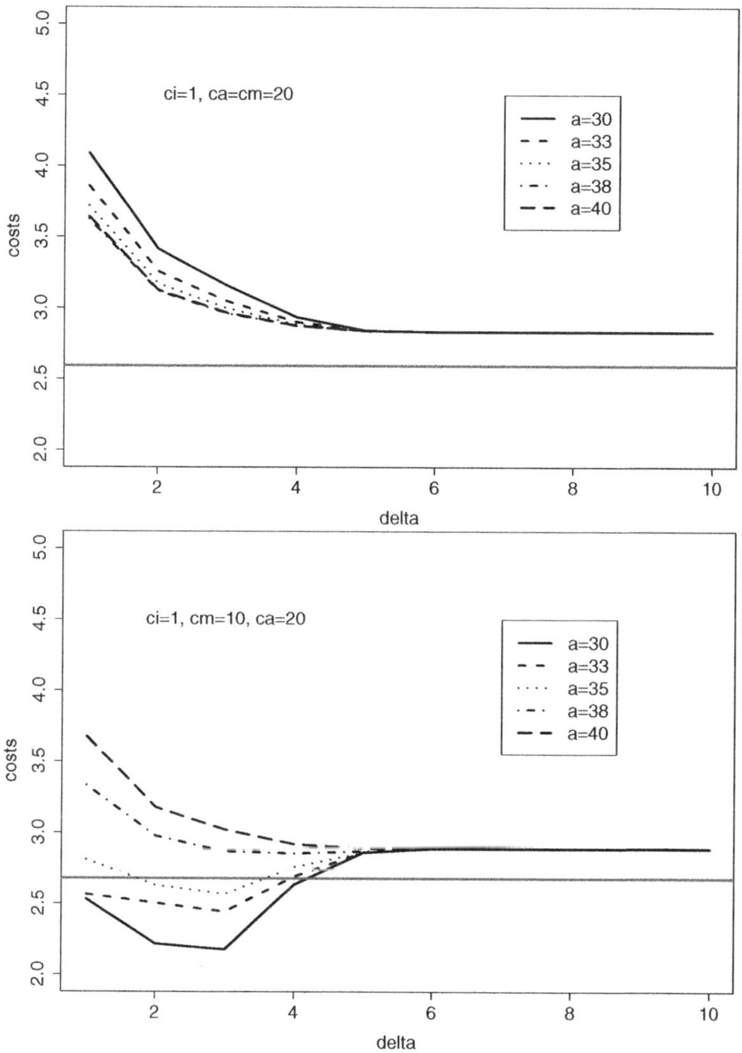


**Figure 21.3.** *Average costs for maintenance and failure for different $a$ and $\delta$*

Kijima considered two types of repair:

$$\text{Kijima Type I:} \quad v_k = \quad v_{k-1} + \xi_k(t_k - t_{k-1})$$

$$\text{Kijima Type II:} \quad v_k = \quad \xi_k(v_{k-1} + (t_k - t_{k-1}))$$

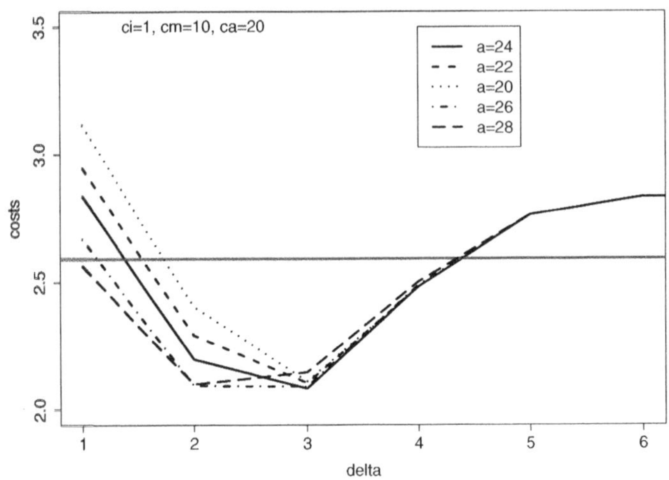


**Figure 21.4.** *Optimal maintenance level and optimal distance between inspections*

In [KAH 07], it was shown that for preventive maintenance the important fact is the age after maintenance. Knowing this age, the degree of repair can be found for both Kijima-type repairs.

Now, we want to apply incomplete maintenance to degradation processes. We consider the degradation process $X(t)$ with maintenance points $\delta$, $2\delta$, .... At each maintenance point, we reset the age of the process in the following manner:

– Kijima Type I: $v_k = v_{k-1} + \xi(t_k - t_{k-1}) = v_{k-1} + \xi\delta = \xi k \delta$

– Kijima Type II: $v_k = \xi(v_{k-1} + t_k - t_{k-1}) = v_{k-1} + \xi\delta = \tau \sum_{l=1}^{k} \xi^k$

At age $v_k$, the degradation is normally distributed with

$$E(X(v_k)) = \mu \cdot v_k \qquad \text{and} \qquad Var(X(v_k)) = \sigma^2 \cdot v_k \,.$$

Therefore, a natural idea is to reduce the degradation level to $\mu \cdot v_k$. In Figure 21.5, the effect of preventive maintenance is illustrated for different Kijima type I degrees of repair.

In a further work, the question of cost optimal preventive incomplete maintenance where costs depend on the degree of repair will be considered.

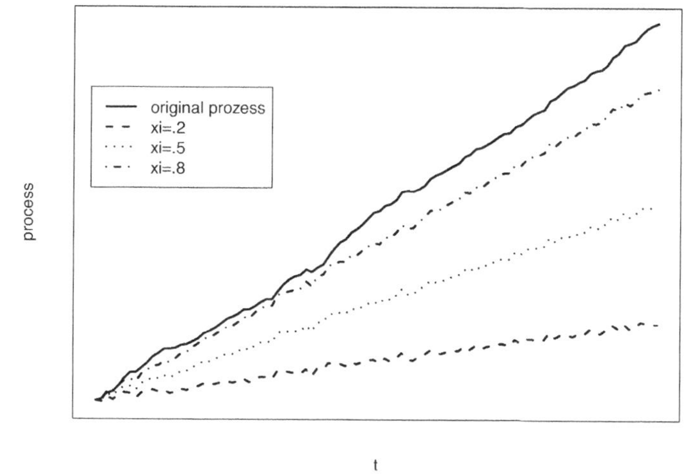


**Figure 21.5.** *Influence of the degree of repair*

## 21.7. Bibliography

[BAG 01] BAGDONAVIČIUS V., NIKULIN M., "Estimation in degradation models with explanatory variables", *Lifetime Data Analysis*, vol. 7, pp. 85–103, 2001.

[BAG 07] BAGDONAVIČIUS V., BIKELIS A., KAZAKEVICIUS V., *et al.*, "Analysis of joint multiple failure mode and linear degradation data with renewals", *Journal of Statistical Planning and Inference*, vol. 137, pp. 2191–2207, 2007.

[DOY 11] DOYEN L., GAUDOIN O., "Modelling and assessment of aging and efficiency of corrective and planned preventive maintenance", *IEEE Transactions on Reliability*, vol. 60, pp. 759–769, 2011.

[DOK 96] DOKSUM K.A., NORMAND S.T., "Models for degradation processes and event times based on gaussian processes", in JEWELL N.P. *et al.* (eds), *Lifetime Data: Models in Reliability and Survival Analysis*, pp. 85–91, 1996. Kluwer academic publishers.

[KAH 94] KAHLE W., "Simultaneous confidence regions for the parameters of damage processes", *Statistical Papers*, vol. 35, pp. 27–41, 1994.

[KAH 98] KAHLE W. LEHMANN A., "Parameter estimation in damage processes: dependent observations of damage increments and first passage time", in KAHLE W. *et al.* (eds), *Advances in Stochastic Models for Reliability, Quality and Safety*, Birkhauser, Boston, pp. 139–152, 1998.

[KAH 07] KAHLE W., "Optimal maintenance policies in incomplete repair models", *Reliability Engineering & System Safety*, vol. 92, pp. 563–565, 2007.

[KIJ 88] KIJIMA M., MORIMURA H., SUZUKI Y., "Periodical replacement problem without assuming minimal repair", *European Journal of Operational Research*, vol. 37, pp. 194–203, 1988.

[KIJ 89] KIJIMA M., "Some results for repairable systems with general repair", *Journal of Applied Probability*, vol. 26, pp. 89–102, 1989.

[LU 93] LU C.J., MEEKER W.Q., "Using degradation measures to estimate a time-to-failure distribution", *Technometrics*, vol. 35, pp. 161–174, 1993.

[MEE 88] MEEKER W.Q., ESCOBAR, L., *Statistical Methods for Reliability Data*, Wiley, New York, 1988.

[VAN 09] VAN NOORTWIJK J.M., "A survey of the application of gamma processes in maintenance", *Reliability Engineering and System Safety*, vol. 94, pp. 2–21, 2009.

[SES 99] SESHADRI V., *The Inverse Gaussian Distribution: statistical theory and applications*, New York: Springer, 1999.

[WHI 95] WHITMORE G.A., "Estimating degradation by a Wiener difusion process subject to measurement error", *Lifetime Data Analysis*, vol. 1, pp. 307–319, 1995.

[WHI 97] WHITMORE G.A., SCHENKELBERG F., "Modelling accelerated degradation data using Wiener difusion with a time scale transformation", *Lifetime Data Analysis*, vol. 3, pp. 27–45, 1997.

[WHI 98] WHITMORE G.A., CROWDER M.J., LAWLESS J.F., "Failure inference from a marker process based on a bivariate Wiener model", *Lifetime Data Analysis*, vol. 4, pp. 229–251, 1998.

# Chapter 22

# Planning Accelerated Destructive Degradation Tests with Competing Risks

Accelerated destructive degradation tests (ADDTs) are widely used in manufacturing industries to obtain timely product reliability information, especially in applications where few or no failures are expected under use conditions in tests of practical length. An ADDT plan specifies the test conditions of accelerating variables, running time and the corresponding allocation of test units to each condition. Usually, variables such as temperature, voltage or pressure can be used as accelerating variables to accelerate degradation of a product. For some applications, however, tests at high-stress test conditions would result in more than one type of failure for test units, called competing risk problems. Careful test planning is important for efficient use of limited resources: test time, test units and test facilities. This chapter describes methods to find unconstrained and constrained optimum test plans for competing risk applications under a given test optimization criterion, such as minimizing the large-sample approximate variance of a failure-time distribution quantile at use conditions. A modified general equivalence theorem (GET) is used to verify the optimality of a given ADDT plan. Generally, an optimum test plan provides insight for constructing a good compromise test plan that tends to be more robust and practical. Monte Carlo simulations are used to provide a visualization of the results that might be obtained from given test plans. The methods are illustrated with an application for an adhesive bond.

---

Chapter written by Ying SHI and William Q. MEEKER.

## 22.1. Introduction

### 22.1.1. *Background*

With the development of new technologies and strong global competition, manufacturers are facing pressure to produce high-reliability products in a shorter time. This has motivated a strong interest in conducting up-front accelerated reliability tests on materials and components while products are being designed. Degradation tests are becoming more and more popular nowadays because degradation data can provide considerably more reliability information, compared to the traditional failure-time data (where time to failure is the response), especially in applications where few or no failures are expected within a long period of time. For most applications, to get reliability information quickly, degradation tests are accelerated by testing at higher than usual levels of accelerating variables such as temperature, voltage or humidity. Information obtained from tests at high levels of the accelerating variables is then extrapolated to the use conditions using a reasonable statistical model, often based on the physics of failure knowledge. For some applications, degradation tests are destructive because the measurement process destroys or changes the physical/mechanical characteristics of test units. An accelerated degradation test with such degradation data is called ADDT.

In some applications, tests at high-stress levels of the accelerating variables can induce failure modes that would not be observed at normal operating conditions. If other failure modes (sometimes known as competing risks) are caused only at high-stress levels of the accelerating variables, and primary interest is focused on one particular failure cause that commonly occurs at normal conditions, failure from other causes can be viewed as a form of *right random censoring* in the data analysis (as long as the other failure modes do not completely dominate the failure mode of interest). Therefore, for such applications, careful test planning is extremely important for efficient use of limited resources: test time, test units and test facilities.

### 22.1.2. *Motivation: adhesive bond C*

The application motivating this work was an adhesive bond (adhesive bond C), which was to be used to attach two components of a product together. The experimental response was strength (in Newtons) of the adhesive bond and the strength deteriorated over time. In particular, the product engineers wanted to estimate the time at which 1% of the product would have an adhesive strength less than 40 N when operating at a room temperature of 25°C (i.e. they wanted to estimate the 0.01 quantile of the failure-time distribution at use conditions). To obtain the degradation information quickly, temperature was used as an accelerating variable for the test. To measure the strength of the test units, engineers applied an increasing force until the two components attached by the adhesive bond broke apart.

Thus, the measurement process for this application was destructive. For the adhesive bond C, two types of failure were observed for the units tested at high-stress test conditions (i.e. high temperature for long running time). In particular, some units failed from *adhesive* failure while others failed from *cohesive* failure. Adhesive failures occurred when one of the components detached from the adhesive bond. Cohesive failures occurred when the adhesive bonds to the material remained intact, but the components came apart due to a loss of cohesion within the layer of the adhesive material. Because of the destructiveness of measurement process, only one type of failure or the other could be observed on each test unit. Because of differences in the two effective activation energy values, the engineers were confident that only adhesive failure would occur on units operating at the normal operating temperatures, even after a long running time. Hence, the primary interest was to estimate the 0.01 quantile of the failure-time distribution for the adhesive failures.

Table 22.1 shows the ADDT plan originally proposed for the adhesive bond C application in order to estimate the 0.01 quantile of the failure-time distribution at use conditions. The 15 units tested at time 0 would have no aging and were to serve as baseline units. A total of 59 additional test units were aged and measured after running according to the test conditions as shown in Table 22.1. Note that no test units were assigned at the highest stress test condition (16 weeks, 70°C), in an attempt to avoid cohesive failures.

| Temperature | Weeks | | | | | | Totals |
|---|---|---|---|---|---|---|---|
| (°C) | 0 | 2 | 4 | 6 | 12 | 16 | |
| 70 | | 4 | 4 | 4 | 4 | 0 | 16 |
| 60 | | 5 | 0 | 5 | 5 | 5 | 20 |
| 50 | | 6 | 0 | 6 | 6 | 5 | 23 |
| — | 15 | | | | | | 15 |
| Total | 15 | 15 | 4 | 15 | 15 | 10 | 74 |

**Table 22.1.** *The original ADDT plan for adhesive bond C*

### 22.1.3. *Related literature*

There is a large amount of literature on planning accelerated tests, especially for accelerated life tests (ALTs) where failure time is the response. [NEL 05a] and [NEL 05b] summarize such test planning work. Some work that is particularly relevant to this chapter is included in the following references. [NEL 90, Chapter 6] describes methods for planning ALTs based on a simple model. [MEE 98, Chapter 20] provides details and examples on how to plan a single-variable ALT. [ESC 95] describes methods for planning ALTs with two or more explanatory variables. There are some important differences between ALTs and accelerated degradation tests because of their different responses. [BOU 94] gives methods for planning *repeated measures* accelerated degradation tests. [SHI 09] derives methods to find good ADDT plans for an important class of destructive degradation model. [JEN 11] investigates a distribution misspecification problem in ADDT and proposes robust test plans that minimize the approximate mean square error of the lifetime quantile estimate. [SHI 12] uses Bayesian methods for ADDT planning under a class of nonlinear degradation models with one accelerating variable. None of these methods has considered planning ADDTs for applications with more than one type of failure.

### 22.1.4. *Overview*

The remaining sections of this chapter are organized as follows. Section 22.2 describes a class of degradation models for ADDT data with more than one type of failure mode and gives expressions for the degradation distribution and quantiles. Section 22.3 illustrates the failure-time distribution and quantiles induced by the degradation models. Section 22.4 discusses the criterion of ADDT planning for applications with more than one type of failure. Section 22.5 gives a constrained optimum ADDT plan and applies a modification of the GET to verify the optimality of this test plan. Section 22.5 also presents a compromise ADDT plan that meets practical constraints. Section 22.6 uses Monte Carlo simulation to provide visualization of the results that might be obtained from given test plans. Section 22.7 contains some concluding remarks and extensions for future research work. The appendix in section 22.8 provides derivations and technical details about the large-sample approximations used in this work.

## 22.2. Degradation models with competing risks

For applications resulting in data with more than one type of failure, the statistical competing risk model, e.g. [DAV 78] and [CRO 01], can be used. In this chapter, we mainly focus on accelerated test planning in the applications with two types of failure where the marginal distribution of time to failure for one of the failure types is of primary interest. The generic responses corresponding to the two types of failure are called the primary response and the competing response, respectively. In an ADDT with competing risks, for each experimental unit, we observe only the minimum of the primary response and the competing response.

### 22.2.1. *Accelerated degradation model for the primary response*

An important class of the degradation models for the primary response at time $t$ and the accelerating variable AccVar is

$$Y = \beta_0 + \beta_1 \exp(\beta_2 x)\tau + \epsilon, \quad [22.1]$$

where $Y = h_d(\text{primary response})$ is transformed degradation, and $\tau = h_t(t)$ and $x = h_a(\text{AccVar})$ are known monotone increasing transformations of time and the accelerating variable, respectively. These functions may be suggested by the physics of failure or determined empirically. $\epsilon$ is the residual deviation that describes unit-to-unit variability. Here, $\epsilon/\sigma \sim \Phi(z)$ , where $\Phi(z)$ is a completely specified cumulative distribution function (cdf), such as the standardized normal cdf, $\Phi_{\text{nor}}(z)$, or the standardized smallest extreme value cdf, $\Phi_{\text{sev}}(z)$. The model parameters $\boldsymbol{\theta} = (\beta_0, \beta_1, \beta_2, \sigma)$ are fixed but unknown, and will be estimated from ADDT data.

Model [22.1] is linear in the sense that for a specified $x$, $\mu(\tau, x) = \beta_0 + \beta_1 \exp(\beta_2 x)\tau$ is linear in $\tau$. The full model in [22.1] is, however, nonlinear in the parameters. The intercept $\beta_0$ is the location parameter of the transformed degradation distribution when $\tau = 0$. The degradation rate of $\mu(\tau, x)$ with respect to $\tau$ at $x$ is $\omega(x) = \beta_1 \exp(\beta_2 x)$. The sign of $\beta_1$ determines whether the degradation is increasing or decreasing over time. For example, if the degradation response is the size of a crack, $\beta_1$ would be expected to be positive. On the other hand, if the degradation response is the strength of an adhesive bond, $\beta_1$ would be expected to be negative.

### 22.2.2. *Accelerated degradation model for the competing response*

The degradation model for the competing response is parallel to the model for the primary response, and is given as

$$Y^{(c)} = \beta_0^{(c)} + \beta_1^{(c)} \exp(\beta_2^{(c)} x)\tau + \epsilon^{(c)},$$

where $Y^{(c)} = h_d(\text{competing response})$ is the transformed competing degradation, $\tau$ and $x$ are the same as in section 22.2.1 and again $\epsilon^{(c)}/\sigma^{(c)} \sim \Phi(z)$. The model parameters for this competing model are $\boldsymbol{\theta}^{(c)} = (\beta_0^{(c)}, \beta_1^{(c)}, \beta_2^{(c)}, \sigma^{(c)})$ and they are also fixed but unknown.

### 22.2.3. *Degradation models for adhesive bond C*

The degradation model for adhesive bond C is an extension of the model derived by Escobar *et al.* [ESC 03] for a similar application without competing risks, adhesive bond B. Product engineers believe that these two types of bond have the same characteristics for the adhesive strength, but for adhesive bond C, cohesive failures can, after a long running time, occur for units tested at high temperature.

For the adhesive bond C application, the degradation models for the adhesive strength and cohesive strength are, respectively,

$$\begin{aligned} Y &= \beta_0 + \beta_1 \exp(\beta_2 x)\tau + \epsilon \qquad \text{and} \\ Y^{(c)} &= \beta_0^{(c)} + \beta_1^{(c)} \exp(\beta_2^{(c)} x)\tau + \epsilon^{(c)}, \end{aligned} \qquad [22.2]$$

where

$$
\begin{aligned}
Y &= \log(\text{Adhesive Strength in Newtons}) \\
Y^{(c)} &= \log(\text{Cohesive Strength in Newtons}) \\
x &= -\frac{11605}{\text{Temperature in }^\circ\text{C} + 273.15} \\
\tau &= \sqrt{\text{Time in Weeks}} \qquad [22.3] \\
(\epsilon/\sigma) &\sim \Phi_{\text{nor}}(z) \quad \text{and} \quad (\epsilon^{(c)}/\sigma^{(c)}) \sim \Phi_{\text{nor}}(\text{z}). \qquad [22.4]
\end{aligned}
$$

For this application, the accelerating variable is temperature and $x$ is the Arrhenius transformed temperature. The denominator in $x$ is temperature on the Kelvin (K) scale and the numerator is the reciprocal of Boltzmann's constant in units of electronvolt per Kelvin (eV/K). Under this parameterization, $\beta_2$ and $\beta_2^{(c)}$ have the interpretations of effective activation energy corresponding to the degradation mechanism for adhesive and cohesive strengths, respectively.

Figure 22.1 is an illustration of degradation distributions for the adhesive strength at 25°C and different values of time for some specific values of the parameters $\beta_0, \beta_1, \beta_2, \sigma$ corresponding to the maximum likelihood (ML) estimates given in [ESC 03]. This figure gives a visual impression of the adhesive strength degradation distribution over time. For the cohesive strength, the degradation distribution should have a similar decreasing trend over time, but at a different degradation rate from the adhesive strength because of differences in the parameters.

Because of the competing risk between two types of failure, we can only observe $\min(Y, Y^{(c)})$. If an adhesive failure occurs, then $Y < Y^{(c)}$; if a cohesive failure occurs, then $Y^{(c)} < Y$. Because the adhesive strength is of primary interest, we consider an observation to be *exact* for an adhesive failure and *right censored* for a cohesive failure. For right-censored observations, the adhesive strength is unobserved but known to be greater than the censored value. Such right-censored observations contain relatively little information about the adhesive strength. Thus, using the competing risk model for test planning should result in a test plan that limits the probability of having such censored observations in the accelerated test.

### 22.2.4. *Degradation distribution and quantiles*

For a given time and accelerating variable level, the cdf of the primary transformed degradation $Y$ is

$$F_Y(y;\tau,x) = \Pr(Y \le y;\tau,x) = \Phi\left[\frac{y-\mu(\tau,x)}{\sigma}\right].$$

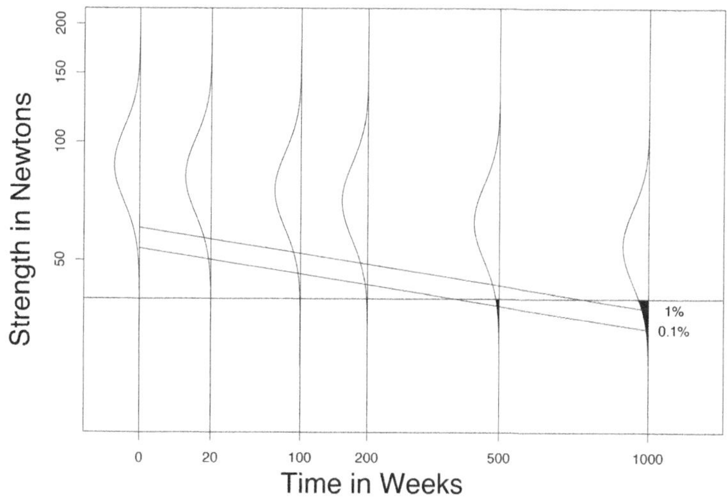


**Figure 22.1.** *Adhesive strength degradation distributions of adhesive bond C as a function of time at* 25°C. *The strength axis is a logarithmic axis and the time axis is a square root axis. The horizontal line at* $\mathcal{D}_{\rm f} = 40$ *N is the threshold for the failure-definition degradation level. At each time, the shaded area below the horizontal line is the failure probability at that specific time*

For the adhesive bond C example, the cdf for $Y$, the logarithm of the adhesive strength at a fixed test condition of time and temperature, can be obtained by replacing $\Phi$ with $\Phi_{\rm nor}$.

The $p$ quantile function for the primary transformed degradation distribution at $(\tau, x)$ is

$$y_p = \beta_0 + \beta_1 \exp(\beta_2 x)\tau + \sigma\Phi^{-1}(p),$$

where $\Phi^{-1}(p)$ is the $p$ quantile of a standard location-scale distribution. Substituting $\Phi^{-1}_{\rm nor}(p)$ for $\Phi^{-1}(p)$, we obtain the $p$ quantile of the transformed degradation (log Newtons) for the adhesive strength in the adhesive bond C application, as shown in Figure 22.1 for $p = 0.01$ and $p = 0.001$.

## 22.3. Failure-time distribution with competing risks

### 22.3.1. *Relationship between degradation and failure*

For some products, there is a gradual loss of performance as usage time increases (e.g. decreasing adhesive strength of an adhesive bond). A "soft failure" [MEE 98, section 13.4] occurs when performance reaches a critical level $\mathcal{D}_{\mathrm{f}}$ after a certain usage time. Because we are mainly interested in evaluating the failure-time distribution for the primary response, the failure-time $T$, for the competing risk applications, is defined as the time when the observed primary degradation crosses the critical level $\mathcal{D}_{\mathrm{f}}$.

### 22.3.2. *Failure-time distribution and quantiles*

For decreasing degradation, when $\beta_1$ is negative, the event of failure-time $T$ being less than $t$ is equivalent to an observed primary degradation response being less than the critical level $\mathcal{D}_{\mathrm{f}}$ at time $t$ (i.e. $Y \leq y_{\mathrm{f}}$, where $y_{\mathrm{f}} = h_d(\mathcal{D}_{\mathrm{f}})$). Then, the failure-time cdf is

$$\begin{aligned} F_T(t;x) &= \Pr(T \leq t) = \Pr(Y \leq y_{\mathrm{f}}) = F_Y(y_{\mathrm{f}};\tau,x) \\ &= \Phi\left[\tfrac{y_{\mathrm{f}}-\mu(\tau,x)}{\sigma}\right] = \Phi\left(\tfrac{\tau-\nu}{\varsigma}\right), \quad \text{for } t \geq 0, \end{aligned} \qquad [22.5]$$

where

$$\nu = \frac{(\beta_0 - y_{\mathrm{f}})\exp(-\beta_2 x)}{\mid \beta_1 \mid} \quad \text{and} \quad \varsigma = \frac{\sigma \exp(-\beta_2 x)}{\mid \beta_1 \mid}.$$

Note that in this model, there is a positive $\Pr(T = 0) = \Phi\left(-\nu/\varsigma\right)$ at $t = 0$. This spike is sometimes called the dead-on-arrival (DOA) probability.

From [22.5], the $p$ quantile of the failure time for decreasing degradation is

$$t_p = \begin{cases} h_t^{-1}\left[\nu + \varsigma\Phi^{-1}(p)\right] & \text{if } p \geq \Phi\left(-\nu/\varsigma\right) \\ 0 & \text{otherwise.} \end{cases}$$

For increasing degradation, when $\beta_1$ is positive, the derivations of the failure-time cdf and quantiles are similar.

## 22.4. Test planning with competing risks

### 22.4.1. *ADDT planning information*

For ADDT planning, there are usually practical constraints. For example, for the adhesive bond C application, there are the following three constraints for the ADDT planning:

– The maximum sample size is 74 test units.

– The maximum temperature that can be used is 70°C. (The engineers involved in the testing felt that higher temperature would cause the degradation models to break down.)

– The maximum time available for testing is 16 weeks.

To do test planning, planning information for the parameters of the given degradation models is required. This is because the test plan evaluation criteria depend on the model parameters. Such planning information could be obtained from previous experiments, past data, engineers' experience, etc. For the adhesive bond C application, planning values for parameters in the primary response degradation model are obtained from the data analysis in [ESC 03] because it is believed that the adhesive degradation properties of bond C are similar to those of bond B, as mentioned in section 22.2.3. That is, $\beta_0^{\square} = 4.471$, $\beta_2^{\square} = 0.6364$, $\sigma^{\square} = 0.158$, and the degradation rate at 50°C is $\omega_{50}^{\square} = -0.1026$. These parameters have clear practical interpretations, which makes them easier to elicit from experts when needed. With the given planning values, we get $\beta_1^{\square} = \omega_{50}^{\square} \exp(-\beta_2^{\square} x_{50}) = -0.1026 \exp(0.6364 \times 35.912) = -8.643 \times 10^8$, where $x_{50} = -11605/(50 + 273.15) = -35.912$ is the transformed temperature of 50°C.

Planning values for parameters in the competing response degradation model are $\beta_0^{(c)\square} = 5.7$, $\beta_2^{(c)\square} = 0.6$, $\omega_{50}^{(c)\square} = -0.23$ and $\sigma^{(c)\square} = 0.2$. This information was obtained from a combination of limited previous experience with cohesive failures in accelerated tests and engineering judgment. In some preliminary experiments, engineers did not find any cohesive failures at test conditions with low temperature and they would not expect to see any cohesive failures at use conditions, even after a long period of running time (i.e. with probability approaching 1, adhesive failures would occur first at use conditions). Cohesive failures only occurred at some high-stress test conditions. Under the specified planning values for the model parameters, Figure 22.2 shows a contour plot of the probability that the cohesive strength is less than adhesive strength (i.e. the right-censored probability) as a function of test conditions within the experimental region, and Figure 22.3 compares the mean degradation responses of adhesive and cohesive strength at various test conditions.

### 22.4.2. *Criterion for ADDT planning with competing risks*

For ADDT planning, the appropriate test planning criterion depends on the purpose of the experiment. For our application, the objective is to estimate $t_p$, a particular quantile of the failure-time distribution at use conditions. A commonly used criterion for test planning is to minimize $\text{Avar}(\widehat{t_p})$, the large-sample approximate variance of the ML estimator of the specified failure-time quantile.

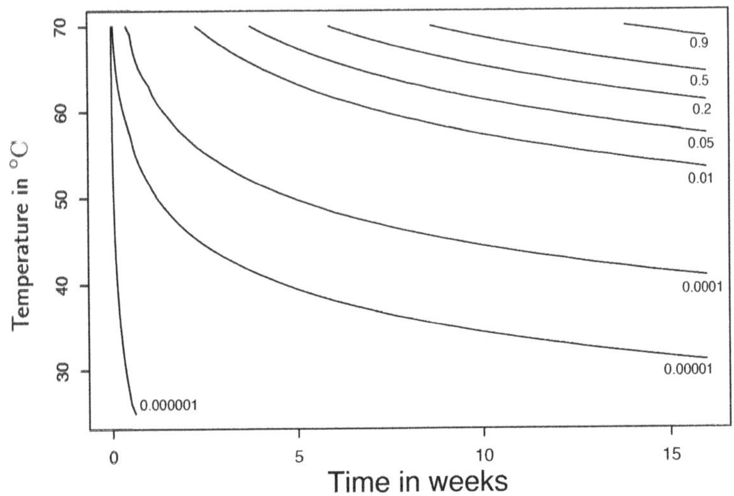


**Figure 22.2.** *The probability of cohesive strength being less than adhesive strength (i.e. probability of a right-censored observation) under specified planning values for the model parameters*

We use $\boldsymbol{v} = (t, \text{AccVar})$ to denote a test condition specifying the time $t$ and the accelerating variable level AccVar. An ADDT plan, denoted by $\boldsymbol{\xi}$, will specify a set of test conditions $\boldsymbol{v}_i$, and the corresponding proportional allocation $\pi_i$ of test units at each $\boldsymbol{v}_i$. If a test plan has $r$ test conditions, then the proportional allocations satisfy $\pi_i > 0$, for each $i = 1, 2, \ldots, r$, and $\sum_{i=1}^{r} \pi_i = 1$.

Because $h_t(t_p)$ is a monotone increasing function of $t_p$, minimizing $\text{Avar}[h_t(\widehat{t_p})]$ is equivalent to minimizing $\text{Avar}(\widehat{t_p})$. Thus, from section 22.8.2, the optimization criterion is equivalent to finding a test plan $\boldsymbol{\xi}$ that maximizes the objective function

$$\Psi(\boldsymbol{\xi}) = -\boldsymbol{c}'[\boldsymbol{I}_{\boldsymbol{\theta},\boldsymbol{\theta}^{(c)}}(\boldsymbol{\xi})]^{-1}\boldsymbol{c}, \qquad [22.6]$$

where $\boldsymbol{c} = \partial h_t(t_p)/\partial\boldsymbol{\theta}$ and $\boldsymbol{I}_{\boldsymbol{\theta},\boldsymbol{\theta}^{(c)}}(\boldsymbol{\xi})$ is the Fisher information matrix for test plan $\boldsymbol{\xi}$. Details are defined in section 22.8. This criterion is known as *c optimality*.

## 22.5. ADDT plans with competing risks

### 22.5.1. *Initial optimum ADDT plan with competing risks*

For applications with no competing risks and a single degradation model like that in [22.1], [SHI 09] presents an ADDT optimum plan structure, illustrated in Figure

22.4. Under the practical constraints of a maximum time $t_M$ and a maximum accelerating variable level $\text{AccVar}_M$, the optimum plan with no competing risks is not unique. That is, there are many plans that provide the same smallest large-sample approximate variance for the ML estimator of a specified failure-time quantile. As shown in Figure 22.4, such optimum plans have some test units allocated at $\boldsymbol{v}_1$ (the baseline test condition) with $t = 0$ (note from the model that the level of AccVar is not a factor at time $t = 0$), some at the $\boldsymbol{v}_2$ test condition with $\text{AccVar}_M$ and a time level larger than a lower boundary $t_L$ and some at the $\boldsymbol{v}_3$ test condition with $t_M$ and an optimized AccVar value. The AccVar value for the $\boldsymbol{v}_3$ test condition and the proportional allocations $\pi_1$, $\pi_2$ of test units are chosen to minimize the large-sample approximate variance for each value of time for $\boldsymbol{v}_2$ between $t_L$ and $t_M$. For the no-competing-risk model, each such plan has the same optimum value for the objective function.

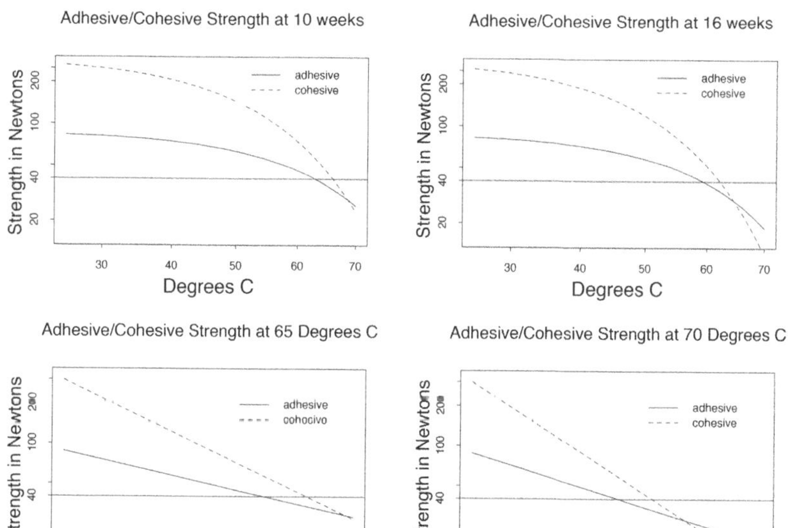


**Figure 22.3.** *The mean degradation responses of adhesive and cohesive strength at various test conditions under specified planning values for the model parameters. The strength axis is a logarithmic axis, the time axis is a square root axis and the temperature axis is an Arrhenius axis*

Generally, in a regression model, if all of the parameters are linear with respect to the explanatory variables, the number of test conditions for the optimum plan will not be any larger than the number of parameters needed to define the response surface. Although no such definite conclusion exists for models that are nonlinear in their parameters, such as our degradation model, numerical experiments for the ADDT model have not found a counter example. Because there are three regression parameters in the degradation model described in [22.1], we expect that a non-degenerate optimum ADDT plan for an application with the competing risk degradation model could also be a three-point plan (i.e. the optimum test plan should have three test conditions). We will check this using the GET.

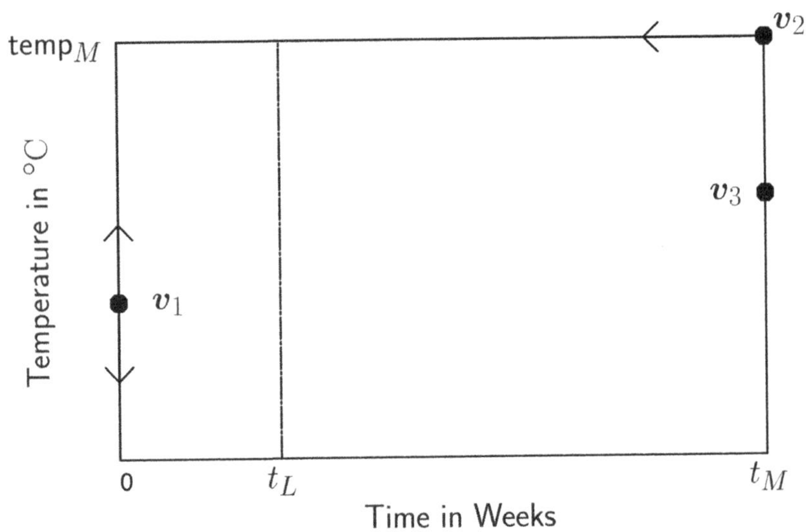


**Figure 22.4.** *Optimum plan structure*

For the adhesive bond C application, which has competing risks, our first attempt to find an optimum plan was to investigate test plans that have a structure similar to that for the no-competing-risk model, as shown in Figure 22.4. For most practical situations in which accelerated tests are used, all the test conditions for an optimum plan are on the boundaries of the experimental region and they spread out as much as possible, providing better estimates of the regression coefficients than closely spaced test conditions. For the competing risk model, the probability of getting a cohesive failure (a right-censored observation, providing relatively little information about adhesive failures) is highest when $\boldsymbol{v}_2$ is in the north east corner so we might expect the optimum $\boldsymbol{v}_2$ to be at an interior point. Hence, we assumed similar $\boldsymbol{v}_1$ and $\boldsymbol{v}_3$ test conditions but explored the properties of test plans with the $\boldsymbol{v}_2$ point moving within a subset of the experimental region. Then, at each fixed test condition for $\boldsymbol{v}_2$, temperature for the $\boldsymbol{v}_3$ and the proportional allocations $\pi_1$, $\pi_2$ of test units were chosen to minimize $\text{Avar}(\widehat{t}_{0.01})$. Figure 22.5 shows the contour plot of the large-sample approximate standard error of $\widehat{t}_{0.01}$, $\text{Ase}(\widehat{t}_{0.01})$, for each optimized plan as the test condition $\boldsymbol{v}_2$ varies within the experimental region, under the usual constraint that proportional allocations for any test condition should be non-negative and that they sum to 1. Figure 22.5 indicates that having $v_2$ on the 70°C boundary with time approximately 3 weeks results in a best test plan with minimum Ase($\widehat{t_{0.01}}$). This optimization result for the competing risk model is different from the result with no-competing risk where many optimum plans are derived by moving the time for the $v_2$ point between a lower boundary $t_L$ and $t_M = 16$ weeks at 70°C. The reason for different optimization conclusions between the two models is that testing at temperature approaching 70°C will tend to cause cohesive failures.

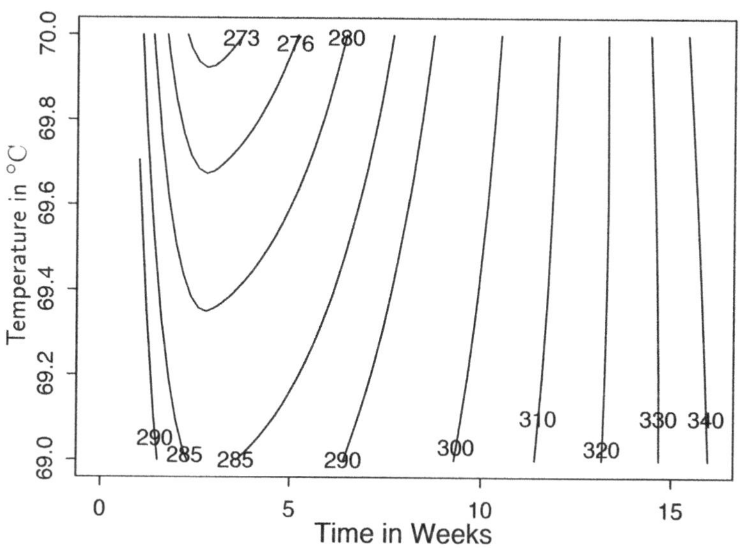


**Figure 22.5.** *Contour plot of the large-sample approximate standard error of $\widehat{t_{0.01}}$ for optimized plans when the test condition $v_2$ moves within the experimental region*

As mentioned before, the boundary of the experimental region is of particular interest. Figure 22.6 shows Ase($\widehat{t_{0.01}}$) and the proportional allocations of test conditions obtained for different optimized test plans, when the point $v_2$ moves along the maximum 70°C temperature line. Because all of the test plans are optimized under the condition that the proportional allocation for each test condition must be non-negative and sum to 1, $\pi_1$ is close to zero for optimized plans when the time for $v_2$ is less than approximately 3 weeks. At the minimum Ase($\widehat{t_{0.01}}$) value 272.06, the corresponding proportional allocation $\pi_1$ is close to zero, which is nearly a degenerate optimum test plan with little practical appeal.

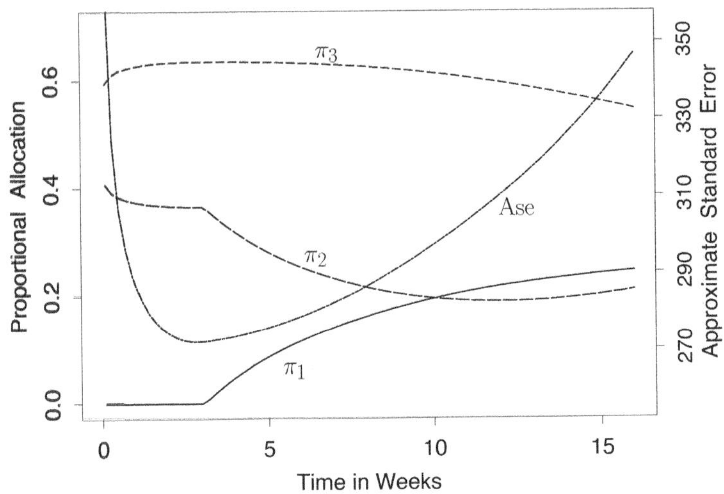


**Figure 22.6.** *Proportional allocations and large-sample approximate standard error of $\widehat{t}_{0.01}$ as a function of time in weeks for $\boldsymbol{v}_2$ showing different optimized plans arising from the constraint that proportional allocation for each test condition should be non-negative and sum to 1*

### 22.5.2. *Constrained optimum ADDT plan with competing risks*

In practical applications, engineers prefer to allocate a certain number of test units at the baseline condition at time 0. For the adhesive bond C application, instead of the initial optimized "degenerate" plan derived in section 22.5.1, it is more meaningful to assign a fixed proportion of test units at the baseline condition and then find a constrained optimum ADDT plan. For purposes of illustration, we assign 10% of test units to the baseline test condition $\boldsymbol{v}_1$. This amount is typical of what engineers have used in previous ADDT experiments that we have seen. The time of condition $\boldsymbol{v}_2$, the temperature of condition $\boldsymbol{v}_3$ and the proportional allocation $\pi_2$ are then chosen to minimize $\text{Avar}(\widehat{t}_{0.01})$. Table 22.2 shows the constrained optimum ADDT plan. The optimum $\text{Ase}(\widehat{t}_{0.01})$ for this test plan is 274.62, only slightly larger than that for the initial optimum degenerate plan.

### 22.5.3. *General equivalence theorem*

Shi *et al.* [SHI 09] use a variation of the GET described by Whittle [WHI 73] to check the optimality of a given ADDT plan without competing risks. For an ADDT with competing risks, the objective function shown in [22.6] has a form that is similar to the one in [SHI 09], except that the expression of $\boldsymbol{I}_{\boldsymbol{\theta},\boldsymbol{\theta}^{(c)}}(\boldsymbol{\xi})$ is different. The derivation of $\boldsymbol{I}_{\boldsymbol{\theta},\boldsymbol{\theta}^{(c)}}(\boldsymbol{\xi})$ for the competing risk model is given in section 22.8. To use

GET appropriately for checking the optimality of an ADDT plan with competing risks, some conditions need to be satisfied. The proof for the satisfaction of GET conditions can be done in a manner that is similar to that presented in [SHI 09, Appendix B.1]. To save space, however, the details of proof are not included here, only the idea of GET for an optimality check is shown. For the GET, the directional derivative, $\Lambda$, of $\Psi$ at $\boldsymbol{\xi}$ and in the direction of an alternative plan $\boldsymbol{\eta}$ is defined as

$$\begin{aligned}\Lambda(\boldsymbol{\xi},\boldsymbol{\eta}) &= \lim_{\delta\to 0+}\frac{\Psi[(1-\delta)\boldsymbol{\xi}+\delta\boldsymbol{\eta}]-\Psi(\boldsymbol{\xi})}{\delta}\\ &= \boldsymbol{c}'[\boldsymbol{I}_{\boldsymbol{\theta},\boldsymbol{\theta}^{(c)}}(\boldsymbol{\xi})]^{-1}\boldsymbol{I}_{\boldsymbol{\theta},\boldsymbol{\theta}^{(c)}}(\boldsymbol{\eta})[\boldsymbol{I}_{\boldsymbol{\theta},\boldsymbol{\theta}^{(c)}}(\boldsymbol{\xi})]^{-1}\boldsymbol{c}-\boldsymbol{c}'[\boldsymbol{I}_{\boldsymbol{\theta},\boldsymbol{\theta}^{(c)}}(\boldsymbol{\xi})]^{-1}\boldsymbol{c},\end{aligned}$$

where $\boldsymbol{c}$, $\boldsymbol{I}_{\boldsymbol{\theta},\boldsymbol{\theta}^{(c)}}(\boldsymbol{\xi})$ and $\boldsymbol{I}_{\boldsymbol{\theta},\boldsymbol{\theta}^{(c)}}(\boldsymbol{\eta})$ are evaluated using the planning values of the model parameters $\boldsymbol{\theta}^{\square}$ and $\boldsymbol{\theta}^{(c)\square}$. Let $\boldsymbol{\xi}_{\boldsymbol{v}}$ be a singular test plan that puts all units at the test condition $\boldsymbol{v}$. Suppose a given test plan $\boldsymbol{\xi}$ has $r$ test conditions, $\boldsymbol{v}_1,\boldsymbol{v}_2,\ldots,\boldsymbol{v}_r$. Then, the plan $\boldsymbol{\xi}$ is optimal if it satisfies $\Lambda(\boldsymbol{\xi},\boldsymbol{\xi}_{\boldsymbol{v}_1}) = \Lambda(\boldsymbol{\xi},\boldsymbol{\xi}_{\boldsymbol{v}_2}) = \ldots = \Lambda(\boldsymbol{\xi},\boldsymbol{\xi}_{\boldsymbol{v}_r}) = 0$ and $\Lambda(\boldsymbol{\xi},\boldsymbol{\xi}_{\boldsymbol{v}}) \leq 0$ for any other singular plan $\boldsymbol{\xi}_{\boldsymbol{v}}$ in the experimental region.

| Test condition | Time (weeks) | Temperature (°C) | Proportional allocation |
|---|---|---|---|
| $\boldsymbol{v}_1$ | 0 | – | 0.1 |
| $\boldsymbol{v}_2$ | 4.04 | 70 | 0.279 |
| $\boldsymbol{v}_3$ | 16 | 53.519 | 0.621 |

Note: "–" indicates that at time 0, the level of temperature has no effect on the model.

**Table 22.2.** *Constrained optimum ADDT plan with competing risks*

For a constrained optimum ADDT plan with a fixed proportion of test units at the baseline condition, a modification to the GET is needed. Instead of allocating all units at a single test condition, the alternative test plan $\boldsymbol{\eta}$ must now put the same proportion of test units at the baseline condition $\boldsymbol{v}_1$ as in the constrained optimum plan, and the remaining proportion of units at a single test condition $\boldsymbol{v}$. Let $\boldsymbol{\xi}_{\boldsymbol{v}_1,\boldsymbol{v}}$ denote a plan with two test conditions, one baseline condition $\boldsymbol{v}_1$ with a fixed proportion of test units, and the other condition $\boldsymbol{v}$ with all the remaining proportion of test units. For a given test plan $\boldsymbol{\xi}$ with a fixed test condition $\boldsymbol{v}_1$ and $r-1$ additional test conditions, $\boldsymbol{v}_2,\boldsymbol{v}_3,\ldots,\boldsymbol{v}_r$, the plan $\boldsymbol{\xi}$ is a constrained optimum plan if and only if it satisfies $\Lambda(\boldsymbol{\xi},\boldsymbol{\xi}_{\boldsymbol{v}_1,\boldsymbol{v}_2}) = \Lambda(\boldsymbol{\xi},\boldsymbol{\xi}_{\boldsymbol{v}_1,\boldsymbol{v}_3}) = \ldots = \Lambda(\boldsymbol{\xi},\boldsymbol{\xi}_{\boldsymbol{v}_1,\boldsymbol{v}_r}) = 0$ and $\Lambda(\boldsymbol{\xi},\boldsymbol{\xi}_{\boldsymbol{v}_1,\boldsymbol{v}}) \leq 0$ for any other single condition $\boldsymbol{v}$ in the experimental region.

For the adhesive bond C application, this modified GET is used to check the optimality of the constrained test plan illustrated in Table 22.2. Figure 22.7 shows the directional derivatives $\Lambda(\boldsymbol{\xi}, \boldsymbol{\xi}_{\boldsymbol{v}_1,\boldsymbol{v}})$ of this constrained optimum plan as a function of temperature and time for the test condition $\boldsymbol{v}$, where $\boldsymbol{\xi}_{\boldsymbol{v}_1,\boldsymbol{v}}$ is a plan that puts 10% of the test units at the baseline condition $\boldsymbol{v}_1$ and the remaining 90% of the test units at the test condition $\boldsymbol{v}$. Observe that, as expected, the directional derivatives are zero at those two unconstrained test conditions $\boldsymbol{v}_2, \boldsymbol{v}_3$ of the constrained optimum plan and less than zero everywhere else. This behavior of the directional derivative curves also implies that the constrained optimum plan is unique.

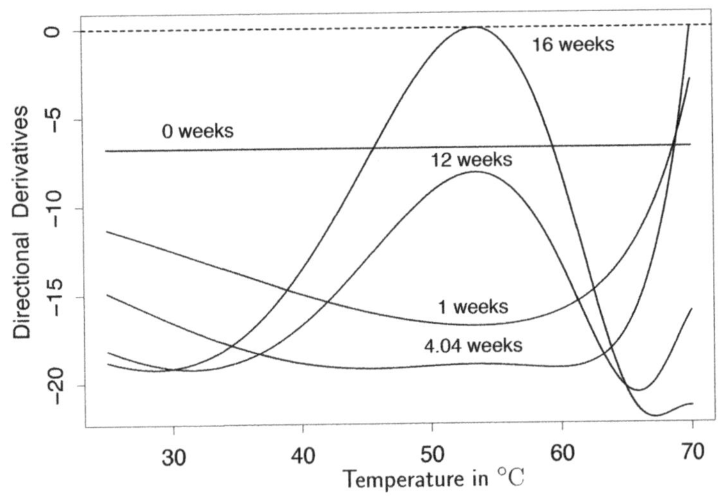


**Figure 22.7.** *Directional derivatives $\Lambda(\boldsymbol{\xi}, \boldsymbol{\xi}_{\boldsymbol{v}_1,\boldsymbol{v}})$ of the constrained optimum plan as a function of temperature and time for the test condition $\boldsymbol{v}$ in the adhesive bond C application. $\boldsymbol{\xi}_{\boldsymbol{v}_1,\boldsymbol{v}}$ is a plan that puts 10% test units at the baseline condition $\boldsymbol{v}_1$ and the remaining 90% test units at the test condition $\boldsymbol{v}$*

### 22.5.4. *Compromise ADDT plan with competing risks*

An optimum plan results in the smallest large-sample approximate variance of the estimated failure-time quantile. An optimum plan, however, provides no information to check the adequacy of the model and tends to be highly sensitive to planning information specification errors [MEE 84]. An optimum plan does, however, provide insight for obtaining good compromise test plans that tend to be more robust and useful in practical applications. A compromise test plan is generally suggested by the optimum plan but uses more test conditions to achieve robustness.

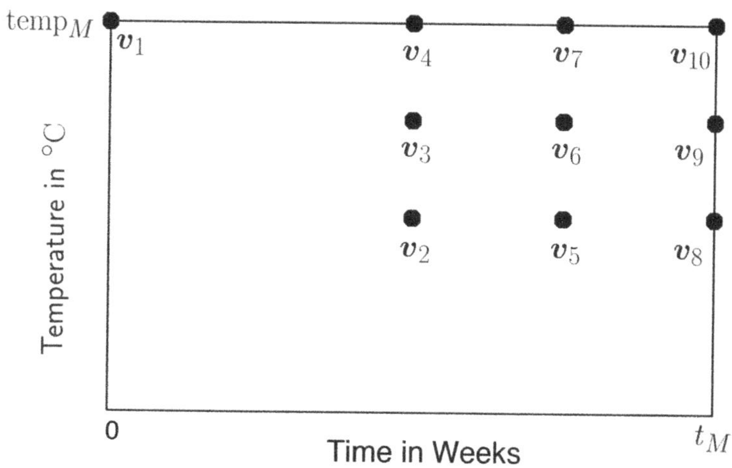


**Figure 22.8.** *Compromise plan structure*

[SHI 09] proposes an optimized compromise plan, whose structure is shown in Figure 22.8, for an adhesive bond application with no competing risks. This compromise plan has 10 test conditions, numbered 1–10, including the baseline condition and nine additional test conditions consisting of equally spaced levels of time and temperature within the experimental region. For practical reasons (a limited number of temperature-controlled test chambers and the test groups of units at one time), the rectangular pattern is preferred by engineers. For our adhesive bond C application with competing risks, we expect to obtain little useful information about adhesive strength by allocating test units at high-stress test conditions because of the high right-censored probabilities at those conditions, as shown in Figure 22.2. For this reason, some of the high-stress test conditions in the compromise plan structure of Figure 22.8 should be omitted to reduce the risk of cohesive failures and to get a better estimate of $\widehat{t}_{0.01}$ for competing risk applications. Table 22.3 compares $\text{Ase}(\widehat{t}_{0.01})$ for various compromise plans obtained by omitting different combinations of the high-stress test conditions. The proportional allocations for all the remaining test conditions in the plan are equal. For each compromise plan, beyond a baseline test condition, there are three equally spaced time levels and three equally spaced temperature levels. The highest time level is fixed at 16 weeks and the highest temperature level is fixed at 70°C. The lowest time level and the lowest temperature level are chosen to minimize $\text{Ase}(\widehat{t}_{0.01})$. The middle time level is the mean of the other two time levels and the middle temperature level is the mean of the other two temperature levels. As can be seen from Table 22.3, the plan after omitting three high-stress test conditions, numbered 7, 9 and 10, appears to be the best optimized compromise plan with competing risks in the sense that it provides the smallest $\text{Ase}(\widehat{t}_{0.01})$. Note that there is a trade-off between the estimation precision and robustness of the test plan. We can obtain a test plan with a smaller $\text{Ase}(\widehat{t}_{0.01})$ by omitting more test conditions, but we have to pay for this by some loss of robustness. We would recommend the last compromise plan in Table 22.3 or a similar compromise plan.

| Omitted conditions | Lowest time (weeks) | Lowest temperature (°C) | Proportional allocation | Ase($\widehat{t}_{0.01}$) |
|---|---|---|---|---|
| None | 8.19 | 50.437 | 1/10 | 389.83 |
| 10 | 7.56 | 50.182 | 1/9 | 372.47 |
| 9, 10 | 10.36 | 52.087 | 1/8 | 373.88 |
| 7, 10 | 7.89 | 48.673 | 1/8 | 368.26 |
| 7, 9, 10 | 9.39 | 51.552 | 1/7 | 364.44 |

**Table 22.3.** *Comparison of different compromise ADDT plans with competing risks*

## 22.6. Monte Carlo simulation to evaluate test plans

Monte Carlo simulation is a useful tool to provide visualization of the results obtained from a test plan. Simulation also provides a check on the adequacy of the large-sample approximations. We do simulations to evaluate the three ADDT plans illustrated above: the original plan in Table 22.1, the constrained optimum plan in Table 22.2 and the last compromise plan in Table 22.3. For the adhesive bond C application, the large-sample approximate standard errors of the estimated 0.01 failure-time quantile, Ase($\widehat{t}_{0.01}$), for these test plans are 396.68, 274.62 and 364.44, respectively. For each of these test plans, a simulation trial consists of a set of 74 observations obtained according to the test plan, the given degradation models and the planning information.

Figure 22.9 shows estimates of the 0.01 failure-time quantile versus temperature for 50 realizations of simulations for each test plan. The longer lines represent the values computed from the planning values, which we call the "true" values. The constrained optimum plan has the narrowest group of simulated lines. The spread of simulated lines for the compromise plan is wider than that for the constrained optimum plan but narrower than that for the original plan. These results are exactly consistent with the estimation precisions of different test plans, based on the large-sample approximate Ase($\widehat{t}_{0.01}$).

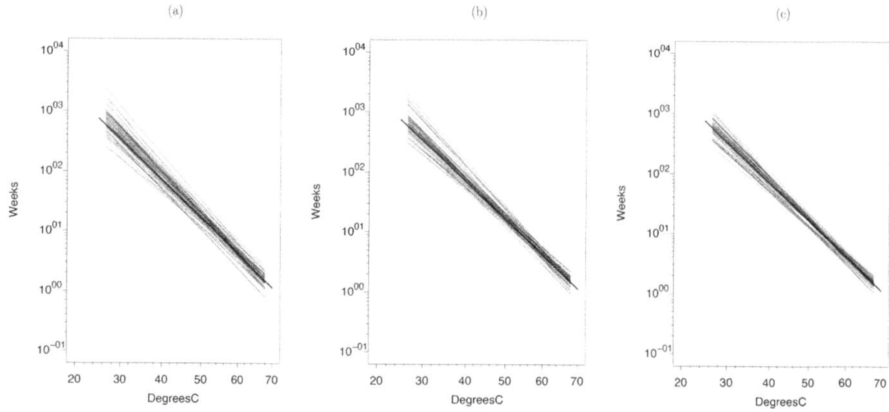


**Figure 22.9.** *Simulation of* 0.01 *failure-time quantile estimates versus temperature for a) the original plan, b) the compromise plan and c) the constrained optimum plan*

## 22.7. Conclusions and extensions

Accelerated destructive degradation testing is an important tool for making reliability inferences and predictions, especially when few or no failures are expected at use conditions within severe testing time constraints. Good ADDT plans can yield significant benefits to industry. However, accelerated tests might result in more than one type of failure for units tested at high-stress test conditions. Because of the competing risks problem, designing an appropriate ADDT plan could give more useful information about the primary degradation responses within the limited test time. The methodology presented in this chapter can be extended in several important directions, suggesting areas for future research. These include the following:

1) Test plans for an accelerated degradation model with multiple accelerating variables (e.g. temperature and humidity) could be developed.

2) As explained in section 22.2.1, the relationship between the location parameter and the transformed time is linear at a fixed accelerating variable level. The work in this chapter could be extended to degradation models that have a nonlinear relationship (such a relationship might, for example, be suggested by a physics of failure model).

3) For some applications, prior knowledge about the failure mechanism might provide information about some of the model parameters. It is important that such prior information be used in test planning as well as analysis. [SHI 12], following methods in [ZHA 06], describes Bayesian methods for ADDT planning.

## 22.8. Appendix: technical details

### 22.8.1. *The Fisher information matrix for ADDT with competing risks*

Because of the competing risks, each obtained observation could be an exact or right random censoring value (for example see [NEL 82, Chapter 5] or [LAW 03, Chapter 9]). [ESC 98] shows in detail how to compute the Fisher information matrix and large-sample approximate covariance matrix of ML estimators for a wide class of parametric models, including models with right random censoring data. The specific approach used here follows the general approach outlined in [ESC 98].

Let $\widehat{\boldsymbol{\theta}}$ be the ML estimator of the parameters $\boldsymbol{\theta} = (\beta_0, \beta_1, \beta_2, \sigma)'$ in the primary response model based on $n$ observations. Under the usual regularity conditions (for example [MEE 98, Appendix B.4]), $\widehat{\boldsymbol{\theta}} \dot{\sim} \text{MVN}(\boldsymbol{\theta}, \boldsymbol{\Sigma})$, where $\boldsymbol{\Sigma} = \boldsymbol{I}^{-1}_{\boldsymbol{\theta},\boldsymbol{\theta}^{(c)}}$, and the Fisher information matrix $\boldsymbol{I}_{\boldsymbol{\theta},\boldsymbol{\theta}^{(c)}}$ with $r$ test conditions is $\boldsymbol{I}_{\boldsymbol{\theta},\boldsymbol{\theta}^{(c)}} = n\sum_{i=1}^{r} \pi_i \boldsymbol{I}_i$. Here, $\boldsymbol{I}_i$ is the contribution of one observation at a specific test condition $\boldsymbol{v}_i = (\tau_i, x_i)$ to $\boldsymbol{I}_{\boldsymbol{\theta},\boldsymbol{\theta}^{(c)}}$. For an ADDT with competing risks, following the approach given in [ESC 98], $I_i$ can be expressed in terms of parameters $\boldsymbol{\theta}$ and $\boldsymbol{\theta}^{(c)}$ as

$$\boldsymbol{I}_i = \frac{1}{\sigma^2}\left[\begin{array}{cc} \left[\int_{-\infty}^{\infty} f_{11}(-\infty, \frac{y-\mu_i}{\sigma})h_i(y)dy\right]\boldsymbol{u}_i\boldsymbol{u}_i' & \left[\int_{-\infty}^{\infty} f_{12}(-\infty, \frac{y-\mu_i}{\sigma})h_i(y)dy\right]\boldsymbol{u}_i \\ \left[\int_{-\infty}^{\infty} f_{12}(-\infty, \frac{y-\mu_i}{\sigma})h_i(y)dy\right]\boldsymbol{u}_i' & \left[\int_{-\infty}^{\infty} f_{22}(-\infty, \frac{y-\mu_i}{\sigma})h_i(y)dy\right] \end{array}\right],$$

where $\mu_i$ is the location parameter for the primary response, defined in section 22.2.1 as $\mu_i = \beta_0 + \beta_1 \exp(\beta_2 x_i)\tau_i$ with $\tau_i = h_t(t_i)$ and $x_i = h_a(\text{AccVar}_i)$. $f_{jk}(-\infty, \frac{y-\mu_i}{\sigma}), jk = 11, 12, 22$ are the elements of the Fisher information matrix, multiplied by $\sigma^2$, for right-censored observations (see [ESC 98] for details). These elements can be computed using the location-scale distribution Fisher information matrix (LSINF) algorithm (see [ESC 94]). Here, $h_i(y) = \frac{1}{\sigma^{(c)}}\phi\left(\frac{y-\mu_i^{(c)}}{\sigma^{(c)}}\right)$ is the pdf of right random censoring point $y$ due to competing risks, where $\mu_i^{(c)} = \beta_0^{(c)} + \beta_1^{(c)} \exp(\beta_2^{(c)} x_i)\tau_i$, and $\boldsymbol{u}_i$ is the vector of partial derivatives of $\mu_i$ with respect to the parameters $(\beta_0, \beta_1, \beta_2)$. That is

$$\boldsymbol{u}_i = \left[\begin{array}{c} \dfrac{\partial \mu_i}{\partial \beta_0} \\ \dfrac{\partial \mu_i}{\partial \beta_1} \\ \dfrac{\partial \mu_i}{\partial \beta_2} \end{array}\right] = \left[\begin{array}{r} 1 \\ \exp(\beta_2 x_i)\tau_i \\ \beta_1\, x_i \exp(\beta_2 x_i)\tau_i \end{array}\right].$$

The Fisher information matrix is evaluated based on planning information for model parameters $\boldsymbol{\theta}^{\Box}$ and $\boldsymbol{\theta}^{(c)\Box}$.

### 22.8.2. *Large-sample approximate variance of* $h_t(\widehat{t}_p)$ *and* $\widehat{t}_p$

When $\widehat{\boldsymbol{\theta}} \dot{\sim} \text{MVN}(\boldsymbol{\theta}, \Sigma)$, for a scalar function $\widehat{g} = g(\widehat{\boldsymbol{\theta}}) \dot{\sim} \text{NOR}[g(\boldsymbol{\theta}), \text{Avar}(\widehat{g})]$, and the delta method gives

$$\text{Avar}(\widehat{g}) = \left[\frac{\partial g(\boldsymbol{\theta})}{\partial \boldsymbol{\theta}}\right]' \Sigma \left[\frac{\partial g(\boldsymbol{\theta})}{\partial \boldsymbol{\theta}}\right],$$

providing the large-sample approximate variance for the ML estimator of a desired function of $\boldsymbol{\theta}$.

$\text{Avar}[h_t(\widehat{t}_p)]$ can be written as a function of $\Sigma$ using delta method and expressed as

$$\text{Avar}[h_t(\widehat{t}_p)] = \boldsymbol{c}'\Sigma\boldsymbol{c} = \boldsymbol{c}'\boldsymbol{I}^{-1}_{\boldsymbol{\theta},\boldsymbol{\theta}^{(c)}}\boldsymbol{c},$$

where $\boldsymbol{c} = \partial h_t(t_p)/\partial\boldsymbol{\theta}$. For decreasing degradation, $h_t(t_p) = \nu + \varsigma\Phi^{-1}(p)$ for $p \geq \Phi\left(-\nu/\varsigma\right)$. In this case, the elements of $\boldsymbol{c}$ are

$$\frac{\partial h_t(t_p)}{\partial \beta_0} = -\frac{1}{\beta_1 \exp(\beta_2 x)} \qquad \frac{\partial h_t(t_p)}{\partial \beta_1} = -\frac{h_t(t_p)}{\beta_1}$$

$$\frac{\partial h_t(t_p)}{\partial \beta_2} = -x\, h_t(t_p) \qquad \frac{\partial h_t(t_p)}{\partial \sigma} = -\frac{\Phi^{-1}(p)}{\beta_1 \exp(\beta_2 x)}.$$

Using the delta method again

$$\text{Avar}(\widehat{t}_p) = \left(\left.\frac{\partial h_t^{-1}(z)}{\partial z}\right|_{h_t(t_p)}\right)^2 \text{Avar}[h_t(\widehat{t}_p)].$$

## 22.9. Bibliography


[BOU 94] BOULANGER M., ESCOBAR L.A., "Experimental design for a class of accelerated degradation tests", *Technometrics*, vol. 36, pp. 260–272, 1994.

[CRO 01] CROWDER M., *Classical Competing Risks*, Chapman and Hall/CRC, Boca Raton, FL, 2001.

[DAV 78] DAVID H.A., MOESCHBERGER M.L., *The Theory of Competing Risks*, Griffin, London, 1978.

[ESC 94] ESCOBAR L.A., MEEKER W.Q., "Fisher information matrix for the extreme value, normal, and logistic distributions and censored data", *Applied Statistics*, vol. 43, pp. 533–540, 1994.

[ESC 95] ESCOBAR L.A., MEEKER W.Q., "Planning accelerated life tests with two or more experimental Factors", *Technometrics*, vol. 37, pp. 411–427, 1995.

[ESC 98] ESCOBAR L.A., MEEKER W.Q., "Fisher information matrices with censoring, truncation, and explanatory variables", *Statistica Sinica*, vol. 8, pp. 221–237, 1998.

[ESC 03] ESCOBAR L.A., MEEKER W.Q., KUGLER D.L., KRAMER L.L., "Accelerated destructive degradation tests: data, models, and analysis", in LINDQVIST B.H., DOKSUM K.A. (eds), *Mathematical and Statistical Methods in Reliability*, pp. 319–337, World Scientific Publishing Company, River Edge, NJ, 2003.

[JEN 11] JENG S., HUANG B., MEEKER W.Q., "Accelerated destructive degradation tests robust to distribution misspecification", *IEEE Transactions on Reliability*, vol. 60, pp. 701–711, 2011.

[LAW 03] LAWLESS J.F., *Statistical Models and Methods for Lifetime Data*, John Wiley & Sons, New York, 2003.

[MEE 84] MEEKER W.Q., "A comparison of accelerated life test plans for weibull and lognormal distributions and type I censoring", *Technometrics*, vol. 26, pp. 157–171, 1984.

[MEE 98] MEEKER W.Q., ESCOBAR L.A., *Statistical Methods for Reliability Data*, John Wiley & Sons, New York, 1998.

[NEL 82] NELSON W., *Applied Life Data Analysis*, John Wiley & Sons, New York, 1982.

[NEL 90] NELSON W., *Accelerated Testing: Statistical Models, Test Plans, and Data Analyses*, John Wiley & Sons, New York, 1990.

[NEL 05a] NELSON W., "A bibliography of accelerated test plans", *IEEE Transactions on Reliability*, vol. 54, pp. 194–197, 2005.

[NEL 05b] NELSON W., "A bibliography of accelerated test plans part II - references", *IEEE Transactions on Reliability*, vol. 54, pp. 370–373, 2005.

[SHI 09] SHI Y., ESCOBAR L.A., MEEKER W.Q., "Accelerated destructive degradation test planning", *Technometrics*, vol. 51, pp. 1–13, 2009.

[SHI 12] SHI Y., MEEKER W.Q., "Bayesian methods for accelerated destructive degradation test planning", *IEEE Transactions on Reliability*, vol. 61, pp. 245–253, 2012.

[WHI 73] WHITTLE P., "Some general points in the theory of optimal experimental design", *Journal of the Royal Statistical Society, Series B*, vol. 35, pp. 123–130, 1973.

[ZHA 06] ZHANG Y., MEEKER W.Q., "Bayesian methods for planning accelerated life tests", *Technometrics*, vol. 48, pp. 49–60, 2006.

Chapter 23

# A New Goodness-of-Fit Test for Shape-Scale Families

In this chapter, a new goodness-of-fit test for validation of shape-scale families of probability distributions is proposed. Asymptotic distribution of the test statistic is found. Approximations of critical values for large and medium size samples are given. Asymptotic variance of the test statistic for widely used families such as Weibull, lognormal and loglogistic is calculated.

## 23.1. Introduction

Suppose that $T$ is a random variable with the cumulative distribution function (CDF) $F(t)$.

Let us consider the hypothesis

$$H : F(t) = F_0((t/\theta)^\nu) \quad \text{for all } t \in \mathbf{R},$$

where $F_0$ is a completely specified CDF, $F_0(t) = 0$ for $t < 0$ and $\theta, \nu$ are unknown parameters.

Several classes of tests for the hypothesis $H$ are used in statistical literature: tests based on differences of observed and expected numbers of failures in intervals of some partition of $[0, \infty)$ (chi-squared tests; see [NIK 73]); tests based on weighted

---

Chapter written by Vilijandas BAGDONAVIČIUS.

differences of non-parametric and parametric of the CDF (modifications of classical Kolmogorov–Smirnov, Cramér–von Mises, Anderson–Darling tests); tests based on the differences of logarithms of order statistics ([MAN 73] and [TIK 81] tests); tests based on empirical correlation coefficients between logarithms of order statistics and their "expected values" (correlation tests; [GER 79]), integrated loglikelihood tests [ZHA 02]. In the case of lognormal distribution, special normality tests (D'Agostino, Geary, Jarque–Berra, Sarkadi, Shapiro–Wilktests) (see [BAG 11]) may be used.

A new test is proposed in this chapter. Asymptotic distribution of the test statistic is found.

## 23.2. The test statistic

Let

$$T_1, ..., T_n$$

be a sample from the distribution $F$. Set

$$S_0(t) = 1 - F_0(t), \quad \Lambda_0(t) = -\ln S_0(t), \quad f_0(t) = F_0'(t),$$

$$\lambda_0(t) = \Lambda_0'(t), \quad G_0(t) = (\Lambda_0(t) - 1)^2, \quad g_0(t) = G_0'(t).$$

Denote by $\hat{\theta}, \hat{\nu}$ the maximum likelihood estimators of the parameters $\theta, \nu$. Note that under hypothesis $H$, the random variables $V_i = (T_i/\theta)^\nu$ are independent and identically distributed (i.i.d.) and each has the distribution function $F_0$, so the random variables $\Lambda_0(V_i)$ are also i.i.d. and each has the standard exponential distribution with the mean 1. The values of $\Lambda_0((T_i/\hat{\theta})^{\hat{\nu}})$ are scattered around 1 under the hypothesis, and a test may be based on the sum of squared differences $\sum_{i=1}^n G_0((T_i/\hat{\theta})^{\hat{\nu}}) = \sum_{i=1}^n (\Lambda_0((T_i/\hat{\theta})^{\hat{\nu}}) - 1)^2$. The median of the standard exponential distribution is $\ln 2$, so an alternative test based on the sum of squared differences $\sum_{i=1}^n (\Lambda_0((T_i/\hat{\theta})^{\hat{\nu}}) - \ln 2)^2$ may be considered.

So we propose a test for the hypothesis $H$ based on the following statistic:

$$\tilde{B} = \frac{1}{\sqrt{n}}(\sum_{i=1}^{n} G_0(\tilde{V}_i) - n), \quad \tilde{V}_i = (T_i/\hat{\theta})^{\hat{\nu}}.$$

For any $n$, the distribution of the random vector $(\tilde{V}_1, ..., \tilde{V}_n)$ is parameter-free; so the distribution of the statistic $\tilde{B}$ is also parameter-free.

## 23.3. The asymptotic distribution of the test statistic

Assumptions: under $H$, we have:

1)

$$(\sqrt{n}(\hat{\theta}-\theta), \sqrt{n}(\hat{\nu}-\nu))^T = \mathbb{1}^{-1}(\theta,\nu)(\frac{1}{\sqrt{n}}\dot{\ell}_\theta, \frac{1}{\sqrt{n}}\dot{\ell}_\nu)^T + o_P(1),$$

where $\dot{\ell}_\theta$ and $\dot{\ell}_\nu$ are the score functions and $\mathbb{1} = [i_{kl}]_{2\times 2}$ is the Fisher information matrix for one observation.

2) $\Delta_i < \infty$, $i = 1, 2, 3$, where

$$\Delta_1 = \int_0^\infty \frac{\partial^2}{\partial\theta^2} G_0((t/\theta)^\nu) dF_0((t/\theta)^\nu),$$

$$\Delta_2 = \int_0^\infty \frac{\partial^2}{\partial\theta\partial\nu} G_0((t/\theta)^\nu) dF_0((t/\theta)^\nu),$$

$$\Delta_3 = \int_0^\infty \frac{\partial^2}{\partial\nu^2} G_0((t/\theta)^\nu) dF_0((t/\theta)^\nu).$$

3) $A_i < \infty$, $i = 1, 2$, where

$$A_1 = \int_0^\infty x g_0(x) dF_0(x), \quad A_2 = \int_0^\infty x \ln x g_0(x) dF_0(x).$$

Assumptions 1–3 are evidently satisfied taking the Weibull, loglogistic or lognormal models. Set $\mathbb{1}^{-1} = [i^{kl}]_{2\times 2}$.

THEOREM 23.1.– Under assumptions 1–3:

$$\tilde{B} \xrightarrow{d} \tilde{b} \sim N(0,\sigma^2), \quad \sigma^2 = 8 - [\frac{\nu^2}{\theta^2} A_1^2 i^{11} - \frac{2}{\theta} A_1 A_2 i^{12} + \frac{1}{\nu^2} A_2^2 i^{22}].$$

PROOF.– The statistic $\tilde{B}$ can be written in the form $\tilde{B} = \tilde{B}_1 + B$, where

$$\tilde{B}_1 = \frac{1}{\sqrt{n}} \sum_{i=1}^n [G_0(\tilde{V}_i) - G_0(V_i)], \quad B = \frac{1}{\sqrt{n}} (\sum_{i=1}^n G_0(V_i) - n).$$

At first, let us consider the statistic $\tilde{B}_1$. Set

$$N(t) = \sum_{i=1}^{n} \mathbf{1}_{\{T_i \le t\}}, \quad \hat{F}(t) = N(t)/n.$$

$\hat{F}(t)$ is the empirical distribution function which has the following property:

$$\sqrt{n}(\hat{F}(t) - F(t)) \xrightarrow{d} V(t)$$

on $D[0,\infty)$, where $V$ is a Gaussian process with

$$\mathbf{E}V(u) = 0, \quad \mathbf{cov}(V(u), V(v)) = F(u)S(v), \quad u \le v.$$

The statistic $\tilde{B}_1$ can be written in the form

$$\begin{aligned}\tilde{B}_1 &= \int_0^\infty [G_0((t/\hat{\theta})^{\hat{\nu}}) - G_0((t/\theta)^{\nu})] d(\sqrt{n}(\hat{F}(t) - F(t))) \\ &\quad + \sqrt{n} \int_0^\infty [G_0((t/\hat{\theta})^{\hat{\nu}}) - G_0((t/\theta)^{\nu})] dF_0((t/\theta)^{\nu}) \\ &= \int_0^\infty [G_0((t/\hat{\theta})^{\hat{\nu}}) - G_0((t/\theta)^{\nu})] dF_0((t/\theta)^{\nu}) + o_P(1),\end{aligned}$$

because

$$\int_0^\infty [G_0((t/\hat{\theta})^{\hat{\nu}}) - G_0((t/\theta)^{\nu})] d(\sqrt{n}(N(t)/n - F(t))) \xrightarrow{P} 0.$$

We have

$$\tilde{B}_1 = \sqrt{n} \int_0^\infty (t/\theta)^{\nu} g_0((t/\theta)^{\nu})[-\frac{\nu}{\theta}\sqrt{n}(\hat{\theta} - \theta)$$

$$+\frac{1}{\nu} \ln(t/\theta)^{\nu} \sqrt{n}(\hat{\nu} - \nu)] dF_0((t/\theta)^{\nu})$$

$$+\Delta_{1n}\sqrt{n}(\hat{\theta} - \theta)^2 + 2\Delta_{2n}\sqrt{n}(\hat{\theta} - \theta)(\hat{\nu} - \nu) + \Delta_{3n}\sqrt{n}(\hat{\nu} - \nu)^2 + o_P(1),$$

$$\Delta_{in} \xrightarrow{P} \Delta_i, \quad i = 1, 2, 3.$$

We have

$$\tilde{B}_1 = -\frac{\nu}{\theta}\sqrt{n}(\hat{\theta}-\theta)^T\frac{1}{n}\sum_{i=1}^{n}V_i g_0(V_i) + \frac{1}{\nu}\sqrt{n}(\hat{\nu}-\nu)\frac{1}{n}\sum_{i=1}^{n}V_i\ln V_i g_0(V_i) + o_P(1).$$

By assumptions 1–3

$$(\sqrt{n}(\hat{\theta}-\theta),\sqrt{n}(\hat{\nu}-\nu))^T = \left(i^{11}\frac{1}{\sqrt{n}}\dot{\ell}_\theta + i^{12}\frac{1}{\sqrt{n}}\dot{\ell}_\nu, i^{12}\frac{1}{\sqrt{n}}\dot{\ell}_\theta + i^{22}\frac{1}{\sqrt{n}}\dot{\ell}_\nu\right)^T + o_P(1).$$

So

$$\tilde{B}_1 = C_1\frac{1}{\sqrt{n}}\dot{\ell}_\theta + C_2\frac{1}{\sqrt{n}}\dot{\ell}_\nu + o_P(1),$$

where

$$C_1 = -\frac{\nu}{\theta}A_1 i^{11} + \frac{1}{\nu}A_2 i^{12}, \quad C_2 = -\frac{\nu}{\theta}A_1 i^{12} + \frac{1}{\nu}A_2 i^{22}.$$

To find the Fisher information matrix $\mathbb{1}$, the loglikelihood function $\ell_i(\theta,\nu)$ for one observation $T_i$ is used. Under hypothesis $H$, the loglikelihood function for the observation is

$$\ell_i(\theta,\nu) = (\nu-1)\ln T_i + \ln\nu - \nu\ln\theta + (\ln f_0)((T_i/\theta)^\nu).$$

The score functions are

$$\dot{\ell}_{i\theta} = -\frac{\nu}{\theta}[1 + (T_i/\theta)^\nu(\ln f_0)'((T_i/\theta)^\nu)],$$

$$\dot{\ell}_{i\nu} = \frac{1}{\nu}[1 + \ln(T_i/\theta)^\nu + (T_i/\theta)^\nu\ln(T_i/\theta)^\nu(\ln f_0)'((T_i/\theta)^\nu)].$$

Set

$$\tilde{B}_1^0 = C_1\frac{1}{\sqrt{n}}\dot{\ell}_\theta + C_2\frac{1}{\sqrt{n}}\dot{\ell}_\nu = \frac{1}{\sqrt{n}}\sum_{j=1}^{n}\xi_{1j},$$

$$\tilde{B} = \frac{1}{\sqrt{n}} \sum_{j=1}^{n} (G_0(V_i) - 1) = \frac{1}{\sqrt{n}} \sum_{j=1}^{n} \xi_{2j},$$

where

$$\xi_{1j} = C_1 \dot{\ell}_{j\theta} + C_2 \dot{\ell}_{j\nu}, \quad \xi_{2j} = G_0(V_j) - 1.$$

The limit distribution of the random vector $(\tilde{B}_1, \tilde{B})^T$ coincides with the limit distribution of the random vector $(\tilde{B}_1^0, \tilde{B})^T$.

Let us find the covariance matrix

$$\Sigma = [\sigma_{ik}]_{2\times 2}, \quad \sigma_{ik} = \mathbf{cov}(\xi_{ij}, \xi_{kj}).$$

We have

$$\begin{aligned}
\sigma_{11} &= \mathbf{Var}(\xi_{11}) = C_1^2 i_{11} + 2C_1 C_2 i_{12} + C_2^2 i_{22} \\
&= \frac{\nu^2}{\theta^2} A_1^2 i^{11} - \frac{2}{\theta} A_1 A_2 i^{12} + \frac{1}{\nu^2} A_2^2 i^{22}, \\
\sigma_{12} = \mathbf{cov}(\xi_{11}, \xi_{21}) &= C_1 \mathbf{cov}(\dot{\ell}_{1\theta}, G_0(V_1)) + C_2 \mathbf{cov}(\dot{\ell}_{1\nu}, G_0(V_1)), \\
\mathbf{cov}(\dot{\ell}_{1\theta}, G_0(V_1)) &= -\frac{\nu}{\theta}[1 + \int_0^\infty tG_0(t)(\ln f_0)'(t) dF_0(t)] \\
&= -\frac{\nu}{\theta}[1 - \int_0^\infty [G_0(t) + tg_0(t)] dF_0(t) = \frac{\nu}{\theta} A_1, \\
\mathbf{cov}(\dot{\ell}_{1\nu}, G_0(V_1)) &= \frac{1}{\nu}[1 + \int_0^\infty G_0(t) \ln t (1 + t(\ln f_0)'(t)) dF_0(t)] \\
&= \frac{1}{\nu}[1 + \int_0^\infty G_0(t) \ln t \, dF_0(t) \\
&\quad - \int_0^\infty [(\ln t + 1) G_0(t) + t \ln t \, g_0(t)] dF_0(t)] \\
&= -\frac{A_2}{\nu}.
\end{aligned}$$

So

$$\sigma_{12} = C_1 \frac{\nu}{\theta} A_1 - C_2 \frac{A_2}{\nu} = -A_1^2 \frac{\nu^2}{\theta^2} i^{11} + \frac{2}{\theta} A_1 A_2 i^{12} - \frac{1}{\nu^2} A_2^2 i^{22} = -\sigma_{11}.$$

Note that under the hypothesis

$$\mu_0 = \mathbf{E}\, G_0(V_i) = \int_0^\infty G_0(u)dF_0(u) = \int_0^\infty (x-1)^2 e^{-x} dx = 1,$$

$$\sigma_{22} = \mathbf{Var}\, G_0(V_i) = \int_0^\infty G_0^2(u)dF_0(u) - \mu_0^2 = \int_0^\infty (x-1)^4 e^{-x} dx - 1 = 8.$$

By the central limit theorem,

$$(\tilde{B}_1^0, B) \xrightarrow{d} (b_1, b) \sim N_2(\mathbf{0}, \Sigma).$$

Finally

$$\tilde{B} = \tilde{B}_1^0 + B + o_P(1) \xrightarrow{d} b_1 + b \sim N(0, \sigma^2), \quad \sigma^2 = \sigma_{11} + 2\sigma_{12} + \sigma_{22} = 8 - \sigma_{11}.$$

The proof is complete.

The theorem implies that if $n$ is large, then the distribution of the test statistic is approximated with the normal law with zero mean and variance $\sigma^2$, which is known when the baseline distribution function $F_0$ is specified.

The second derivatives of the loglikelihood function are

$$\ddot{\ell}_{i\theta^2} = \frac{\nu^2}{\theta^2} V_i^2 (\ln f_0)''(V_i) + \frac{\nu}{\theta^2} + \frac{\nu(\nu+1)}{\theta^2} V_i (\ln f_0)'(V_i),$$

$$\ddot{\ell}_{i\theta\nu} = -\frac{1}{\theta} V_i^2 \ln V_i (\ln f_0)''(V_i) - \frac{1}{\theta} V_i (\ln f_0)'(V_i) - \frac{1}{\theta} V_i \ln V_i (\ln f_0)'(V_i) - \frac{1}{\theta},$$

$$\ddot{\ell}_{i\nu^2} = \frac{1}{\nu^2} V_i^2 \ln^2 V_i (\ln f_0)''(V_i) + \frac{1}{\nu^2} V_i \ln^2 V_i (\ln f_0)'(V_i) - \frac{1}{\nu^2}.$$

The elements of the Fisher information matrix $\mathbb{1}$ are $i_{11} = -\mathbf{E}\dot{\ell}_{i\theta^2}$, $i_{12} = -\mathbf{E}\dot{\ell}_{i\theta\nu}$, $i_{22} = -\mathbf{E}\dot{\ell}_{i\nu^2}$. So

$$i_{11} = -\frac{\nu^2}{\theta^2} \int_0^\infty x[x(\ln f_0)''(x) + (\ln f_0)'(x)]dF_0(x),$$

$$i_{12} = \frac{1}{\theta}\{1 + \int_0^\infty x[x \ln x (\ln f_0)''(x) + (1 + \ln x)(\ln f_0)'(x)]dF_0(x)\},$$

$$i_{22} = \frac{1}{\nu^2}\{1 - \int_0^\infty x \ln^2 x [x(\ln f_0)''(x) + (\ln f_0)'(x)]dF_0(x)\}.$$

## 23.4. The test

The test statistic $\mathcal{Z}$ is defined as follows:

$$\mathcal{Z} = \frac{1}{\sigma\sqrt{n}}(\sum_{i=1}^{n} G_0((T_i/\hat{\theta})^{\hat{\nu}}) - n).$$

As mentioned earlier, for any $n$ the distribution of the random vector $(\tilde{V}_1, ..., \tilde{V}_n)$ is parameter-free; so the distribution of the test statistic is also parameter-free. For sample size $n$, denote by $b_\alpha(n)$ the $\alpha$ critical value of the statistic $\mathcal{B}$ under $H$. The critical values are simply found by simulation.

Hypothesis $H$ is rejected with the significance level $\alpha$ if $\mathcal{Z} < b_{1-\alpha/2}(n)$ or $\mathcal{Z} > b_{\alpha/2}(n)$.

We proved that $\mathcal{Z} \xrightarrow{d} Z \sim N(0,1)$. So for large $n$, we can use the approximations

$$b_\alpha(n) \approx z_\alpha,$$

where $z_\alpha$ is the $\alpha$ critical value of the standard normal distribution.

Taking into account that the sum $\sum_{i=1}^{n} G_0((T_i/\hat{\theta})^{\hat{\nu}})$ takes positive values and $z_\alpha \approx (\chi^2_\alpha(n) - n)/\sqrt{2n}$, we may use the approximation

$$b_\alpha(n) \approx (\chi^2_\alpha(n) - n)/\sqrt{2n},$$

which may be more accurate than the approximation $b_\alpha(n) \approx z_\alpha$ for intermediate values of $n$.

## 23.5. Weibull distribution

In the case of the Weibull distribution

$$S_0(t) = e^{-t}, \quad \Lambda_0(t) = t, \quad f_0(t) = e^{-t}, \quad \lambda_0(t) = 1,$$

$$G_0(t) = (t-1)^2, \quad g_0(t) = 2(t-1), \quad (\ln f_0)'(t) = -1, \quad (\ln f_0)''(t) = 0,$$

$$A_1 = 2\int_0^\infty x(x-1)e^{-x}dx = 2,$$

$$A_2 = \int_0^\infty x(x-1)\ln x e^{-x}dx = 2[1 + \Gamma'(2)],$$

$$\mathbb{1} = \begin{pmatrix} \nu^2/\theta^2 & -\Gamma'(2)/\theta \\ -\Gamma'(2)/\theta & [1+\Gamma''(2)]/\nu^2 \end{pmatrix},$$

$$\mathbb{1}^{-1} = \frac{1}{det}\begin{pmatrix} \theta^2[1+\Gamma''(2)]/\nu^2 & \theta\Gamma'(2) \\ \theta\Gamma'(2) & \nu^2 \end{pmatrix}, \quad det = 1+\Gamma''(2) - (\Gamma'(2))^2.$$

So

$$\sigma_{11} = \frac{4\theta^2}{\nu^2}[1+\Gamma''(2) - 2(1+\Gamma'(2))\Gamma'(2) + (1+\Gamma'(2))^2)/det = 4 + \frac{4}{det},$$

$$\sigma^2 = 4 - \frac{4}{det} \approx 1.5682916.$$

The test statistic is

$$\mathcal{B} = \frac{1}{\sigma\sqrt{n}}(\sum_{i=1}^{n}(\tilde{V}_i - 1)^2 - n).$$

## 23.6. Loglogistic distribution

In the case of the loglogistic distribution

$$S_0(t) = \frac{1}{1+t}, \quad \Lambda_0(t) = \ln(1+t), \quad f_0(t) = \frac{1}{(1+t)^2}, \quad \lambda_0(t) = \frac{1}{1+t},$$

$$G_0(t) = (\ln(1+t) - 1)^2, \quad g_0(t) = \frac{2}{1+t}(\ln(1+t) - 1),$$

$$(\ln f_0)'(t) = -\frac{2}{1+t}, \quad (\ln f_0)''(t) = \frac{2}{(1+t)^2}.$$

$$A_1 = 2\int_0^\infty \frac{x}{(1+x)^3}[\ln(1+x) - 1]dx$$

$$= 2\int_0^\infty \ln(1+x) - 1]d(-\frac{1}{x+1} + \frac{1}{2(x+1)^2})-$$

$$2(1-1/2) = 2\int_0^\infty (\frac{1}{(x+1)^2} - \frac{1}{2(x+1)^3})dx - 1 = \frac{1}{2},$$

$$A_2 = 2\int_0^{\infty} \frac{x}{(1+x)^3}[\ln(1+x)-1]\ln x dx \approx 2.144934066,$$

$$\mathbb{1} = \begin{pmatrix} \nu^2/(3\theta^2) & 0 \\ 0 & d/\nu^2 \end{pmatrix},$$

$$\mathbb{1}^{-1} = \begin{pmatrix} 3\theta^2/\nu^2 & 0 \\ 0 & \nu^2/d \end{pmatrix}, \quad d = 1 + \frac{2}{3}(\Gamma''(2) - (\Gamma'(2))^2).$$

So

$$\sigma_{11} = \frac{\nu^2}{\theta^2}\frac{1}{4}\frac{3\theta^2}{\nu^2} + \frac{1}{\nu^2}A_2^2\frac{\nu^2}{d} = \frac{3}{4} + \frac{A_2^2}{d} \approx 3,96740110,$$

$$\sigma^2 = 8 - \sigma_{11} \approx 4,03259890.$$

The test statistic is

$$Z = \frac{1}{\sigma\sqrt{n}}(\sum_{i=1}^{n}[\ln(1+\tilde{V}_i)-1] - n).$$

## 23.7. Lognormal distribution

$$f_0(t) = \varphi(\ln t)/t, \quad (\ln f_0)'(t) = -\frac{1+\ln t}{t}, \quad (\ln f_0)''(t) = \frac{\ln t}{t^2},$$

$$G_0(t) = [1+\ln(1-\Phi(\ln t))]^2, \quad g_0(t) = \frac{2[1+\ln(1-\Phi(\ln t))]}{1-\Phi(\ln t)}\varphi(\ln t)\frac{1}{t},$$

where $\varphi(x)$ and $\Phi(x)$ are the density function and the CDF of the standard normal distribution, respectively

$$\mathbb{1} = \begin{pmatrix} \nu^2/\theta^2 & 0 \\ 0 & 2/\nu^2 \end{pmatrix}, \quad \mathbb{1}^{-1} = \begin{pmatrix} \theta^2/\nu^2 & 0 \\ 0 & \nu^2/2 \end{pmatrix},$$

$$A_1 = -2\int_{-\infty}^{\infty} \varphi^2(x)\frac{1+\ln(1-\Phi(x))}{1-\Phi(x)})dx \approx 1.19160993,$$

$$A_2 = -2\int_{-\infty}^{\infty} x\varphi^2(x)\frac{1+\ln(1-\Phi(x))}{1-\Phi(x)}dx \approx 2.70710729.$$

So

$$\sigma_{11} = A_1^2 + \frac{A_2^2}{2} \approx 5,08414917,$$

$$\sigma^2 = 8 - \sigma_{11} \approx 2,91585083.$$

The test statistic is

$$\mathcal{Z} = \frac{1}{\sigma\sqrt{n}}(\sum_{i=1}^{n}[\ln(1-\Phi(\ln\tilde{V}_i))+1]^2 - n).$$

## 23.8. Bibliography

[BAG 11] BAGDONAVIČIUS V., KRUOPIS J., NIKULIN M., *Non-Parametric Tests for Complete Data*, ISTE, London, John Wiley and Sons, New York, 2011.

[GER 79] GERLACH B., "A consistent correlation-type goodness-of-fit test with application to the two-parameter Weibull distribution", *Series Statistics,* vol. 10, pp. 427–452, 1979.

[MAN 73] MANN M.R., SCHEUER E.M., FERTIG K.W., "A new goodness-of-fit test for the two parameter Weibull or extreme- value distribution with unknown parameters", *Communications in Statistics,* vol. 2, pp. 283–400, 1973.

[NIK 73] NIKULIN M.S., "Chi-square test for continuous distributions with shift and scale parameters", *Theory of Probability and Its Applications,* vol. 18, pp. 559–568, 1973.

[TIK 81] TIKU M.L., SINGH M., "Testing the two parameter Weibull distribution", *Communications in Statistics Part A.,* vol. 10, pp. 907–917, 1981.

[ZHA 02] ZHANG J., "Powerful goodness-of-fit tests based on the likelihood ratio", *Journal of Royal Statistical Society B,* vol. 64, pp. 281–294, 2002.

Chapter 24

# Time-to-Failure of Markov-Modulated Gamma Process with Application to Replacement Policies

## 24.1. Introduction

One of the most popular stochastic processes used to model degradation of a device in reliability theory is the gamma process. Briefly, a stochastic process is a stationary gamma process if, and only if, its increments are stationary, independent and gamma distributed.

In several papers, covariates or random effects have been incorporated into a gamma process in order to take into account the environment or the individual heterogeneity. For instance, random effects have been considered by Lawless and Crowder [LAW 05] assuming that the scale parameter is random and may depend on covariates. Bagdonavičius and Nikulin [BAG 01] proposed an accelerated life model by scaling the time index by time-dependent covariates. However, (time-dependent or not) covariates are all assumed to be deterministic.

Here, we propose and study a gamma process integrating covariates, which evolves according to a Markov jump process (which is assumed to be independent of the underlying gamma processes). For the sake of simplicity, we restrict ourselves to a two-state Markov process (or binary Markov process), but it can be extended to a multistate Markov process. This Markov process represents the environment in

Chapter written by Christian PAROISSIN and Landy RABEHASAINA.

which the device is used. For instance, assume that the device could be used under nominal stress (state 0) or accelerated stress (state 1). When the transition rate from 1 to 0 is null, it turns out to be a special case of the Saassouh *et al.* [SAA 07] for which we will assume an exponential duration in the nominal state. Another related model is the one studied by Zhao *et al.* [ZHA 10b] (see also [ZHA 10a]). Related models have been considered earlier (see [SIN 95], for instance).

First, we will define the model we propose, also providing some basic properties. Such model is called Markov-modulated gamma process or Markov regime-switching gamma process. In section 24.3, we will consider the time-to-failure distribution for the Markov-modulated gamma process. Finally, we will also study numerically two replacement policies associated with this degradation model.

## 24.2. Degradation model

In the proposed model, we consider that the degradation process depends on binary covariates, i.e. the covariates can take two values, 0 or 1. However, it should be easily extended to the case of finite state space. Here, the two states, 0 and 1, may represent, respectively, the normal condition of a device and a higher level of solicitation.

### 24.2.1. *Covariate process*

The covariates (or solicitation conditions) are modeled by a binary Markov process $(J(t))$. We will denote by $\lambda$ (respectively, $\mu$) the rate from state 0 to state 1 (respectively, from state 1 to state 0). Hence, the infinitesimal generator $Q$ is given by

$$Q = \begin{pmatrix} -\lambda & \lambda \\ \mu & -\mu \end{pmatrix}.$$

We recall that a Markov process is characterized by its infinitesimal generator and its initial distribution $\nu$. For any $t \geqslant 0$, the transition matrix between instants 0 and $t$ is given by $P_t = \exp(tQ)$ and thus the distribution of $J(t)$ is $\nu \exp(tQ)$. For any $t \geqslant 0$

$$P_t = \frac{1}{\lambda+\mu}\begin{pmatrix} \mu & \lambda \\ \mu & \lambda \end{pmatrix} + \frac{e^{-(\lambda+\mu)t}}{\lambda+\mu}\begin{pmatrix} \lambda & -\lambda \\ -\mu & \mu \end{pmatrix}.$$

It follows that its unique stationary distribution $\pi$ is equal to

$$\pi = \begin{pmatrix} \frac{\mu}{\lambda+\mu} \\ \frac{\lambda}{\lambda+\mu} \end{pmatrix}.$$

In the general context, we have to assume that $(J(t))$ is an homogeneous, irreducible and recurrent Markov process. Most of the time, the transition matrix cannot be computed explicitly.

### 24.2.2. *Degradation process*

The degradation process $(D(t))$ that we propose is a non-homogeneous gamma process with shape function $(\eta(t))$ and scale parameter $\xi$. The shape function is piecewise-linear, depending on the covariate process $(J(t))$ as follows

$$\forall n \in \mathbb{N}, \quad \eta(t) = \begin{cases} \eta(\tau_{2n}) + \alpha_0(t - \tau_{2n}) & \text{if } \tau_{2n} < t \leqslant \tau_{2n+1}, \\ \eta(\tau_{2n+1}) + \alpha_1(t - \tau_{2n+1}) & \text{if } \tau_{2n+1} < t \leqslant \tau_{2n+2}, \end{cases}$$

where $(\tau_n)_{n \geqslant 0}$ are the instants of jumps of $(J(t))$, with the convention that $\tau_0 = 0$ and $\eta(0) = 0$. The sequence $(\tau_n)_{n \geqslant 0}$ is an alternated renewal process.

This process can also be described by considering its increments and gamma processes in the non-modulated cases. Let $(D^{(0)}(t))$ (respectively, $(D^{(1)}(t))$) be the homogeneous gamma process with shape parameter $\alpha_0$ (respectively, $\alpha_1$) and scale parameter $\xi$. We have the following two identities, conditionally to $(\tau_n)_{n \geqslant 0}$,

$$\forall t \in (\tau_{2n}, \tau_{2n+1}], \quad D(t) - D(\tau_{2n}) \overset{(d)}{=} D^{(0)}(t - \tau_{2n})$$

and

$$\forall t \in (\tau_{2n+1}, \tau_{2n+2}], \quad D(t) - D(\tau_{2n+1}) \overset{(d)}{=} D^{(0)}(\tau_{2n+1} - t),$$

where $\overset{(d)}{=}$ stands for equality in distribution.

In Figure 24.1, a simulation of the degradation process is plotted. The curve with dotted lines (respectively, full lines) corresponds to a path of the gamma process under nominal stress (respectively, under high stress) while the curve with dashed lines corresponds to a path of the Markov-modulated gamma process. Light gray represents the periods of nominal stress while dark gray represents the periods of high stress.

## 24.3. Time-to-failure distribution

Time-to-failure is defined as the time to reach a fixed level $c > 0$ by a Markov-modulated gamma process

$$T_c = \inf \{t \geqslant 0 \, ; \, D(t) \geqslant c\} .$$

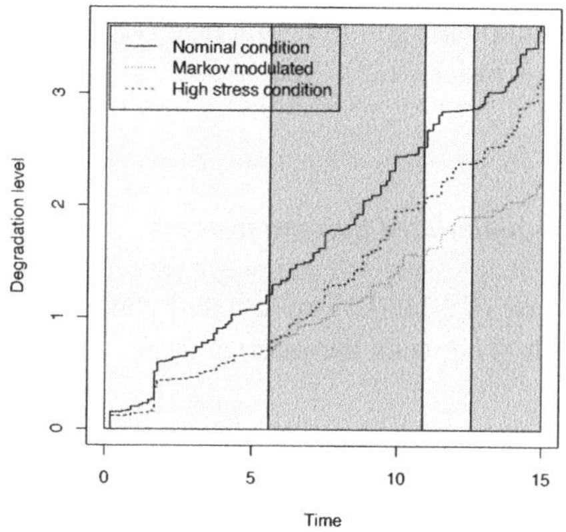


**Figure 24.1.** *Simulation of sample paths*

Since $(D(t))$ has increasing paths, it follows that

$$\forall t \geqslant 0\,, \quad \mathbb{P}[T_c > t] = \mathbb{P}[D(t) < c].$$

Thus, it follows that it is sufficient to study the distribution of $D(t)$ for any $t \geqslant 0$. We first recall some facts about the time-to-failure for the (non-modulated) gamma process. Then, we will consider the modulated case.

### 24.3.1. *Case of a non-modulated gamma process*

First, we consider the distribution of the time-to-failure for a non-modulated gamma process. According to Park and Padgett [PAR 04], the cumulative distribution function (cdf) of $T_c$ is, for any $t \geqslant 0$,

$$F_{T_c}(t) = \mathbb{P}[D(t) \geqslant c] = \frac{\Gamma(\alpha t, c/\xi)}{\Gamma(\alpha t)}, \qquad [24.1]$$

where $\Gamma(\cdot, \cdot)$ is the upper incomplete gamma function. The ratio $F_{T_c}$ is also called normalized upper incomplete gamma function. The probability density function (pdf) of $T_c$ is, for any $t \geqslant 0$,

$$f_{T_c}(t) = \alpha \left( \Psi(\alpha t) - \log\left(\frac{c}{\xi}\right)\right) \frac{\gamma(\alpha t, c/\xi)}{\Gamma(\alpha t)} + \frac{1}{\alpha t^2 \Gamma(\alpha t)} \left(\frac{c}{\xi}\right)^{\alpha t} {}_2F_2(\alpha t, \alpha t; \alpha t + 1, \alpha t + 1; -c/\xi),$$

where $\Psi$ is the di-gamma function (or logarithmic derivative of the gamma function), $\gamma(\cdot,\cdot)$ is the lower incomplete gamma function and ${}_2F_2$ is the generalized hypergeometric function of order $(2,2)$. See [ABR 72] or [GRA 65] for details on special functions.

### 24.3.2. *Case of a Markov-modulated gamma process*

Let us reconsider the case of a Markov-modulated gamma process. Consider $\Delta_{[0,t]}$ the occupation time of state 0 between the interval $[0,t]$

$$\Delta_{[0,t]} = \int_0^t \mathbb{I}_{J(u)=0} \mathrm{d}u.$$

Such random variables have already been studied in the literature (see [KOV 10] and references therein). In this chapter, Kovchegov *et al.* [KOV 10] obtained a closed-form expression of the Laplace transform of the pdf of $\Delta_{[0,t]}$ for any Markov jump process. For the particular two-state Markov process, the generalized pdf was initially derived by Pedler [PED 71]: for all $u \in [0,t]$

$$\begin{aligned} f_{\Delta_{[0,t]}}(u)\mathrm{d}u &= e^{-\lambda t}\delta_t(u) \\ &+ e^{-\lambda u}e^{-\mu(t-u)}\left[\lambda I_0\left(2\sqrt{\lambda\mu u(t-u)}\right) + \sqrt{\frac{\lambda\mu u}{t-u}}I_1\left(2\sqrt{\lambda\mu u(t-u)}\right)\right]\mathrm{d}u, \end{aligned}$$

where $\delta_t$ is the Dirac mass at point $t$ and $I_r$ is the modified Bessel function of the kind

$$I_r(z) = \sum_{k=0}^{\infty} \frac{1}{k!\Gamma(k+r+1)}\left(\frac{z}{2}\right)^{2k+r}, \quad \text{for } r > -1.$$

Hence, the distribution of $\Delta_{[0,t]}$ can be viewed roughly speaking as a mixture of a Dirac distribution (corresponding to the event that no jump occurs between 0 and $t$) and an absolutely continuous distribution.

The independent increment property of the gamma process, as well its independence from the Markov process, implies that we have equality in distribution

$$D(t) \stackrel{(d)}{=} D^{(0)}(\Delta_{[0,t]}) + D^{(1)}(t - \Delta_{[0,t]}).$$

Thus, after conditioning on $\Delta_{[0,t]}$, we can decompose the integral as previously and get:

$$\mathbb{P}[T_c \geqslant t] = \int_0^t \left[1 - \frac{\Gamma(\alpha_0 u + \alpha_1(t-u), c/\xi)}{\Gamma(\alpha_0 u + \alpha_1(t-u))}\right] f_{\Delta_{[0,t]}}(u)\mathrm{d}u, \qquad [24.2]$$

using the fact that the convolution of gamma distributions with the same scale parameter is still a gamma distribution. It seems that there is no closed expression for the survival function of $T_c$. However, numerical computations can be easily done, since Bessel functions are implemented in many software or languages.

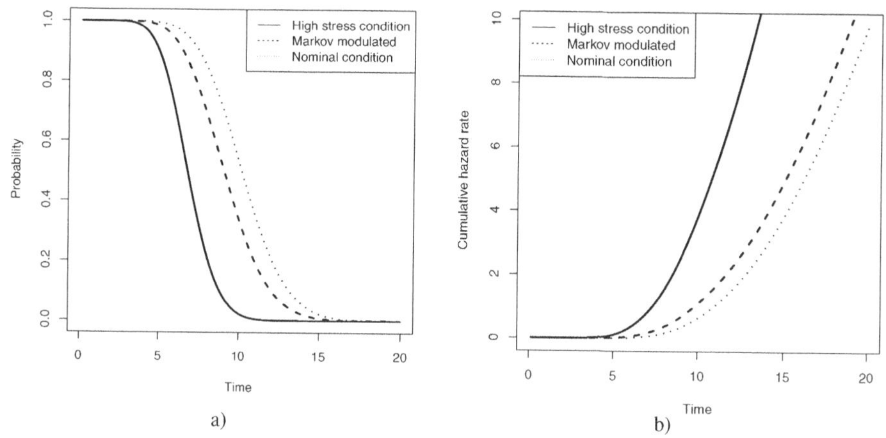


**Figure 24.2.** *Survival functions a) and cumulative hazard rates b)*

Computations were carried out in order to compare time-to-failure distributions. The parameters were set as follows: $\lambda = \mu = 1/10$, $\xi = 0.1$, $\alpha_0 = 2$, $\alpha_1 = 3$ and $c = 2$. In Figure 24.2, we represent, respectively, survival function and cumulative hazard rate of time-to-failures. The curves with dotted lines correspond to the case when the covariate is always equal to 0 (i.e. when $\mu = 0$) while the curves with full lines correspond to the case when the covariate is always equal to 1 (i.e. when $\lambda = 0$). The curves with dashed lines correspond to the modulated case.

### 24.3.3. *Stochastic comparison*

Let us denote by $T_c^{(0)}$ the time-to-failure when $\mu = 0$ (deterioration only under the nominal mode) and by $T_c^{(1)}$ the time-to-failure when $\lambda = 0$ (deterioration only under the high-stress mode). The following quite natural result establishes the stochastic orders (in the usual sense) between the three times-to-failure.

THEOREM 24.1.– We have the following relations

$$T_c^{(1)} \preceq_{st} T_c \preceq_{st} T_c^{(0)}.$$

PROOF.– We have to show that, for all $t \geqslant 0$, we have $\mathbb{P}[T_c^{(1)} > t] \leqslant \mathbb{P}[T_c > t] \leqslant \mathbb{P}[T_c^{(0)} > t]$. Let us first note that, from equation [24.1], $\mathbb{P}[T_c \leqslant t]$ is an increasing function in $\alpha$: this can be easily observed as it is an increasing function in $t$ (being a cdf) and because of obvious symmetry of $\alpha$ and $t$. This, in particular, entails that $\Gamma(., c/\xi)/\Gamma(.)$ is an increasing function, and, since $\alpha_0 \leqslant \alpha_1$, this also proves that $T_c^{(1)} \preceq_{st} T_c^{(0)}$. Let us now prove that $\mathbb{P}[T_c > t] \leqslant \mathbb{P}[T_c^{(0)} > t]$, the same method applying for proving that $\mathbb{P}[T_c^{(1)} > t] \leqslant \mathbb{P}[T_c > t]$. By using the same argument as before, and by the remark that $\Gamma(., c/\xi)/\Gamma(.)$ is increasing, we have from equation [24.2] that

$$\begin{aligned}
\mathbb{P}[T_c > t] &= \int_0^t \left[1 - \frac{\Gamma(\alpha_0 u + \alpha_1(t-u), c/\xi)}{\Gamma(\alpha_0 u + \alpha_1(t-u))}\right] f_{\Delta_{[0,t]}}(u)\mathrm{d}u \\
&\leqslant \int_0^t \left[1 - \frac{\Gamma(\alpha_0 u + \alpha_0(t-u), c/\xi)}{\Gamma(\alpha_0 u + \alpha_0(t-u))}\right] f_{\Delta_{[0,t]}}(u)\mathrm{d}u \\
&= \int_0^t \left[1 - \frac{\Gamma(\alpha_0 t, c/\xi)}{\Gamma(\alpha_0 t)}\right] f_{\Delta_{[0,t]}}(u)\mathrm{d}u \\
&= \mathbb{P}[T_c^{(0)} > t],
\end{aligned}$$

which completes the proof.

As is well known, the above result implies the following order between mean time to failures (MTTF)

$$\mathbb{E}[T_c^{(1)}] \leqslant \mathbb{E}[T_c] \leqslant \mathbb{E}[T_c^{(0)}].$$

Using the approximation of the MTTF (in the non-modulated case) obtained by Bérenguer *et al.* [BEŔ 03], we can deduce lower and upper bounds for the MTTF in the modulated case (provided that $c/\xi$ is large enough)

$$\frac{1}{\alpha_1}\left(\frac{c}{\xi} + \frac{1}{2}\right) \lessapprox \mathbb{E}[T_c] \lessapprox \frac{1}{\alpha_0}\left(\frac{c}{\xi} + \frac{1}{2}\right).$$

We illustrate these inequalities by considering the same simulations as previously. Results are reported in Table 24.1.

| MTTF | $\mathbb{E}[T_c^{(1)}]$ | $\mathbb{E}[T_c]$ | $\mathbb{E}[T_c^{(0)}]$ |
|---|---|---|---|
| Numerical computation | 6.83 | 8.39 | 10.25 |
| Approximation | 6.83 | - | 10.25 |

**Table 24.1.** *Approximations and bounds for the MTTF*

More precisely, we have an explicit relationship between survival functions $S_{T_c^{(0)}}$ and $S_{T_c^{(1)}}$, respectively, of $T_c^{(0)}$ and $T_c^{(1)}$. Indeed, from their expression (see equation [24.1]), we have

$$\forall t \geqslant 0, \quad S_{T_c^{(1)}}(t) = S_{T_c^{(0)}}(\gamma t),$$

with $\gamma = \alpha_1/\alpha_0 > 1$. So, we have $\mathbb{E}[T_c^{(1)}] = \gamma^{-1}\mathbb{E}[T_c^{(0)}]$. It also follows that their pdfs $f_{T_c^{(0)}}$ and $f_{T_c^{(1)}}$ satisfy

$$\forall t \geqslant 0, \quad f_{T_c^{(1)}}(t) = \gamma f_{T_c^{(0)}}(\gamma t)$$

and at least for their failure (or hazard) rate functions $h_{T_c^{(0)}}$ and $h_{T_c^{(1)}}$ satisfy

$$\forall t \geqslant 0, \quad h_{T_c^{(1)}}(t) = \gamma h_{T_c^{(0)}}(\gamma t).$$

Abdel-Hameed [ABD 75] has proved that $T_c^{(0)}$ (and also $T_c^{(1)}$) has an increasing failure rate. Then, for any $t > 0$, $h_{T_c^{(1)}}(t) > h_{T_c^{(0)}}(t)$.

## 24.4. Replacement policies

In this section, we consider two replacement policies: the block replacement policy and the age replacement policy. For more details on these policies (and for others), one can refer, respectively, to Chapters 5 and 3 in [NAK 05]. These two policies depend on a single parameter $\delta$, the delay between two consecutive inspections and induce a certain cost. Thus, one can be interested in determining the optimal inter-inspection delay $\delta_\star$. To determine this, we consider the asymptotic cost per unit of time defined as follows

$$C(\delta) = \lim_{t\to\infty} \frac{C_t(\delta)}{t},$$

where $C_t(\delta)$ is the cost over the time interval $[0, t]$ when the device is inspected at $(k\delta)_{k\in\mathbb{N}}$. As is well known, according to the renewal theory, we have

$$C(\delta) = \frac{\text{Expected cost over a cycle}}{\text{Expected cycle length}}.$$

### 24.4.1. *Block replacement policy*

We assume that the degradation level of a device can be measured only during inspections (i.e. no continuous monitoring) and that, at each replacement, the device is replaced by a new device or is perfectly repaired (AGAN), the replacement/repair duration being negligible. Moreover, replacement occurs only after an inspection (in particular there is no replacement at times-to-failure). Such scheme is the so-called block replacement policy.

Hence, there exist two different costs: the cost $c_r$ for replacing the device by a new device and the unavailability cost $c_u$. We have

$$C_{brp}(\delta) = \frac{\mathbb{E}[c_r + c_u(\delta - T_c)_+]}{\delta} = \frac{c_r + c_u \int_0^\delta F_{T_c}(u)\mathrm{d}u)}{\delta},$$

where $F_{T_c}$ is the cumulative distribution of $T_c$. We have

$$\lim_{\delta\to 0} C_{brp}(\delta) = +\infty \quad \text{and} \quad \lim_{\delta\to+\infty} C_{brp}(\delta) = c_u.$$

Let us denote $\delta_\star := \text{argmin}_{\delta>0} C_{brp}(\delta)$. Differentiating the above expression of the cost function, $\delta_\star$ is the solution of the following equation with respect to $t$

$$\phi_{brp}(t) = \int_0^t u f_{T_c}(u)\mathrm{d}u = \mathbb{E}[T_c \mathbf{1}_{T_c \leqslant t}] = \frac{c_r}{c_u} (\leqslant 1).$$

The solution depends on the two costs $c_r$ and $c_u$ only through their ratio (and so does not depend on the monetary unit). Since $\phi_{brp}$ is an increasing function toward $\mathbb{E}[T_c]$, it follows that $\delta_\star$ is finite if $\mathbb{E}[T_c] > c_r/c_u$ and is infinite otherwise.

The stochastic order properties between times-to-failure shown previously induce that the same relationship also holds for the cost functions

$$\forall \delta > 0, \quad C_{brp}^{(0)}(\delta) \leqslant C_{brp}(\delta) \leqslant C_{brp}^{(1)}(\delta),$$

where $C_{brp}^{(0)}$ and $C_{brp}^{(1)}$ denote the cost function when the device is, respectively, under nominal stress and high stress. However, this order is no longer preserved for functions $\phi_{brp}^{(0)}$, $\phi_{brp}$ and $\phi_{brp}^{(1)}$. In fact, we can prove that at least $\phi_{brp}^{(0)}$ and $\phi_{brp}^{(1)}$ are crossing each other. This comes from a more general result for uni-modal distributions (using similar notations as above).

THEOREM 24.2.– Let $X$ and $Y$ be the two positive random variables such that $X \preceq_{st} Y$ and such that distributions $X$ and $Y$ are uni-modal with respective mode $m_X \neq 0$

and $m_Y$ with $m_X < m_Y$ and with continuous pdfs. Then, there exists a unique $\hat{\delta}$ such that $\phi_X(t) > \phi_Y(t)$ for any $t < \hat{\delta}$, $\phi_X(\hat{\delta}) = \phi_Y(\hat{\delta})$ and $\phi_X(t) < \phi_Y(t)$ for any $t > \hat{\delta}$.

PROOF.– From stochastic order between $X$ and $Y$, it follows that there exists a unique $\hat{z} > m_X$ such that $f_X(t) > f_Y(t)$ for $t < \hat{z}$, $f_X(\hat{z}) = f_Y(\hat{z})$ and $f_X(t) < f_Y(t)$ for $t > \hat{z}$.

We will now prove the existence and the uniqueness of the crossing point $\hat{\delta}$ as defined above. Recall that $\phi_X(t) = \int_0^t u f_X(u) du$ and so $\phi'_X(t) = t f_X(t)$ (and similarly for $Y$). Existence of $\hat{\delta}$ follows from the two following remarks. On the one hand, since $f_X(t) > f_Y(t)$ for $t < \hat{z}$ and from the expressions of $\phi'_X$ and $\phi'_Y$, we have for all $t < \hat{z}$

$$\phi_X(t) = \int_0^t \phi'_X(u) du > \int_0^t \phi'_Y(u) du = \phi_Y(t),$$

using the fact that $\phi_X(0) = \phi_Y(0) = 0$. On the other hand, $\phi_X$ and $\phi_Y$ are increasing functions with respective limits $\mathbb{E}[X]$ and $\mathbb{E}[Y] > \mathbb{E}[X]$ as $t$ tends to infinity. As a result, $\hat{\delta}$ exists and moreover $\hat{\delta} \geqslant \hat{z}$. To prove the uniqueness of $\hat{\delta}$, let us introduce the function $\Phi$ given for any $t > 0$ by $\Phi(t) = \phi_X(t) - \phi_Y(t)$. As shown previously, $\Phi(\hat{z}) > 0$, $\Phi(\infty) = \mathbb{E}[X] - \mathbb{E}[Y] < 0$ and at least $\Phi$ is decreasing on $[\hat{z}, \infty[$. Since we have, for all $t > \hat{z}$,

$$\Phi'(t) = \phi'_X(t) - \phi'_Y(t) = t(f_X(t) - f_Y(t)) < 0,$$

then $\Phi$ is strictly decreasing over $]\hat{z}, \infty[$ and thus there exists a unique solution to the equation $\Phi(t) = 0$.

We conjecture the existence and uniqueness of the mode for $T_c^{(0)}$ (and for $T_c^{(1)}$), which is supported by various numerical computations. Let $m_0$ and $m_1$ be the mode of the distributions of $T_c^{(0)}$ and $T_c^{(1)}$. From the relationship between pdfs of $T_c^{(0)}$ and $T_c^{(1)}$, we have

$$\forall t \geqslant 0, \quad f'_{T_c^{(1)}}(t) = \gamma^2 f'_{T_c^{(0)}}(\gamma t),$$

and so we get

$$0 = f'_{T_c^{(1)}}(m_1) = \gamma^2 f'_{T_c^{(0)}}(\gamma m_1) = f'_{T_c^{(0)}}(\gamma m_0)$$

and thus $m_0 = \gamma m_1 > m_1$. Then, applying the above result, we have that there exists a unique crossing point $\hat{\delta}$ for functions $\phi_{brp}^{(0)}$ and $\phi_{brp}^{(1)}$. As a result, we conclude that

$$\begin{cases} \hat{\delta} \leqslant \delta_\star^{(0)} \leqslant \delta_\star^{(1)} & \text{if } \phi_{brp}^{(0)}(\hat{\delta}) = \phi_{brp}^{(1)}(\hat{\delta}) \leqslant c_r/c_u \\ \delta_\star^{(1)} \leqslant \delta_\star^{(0)} \leqslant \hat{\delta} & \text{otherwise} . \end{cases}$$

To conclude, we provide a numerical illustration of this problem. We consider the same parameters value as above and, in addition, we set $c_r = 100$ (replacement cost) and $c_u = 500$ (unavailability cost). In Figure 24.3, we have plotted the three cost functions. We obtained that $\delta_\star^{(1)} = 4.49$, $\delta_\star = 5$ and $\delta_\star^{(0)} = 6.41$ (points in the figure indicate minimum of the three cost functions).

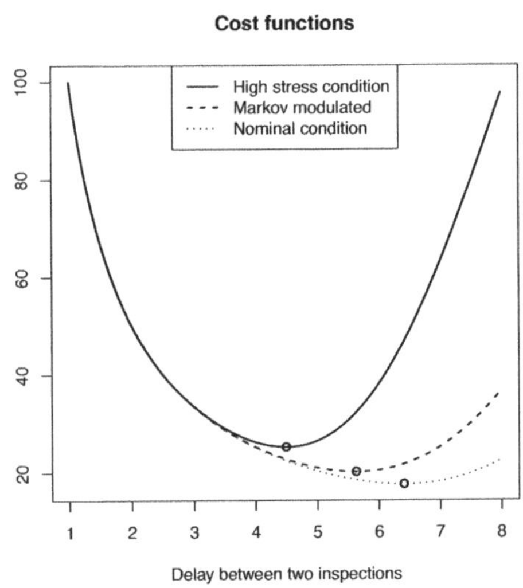


**Figure 24.3.** *Block replacement policy–cost functions and optimal parameter values*

### 24.4.2. *Age replacement policy*

Here, we consider another policy, the age replacement policy. A device is replaced either at failure or at age $\delta$ if no failure occurs between 0 and $\delta$ ($\delta = \infty$ means that replacement occurs only at failure). Hence, there are two different costs: $c_f$, the cost incurred by the replacement of a failed device, and $c_{nf} < c_f$, the cost incurred by the replacement of a non-failed device. We have

$$C_{arp}(\delta) = \frac{c_f F_{T_c}(\delta) + c_{nf} S_{T_c}(\delta)}{\int_0^\delta S_{T_c}(u)du}.$$

When $\delta$ tends to infinity, since it corresponds to the case where replacement occurs only at failure, it can easily be seen that $C_{arp}(\delta)$ tends to $\frac{c_f}{\mathbb{E}[T_c]} := C_{arp}(\infty)$.

From the stochastic order properties between times-to-failure shown previously and since we have assumed that $c_{nf} < c_f$, it follows that the same relationship also holds for the cost functions

$$\forall \delta > 0, \quad C_{arp}^{(0)}(\delta) \leqslant C_{arp}(\delta) \leqslant C_{arp}^{(1)}(\delta),$$

where $C_{arp}^{(0)}$ and $C_{arp}^{(1)}$ denote the cost functions of the age replacement policy when the device is, respectively, under nominal stress and high stress. Moreover, from the comparison between MTTF, it follows that

$$C_{arp}^{(0)}(\infty) \leqslant C_{arp}(\infty) \leqslant C_{arp}^{(1)}(\infty).$$

Let us denote $\delta_\star := \text{argmin}_{\delta>0} C_{arp}(\delta)$. The order between cost functions cannot be used here as well to deduce the existence (or not) of an order between $\delta_\star$, $\delta_\star^{(0)}$ and $\delta_\star^{(1)}$. If $T_c$ has a strictly increasing and continuous failure rate and if $h_{T_c}(\infty) > c_f/(\mathbb{E}[T_c](c_f - c_{nf}))$, then $\delta_\star$ is the unique and finite solution of the equation (see theorem 3.2 in [NAK 05])

$$\phi_{T_c}(t) := h_{T_c}(t) \int_0^t S_{T_c}(u)du - F_{T_c}(t) = \frac{c_{nf}}{c_f - c_{nf}}.$$

Nakagawa [NAK 05, p. 74] has proved that the provided $\phi_{T_c}$ function is strictly increasing (provided that $h_{T_c}$ is continuous and strictly increasing). This result implies that provided $\delta_\star^{(1)} < \delta_\star^{(0)}$, we can prove that $h_{T_c^{(0)}}$ satisfies the above assumptions (Abdel-Hameed [ABD 75] has proved its increasingness). Indeed, from distributional relationships between $h_{T_c^{(0)}}$ and $h_{T_c^{(1)}}$, we deduce that

$$\forall t \geqslant 0, \quad \phi_{T_c^{(1)}}(t) = \phi_{T_c^{(0)}}(\gamma t) > \phi_{T_c^{(0)}}(t),$$

from the increasingness of $\phi_{T_c^{(0)}}$ and since $\gamma > 1$. As a result, $\delta_\star^{(1)} < \delta_\star^{(0)}$

To conclude, we provide a numerical illustration. We consider the same parameters value as above and, in addition, we set $c_{nf} = 100$ (replacement cost of a non-failed component) and $c_f = 500$ (replacement cost of a failed component). In Figure 24.4, we have plotted the three cost functions and we obtain that $\delta_\star^{(1)} = 4.32$, $\delta_\star = 4.69$ and $\delta_\star^{(0)} = 6.47$ (points in the figure indicate minimum of the three cost functions).

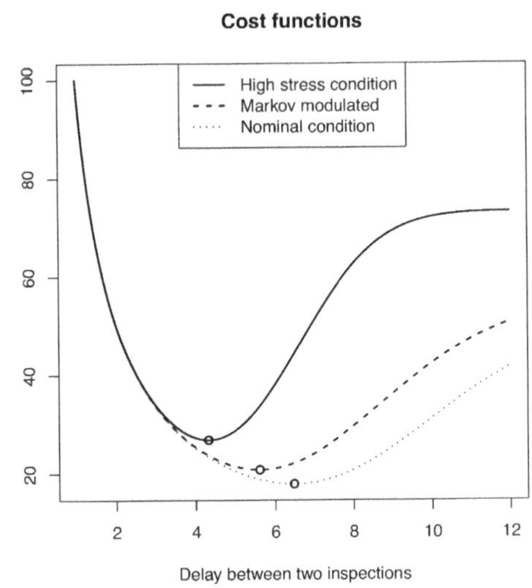


**Figure 24.4.** *Age replacement policy-cost functions and optimal parameter values*

## 24.5. Conclusion

In this chapter, we have studied the time-to-failure $T_c$ of a Markov-modulated gamma process. Properties of $T_c$ are quite difficult and we have used stochastic orders with $T_c^{(0)}$ and $T_c^{(1)}$ to establish some facts. We have also considered two replacement policies, namely the block replacement policy and the age replacement policy. Despite the fact that in both cases orders are preserved for the cost functions, the same does not necessarily hold for the optimal delays between successive replacement.

## 24.6. Acknowledgment

This work was supported by the French National Agency of Research (ANR) (project AMMSI, number ANR-2011-BS01-021).

## 24.7. Bibliography

[ABD 75] ABDEL-HAMEED M., "A gamma wear process", *IEEE Transactions on Reliability*, vol. 24, pp. 152–153, 1975.

[ABR 72] ABRAMOWITZ M., STEGUN I. (eds), *Handbook of Mathematical Functions with Formulas, Graphs, and Mathematical Tables*, National Bureau of Standards, 1972.

[BAG 01] BAGDONAVIČIUS V., NIKULIN M., "Estimation in degradation models with explanatory variables", *Lifetime Data Analysis*, vol. 7, pp. 82–103, 2001.

[BEŔ 03] BÉRENGUER C., GRALL A., DIEULLE L., *et al.*, "Maintenance policy for a continuously monitored deteriorating system", *Probability in Enginering and Informational Sciences*, vol. 17, no. 2, pp. 235–250, 2003.

[GRA 65] GRADSHTEYN I., RYZHIK I., *Table of Integrals, Series and Products*, Academic Press, 1965.

[KOV 10] KOVCHEGOV Y., MEREDITH N., NIR E., "Occupation times and Bessel densities", *Statistics and Probability Letters*, vol. 80, pp. 104–110, 2010.

[LAW 05] LAWLESS J., CROWDER M., "Accelerated degradation models for failure based on geometric Brownian motion and gamma process", *Lifetime Data Analysis*, vol. 11, pp. 511–527, 2005.

[NAK 05] NAKAGAWA T., *Maintenance Theory of Reliability*, Springer-Verlag, London, 2005.

[PAR 04] PARK C., PADGETT W., "Covariates and random effects in a gamma process model with application to degradation and failure", *Lifetime Data Analysis*, vol. 10, pp. 213–227, 2004.

[PED 71] PEDLER P., "Occupation times for two-state Markov chains", *Journal of Applied Probability*, vol. 8, pp. 381–390, 1971.

[SAA 07] SAASSOUH B., DIEULLE L., GRALL A., "On-line maintenance policy for a deteriorating system with random change of mode", *Reliability Enginering and System Safety*, vol. 92, pp. 1677–1685, 2007.

[SIN 95] SINGPURWALLA N., "Survival in dynamic environments", *Statistical Science*, vol. 10, no. 1, pp. 96–103, 1995.

[VAN 09] VAN NOORTWIJK J., "A survey of the application of gamma processes in maintenance", *Reliability Enginering and System Safety*, vol. 94, pp. 2–21, 2009.

[ZHA 10a] ZHAO X., FOULADIRAD M., BÉRENGUER C., "Residual-based inspection/replacement policy for a deteriorating system with Markovian covariates", *Industrial Engineering and Engineering Management (IEEM)*, IEEE, *International Conference*, Macao, pp. 636–640, 7–12 December 2010.

[ZHA 10b] ZHAO X., FOULADIRAD M., BÉRENGUER C., *et al.*, "Condition-based inspection/replacement policies for monotone deteriorating systems with environmental covariates", *Reliability Enginering and System Safety*, vol. 95, no. 8, pp. 921–934, 2010.

# Chapter 25

# Calculation of the Redundant Structure Reliability for Aging-type Elements

This chapter discusses the calculation of the reliability of redundant structures, taking into account the aging elements. The operating practice of modern industrial plants is such that both the basic element and elements from the spare subject are affected by the process. Failed objects have to be renewed. After being repaired they return to spare elements. It should be noted that when the repair is finished, usually a partial renewal of capacity occurs. An object produces a part of the resource during the previous operation and full renewal does not takes place. The problem of determining spare element composition taking into account renewal and working out of a certain part of their resource is solved by simulation.

## 25.1. Introduction

Organizing the operation of industrial objects, especially high-risk objects such as nuclear power plants (NPP), high demand is on the safety and reliability of their operation. One way to improve reliability is to plan preventive maintenance, functional testing of objects and the creation of sets of spare elements for the rapid replacement of faulty equipment.

In this chapter we consider the calculation of reliability issues of restorable equipment taking into account the availability of spare elements and the determination of the required number of spare elements that guarantee support of the

---

Chapter written by Alexandr ANTONOV, Alexandr PLYASKIN and Khizri TATAEV.

specified reliability coefficients. The questions of calculation of the equipment reliability taking into account spare parts and of determination of its optimal composition have been considered by both Russian and foreign specialists.

We note here one feature that is characteristic of a large amount of equipment in service in various industries. This feature is that the equipment has a long operating time. Often it exceeds a resource or life-time that is established for it in the regulations. The objects operating in many industries were put into operation during the Soviet era. Replacement of the equipment is a slow process. As a result, we can assume that aging process takes place in this equipment due to wear and tear of materials and degradation processes occurring within the product. In this connection, there is the problem of calculation of the equipment reliability with spare parts, taking into account the aging elements.

## 25.2. The operation process of the renewal and repaired products

Let us consider the method of calculation of the reliability characteristics of renewal and repaired object with $n$ spare elements. The strategy of the element operation is described below. The item is in good condition at the initial time.

The element fails with the intensity $\lambda$(t). The element is replaced by a reserve one at the time of failure. The intensity of the replacement element is $\mu$(t). The faulty element is sent in for repair. The element is considered to be renewed after repair and proceeds into the reserve. The intensity of the repair is $\nu$(t). If available elements in the reserve are left, then a failure occurs. The above strategy of operation can be represented by the graph shown in Figure 25.1.

It is necessary to assess the reliability of the system (availability factor, the probability of system failure due to lack of spare elements) for the reviewed strategy and to determine the required number of spare elements that provide a given level of system availability.

We denote the state of the object on the graph by two symbols $(k, i)$, where the first character indicates the number of spare elements, $k = 0, \ldots, n$, and the second character $i$ indicates the state of a basic element: $i = 1$, the element is operational, and $i = 0$, element is not operational.

Consider the operation of the object with spare elements in more detail. An element is with probability 1 in the state ($n$, 1) in the beginning (there are $n$ available spare elements, the object is operational). The element changes its state ($n$, 0) at a random moment in time with the intensity of failure $\lambda$(t) ($n$ spare elements, the object is in a failure state, replacing of the element is beginning).

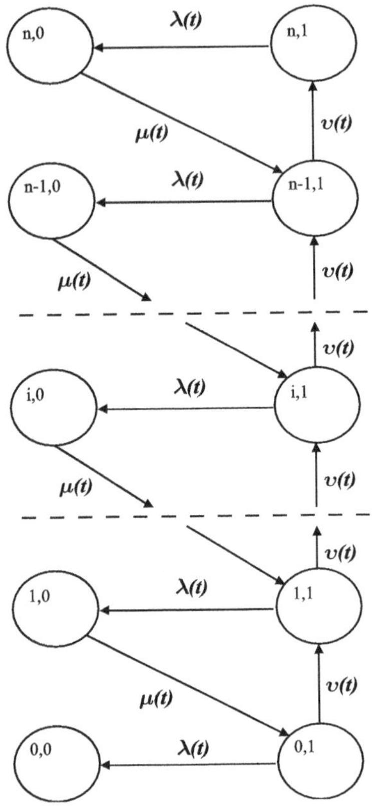


**Figure 25.1.** *Transition graph of renewal and repaired element*

| | |
|---|---|
| $\lambda$(t) | Failure rate |
| $\mu$(t) | Intensity of the replacement |
| $\nu$(t) | Intensity of the repair |
| $n$ | Number of elements |

**Table 25.1.** *Symbols*

The object passes into a $(n-1, 1)$ state with the intensity of recovery $\mu$(t) $(n-1)$ spare elements, the object is operational). From this state, transitions to state $(n, 1)$ with an intensity of recovery $\nu$(t) (repair is finished, $n$ elements are in reserve again) or to state $(n-1, 0)$ with intensity $\lambda$(t) (repairs are not completed until the next failure) are possible, and so on. Getting into a condition (0,0), the object stops working and it is in a condition of refusal till the moment of replenishment of spare parts.

The calculation of the probability of finding each of the intermediate states $P_{i,j}(t)$, where $i = 1, n, j = 0, 1$, is of interest in this problem. The problem of determining the probability of falling into a state $P_{0,0}(t)$ characterized by the fact that the basic element is failed and there are no spare elements is very important. The above problem will be solved by methods of simulation. To describe the dependence of the intensity of the element failures, we use the model of the geometric process.

## 25.3. The model of the geometric process

Let us consider the model describing the variation of object reliability characteristics and taking into account incomplete repair of operability after failure [ANT 07, FIN 93, LAM 88, LAM 03]. The behavior of complex systems is well described by this model.

Let us consider the following strategy of the object operation. The object is functioning properly for a random time. It is restored after the failure. It is understood that the restoration is not complete. Incompleteness of restoration is characterized by the coefficient $\gamma$. As a result of incomplete repair, the operating time of the renewal object is reduced (by probability) by $\gamma$ times in comparison with the previous operation phase:

$$\xi_2^d = \gamma\xi_1, ...\xi_n^d = \gamma^{n-1}\xi_1, 0 < \gamma \leq 1.$$

Mathematical dependence between distribution functions of the operating time between failures of the restored object (taking into account incomplete repair) can be expressed as:

$$F_{\xi_2}(t) = F_{\xi_1}\left(\frac{t}{\gamma}\right), ..., F_{\xi_n(t)} = F_{\xi_1}\left(\frac{t}{\gamma^{n-1}}\right),$$

where $F_{\xi_i}(t)$ is the distribution function of the operating time between failures of $(i-1)$ times the restored object and $\gamma$ is the coefficient of incomplete restoration). Then, distribution densities are related through the following equation:

$$f_{\xi_n}(t) = \frac{1}{\gamma^{n-1}} f_{\xi_1}\left(\frac{t}{\gamma^{n-1}}\right).$$

We define the inverse value of the coefficient of incomplete restoration $\alpha = \frac{1}{\gamma}$ and call it the degradation coefficient. The degradation coefficient $\alpha$ is an average value that reflects the accumulated process of damages and defects and indirectly characterizes the process of gradual material weariness, physical ageing, wear ability, corrosion, etc.

In some cases, $\alpha$ can be understood as a factor, reflecting the increased load on the object due to variable operating conditions.

DEFINITION 25.1.– Let $\{\xi_i\}, i \geq 1$ be the sequence of independent random variables. Each $\xi_i$ corresponds to operating time between failures of the object with the distribution function $F_{\xi_i}(t)$ generated by the distribution $F(t)$ as follows:

$$F_{\xi_i}(t) = F\left(\frac{t}{\gamma^{i-1}}\right), i = 1, 2, ...,$$

where $\gamma$ is a positive constant. Then, the sequence $\{\xi_i\}, i \geq 1$ is called geometric process.

Let us define the expression that establishes the relationship between the failure rate at the initial stage of operation and the failure rate after the $(n-1)$th failure. By definition, the failure rate can be defined by:

$$\lambda(t) = \frac{f(t)}{1 - F(t)}.$$

Then, the expression of the failure rate of $(n-1)$ times restored object can be written as:

$$\lambda_{\xi_n}(t) = \frac{f_{\xi_n}(t)}{1 - F_{\xi_n}(t)} = \frac{\frac{1}{\gamma^{n-1}} f_{\xi_1}\left(\frac{1}{\gamma^{n-1}}\right)}{1 - F_{\xi_1}\left(\frac{t}{\gamma^{n-1}}\right)} = \frac{1}{\gamma^{n-1}} \lambda_{\xi_1}\left(\frac{t}{\gamma^{n-1}}\right), \qquad [25.1]$$

Thus, after each restoration, the failure rate becomes $\frac{1}{\gamma}$ times more than the failure rate during the previous time interval. The timescale of the process also changes.

## 25.4. Task solution

The simulation process is organized in accordance with the description provided in section 25.2. The object is available at the initial time. Further, the failure of the object is occurred in the random time. The failed element is removed from operation, is sent being repaired and the element from the spare set is put into its place. After being repaired, object is returned into the spare elements and is put at the end of the queue to use. That is, the next time it will be installed in the system will be after all the spare elements that are before it have worked out to failure. The operation process of the installation of a system, failure, repair and return of spare parts for all other elements are organized in a similar way.

Note that the object has the failure intensity $\lambda_{\xi_1}(t)$ in the first cycle, intensity varies according to the expression after the first failure and is $\lambda_{\xi 2}(t)$, and after the $i$th failure, it will be equal to $\lambda_{\xi_{i+1}}(t)$.

We represent the results of test calculations, which were carried out using the described method. The following values were used as the initial data for calculation:

– failure intensity of the element is $\lambda$ (t) = 0.001 × 1/h;

– the intensity of the element replacement is $\mu$(t) = 1 × 1/h;

– the intensity of repair is $\nu$(t) = 0.1 × 1/h.

The number of simulations is $N = 10^6$. Operating time of the redundant structure is equal to 16,000 h.

At the first stage, we will carry out calculations of changing of an average operating time of the element depending on the number of renewals. In this case, we change the coefficient of degradation. The calculations considered a structure with a basic element and three spares. Figure 25.2 shows plots of the average time to failure of the element depending on the number of renewals of the elements for different values of degradation coefficient (WT = average time to failure, RC = recovery time, $\alpha$ = degradation coefficient).

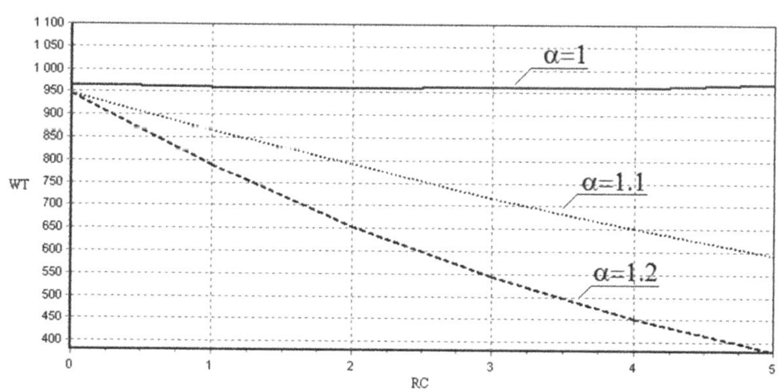


**Figure 25.2.** *Changing the average time to failure of the object depending on the number of renewals*

Several lines of the average time to failure of the object for different $\alpha$ coefficients are shown in the figure.

We can see that the graph's average time to failure does not change for elements without taking into account the aging elements and is equal to 1,000 h, a situation characterized by the fact that the greater the degradation coefficient is, the faster the average time to failure will decrease for all other graphs. Availability coefficient to 16,000 h of operation work was calculated for other parameter. The calculation results are presented in Table 25.2.

| Degradation coefficient $\alpha$ | 1 | 1.05 | 1.1 | 1.15 | 1.2 |
|---|---|---|---|---|---|
| Availability coefficient | 0.999 | 0.9989 | 0.9988 | 0.9986 | 0.9979 |

**Table 25.2.** *Calculation results*

The results show that the availability coefficient decreases as the degradation coefficient increases.

The proposed mathematical model can be used to solve the problem of calculating the required number of spare elements that guarantee the achievement of specified reliability parameters on the considered time interval of operation of the structure. We take the availability coefficient as an indicator of the reliability. It is necessary to provide a value of availability coefficient not less than 0.99 of the time of 16,000 h. The results of these calculations are presented in Table 25.3.

| Degradation coefficient $\alpha$ | 1 | 1/1.05 | 1/1.1 | 1/1.15 | 1/1.2 |
|---|---|---|---|---|---|
| The number of spare elements to ensure specified requirements | 2 | 2 | 2 | 2 | 3 |
| Availability coefficient | 0.999 | 0.9989 | 0.9988 | 0.9986 | 0.9979 |

**Table 25.3.** *Calculation results*

As can be seen from Table 25.3, the structure with one basic and two spare element requirements specified by the value of the availability coefficient satisfies for degradation coefficients $\alpha$= 1 – 1. 15, and three spare elements are for $\alpha$= 1. 2.

## 25.5. Conclusion

Thus, we can say that a model for calculating the reliability of the redundant structure is presented in this chapter, taking into account the aging elements. It has been suggested that either a basic element or spare elements spend part of their resource in the operation process. The result is a partial renewal of the system availability. Investigations of the behavior of the system parameters depending on the values of the degradation were done in the test examples. It was shown that the developed model can be used to calculate the required number of spare elements that guarantee the achievement of specified parameters of the reliability in the considered time interval of the structure operation.

## 25.6. Bibliography

[ANT 07] ANTONOV A.V., POLIAKOV A.A., "One statistical age-dependent reliability model in operating of nuclear power plant equipment", *International Conference on Mathematical Methods in Reliability: Methodology, Practice and Interference (MMR' 2007)*, Glasgow, GB, 1–4 July 2007.

[FIN 93] FINKELSTEIN M., "A scale model of general repair", *Microelectronics and Reliability*, pp. 41–46, 1993.

[LAM 88] LAM Y., "Geometric process and the replacement problem", *Acta Mathematicae Applicatae Sinica*, pp. 366–382, 1988.

[LAM 03] LAM Y., "Some limit theorems in geometric processes", *Acta Mathematicae Applicatae Sinica*, pp. 405–416, 2003.

# Chapter 26

# On Engineering Risks of Complex Hierarchical Systems Analysis

In this chapter, the notion of "risk" is discussed. Some new characteristics of engineering risk are proposed and the methods of their analysis for complex hierarchical systems are considered.

## 26.1. Introduction

During the last few years, the term *risk* became very popular in various fields: mathematics, engineering, economics, environment, management, biology, medicine, etc. However, different authors as well as national and international documents treat this term in different senses (see, for example, [RYK 99; RYK 13]). In some special books devoted to risk analysis (e.g. [BED 01, GRA 01, SIN 06, SOL 06]), there is no definition of risk at all. Thus, the situation in this area looks like a situation in reliability theory more than half a century ago, when different specialists used the term *reliability* in different senses and understood reliability as having different characteristics. The situation in this area has been fixed due to B.V. Gnedenko, who in 1965 ([GNE 69, p. 9]) proposed the following definition of reliability, which is recognized now by most of the specialists: "The Reliability Theory is a general scientific discipline that deals with general methods and tools for: projecting, producing, accepting, transportation, and exploitation of items and systems in order to provide the necessary (or maximal) efficiency during life time, as well as for

Chapter written by Vladimir RYKOV.

calculation of the system's quality characteristics using known quality characteristics of their parts".

Historically, among mathematicians and actuaries, the term "risk" arose in the framework of *ruin models* [CRA 30, CRA 55, GRA 91]. From another side, among engineers, the notion of "risk" is used in the framework of reliability theory and it means the *losses* or *damages* for people, the environment and finance [HEN 92]. There is one more discussion of "risk" in financial mathematics, where it is used for comparison of random variables (RVs) and their distributions [ROT 94] in the framework of utility theory [VON 53]. Therefore, now there exist at least three different risk theories:

– insurance risk theory, which deals with the *ruin problems*;

– engineering risk theory, which deals with investigation of the *consequences* (loses, damages) generated by *failures* of complex non-reliable systems;

– business risk theory, which considers the problems of RVs and their distributions comparison in the framework of *utility theory*.

In this situation, the strong notion of *risk* and the methods for its measurement and assessment, which could be used for interdisciplinary connections are needed.

Examples of different risk situations in habitual life, engineering and anthropogenic, financial and business, natural and ecological, social and medical risks, as well as insurance of risks and risk of insurance [RYK 13] show that risk arises as an *uncertain event* (or *family of events*), which is distributed in *time* and leads to some *damages* or *losses* that are measured in specific units (in some cases with money but also with other units). This analysis allows us to propose some general definitions and mathematical models of risk.

## 26.2. Risk definition and measurement

From the probabilistic point of view, *risk* is an *event* or *sequence of events* under some probabilistic space $(\Omega, \mathcal{F}, \mathbf{P})$, which is accompanied with two-dimensional RV $(T, Y)$ or sequence of two-dimensional RVs $\{(S_n, Y_n),\ n = 0, 1, \ldots\}$ (a marked point process), which we will refer to as a *risk process*. In this case, the main characteristics of any risk are:

– its two-dimensional cumulative distribution function (CDF) referred to as *risk distribution*

$$F(t, y) = \mathbf{P}\{T \leq t,\ Y \leq y\},$$

– or distribution of a risk process $(S_n, Y_n,\ n = 0, 1, \dots)$ jointly with appropriate characteristics of these distributions.

This definition allows us to consider all risk situations as its special cases. For example, in the *insurance risk* analysis, the main mathematical problems consist of the investigation of the *reserve process*:

$$Y(t) = u + ct - \sum_{1 \le i \le N(t)} Y_i \quad \text{with } N(t) = \sum_{n \ge 0} 1_{\{S_n \le t\}},$$

calculation of the time to ruin distribution

$$F_S(t) = \mathbf{P}\{S \le t\} \quad \text{with } S = \inf\{t : Y(t) < 0\},$$

and evaluation of its defect–*ruin probability*:

$$\psi(u) = F_S(\infty) = \mathbf{P}\{T < \infty\}.$$

Financial risk models deal with the comparison of these two-dimensional RV $(T, Y)$ in the sense of their expected utility or regret [ROT 94].

## 26.3. Engineering risk

In this chapter, we deal with *engineering risks* of complex technical systems or biological objects, which are usually connected with their failures. Therefore, their analysis is based on the study of appropriate objects' reliability, and so far the term "failure" is used as a synonym for "risk event". Up-to-date complex systems have a hierarchical structure, which should be taken into account for their reliability and risk analysis. The main problem in engineering risk analysis (ERA) consists of modeling risk events and their characteristics $(T, Y)$. From the probabilistic point of view, the problem consists of the description of risk situation evolution as some stochastic process on a risk probabilistic space $(\Omega, \mathcal{F}, \mathbf{P})$. It seems very simple, however, that any risk situation is unique and should be investigated separately.

For a description of complex hierarchical systems, their subsystems and elements, it is convenient to denote elements of a system by a vector index $\mathbf{i} = (i_1, i_2, \dots, i_r)$, its subsystems of $k$th level by a truncated vector $\mathbf{i}_k = (i_1, i_2, \dots, i_k)$ and the $j$th subsystem of the lower level subsystem $\mathbf{i}_k$ by $j(\mathbf{i}_k)$. Here, $i_1$ is a number of the first subsystem of whole system, $i_2$ is a number of its subsystem (i.e. subsystem of the second level) that contains this element etc., $i_r$ is the number of an element of the subsystem $\mathbf{i}_{r-1}$ and $r$ is its hierarchical level which we will refer to as its *rank*. In order to indicate the path for risk situation evolution up to $k$th subsystem, the notation

$\vec{\mathbf{i}}_k = (i_r, i_{r-1}, \ldots, i_k)$ will be used, which for $k = 0$ shows the appropriate path for the whole system.

A very useful tool for risk analysis consists of a risk tree construction and analysis (see, for example, [HEN 92]), where some special symbols for events and their connections (gates) as well as their functional (algebraic) representations have also been proposed. For hierarchical systems, a risk tree follows the system structure.

The structural variables of elements (for $k = r$), the subsystem and the whole system (for $k = 0$) are denoted as $x_{\mathbf{i}_k}$ and their appropriate states at time $t$ as $X_{\mathbf{i}_k}(t)$:

$$X_{\mathbf{i}_k}(t) = \begin{cases} 0, & \text{if subsystem } \mathbf{i}_k \text{ is in up state in time } t, \\ 1, & \text{if subsystem } \mathbf{i}_k \text{ is in down state in time } t. \end{cases}$$

The *risk structural functions* are denoted by $\phi_{\mathbf{i}_k}(x_{(\mathbf{i}_k,1)}, \ldots x_{(\mathbf{i}_k, n(\mathbf{i}_k))})$ for any subsystems $(k = 0, 1, \ldots, r)$ and with the help of algebraic representation of different gates [HEN 92], they allow us to calculate step by step the structure function of the whole system:

$$\phi_{\mathbf{i}_k}(x_{(\mathbf{i}_k,1)}, \ldots x_{(\mathbf{i}_k, n(\mathbf{i}_k))}) = x_{\mathbf{i}_{k-1}} \quad (k = 1, 2, \ldots, r).$$

It is necessary to note that risk analysis depends on the problem setting: long-time or short-time risk model investigation and accessible information about the object. For the long-time (general) study of risk models according to the given risk definition, the initial information should be done in terms of:

– CDF time to initial risk event (lifelength of $i$th element for engineering risks) and appropriate loss size:

$$F_{\mathbf{i}}(t) = \mathbf{P}\{T_{\mathbf{i}} \le t\}, \quad G_{\mathbf{i}}(y) = \mathbf{P}\{Y_{\mathbf{i}} \le y\}, \qquad [26.1]$$

– distributions of delay time and additional loss value $(T_{j(\mathbf{i}_k)}, Y_{j(\mathbf{i}_k)})$ for the failure evolution along the $j$th path in the subsystem $\mathbf{i}_k$:

$$F_{j(\mathbf{i}_k)}(t) = \mathbf{P}\{T_{j(\mathbf{i}_k)} \le t\}, \quad G_{j(\mathbf{i}_k)}(y) = \mathbf{P}\{Y_{j(\mathbf{i}_k)} \le y\}. \qquad [26.2]$$

As a result of the ERA, the following characteristics should be found:

– CDF time to risk event occurrence (unreliability function) $F_{\mathbf{i}_k}(t) = \mathbf{P}\{T_{\mathbf{i}_k} \le t\}$ or appropriate reliability function $R_{\mathbf{i}_k}(t) = 1 - F_{\mathbf{i}_k}(t) = \mathbf{P}\{T_{\mathbf{i}_k} > t\}$ and CDF of appropriate loss size $G_{\mathbf{i}_k}(y) = \mathbf{P}\{Y_{\mathbf{i}_k} \le y\}$ for subsystems $k = 1, \ldots, r$ and whole system for $k = 0$ as well as its characteristics mean value, variance, etc.;

– unavailability functions (state probabilities at time $t$) $Q_{\mathbf{i}_k}(t) = \mathbf{P}\{X_{\mathbf{i}_k}(t) = 1\}$ for subsystems $k = 1, \ldots, r$ and whole system for $k = 0$;

– probabilities $q_{\vec{\mathbf{i}}_k}$ of a risk event evolution along different paths $\vec{\mathbf{i}}_k$ of subsystems and the whole system;

– mostly dangerous paths of risk evolution with respect to different criteria;

– sensitivity analysis of risk characteristics to distribution shape and small changes of initial information parameters.

## 26.4. Risk characteristics for general model calculation

Risk tree construction and risk structure function calculation along with the initial information about the model allow us to calculate all the above-mentioned risk characteristics and produce an appropriate risk analysis. We consider here these problems in more detail.

### 26.4.1. *Lifelength and appropriate loss size CDF*

Observation that each vector $\mathbf{i} = (i_1, i_2, \ldots, i_r)$ unequally determines system element and truncated vector $\mathbf{i}_k = (i_1, i_2, \ldots, i_k)$ unequally determines appropriate subsystem which allows us to calculate the lifelength CDF for each subsystem and the whole system.

It is necessary to take into account that for subsystems with dependent elements (for example, for reparable subsystems) the probability state distributions (unavailability functions) $Q_{\mathbf{i}_k}(t)$ and lifelength CDF (unreliability functions) $F_{\mathbf{i}_k}(t)$ should be calculated with special models. Therefore, reparable subsystems should be specially fixed during the risk tree construction. We will not detail this problem here. However, if these functions are calculated, further analysis goes to the following procedure:

– Taking into account that failure and appropriate loss size for subsystem $\mathbf{i}_k$ along the path $j$ are

$$T_{(\mathbf{i}_k(j))} = T_{(\mathbf{i}_k, j)} + T_{j(\mathbf{i}_k)}, \quad Y_{(\mathbf{i}_k(j))} = Y_{(\mathbf{i}_k, j)} + Y_{j(\mathbf{i}_k)} \quad (k = r, \ldots 1) \; [26.3]$$

their CDF under assumption about independency of summands represents in terms of convolution

$$F_{(\mathbf{i}_k(j))}(t) = F_{(\mathbf{i}_k, j)} * F_{j(\mathbf{i}_k)}(t), \quad G_{(\mathbf{i}_k(j))}(t) = G_{(\mathbf{i}_k, j)} * G_{j(\mathbf{i}_k)}(t). \qquad [26.4]$$

– To calculate the unreliability function $F_{\mathbf{i}_k}(t)$ for subsystem $\mathbf{i}_k$ following the rules of reliability theory [GER 00, HEN 92, RYK 01], we should substitute into the structure function of this subsystem appropriate unreliability functions for the subsystems of lower level taking into account possible delay in risk evolution along any path $j$ and the failure evolution for reparable subsystems,

$$F_{\mathbf{i}_k}(t) = \phi_{\mathbf{i}_k}(F_{\mathbf{i}_k(1)}(t), \ldots, F_{\mathbf{i}_k(n(\mathbf{i}_k))}(t)). \qquad [26.5]$$

Analogously, we should calculate the unavailability functions $Q_{(\mathbf{i}_k(j))}(t)$,

$$Q_{\mathbf{i}_k}(t) = \phi_{\mathbf{i}_k}(Q_{\mathbf{i}_k(1)}(t), \ldots, Q_{\mathbf{i}_k(n(\mathbf{i}_k))}(t)). \qquad [26.6]$$

– For appropriate calculations, we should fulfill beginning from the lowest level up to the subsystem at which we want localize the risk.

This approach allows us to decompose a complex risk event into more simple events that allow a more accurate evaluation of its characteristics using statistical methods or expert evaluation methods.

However, note that the above-mentioned detailed information is practically unavailable and moreover is not needed for short-time risk models investigation. In this case, it is usually required to limit ourselves to the calculation of only some of its moments that essentially simplify the analysis (see section 26.5).

### 26.4.2. *Probability of risk event evolution*

In order to calculate the probability $q_{j(\mathbf{i}_k)}$ of risk event evolution for subsystem $\mathbf{i}_k$ due to failure of its $j$th subsystem for gates "AND" and "OR", note that for gate "AND" the event "failure of a subsystem $\mathbf{i}_k$ due to failure of its $j$th subsystem" is

$$\{\max_{1 \leq l \leq n(\mathbf{i}_k)} T_{(\mathbf{i}_k,l)} = T_{(\mathbf{i}_k,j)}\},$$

and for gate "OR" it is

$$\{\min_{1 \leq l \leq n(\mathbf{i}_k)} T_{(\mathbf{i}_k,l)} = T_{(\mathbf{i}_k,j)}\}.$$

From here, we can find

– for gate "AND"

$$\begin{aligned} q_{j(\mathbf{i}_k)} &= \mathbf{P}\{\max_{1 \leq l \leq n(\mathbf{i}_k)} T_{(\mathbf{i}_k,l)} = T_{(\mathbf{i}_k,j)}\} = \\ &= \int_0^\infty \left[\prod_{1 \leq l \leq n(\mathbf{i}_k),\, l \neq j} F_{(\mathbf{i}_k,l)}(u)\right] F_{(\mathbf{i}_k,l)}(du); \end{aligned} \qquad [26.7]$$

– and for gate "OR"

$$q_{j(\mathbf{i}_k)} = \mathbf{P}\{\min_{1\leq l\leq n(\mathbf{i}_k)} T_{(\mathbf{i}_k,l)} = T_{(\mathbf{i}_k,j)}\} = \\ = \int_0^\infty \left[\prod_{1\leq l\leq n(\mathbf{i}_k),\, l\neq j} (1 - F_{(\mathbf{i}_k,l)}(u))\right] F_{(\mathbf{i}_k,l)}(du). \qquad [26.8]$$

Corresponding values $q_{j(\mathbf{i}_k)}$ are conditional probabilities of subsystem $\mathbf{i}_k$ failure due to its $j$th component. The subsystem failure along any path can be represented with the help of gate "PRIORITY OR" (really, alternative event "system work" represented with gate "PRIORITY AND"). Therefore, unconditional failure subsystems $\mathbf{i}_k,\ \ k = r, \ldots, 1$, probabilities $q(\vec{\mathbf{i}}_k)$ along the path $\vec{\mathbf{i}} = (i_r, \ldots, i_k)$ are

$$q(\vec{\mathbf{i}}_k) = q(i_r, \ldots, i_k) = 1 - (1 - q_{i_r(\mathbf{i}_{r-1})}) \cdots (1 - q_{i_k(\mathbf{i}_{k-1})}) = \\ = 1 - \prod_{r\geq l\geq k} (1 - q_{i_l(\mathbf{i}_{l-1})}). \qquad [26.9]$$

These probabilities are also used for the calculation of risk time evolution along any path and appropriate loss size as well as for finding mostly dangerous (in different senses) paths of risk evolution.

### 26.4.3. *Lifelength and loss moments*

The numerical risk characteristics are calculated using the usual formulas:

– for mean value and variance of time-to-risk t for any subsystem and the whole system for $k = 0$

$$\mu_T(\mathbf{i}_k) = \mathbf{E}[T_{\mathbf{i}_k}] = \int_0^\infty (1 - F_{\mathbf{i}_k}(t))dt, \qquad [26.10]$$

$$\sigma_T^2(\mathbf{i}_k) = \mathbf{Var}[T_{\mathbf{i}}] = \int_0^\infty (t - \mu_T(\mathbf{i}_k))^2 F_{\mathbf{i}_k}(dt), \qquad [26.11]$$

– and for mean value and variance of losses for any subsystem and the whole system

$$\mu_Y(\mathbf{i}_k) = \mathbf{E}[Y_{\mathbf{i}_k}] = \int_0^\infty (1 - G_{\mathbf{i}_k}(y))dy, \qquad [26.12]$$

$$\sigma_Y^2(\mathbf{i}_k) = \mathbf{Var}[Y_{\mathbf{i}_k}] = \int_0^\infty (y - \mu_Y(\mathbf{i}_k))^2 G_{\mathbf{i}_k}(dy). \qquad [26.13]$$

For optimization of the risk evolution with respect to different criteria (see the next section), one needs to calculate the lifelength and loss means for different paths of the risk evolution. If path of risk evolution $\vec{\mathbf{i}}_k = (i_r, \dots, i_k)$ is fixed, then the lifelength and the loss for subsystem $\mathbf{i}_k$ are determined only with delays (including those for reparable subsystems) and additional losses along this path

$$T_{\vec{\mathbf{i}}_k} = T_{\mathbf{i}} + T_{i_r(\mathbf{i}_{r-1})} + \cdots + T_{i_k(\mathbf{i}_{k+1})}, \qquad [26.14]$$

$$Y_{\vec{\mathbf{i}}_k} = Y_{\mathbf{i}} + Y_{i_r(\mathbf{i}_{r-1})} + \cdots + Y_{i_k(\mathbf{i}_{k+1})}. \qquad [26.15]$$

Under assumption about their independency, appropriate losses CDF for risk localization in subsystem $\mathbf{i}_k$ is calculated with the help of convolution as

$$G_{\mathbf{i}_k}(t) = G_{i_r,} * G_{\mathbf{i}_{r-1}(i_r)} * \cdots * G_{\mathbf{i}_{k+1}(i_k)}(t), \qquad [26.16]$$

and moments are summing,

$$\mu_T(\vec{\mathbf{i}}_k) = \mathbf{E}[T_{\mathbf{i}}] + \sum_{r \geq l \geq k} \mathbf{E}[T_{i_{l-1}(\mathbf{i}_l)}] = \mu_T(\mathbf{i}) + \sum_{r \geq l \geq k} \mu_T(i_{l-1}(\mathbf{i}_l)), \qquad [26.17]$$

$$\sigma_T^2(\vec{\mathbf{i}}_k) = \mathbf{Var}[T_{\mathbf{i}}] + \sum_{r \geq l \geq k} \mathbf{Var}[T_{i_{l-1}(\mathbf{i}_l)}] = \sigma_T^2(\mathbf{i}) + \sum_{r \geq l \geq k} \sigma_T^2(i_{l-1}(\mathbf{i}_l)), \qquad [26.18]$$

$$\mu_Y(\vec{\mathbf{i}}_k) = \mathbf{E}[Y_{\mathbf{i}}] + \sum_{r \geq l \geq k} \mathbf{E}[Y_{i_{l-1}(\mathbf{i}_l)}] = \mu_Y(\mathbf{i}) + \sum_{r \geq l \geq k} \mu_Y(i_{l-1}(\mathbf{i}_l)), \qquad [26.19]$$

$$\sigma_Y^2(\vec{\mathbf{i}}_k) = \mathbf{Var}[Y_{\mathbf{i}}] + \sum_{r \geq l \geq k} \mathbf{Var}[Y_{i_{l-1}(\mathbf{i}_l)}] = \sigma_Y^2(\mathbf{i}) + \sum_{r \geq l \geq k} \sigma_Y^2(i_{l-1}(\mathbf{i}_l)). \qquad [26.20]$$

One should calculate these characteristics beginning from the lowest level $r$ up to the subsystem $\mathbf{i}_k$ for which appropriate risk is localized or for the whole system, when $k = 0$.

Appropriate mean values of lifelength and loss for subsystem $\mathbf{i}_k$ can be found with the help of weighing appropriate means $\mu_T(\vec{\mathbf{i}}_k)$, $\mu_Y(\vec{\mathbf{i}}_k)$ along different paths with probabilities $q_{\vec{\mathbf{i}}_k}$ of risk situation evolution along this path,

$$\mu_T(\mathbf{i}_k) = \sum_{\vec{\mathbf{i}}_k} q_{\vec{\mathbf{i}}_k} \mu_T(\vec{\mathbf{i}}_k), \quad \mu_Y(\mathbf{i}_k) = \sum_{\vec{\mathbf{i}}_k} q_{\vec{\mathbf{i}}_k} \mu_Y(\vec{\mathbf{i}}_k)$$

and they should coincide with those calculated accordingly to [26.10] and [26.12].

A very useful and essential result of a risk tree analysis is the possibility of risk analysis in terms of *sensitivity* of risk characteristics to the shapes and parameters of the risk component distributions.

### 26.4.4. *Mostly dangerous paths of risk event evolution and sensitivity analysis*

One of the important problems of ERA consists of finding the mostly dangerous paths of risk event evolution with respect to different criteria:

– probability of risk event maximization;

– event mean time maximization;

– mean loss of risk minimization.

These problems have to be solved by optimization of appropriate functions [26.9], [26.17] and [26.19] with respect to their arguments.

Maximization of function [26.9] with respect to all possible paths of risk evolution gives the first solution. Taking into account the form of expression [26.9], we can see that it attains the maximum when

$$\prod_{1 \le l \le k} (1 - q_{i_l(\mathbf{i}_{l-1})}) \Longrightarrow \min$$

Taking the logarithms in the last expression, we find

$$\ln \prod_{1 \le l \le k} (1 - q_{i_l(\mathbf{i}_{l-1})}) = \sum_{1 \le l \le k} \ln(1 - q_{i_l(\mathbf{i}_{l-1})}).$$

The last expression attains its minimal value for maximal values of probabilities $q_{i_l(\mathbf{i}_{l-1})}$ beginning from the lowest level. Following this rule, we find the mostly dangerous paths of the risk event evolution (with respect to its probability) for any subsystem and the whole system for $k = 0$.

$$i_r^*(\mathbf{i}_{r-1}^*), \ \ldots, i_k^*(\mathbf{i}_{k-1}).$$

By collecting these indices, we find the mostly dangerous path of risk evolution with respect its probability

$$\vec{\mathbf{i}}_k^*(q) = (i_r^*(\mathbf{i}_{r-1}^*),\ \ldots,\ i_k^*(\mathbf{i}_{k-1}^*)), \qquad [26.21]$$

and the risk probability along this path,

$$q(\vec{\mathbf{i}}_k^*) = 1 - \prod_{k \le l \le r} (1 - q_{i_l^*(\mathbf{i}_{l-1}^*)}^*). \qquad [26.22]$$

Concerning mostly dangerous paths of risk evolution with respect to time and losses of its evolution, we should optimize corresponding expressions [26.17] and [26.19] with respect to their arguments,

$$\mu_T^*(\mathbf{i}_k) = \min_{\vec{\mathbf{i}}_k \in \mathcal{I}_k} \mu_T(\vec{\mathbf{i}}_k), \quad \vec{\mathbf{i}}_k^*(T) = \operatorname{argmin}_{\vec{\mathbf{i}}_k \in \mathcal{I}_k} \mu_T(\vec{\mathbf{i}}_k) \qquad [26.23]$$

$$\mu_Y^*(\mathbf{i}_k) = \max_{\vec{\mathbf{i}}_k \in \mathcal{I}_k} \mu_Y(\vec{\mathbf{i}}_k), \quad \vec{\mathbf{i}}_k^*(Y) = \operatorname{argmax}_{\vec{\mathbf{i}} \in \mathcal{I}} \mu_Y^*(\vec{\mathbf{i}}_k). \qquad [26.24]$$

Note that there are some special mathematical programming problems.

It is quite impossible to find the precise risk component distributions and their parameters. Therefore, an important part of ERA consists of *risk sensitivity analysis*. This means as follows:

– to check continuous dependence of output risk characteristics on their input parameters;

– to investigate insensitivity of mostly dangerous risk evolution paths $\vec{\mathbf{i}}_k^*(q)$, $\vec{\mathbf{i}}_k^*(T)$, $\vec{\mathbf{i}}_k^*(Y)$ from the shape of its initial CDFs;

– to study dependence of mostly dangerous paths of risk evolution $\vec{\mathbf{i}}_k^*(q)$, $\vec{\mathbf{i}}_k^*(T)$, $\vec{\mathbf{i}}_k^*(Y)$ with respect to different criteria on parameters of initial information about risk model.

An analytical investigation of this problem is a challenging task. However, it can be partially solved numerically by using the above algorithms or by simulation methods, using appropriate computer tools.

## 26.5. Risk analysis for short-time risk models

In most applications, the detailed information given above, which is needed for general model investigation, is not accessible. Moreover, for short-time risk model investigation, it is not needed. For these models, appropriate initial information is

– probabilities $q_{\mathbf{i}} = \mathbf{P}\{T_{\mathbf{i}} \leq t_0\} = F_{\mathbf{i}}(t_0)$ for the failure arising during a given time and appropriate numerical risk characteristics;

– conditional mean $\mu_T(\mathbf{i})$ and variance $\sigma_T^2(\mathbf{i})$ of time to risk event arising;

– conditional mean $\mu_T(j(\mathbf{i}_k))$ and variance $\sigma_T^2(j(\mathbf{i}_k))$ for the risk evolution delay time along the $j$th path to subsystem $\mathbf{i}_k$;

– conditional mean values $\mu_Y(\mathbf{i})$ and variances $\sigma_Y^2(\mathbf{i}))$ of damage sizes from the $\mathbf{i}$th element failure;

– conditional mean values $\mu_Y(j(\mathbf{i}_k))$ and variances $\sigma_Y^2(j(\mathbf{i}_k))$ additional damage sizes from risk evolution along the $j$th path to subsystem $\mathbf{i}_k$.

If the risk tree is constructed, the risk structure function is determined, and appropriate characteristics for subsystems with dependent elements (including those with restoration) are calculated, the appropriate short-time risk analysis consists of the calculation of:

– probabilities $q(\mathbf{i}_k)$ risk event arising for subsystem $\mathbf{i}_k$ (including those for the whole system for $k = 0$);

– conditional moments $\mu_T(\mathbf{i}_k)$, $\sigma_T^2(\mathbf{i}_k)$ time to risk events occurrence and conditional moments $\mu_Y(\mathbf{i}_k)$, $\sigma_Y^2(\mathbf{i}_k)$ of appropriate losses.

Calculation of the probabilities of risk event arising $q(\mathbf{i}_k)$ for subsystem $\mathbf{i}_k$ (including those for the whole system for $k = 0$) follows from the same algorithm (see formulas [26.5]) as for the calculation of unreliability function $F_{\mathbf{i}_k}(t)$ in section 26.4,

$$q(\mathbf{i}_{k-1}) = \phi_{\mathbf{i}_k}(q(\mathbf{i}_k, 1), \ldots, q(\mathbf{i}_k, n(\mathbf{i}_k))). \qquad [26.25]$$

Thus, the calculation of the most dangerous path with respect to maximal probability of risk event evolution for any subsystem follows from the same algorithm as before (see section 26.4.4).

Because the risk event destination time and appropriate losses for any subsystem and the whole system are accumulated along any path $\vec{\mathbf{i}}_k = (i_r, i_r - 1, \ldots, i_k)$ according to formulas [26.14–26.15], their calculation follows from formulas [26.17–26.20] using given information about appropriate values. These results along with given initial information allow us to find the mostly dangerous paths of risk development with respect to minimal mean time and maximal mean losses using formulas [26.23] and [26.24].

## 26.6. Conclusion

Some new risk characteristics of complex hierarchical systems are proposed. Their investigation requires special computer tools. There are many different computer tools used for special cases of risk analysis. For reference, see, e.g., [HEN 92], [BED 06] and [SOL 06].

One of the most developed tools is RELEX (see [REL XX]), which allows risk tree construction, marking with initial information and analysis with direct calculation or with simulation. Nevertheless, it does not contain possibilities for analysis risk trees with respect to:

– mostly dangerous paths of the risk event evolution;

– insensitivity of risk characteristics to the shape of initial distributions and their stability to small changes of initial parameters.

One of the main problems of the proposed approach is the risk tree marking. The estimation of times to all initial and intermediate events and appropriate damages should be done by the usual theoretical or statistical methods. In spite of the above proposed approach destined to the engineering risk models, it can also be used in other types of risk model. For example, any *business-plan* design can be considered as a risk tree construction. However, its marking is fulfilled by experts representing their *subjective* knowledge about the situation. It is not to be confused with the *objective* risk analysis that is based on statistically estimated probabilities. It is not the objective of this chapter to consider some special statistical methods for input risk characteristic analysis, or to give special methods for their expert analysis.

## 26.7. Bibliography

[BED 01] BEDFORD T., COOKE K., *Probabilistic Risk Analysis: Foundations and Methods*, Cambridge University Press, p. 393, 2001.

[CRA 30] CRAMER H., *On the Mathematical Risk Theory*, Skandia Jubilee Volume, Stockholm, 1930.

[CRA 55] CRAMER H., *Collective Risk Theory*, Skandia Jubilee Volume, Stockholm, 1955.

[GNE 69] GNEDENKO B.V., BELYAEV Y.K., SOLOV'EV A.D., *Mathematical Methods of Reliability Theory*, Academic Press, New York, 1969. [Translation from Russian ed. Darlow R.E.]

[GER 00] GERTSBAKH I., *Reliability Theory with Applications to Preventive Maintenance*, Springer, p. 219, 2000.

[GRA 91] GRANDEL J., *Aspects of Risk Theory*, Springer-Verlag, New York/Berlin/London, 1991.

[HEN 92] HENLEY E.J., KUMAMOTO H., *Probabilistic Risk Assessment. Reliability Engineering, Design, and Analysis*, IEEE Press, New York, p. 568, 1992.

[REL] RELIABILITY EXCELLENCE (RELEX), Software solutions, available at http://www.relex.se.

[ROT 94] ROTAR V.I., BENING V.E., "Introduction into mathematical insurance theory", *Review of Applied and Industrial Mathematics*, vol. 1, no. 5, pp. 698–779, 1994. [In Russian].

[RYK 99] RYKOV V.V., YASTREBENETSKY M.A., "Risk: notion and measurement", *Proceedings of the International Conference Probabilistic Analysis of Rare Events: Theory and Problems of Safety, Insurance and Ruin*, Riga Avion University, pp. 63–70, 1999. [In Russian].

[RYK 01] RYKOV V.V., *Reliability of Technical Systems and Technogeneous Risk*, Gubkin Russian State University of Oil and Gas, Moscow, pp. 165, 2001. [In Russian].

[RYK 13] RYKOV V.V., *Reliability, Safety, Risk: Methodological Aspects*, Gnedenko Volume, 2013. [In press].

[SIN 06] SINGPURVALA N., *Reliability and Risk. A Bayesian Perspective*, John Wiley & Sons, Chichester, p. 371, 2006.

[SOL 06] SOLOJENTSEV E.D., *Scenario Logic and Probabilistic Management of Risk in Business and Engineering*, Business-Pressa, Saint-Petersburg, p. 537, 2006.

[VON 53] VON NEUMANN J., MORGENSTERN O., *Theory of Games and Economic Behavior*, Princeton University Press, p. 707, 1953.

# List of Authors

Annick ALPÉROVITCH
University Pierre and Marie Curie
Paris
France

Emmanouil ANDROULAKIS
National Technical University of Athens
Greece

Alexandr ANTONOV
Obninsk State Technical University for Nuclear Power Engineering
Russia

Vilijandas BAGDONAVIČIUS
Vilnius University
Lithuania

Kamal BOUKHETALA
Science and Technology University
Bab-Ezzouar
Algeria

Salim BOUZEBDA
Laboratoire de Mathématiques Appliquées de Compiègne (LMAC)
University of Technology of Compiègne (UTC)
France

Michel BRONIATOWSKI
University Pierre and Marie Curie
Paris
France

Elodie BRUNEL
Montpellier 2 University
France

Virgile CARON
Laboratoire de Statistique Théorique et Appliquée (LSTA)
University Pierre and Marie Curie
Paris
France

Ekaterina CHIMITOVA
Novosibirsk State Technical University
Russia

Fabienne COMTE
University Paris Descartes
France

Vincent COUALLIER
Institut de Mathématiques de Bordeaux
University of Bordeaux
France

Paul DEHEUVELS
Laboratoire de Statistique Théorique et Appliquée (LSTA)
University Pierre and Marie Curie
Paris
France

Jean-François DUPUY
INSA
Rennes
France

Florence GEORGE
Department of Mathematics and Statistics
Florida International University
USA

Léo GERVILLE-RÉACHE
IMB
University of Bordeaux
France

Shulamith T. GROSS
Statistical Consulting Laboratory
Bernard M. Baruch College
City University of New York
USA

Agathe GUILLOUX
LSTA, University of Paris 6
Saint-Antoine Research Centre
Paris
France

Sneh GULATI
Department of Mathematics and Statistics
Florida International University
USA

Isaac Wu HONG-DAR
Institute of Statistics
National Chung-Hsing University
Taiwan

Catherine HUBER-CAROL
University Paris Descartes
France

Zhezhen JIN
Department of Biostatistics
Columbia University
New York
USA

Waltraud KAHLE
Otto-von-Guericke-University
Magdeburg
Germany

Alex KARAGRIGORIOU
Department of Mathematics & Statistics
University of Cyprus
Nicosia
Cyprus

Christos KOUKOUVINOS
Department of Mathematics
National Technical University of Athens
Greece

Yury A. KUTOYANTS
Laboratoire de Statistique et Processus
University of Maine
France

Henning LÄUTER
Potsdam University
Germany

Boris LEMESHKO
Novosibirsk State Technical University
Russia

Stanislav LEMESHKO
Novosibirsk State Technical University
Russia

Nikolaos LIMNIOS
Laboratoire de Mathématiques Appliquées de Compiègne (LMAC)
University of Technology of Compiègne (UTC)
France

Olivier LOPEZ
Laboratoire de Mathématiques Appliquées
University Pierre and Marie Curie
Paris
France

Andreas MAKRIDES
Department of Mathematics & Statistics
University of Cyprus
Nicosia
Cyprus

Gennady MARTYNOV
Institute for Information Transmission Problems
Russian Academy of Sciences and Higher School of Economics
Moscow
Russia

William Q. MEEKER
Department of Statistics
Center for Nondestructive Evaluation
Iowa State University
USA

Mounir MESBAH
Laboratoire de Statistique Théorique et Appliquée (LSTA)
University Pierre and Marie Curie
Paris
France

Amel MEZAOUER
Faculty of Sciences
Saad Dahleb University
Blida
Algeria

Mikhail NIKULIN
IMB
University Bordeaux Segalen
France

Christian PAROISSIN
University of Pau and Pays de l'Adour
France

Petr PHILONENKO
Novosibirsk State Technical University
Russia

Alexandr PLYASKIN
Obninsk State Technical University for Nuclear Power Engineering
Russia

Sergey POSTOVALOV
Novosibirsk State Technical University
Russia

Landy RABEHASAINA
University of Franche Comté
Besancon
France

Andrey ROGOZHNIKOV
Novosibirsk State Technical University
Russia

Vladimir RYKOV
Gubkin Russian State University of Oil and Gas
Moscow
Russia

Philippe SAINT-PIERRE
Laboratoire de Statistique Théorique et Appliquée (LSTA)
University Pierre and Marie Curie
Paris
France

Ying SHI
Dept. of Statistics
Iowa State University
Ames
USA

Valentin SOLEV
Mathematical Institute RAS
Fontanka
St. Petersburg
Russia

Khizri TATAEV
Obninsk State Technical University for Nuclear Power Engineering
Russia

Xuan Quang TRAN
IMB
University Bordeaux Segalen
France

Filia VONTA
Department of Mathematics
National Technical University of Athens
Greece

Li ZHOU
Laboratoire de Statistique et Processus
University of Maine
France

# Index

## F



## G, H



## I



## K



## L



## M



## N, O



## P, Q



## R



## S, T

**U, W, Z**

Printed and bound by CPI Group (UK) Ltd, Croydon, CR0 4YY

13/09/2026

14961082-0003